INTRODUCTION TO DIFFERENTIAL EQUATIONS

with Dynamical Systems

INTRODUCTION TO DIFFERENTIAL EQUATIONS

with Dynamical Systems

Stephen L. Campbell and
Richard Haberman

PRINCETON UNIVERSITY PRESS

PRINCETON AND OXFORD

Published by Princeton University Press, 41 William Street, Princeton, New Jersey 08540

In the United Kingdom: Princeton University Press, 3 Market Place, Woodstock, Oxfordshire OX20 1SY

Library of Congress Control Number: 2007936213

ISBN 978-0-691-12474-2

British Library Cataloging-in-Publication Data is available

This book has been composed in Times Roman with ITC Eras Display

Printed on acid-free paper. ∞

press.princeton.edu

Printed in the United states of America

10 9 8 7 6 5 4

CONTENTS

CHAPTER 2

Linear Second- and Higher-Order Differential Equations 96

CHAPTER 3

The Laplace Transform 197

CHAPTER 4

An Introduction to Linear Systems of Differential Equations and Their Phase Plane 265

PREFACE

OVERVIEW

We have attempted to write a concise modern treatment of differential equations emphasizing applications and containing all the core parts of a course in differential equations. A semester or quarter course in differential equations is taught to most engineering students (and many science students) at all universities, usually in the second year. Some universities have an earlier brief introduction to differential equations and others do not. Some students will have already seen some differential equations in their science classes. We do not assume any prior exposure to differential equations.

The core of the syllabus consists of Chapters 1 and 2 on linear differential equations. The use of Laplace transforms to solve differential equations is described in Chapter 3 since many engineering faculty use them. Series solutions of differential equations used to be part of the course, but the trend is to not do this, and we have decided to omit series. By doing so, we are communicating that linear and nonlinear systems are more important in a differential equations course for the future. Our book is also less expensive without series.

Most universities do some systems in their differential equations course. Some do a little, some do a lot. We have tried to present systems in an elementary introductory way, so that the beginning student understands the material, and also in a flexible way, so that universities that only spend two weeks on systems will be satisfied, and those that spend more will also have options. We also present the phase plane for linear systems in an elementary way. Most universities will want to do systems because present technology makes them very graphically exciting to students and faculty. Each university will use technology in its own unique way. We do not provide expensive software because the software would be useful for only a short portion of the standard differential equations course. Similarily, while numerical simulation is important, it is easily done with numerous software packages, so that only a brief introduction is needed in a first course.

Our book discusses standard topics for differential equations in the standard way, so that it can easily be adopted by most universities. In addition, we have focused on the essential areas, so that the text is concise. Our book is unique in the way that some of the essential topics are discussed, as we now describe. The most important concept for students in our text is to understand linear differential equations and how to solve or analyze them in an elementary way. We have attempted to make the entire book easy for students to read.

CHAPTER 1

Some of the presentations in Chapter 1 are unique and intended to simplify the overall linearity concepts. The key sections are very early in the text. Section 1.6 on first-order linear differential equations discusses all the concepts of linearity for first-order equations, and confirms the form of the general solution using an integrating factor. It shows that the general solution is a particular solution plus a constant times the homogeneous solution. The section is not unique, but it is very well written and easy for students to understand. Ideas of homogeneous and nonhomogeneous equations are introduced early. The unique section is the short Section 1.7 on elementary methods to solve first-order differential equations with constant coefficients and constant input. By our approach, students are taught that differential equations are sometimes easy, not hard and mysterious. All differential equations will be easier after beginning with our approach.

First-order differential equations have many important applications. Sections 1.8–1.9 discusses population growth, radioactive decay, Newton's law of cooling, and mixture problems. All the differential equations that appear in these applications have constant coefficients. In other books these equations are solved by an integrating factor, and students sometimes are confused and get the wrong answer. In our presentation, the elementary method of Section 1.7 is used to solve all the application problems, so that the student believes that the solutions to differential equation are applicable and sometimes easy. All students can benefit by being introduced to differential equations in this way.

CHAPTER 2

Chapter 2 covers linear second- and higher-order differential equations. This is the most important chapter in our book and in most courses in differential equations. The necessary theory in this chapter is difficult in all books. Our students will not be confused here because they have been introduced to the linearity idea in Chapter 1, where enough examples have been done for them already to understand some of the ideas. In Section 2.1, we immediately introduce the idea that the general solution is a particular solution plus a linear combination of homogeneous solutions. We discuss the Wronskian and its relationship to the initial value problem in Section 2.2. Section 2.3 presents reduction of order, so that it we can use it in Section 2.4 to obtain homogeneous solutions of constant coefficient differential equations. In Section 2.5, the vibrations of spring-mass systems are discussed with great care. The differential equations are formulated using Newton's law. Detailed presentations are given of mechanical vibration with no damping, along with presentations of the three cases of damped mechanical vibrations, especially appropriate for science and engineering students. Section 2.6 presents the method of undetermined coefficients with greater clarity than most books. Section 2.7 is a carefully written presentation of forced vibrations including a very detail oriented discussion of mechanical resonance. Section 2.8 is a nice separate section devoted to linear electric circuits, including a derivation from first principles of RLC (resistors, inductor, capacitors) circuits and the phenomena of electrical resonance. Section 2.9 is a short presentation of Euler equations. Chapter 2 finishes with the method of variation of parameters (including a more advanced presentation for higher-order systems).

CHAPTER 3

Chapter 3 discusses how to solve differential equations using Laplace Transforms. Students learn how to use a table. We present both a short and long table, so that instructors can provide the type they wish. In Section 3.1, we emphasize the use of two theorems for Laplace transforms, multiplying by an exponential and multiplying by t. In Section 3.2 we develop the systematic inverse Laplace transform based on roots, quadratics, and partial fractions. We solve many differential equations using Laplace transforms in Section 3.3. To illustrate the science and engineering orientation of our presentation, we do a resonance example. More in-depth discussion of Laplace transforms is possible. For many, the inclusion of discontinuous forcing with Heaviside step functions (Section 3.4), periodic forcing (Section 3.5), the convolution theorem (Section 3.6), and impulsive forcing using the delta function (Section 3.7) are highly recommended.

CHAPTERS 4–6

Chapters 4–6 cover linear and nonlinear systems, an introduction to so-called dynamical systems. The material on linear and nonlinear systems has great flexibility. Chapter 4 includes a very careful introduction to linear systems of differential equations (two first-order linear equations with two unknowns) and the phase plane for these linear systems. It is our anticipation that most students in an elementary differential equations course will at least cover Chapter 4. Chapter 5 is a very brief chapter discussing mostly nonlinear first-order differential equations. It introduces equilibrium, stability, and one-dimensional phase lines. Chapter 6 discusses nonlinear systems of differential equations in the plane. It has been written to require Chapter 4 but not Chapter 5. Some instructors may wish to discuss Chapter 5 before Chapter 6, some may wish to go directly from Chapter 4 to Chapter 6. Equilibrium, linear stability, and phase plane analysis are carefully presented. These chapters have been written so that instructors can discuss as much of these three chapters as they wish. To keep the presentation relatively simple and the cost relatively low, only systems in the plane are discussed.

CHAPTER 4

Chapter 4 introduces linear systems of two differential equations and their phase planes. Section 4.2 shows in an elementary way how solving a linear system of differential equations reduces to solving a linear system of algebraic equations involving a 2×2 matrix. It is not assumed that students have seen matrices before. Section 4.2 introduces eigenvalues and their corresponding eigenvectors for 2×2 systems in an elementary way with many examples. The emphasis is on finding the general solution to the linear system of differential equations. Careful elementary presentations based on examples are discussed for the cases where the eigenvalues are real and distinct and where they are complex. Only an elementary motivational example is given if the eigenvalues are real and repeated. In Section 4.3, the phase plane is introduced but only for linear systems. The presentation is elementary. The first examples are of elementary systems of differential equations that are not coupled; we then discuss examples that require eigenvalues and eigenvectors. Many examples of 2×2 systems of differential equations with real distinct eigenvalues are given, including a systematic discussion of the phase plane associated with stable and unstable nodes and saddle points. Detailed examples of the phase plane for the case of stable and unstable spirals and centers are also given. Some courses will not go any further than Chapter 4.

CHAPTER 5

This innovative chapter on mostly nonlinear first-order differential equations is optional and not required for Chapter 6. In Section 5.2, the concept of equilibrium and linearized stability analysis is introduced in a very elementary context of first-order nonlinear differential equations. In Section 5.3, the very elementary (but very powerful) method of one-dimensional phase lines for first-order nonlinear problems is introduced to determine the stability of the equilibrium and other qualitative behavior of the solution of the differential equation. An application to an elementary nonlinear biological growth model is presented in Section 5.4.

CHAPTER 6

In Chapter 6, we give a comprehensive presentation of nonlinear systems of differential equations in the plane. It is anticipated that different universities will cover varying amounts of this chapter. In Section 6.2, we discuss equilibriums, and we show that the stability of an equilibrium is determined from the eigenvalues of the linearization (Jacobian) matrix. Furthermore, we show that the phase plane for the nonlinear system of differential equations is usually approximated near each equilibrium by the phase plane of a linear system. Thus, the linear systems studied in Chapter 4 apply to nonlinear systems near equilibria. Section 6.3 discusses in depth competing species and predator-prey population models. Section 6.4 discusses the mechanical system corresponding to a nonlinear pendulum. For such conservative systems, we derive conservation of energy, and we show how to obtain periodic solutions and the phase plane using the potential energy.

APPLICATIONS

Many fundamental problems in biological and physical sciences and engineering are described by differential equations. We believe that many problems of future technologies will be described in the same way. Physical problems have motivated the development of much of mathematics, and this is especially true of differential equations. In this book, we study the interactions between mathematics and physical problems. Thus, in our presentation, we devote some effort to mathematical modeling, deriving the governing differential equations from physical principles. We take four major applications and in most cases carry them throughout. They are population growth, mixing problems, mechanical vibrations, and electrical circuits. In this way, the student has the chance to understand the physical problem.

EXERCISES

Differential equations cannot be learned by reading the text alone. We have included a large number of problems of various types and degrees of difficulty. Most are straightforward illustrations of the ideas in each section. The answers are provided in the back for all the odd problems. A student solutions manual exists in which the solutions to the odd exercises are carefully worked out. Frequently, even problems are similar to neighboring odd problems so that the answer to an odd problem can often be used for guidance. An instructor's manual exists with the answers (and some discussion) to all the exercises. Exercises have been class tested.

TECHNOLOGY

The increasing availability of technology (including graphing and programmable calculators, computer algebra systems, and powerful personal computers) has caused some to question the existing syllabi in university courses in differential equations. However, we believe that the importance of applications will continue to motivate the study of differential equations. This book has been written with that in mind. Courses with a strong emphasis on applications can use our book. In addition, those that wish a greater presence of technology can use our book supplemented by increasingly available web based resources or computational supplements.

ACKNOWLEDGMENTS

Many people have had input into this book, including an excellent group of reviewers, users of an earlier edition by both of us, users of an even earlier edition by one of us, and an experienced editorial staff. However, two groups have especially influenced us: the many students to whom we have taught differential equations and the colleagues with whom we have worked in differential equations, electrical and mechanical engineering.

Finally, the book could not have written without the support and understanding of our wives, Gail and Liz, and our children (two each) Matthew and Eric and Ken and Vicki.

Stephen L. Campbell
Richard Haberman

CHAPTER 1

First-Order Differential Equations and Their Applications

1.1 Introduction to Ordinary Differential Equations

Differential equations are found in many areas of mathematics, science, and engineering. Students taking a first course in differential equations have often already seen simple examples in their mathematics, physics, chemistry, or engineering courses. If you have not already seen differential equations, go to the library or Web and glance at some books or journals in your major field. You may be surprised to see the way in which differential equations dominate the study of many aspects of science and engineering.

Applied mathematics involves the relationships between mathematics and its applications. Often the type of mathematics that arises in applications is differential equations. Thus, the study of differential equations is an integral part of applied mathematics. Applied mathematics is said to have three fundamental aspects, and this course will involve a balance of the three:

1. The **modeling process** by which physical objects and processes are described by physical laws and mathematical formulations. Since so many physical laws involve rates of change (or the derivative), differential equations are often the natural language of science and engineering.
2. The **analysis** of the mathematical problems that are posed. This involves the complete investigation of the differential equation and its solutions, including detailed numerical studies. We will say more about this shortly.
3. However, the mathematical solution of the differential equation does not complete the overall process. The **interpretation** of the solution of the differential equation in the context of the original physical problem must be given, and the implications further analyzed.

Using the **qualitative** approach, we determine the behavior of the solutions without actually getting a formula for them. This approach is somewhat similar to the curve-sketching process in introductory calculus where we sketch curves by drawing maxima, minima, concavity changes, and so on. Qualitative ideas will be discussed in Section 1.4 and Chapters 4, 5, and 6.

Using the **numerical** approach, we compute estimates for the values of the unknown function at certain values of the independent variable. Numerical methods are extremely important and for many difficult problems are the only practical approach. We discuss numerical methods in Section 1.5. The safe and effective use of numerical methods requires an understanding of the basic properties of differential equations and their solutions.

Most of this book is devoted to developing **analytical** procedures, that is, obtaining explicit and implicit formulas for the solutions of various ordinary differential equations. We present a sufficient number of applications to enable the reader to understand how differential equations are used and to develop some feeling for the physical information they convey.

Asymptotic and **perturbation** methods are introduced in more advanced studies of differential equations. Asymptotic approximations are introduced directly from the exact analytic solutions in order to get a better understanding of the meaning of the exact analytic solutions. Unfortunately, many problems of physical interest do not have exact solutions. In this case, in addition to the previously mentioned numerical methods, approximation methods known as perturbation methods are often useful for understanding the behavior of differential equations.

In this book, we use the phrase "differential equation" to mean an ordinary differential equation (or a system of ordinary differential equations. An **ordinary differential equation** is an equation relating an unknown function of one variable to one or more functions of its derivatives. If the unknown x is a function of t, $x = x(t)$, then examples of ordinary differential equations are

$$\begin{aligned} \frac{dx}{dt} &= t^7 \cos x, \\ \frac{d^2x}{dt^2} &= x\frac{dx}{dt}, \\ \frac{d^4x}{dt^4} &= -5x^5. \end{aligned} \tag{1}$$

The **order** of a differential equation is the order of the highest derivative of the unknown function (dependent variable) that appears in the equation. The differential equations in (1) are of first, second, and fourth order, respectively. Most of the equations we shall deal with will be of first or second order.

In applications, the dependent variables are frequently functions of time, which we denote by t. Some applications such as

1. population dynamics,
2. mixture and flow problems,
3. electronic circuits,
4. mechanical vibrations and systems

are discussed repeatedly throughout this text. Other applications, such as radioactive decay, thermal cooling, chemical reactions, and orthogonal trajectories, appear only as illustrations of more specific mathematical results. In all cases, modeling, analysis, and interpretation are important.

Let us briefly consider the following motivating population dynamics problem.

Example 1.1.1 *Population Growth Problem*

Assume that the population of Washington, DC, grows due to births and deaths at the rate of 2% per year and there is a net migration into the city of 15,000 people per year. Write a mathematical equation that describes this situation.

• SOLUTION. We let

$$x(t) = \text{population as a function of time } t. \tag{2}$$

From calculus,

$$\frac{dx}{dt} = \text{rate of change of the population.} \tag{3}$$

In this example, 2% growth means 2% of the population $x(t)$. Thus, the population of Washington, DC satisfies

$$\frac{dx}{dt} = 0.02x + 15{,}000. \tag{4}$$

♦

Equation (4) is an example of a differential equation, and we develop methods to solve such equations in this text. We will discuss population growth models in more depth in Section 1.8 and Chapters 5 and 6.

In a typical application, physical laws often lead to a differential equation. As a simple example, we will consider later the vertical motion $x(t)$ of a constant mass. **Newton's law** says that the force F equals the mass m times the acceleration $\frac{d^2x}{dt^2}$:

$$m\frac{d^2x}{dt^2} = F. \tag{5}$$

If the forces are gravity $-mg$ and a force due to air resistance proportional to the velocity $\frac{dx}{dt}$, then the position satisfies the second-order differential equation

$$m\frac{d^2x}{dt^2} = -mg - c\frac{dx}{dt}, \tag{6}$$

where c is a proportionality constant determined by experiments. However, if in addition the mass is tied to a spring that exerts an additional force $-kx$ satisfying Hooke's law, then we will show that the mass satisfies the following second-order differential equation:

$$m\frac{d^2x}{dt^2} = -mg - kx - c\frac{dx}{dt}. \tag{7}$$

In the main body of the book, we will devote some effort to the modeling process by which these differential equations arise, as well as learning how to solve these types of differential equations easily.

We will also see that a typical electronic circuit with a resistor, capacitor, and inductor can often be modeled by the following second-order differential equation:

$$L\frac{d^2i}{dt^2}+R\frac{di}{dt}+\frac{1}{C}i=f(t). \tag{8}$$

Here the unknown variable is $i(t)$, the time-dependent current running through the circuit. We will present a derivation of this differential equation, and you will learn the meaning of the three positive constants R, L, and C (this is called an RLC circuit). For example, R represents the resistance of a resistor, and we will want to study how the current in the circuit depends on the resistance. The right-hand side $f(t)$ represents something that causes current in the circuit, such as a battery.

Physical problems frequently involve systems of differential equations. For example, we will consider the salt content in two interconnected well-mixed lakes, allowing for some inflow, outflow, and evaporation. If $x(t)$ represents the amount of salt in one of the lakes and $y(t)$ the amount in the other, then under a series of assumptions described in the book, the following coupled system of differential equations is an appropriate mathematical model:

$$\begin{aligned}\frac{dx}{dt}&=\frac{1}{2}+2\frac{y}{100}-3\frac{x}{100},\\ \frac{dy}{dt}&=3\frac{y}{100}-\frac{5}{2}\frac{y}{100}.\end{aligned} \tag{9}$$

1.2 The Definite Integral and the Initial Value Problem

This chapter is concerned with first-order differential equations, in which the first derivative of a function $x(t)$ depends on the independent variable t and the unknown solution x. If $\frac{dx}{dt}$ is given directly in terms of t and x, the differential equation has the form

$$\frac{dx}{dt}=f(t,x). \tag{1}$$

In later sections, we will discuss a number of applications of first-order equations such as growth and decay problems for populations, radioactive decay, thermal cooling, mixture problems, evaporation and flow, electronic circuit theory, and several others. In these applications, some care is given to the development of mathematical models.

A **solution** of the differential equation (1) is a function that satisfies the differential equation for all values t of interest:

$$\frac{dx}{dt}(t)=f(t,x(t)) \qquad \text{for all } t.$$

Example 1.2.1 *Showing That a Function Is a Solution*

Verify that $x = 3e^{t^2}$ is a solution of the first-order differential equation

$$\frac{dx}{dt} = 2tx. \tag{2}$$

• SOLUTION. We substitute $x = 3e^{t^2}$ in both the left- and right-hand sides of (2). On the left we get $\frac{d}{dt}(3e^{t^2}) = 2t(3e^{t^2})$, using the chain rule. Simplifying the right-hand side, we find that the differential equation (2) is satisfied

$$6te^{t^2} = 6te^{t^2},$$

which holds for all t. Thus $x = 3e^{t^2}$ is a solution of the differential equation (2). ♦

Before beginning our general development of first-order equations in Section 1.4, we will discuss some differential equations that can be solved with direct integration. These special cases will be used to motivate and illustrate some of the later development.

1.2.1 The Initial Value Problem and the Indefinite Integral

The simplest possible first-order differential equation arises if the function $f(t, x)$ in (1) does not depend on the unknown solution, so that the differential equation is

$$\frac{dx}{dt} = f(t). \tag{3}$$

Solving (3) for x is just the question of antidifferentiation in calculus. Although (3) can be solved by an integration, we can learn some important things concerning more general differential equations from it that will be useful later.

Example 1.2.2 *Indefinite Integration*

Consider the simple differential equation,

$$\frac{dx}{dt} = t^2. \tag{4}$$

By an integration, we obtain

$$x = \frac{1}{3}t^3 + c, \tag{5}$$

where c is an arbitrary constant. From this example, we see that differential equations usually have many solutions. We call (5) the **general solution** of (4), since it is a formula that gives all solutions. Because x depends on t, sometimes we use the notation $x(t)$. Often, especially in applications, we are interested in a specific solution of the differential equation to satisfy some additional condition. For example, suppose we are given that $x=7$ at $t=2$. This is written mathematically as $x(2)=7$ and called an **initial condition**. By letting $t=2$ and $x=7$ in (5), we get

$$7=\frac{8}{3}+c,$$

so that the constant c is determined to be $c=\frac{13}{3}$. Thus, the unique solution of the differential equation that satisfies the given initial condition is

$$x=\frac{1}{3}t^3+\frac{13}{3}. \qquad \blacklozenge$$

More generally, we might wish to solve the differential equation

$$\frac{dx}{dt}=f(t), \tag{6}$$

subject to the initial condition

$$x(t_0)=x_0. \tag{7}$$

We introduce the symbol t_0 for the value of t at which the solution is given. Often $t_0=0$, but in the previous example, we had $t_0=2$ and $x_0=7$. We refer to (6) with the initial condition (7) as the **initial value problem** for the differential equation. We can always write a formula for the solution of (6) using an indefinite integral

$$x=\int f(t)dt+c. \tag{8}$$

Equation (5) is a specific example of (8). If we can obtain an explicit indefinite integral (antiderivative) of $f(t)$, then this initial value problem can be solved like Example 1.2.1. Explicit integrals of various functions $f(t)$ may be obtained by using any of the various techniques of integration from calculus. Tables of integrals or symbolic integration algorithms such as MAPLE or Mathematica that are available on more sophisticated calculators, personal computers, or larger computers may be used. However, if one cannot obtain an explicit integral, then it may be difficult to use (8) directly to satisfy the initial conditions.

1.2.2 The Initial Value Problem and the Definite Integral

A definite integral should usually be used to solve the differential equation $\frac{dx}{dt}=f(t)$ if an explicit integral is not used. The result can automatically incorporate the given initial condition $x(t_0)=x_0$. If both sides of the differential equation (6) are integrated with respect to t from t_0 to t, we get

$$\int_{t_0}^{t}\frac{dx}{d\bar{t}}d\bar{t}=\int_{t_0}^{t}f(\bar{t})d\bar{t},$$

where we have introduced the dummy variable $\bar{t}$. The left-hand side equals $x(\bar{t})\,|_{\bar{t}=t_0}^{\bar{t}=t}=x(t)-x(t_0)$, since the antiderivative of $\frac{dx}{dt}$ is x. We obtain the same result canceling

$\bar{t}$, since $\frac{dx}{dt}d\bar{t} = dx$. Thus, we have

$$x(t) - x(t_0) = \int_{t_0}^{t} f(\bar{t})d\bar{t}, \quad \text{or equivalently,}$$
$$x(t) = x(t_0) + \int_{t_0}^{t} f(\bar{t})d\bar{t}. \tag{9}$$

Note that $x(t_0) = x_0$ since $\int_{t_0}^{t_0} f(\bar{t})d\bar{t} = 0$. Any dummy variable of integration may be used. We have chosen $\bar{t}$.

Example 1.2.3 *Example with a Definite Integral*

Solve the differential equation

$$\frac{dx}{dt} = e^{-t^2}, \tag{10}$$

subject to the initial condition $x(3) = 7$.

● SOLUTION. The function e^{-t^2} does not have any explicit antiderivative. Thus, if we want to solve (10), we use definite integration from $t_0 = 3$, where $x_0 = 7$. Then (9) is

$$x - 7 = \int_{3}^{t} e^{-\bar{t}^2} d\bar{t},$$

and the solution of the initial value problem is

$$x = 7 + \int_{3}^{t} e^{-\bar{t}^2} d\bar{t}. \tag{11}$$

The function e^{-t^2} is important in probability, since $(1/\sqrt{2\pi})e^{-t^2/2}$ is the famous normal curve. ♦

There are many situations in which it is desirable to use a definite integral rather than an explicit antiderivative:

1. It is sometimes difficult to obtain an explicit antiderivative, and a formula like (11) suffices.
2. For some $f(t)$ (as in the previous example), it is impossible to obtain an explicit antiderivative in terms of elementary functions.
3. The function $f(t)$ might be expressed only by some data, in which case the definite integral (9) represents the area under the curve $f(t)$ and can be evaluated by an appropriate numerical integration method such as Simpson's or the trapezoid rule.

It is difficult to give general advice valid for all problems. If an integral is an elementary integral, then the explicit integral should be used. However, what is elementary to one person is not necessarily elementary to another. With the wide availability of computers, a definite integral can usually be evaluated by a numerical integration.

General Solution Using a Definite Integral

Alternatively, in solving for the general solution of

$$\frac{dx}{dt} = f(t), \tag{12}$$

we may use the definite integral starting at any point a. In this case, we obtain the general solution of (12) as

$$x = \int_a^t f(\bar{t})d\bar{t} + c. \tag{13}$$

There appear to be two arbitrary constants, c and the lower limit of the integral a in (13). However, it can be shown that this is equivalent to one arbitrary constant. In practice, the lower limit a is often chosen to be the initial value of t so that $a = t_0$. To see that (13) actually solves the differential equation (12), it is helpful to recall the **fundamental theorem of calculus**:

$$\frac{d}{dt}\left(\int_a^t f(\bar{t})d\bar{t}\right) = f(t). \tag{14}$$

1.2.3 Mechanics I: Elementary Motion of a Particle with Gravity Only

Elementary motions of a particle are frequently described by differential equations. Simple integration can sometimes be used to analyze these elementary motions. For the one-dimensional vertical motion of a particle, we recall from calculus that

$$\begin{aligned} \text{Position} &= x(t), \\ \text{Velocity} &= v(t) = \frac{dx}{dt}, \\ \text{Acceleration} &= a(t) = \frac{dv}{dt} = \frac{d^2x}{dt^2}. \end{aligned} \tag{15}$$

Newton's law of motion ($ma = F$) will yield a differential equation

$$m\frac{d^2x}{dt^2} = F\left(x, \frac{dx}{dt}, t\right), \tag{16}$$

where m is the mass and F is the sum of the applied forces, and we have allowed the forces to depend on position, velocity, and time. Equation (16) is a second-order differential equation, which we will study later in the book.

There are no techniques for solving (16) in all cases. However, (16) can be solved by simple integration if the force F does not depend on x and $\frac{dx}{dt}$. As an example, suppose that the only force on the mass is due to gravity. Then it is known that $F = -mg$, where g is the acceleration due to gravity. The minus sign is introduced because gravity acts downward, toward the surface of the earth. Here we are taking the coordinate system so that x increases toward the sky. The magnitude of the force due to gravity, mg, is called the weight of the body. Near the surface of the planet earth, g is approximately $g = 9.8$ m/s^2 in the mks system used by most of the world ($g = 32$ ft/s^2 when feet are used as the unit of length instead of meters). If we assume that we are interested in a mass that is located sufficiently near the surface of the earth, then g can be approximated by this constant. With the only force being gravity, (16) becomes

$$m\frac{d^2x}{dt^2} = -mg,$$

or equivalently,

$$\frac{d^2x}{dt^2} = -g, \tag{17}$$

since the mass m cancels. Integrating (17) yields

$$\frac{dx}{dt} = -gt + c_1, \tag{18}$$

where c_1 is an arbitrary constant of integration. We assume that the velocity at $t = 0$ is given and use the notation v_0 for this initial velocity. Since

$$v(t) = -gt + c_1$$

from (18), evaluating (18) at $t = 0$ gives $v_0 = c_1$. Thus, the velocity satisfies

$$\frac{dx}{dt} = -gt + v_0. \tag{19}$$

The position can be determined by integrating the velocity (19) to give

$$x = -\frac{1}{2}gt^2 + v_0 t + c_2, \tag{20}$$

where c_2 is a second integration constant. We also assume that the position x_0 at $t = 0$ is given initially. Then, evaluating (20) at $t = 0$ gives $x_0 = c_2$, so that

$$x = -\frac{1}{2}gt^2 + v_0 t + x_0. \tag{21}$$

Equations (19) and (21) are well-known formulas in physics and are used to solve for quantities such as the maximum height of a thrown object. We do not recommend memorizing (19) or (21). Instead, they should be derived in each case from the differential equation (17). If the applied force depends only on time and is not constant, then the formulas for velocity and position may be obtained by integration. If the applied force depends on other quantities, solving the differential equation is not so simple. We describe some more difficult problems in Section 1.11 and Chapter 2.

Example 1.2.4 *Motion with Gravity*

Suppose a ball is thrown upward from ground level with velocity v_0 and the only force is gravity. How high does the ball go before falling back toward the ground?

• SOLUTION. The differential equation (as before) is (17):

$$\frac{d^2x}{dt^2} = -g. \tag{22}$$

The initial conditions are that $x = 0$ and $\frac{dx}{dt} = v_0$ at $t = 0$. By successive integrations of (22) and by applying the initial conditions, we obtain

$$\frac{dx}{dt} = -gt + v_0 \tag{23}$$

and

$$x = -\frac{1}{2}gt^2 + v_0 t. \tag{24}$$

From (24), the height is known as a function of time. To determine the maximum height, we must first determine the time at which the ball reaches this height. From calculus, the maximum of a function $x = x(t)$ occurs at a critical point where $\frac{dx}{dt} = 0$. At the maximum height, the ball has stopped rising and has not started to fall, so the velocity is zero. Thus, the time of the maximum height is determined from (23):

$$0 = -gt + v_0, \quad \text{or equivalently,} \quad t = \frac{v_0}{g}. \tag{25}$$

When this time (25) is substituted into (24), a formula for the maximum height y is obtained:

$$y = -\frac{1}{2}g\left(\frac{v_0}{g}\right)^2 + v_0\frac{v_0}{g} = \frac{v_0^2}{g}\left(-\frac{1}{2} + 1\right) = \frac{v_0^2}{2g}. \tag{26}$$

♦

Example 1.2.5 *Car Braking*

Suppose that a car is going 76 m/s when brakes are applied at $t = 2$ s. Suppose that the nonconstant deceleration is known to be $a = -12t^2$. Determine the distance the car travels.

• SOLUTION. Let x measure the distance traveled once the brakes are applied. The differential equation is

$$\frac{d^2x}{dt^2} = -12t^2. \tag{27}$$

By integrating (27) we obtain

$$\frac{dx}{dt} = -4t^3 + c_1. \tag{28}$$

By integrating (28) again, we obtain

$$x = -t^4 + c_1 t + c_2. \tag{29}$$

The initial conditions are $x = 0$ and $\frac{dx}{dt} = 76$ when $t = 2$. The initial velocity applied to (28) gives

$$76 = -4(2)^3 + c_1 = -32 + c_1,$$

so that $c_1 = 108$. The initial position applied to (29) gives

$$0 = -2^4 + c_1 t + c_2 = -16 + 216 + c_2,$$

and $c_2 = -200$. To determine the distance the car travels, we note that the car stops when the velocity is zero. That is,

$$\frac{dx}{dt} = -4t^3 + 108 = 0,$$

so that the stopping time is $t^3 = 27$ or $t = 3$. Then the distance traveled at time $t = 3$ is given by substituting $t = 3$ into (29):

$$x = -3^4 + 108(3) - 200 = -81 + 324 - 200 = 43 \text{ m}.$$ ♦

Exercises

In Exercises 1–8, verify that the given function x is a solution of the differential equation.

1. $x = 2e^{3t} + 1$, $\dfrac{dx}{dt} = 3x - 3$.
2. $x = t^2$, $\dfrac{dx}{dt} = 2\dfrac{x}{t}$.
3. $x = t - 1$, $\dfrac{dx}{dt} = \dfrac{x}{t-1}$.
4. $x = t^4$, $\dfrac{dx}{dt} = 4\dfrac{x^2}{t^5}$.
5. $x = e^{t^2}$, $\dfrac{dx}{dt} = 2tx$.
6. $x = 3e^{4t} + 2$, $\dfrac{dx}{dt} = 4x - 8$.
7. $x = e^{-2t}$, $\dfrac{dx}{dt} = -2e^{2t}x^2$.
8. $x = t + 3$, $\dfrac{dx}{dt} = \dfrac{x-3}{t}$.

In Exercises 9–16, find the general solution. If you cannot find an explicit integral solution, use a definite integral. Exercises with an asterisk (*) require definite integrals.

9. $\dfrac{dx}{dt} = 3e^t$.

10. $\dfrac{dx}{dt} = 8e^{-t}$.

11. $\dfrac{dx}{dt} = -5\cos 6t$.

12. $\dfrac{dx}{dt} = 2\sin 7t$.

13.* $\dfrac{dx}{dt} = 8\cos(t^{-1/2})$.

14.* $\dfrac{dx}{dt} = 4\sin(t^{-1/3})$.

15.* $\dfrac{dx}{dt} = \ln(4 + \cos^2 t)$.

16.* $\dfrac{dx}{dt} = e^{t^2}$.

In Exercises 17–22, find the solution of the initial value problem as an explicit or definite integral* as appropriate.

17. $\dfrac{dx}{dt} = t^4$ with $x(2) = 3$.

18. $\dfrac{dx}{dt} = t^{3/2}$ with $x(3) = 7$.

19.* $\dfrac{dx}{dt} = \dfrac{\ln t}{4 + \cos^2 t}$ with $x(2) = 5$.

20.* $\dfrac{dx}{dt} = \dfrac{\sin t}{5 + \ln t}$ with $x(8) = 2$.

21.* $\dfrac{dx}{dt} = \dfrac{e^t}{1 + t}$ with $x(1) = 3$.

22.* $\dfrac{dx}{dt} = \cos(t^{-3})$ with $x(2) = 4$.

In Exercises 23–27, assume that when an automobile brakes, there is a constant deceleration. (A more physically accurate model would have acceleration depend on velocity.)

23. At an accident scene, the police investigator is attempting to determine from the rubber marks on the road how fast the car was going. Suppose it is known that this particular car brakes with a deceleration of 15 m/s^2. At what velocity was the car going at the moment it applied its brakes, if the car traveled 75 m before it stopped?
24. Repeat Exercise 23 with the observation that the car traveled 75 m before stopping, but assume that the car decelerates at only 10 m/s^2.
25. An overly cautious traveling distance between you and the car in front would be the distance it takes you to stop. At a speed of 60 km/h, how far does a car travel if it decelerates at 2500 km/h^2?
26. Referring to Exercise 25, with the same deceleration, how far does a car travel at 120 km/h?

27. Referring to Exercise 25, with the same deceleration, determine how far a car travels as a function of its speed before braking.
28. Suppose a car is going 50 km/h when the brakes are applied at $t = 0$. Determine the distance the car travels. Suppose the nonconstant deceleration is known to be $a = -6t$.
29. Suppose a car is going 50 km/h when the brakes are applied at $t = 2$ h. Determine the distance the car travels. Suppose the nonconstant deceleration is known to be $a = -6t$.
30. How long must you slide a book in order for it to fall off a 15 m table if the book decelerates such that $a = -5$ m/s^2?
31. Assume that a snowplow of width w meters moves forward along a road so that it clears a constant volume of snow per hour, Q m^3/h. Assume that the snow started falling at 8 A.M. ($t = 0$) and falls at a constant rate of c m/h.
 a) Show that $dx/dt = 1/(kt)$, where $x(t)$ is the position of the snowplow and $k = wc/Q$.
 b) Where will the snowplow be at noon, if the snowplow did not start moving until 11 A.M.?
32. A toy rocket is fired upward from ground level at $t = 2$ s. Determine the maximum height of the rocket if the velocity at $t = 2$ is 76 m/s. Assume that the nonconstant acceleration of the rocket is known to be $a = -12t^2$.
33. At what velocity should a ball be thrown upward if it is to reach a maximum height of 100 m?
34. At what velocity should a ball be thrown upward if it is to reach its maximum height in 10 s?
35. Suppose a brick is dropped from a construction tower of height 200 m. How long will it take the brick to fall?
36. The general solution of (10) can be written $x = \int_0^t e^{-\bar{t}^2} d\bar{t} + c$. Determine c so that x satisfies the initial condition $x(3) = 7$, and show that (11) is valid.
37. The general solution of (12) can be written $x = \int_0^t f(\bar{t}) d\bar{t} + c$. Determine c so that x satisfies the initial condition $x(t_0) = x_0$, and show that (13) is valid.

1.3 First-Order Separable Differential Equations

Another class of first-order differential equations that can be solved by integration is the separable differential equations. A first-order differential equation $\frac{dx}{dt} = f(t, x)$ is called **separable** if the function $f(t, x)$ can be written as a product of a function of t and a function of x:

$$\frac{dx}{dt} = h(t)g(x). \tag{1}$$

There are two ways to describe solving (1). They both end up performing the same calculations. One is to divide (1) by $g(x)$ to get

$$\frac{1}{g(x)} \frac{dx}{dt} = h(t). \tag{2}$$

Integrating both sides with respect to t gives

$$\int \frac{1}{g(x)}\frac{dx}{dt}dt = \int h(t)dt.$$

By the chain rule for integration, this equation is the same, using $x = x(t)$, as

$$\int \frac{1}{g(x)}dx = \int h(t)dt. \tag{3}$$

Alternatively, the variables t and x in (1) can be separated using differentials:

$$\frac{dx}{g(x)} = h(t)dt. \tag{4}$$

Integration of (4) then gives (3).

The earlier examples $\frac{dx}{dt} = f(t)$ of Section 1.2 as well as differential equations in the form $\frac{dx}{dt} = g(x)$ are separable.

Note that both (2) and (4) assume that $g(x) \neq 0$ for every x. If $x = a$ is a constant such that $g(a) = 0$, then an easy calculation shows that $x = a$ is a constant solution of (1):

> If $g(a) = 0$, then $x = a$ is a constant solution of $\dfrac{dx}{dt} = h(t)g(x)$.

These constant solutions are sometimes lost in integrating (4), so it is necessary to check if they are all accounted for in the solutions.

Example 1.3.1 *Separable Equation*

The first-order differential equation

$$\frac{dx}{dt} = x^2 \cos t \tag{5}$$

is separable. Separating variables gives

$$\int \frac{dx}{x^2} = \int \cos t dt.$$

Hence, by integration,

$$-\frac{1}{x} = \sin t + c. \tag{6}$$

Solving for x yields the solutions

$$x = \frac{-1}{\sin t + c}. \tag{7}$$

Since we divided by x^2, the calculation assumed that $x \neq 0$. Note that $x = 0$ satisfies the differential equation (5) but is not included in the solutions (7) for any finite value of c. The solutions of (5) thus consist of (7) and $x = 0$.

Suppose that we also had the initial condition $x(1)=4$. Letting $t=1$ and $x=4$ in (6), we would determine that $c=-\frac{1}{4}-\sin 1$, so that the solution of the initial value problem would be

$$x=\frac{-1}{\sin t-(1/4)-\sin 1}.$$

Note that it is easier to determine c by applying the initial condition to (6) than to (7). ♦

Some separable differential equations (see the next example) are more difficult because obtaining integrals may not be as easy as in the previous example.

Example 1.3.2 *Separable Example Using Partial Fractions*

The differential equation

$$\frac{dx}{dt}=x(1-x) \tag{8}$$

is called the logistic equation and is very important in the field of population dynamics. We will solve it by separation. We will analyze it again in a different manner in Chapter 5.

• SOLUTION. Equation (8) is separable:

$$\frac{dx}{x(1-x)}=dt, \tag{9}$$

so that

$$\int\frac{dx}{x(1-x)}=\int dt. \tag{10}$$

The integral on the left can be evaluated in several ways. Using partial fractions, for example, we get

$$\frac{1}{x(1-x)}=\frac{A}{x}+\frac{B}{x-1}.$$

Multiplying by $x(x-1)$ yields

$$-1=A(x-1)+Bx. \tag{11}$$

Since $x(x-1)$ has simple linear factors x and $x-1$, we may evaluate (11) at the roots, $x=0$ and $x=1$, to obtain $A=1$ and $B=-1$. Thus, (10) becomes

$$\int\frac{1}{x}dx-\int\frac{1}{x-1}dx=\int dt.$$

Carrying out the integrations gives $\ln|x|-\ln|x-1|=t+c$, or equivalently,

$$\ln\frac{|x|}{|x-1|}=t+c. \tag{12}$$

Keeping the absolute value is proper. Exponentiation of both sides of (12) yields

$$\frac{|x|}{|x-1|}=e^{t+c}=e^{c}e^{t},$$

using properties of the exponential. Here e^c is an arbitrary positive constant. The absolute value signs can be eliminated since

$$\left|\frac{x}{x-1}\right| = e^c e^t \quad \text{implies} \quad \frac{x}{x-1} = \pm e^c e^t.$$

Thus we have

$$\frac{x}{x-1} = ke^t,$$

where $k = \pm e^c$ is an arbitrary constant (positive or negative). We then multiply by $x-1$ and solve for x to get the solution

$$x = \frac{ke^t}{-1 + ke^t}. \tag{13}$$

Since $x(x-1) = 0$ if $x = 0$ or $x = 1$, it follows that $x = 0$ and $x = 1$ are solutions of (8). The solution $x = 0$ corresponds to $k = 0$. However, $x = 1$ is not included in (13) unless we allow $k = \infty$. Thus, the solutions of (8) are (13) and $x = 1$. The general solution involves one arbitrary constant k, which can be determined from any initial conditions. ♦

These algebraic steps involved in obtaining the general solution may be difficult for some students. For the integration step, tables of integrals will be helpful for some. It is not easy to understand the behavior of the solution of the differential equation (8) from this general solution (13). In particular, the cases $k < 0$, $0 < k < 1$, and $k > 1$ turn out to have different behavior.

Philosophically, we think of solving separable equations as being very easy, although this example shows that in practice there can be substantial complications. A different method for analyzing and understanding qualitatively the solution of differential equations such as (8) is introduced in Chapter 5.

Equation (12) is an **implicit solution** of (8), since it does not directly give x in terms of t. Equation (13) is an **explicit solution**, since x is explicitly in terms of t. For many problems, it is impossible to solve the implicit solution to get a formula for x. However, given values of c and t, we can usually solve the implicit equation numerically, say by Newton's method or bisection, to determine the value of x. Thus, the implicit solution is still useful.

1.3.1 Using Definite Integrals for Separable Differential Equations

Definite integrals can also be used on the separable differential equation $\frac{dx}{dt} = h(t)g(x)$. Suppose that the initial condition is $x(t_0) = x_0$. Then using definite integrals we get

$$\int_{t_0}^{t} \frac{1}{g(\bar{y})} \frac{d\bar{x}}{d\bar{t}} d\bar{t} = \int_{t_0}^{t} h(\bar{t}) d\bar{t}$$

or

$$\int_{x_0}^{x} \frac{d\bar{x}}{g(\bar{x})} = \int_{t_0}^{t} h(\bar{t}) d\bar{t}. \tag{14}$$

We consider (14) to represent the solution to the separable differential equation. Often, a more explicit solution is desired. Whether a simpler representation of the solution of a specific separable differential equation may be obtained depends on the question from calculus of evaluating the integrals.

Example 1.3.3 *Separable with Definite Integral*

The general solution of the initial value problem for (8) can be represented by a definite integral:

$$\int_{x_0}^{x} \frac{d\overline{x}}{\overline{x}(1-\overline{x})} = t - t_0.$$

This can be shown to be equivalent to (13). (Note to students: You will have to ask your instructor whether this approach will be acceptable.) ♦

Exercises

In Exercises 1–14, solve the differential equation. Definite integrals are necessary in problems with an asterisk (*). Some exercises are in terms of variables other than x, t.

1. $\dfrac{dx}{dt} = \dfrac{x+1}{t}$.
2. $\dfrac{dr}{d\theta} = \dfrac{r^2+r}{\theta}$.
3. $\dfrac{dx}{dt} = e^t$.
4. $\dfrac{dx}{dt} = e^{t+x}$.
5. $\dfrac{dx}{dt} = tx + 4x + 3t + 12$.
6. $\dfrac{du}{dt} = \dfrac{u^2+4}{t^2+4}$.
7. $\dfrac{dx}{dt} = 3$.
8. $\dfrac{dx}{dt} = t^2x^2$.
9. $\dfrac{dx}{dt} = x^5,\ x(2) = 1$.
10. $\dfrac{dx}{dt} = tx - 2x + t - 2$.

11.* $\dfrac{dx}{dt} = x^2\cos(t^2),\ x(0) = 1$.

12.* $\dfrac{dx}{dt} = x^4\cos(t^{-1/2})$.

13.* $\dfrac{dx}{dt} = t\cos(x^{-1/2}),\ x(1) = 2$.

14.* $\dfrac{dx}{dt} = e^{t^2+x^2}$, $x(2) = 4$

In Exercises 15–18, solve the differential equation.

15. $\dfrac{du}{dt} = \dfrac{t^2+1}{u^2+4}$, $u(0) = 1$.

16. $\dfrac{dr}{d\theta} = \sin r$.

17. $\dfrac{dx}{dt} = t^2x^2 + x^2 + t^2 + 1$, $x(0) = 2$.

18. $\dfrac{du}{dt} = u^3 - u$.

The differential equations in Exercises 19–21 require partial fractions.

19. $\dfrac{dx}{dt} = x(x-1)$.

20. $\dfrac{dx}{dt} = x^2(x-1)^2$.

21. $\dfrac{dx}{dt} = (x-1)(x-2)^2$.

In Exercises 22–28, a first-order differential equation is written in differential form. Solve the differential equation.

22. $x^2 dt + t^3 dx = 0$.

23. $(tx + x)dt + (tx + t)dx = 0$.

24. $r \sin\theta dr + \cos\theta d\theta = 0$.

25. $(t^2 - 4)dz + (z^2 - 9)dt = 0$.

26. $(x^2 + x)dt + (t^3 + 4t^2)dx = 0$.

27. $e^{t+x} dt + e^{2t-3x} dx = 0$.

28. $te^x dt + xe^{-t} dx = 0$.

29. Show that if we make the substitution $z = at + bx + c$ and $b \neq 0$, then the differential equation

$$\frac{dx}{dt} = f(at + bx + c)$$

is changed to a differential equation in z and t that can be solved by separation of variables.

In Exercises 30 through 34, solve the differential equation using Exercise 29.

30. $\dfrac{dx}{dt} = (t + x)^2$.

31. $\dfrac{dx}{dt} = (t + 4x - 1)^2$.

32. $\dfrac{dx}{dt} = \tan(-t + x + 1) + 1$.

33. $\dfrac{dx}{dt} = e^{t+x}(t+x)^{-1} - 1.$

34. $\dfrac{dx}{dt} = \dfrac{t+x+2}{t+x+1}.$

1.4 Direction Fields

We can often get a good pictorial or graphical idea of what the solutions $x(t)$ of a first-order differential equation,

$$\frac{dx}{dt} = f(t, x), \tag{1}$$

look like without actually solving the differential equation. The method we describe is based on tangent lines and is easily implemented on a computer. Software that does this is readily available and is part of most general mathematical software such as MATLAB, Scilab, MAPLE, or Mathematica. It is highly recommended that students take advantage of the computer facilities that are available to them.

We are interested in graphing x as a function of t, where $x(t)$ satisfies the differential equation (1). We first set up a t, x coordinate system, and note that at any point (t, x) the slope $\frac{dx}{dt}$ of the tangent line to the solution x at that point is given by the differential equation $\frac{dx}{dt} = f(t, x)$. Through each point (t, x), we plot a short dash (line segment) with slope given by the differential equation, as shown for one point in figure 1.4.1. Continuing in this manner at other points, we can build up a picture. The resulting plot is sometimes referred to as a sketch of the **direction field** or **vector field** of the solutions.

Example 1.4.1 *Slope at a Point*

Suppose the differential equation is

$$\frac{dx}{dt} = tx^2.$$

Find the slope of the tangent lines to the solution through $t = 1$, $x = 2$ and the solution through $t = -1$, $x = -3$, and plot short segments of both tangent lines.

• SOLUTION. The differential equation says that the slope at (t, x) is tx^2. Thus the slope at $(1, 2)$ is $1(2)^2 = 4$ and the slope at $(-1, -3)$ is $(-1)(-3)^2 = -9$. These points and slopes are graphed in figure 1.4.2. ♦

This example suggests a procedure that is fairly easy to implement on a computer with plot or graphic capabilities. The points would normally be chosen systematically on some grid in the t, x-plane. At each grid point (t_i, x_i), a short line segment is drawn with slope $f(t_i, x_i)$ given by the differential equation. The line segment is the tangent line at that point to the solution going through that point. We can connect the line segments in a smooth way to approximate the solution of the differential equation that

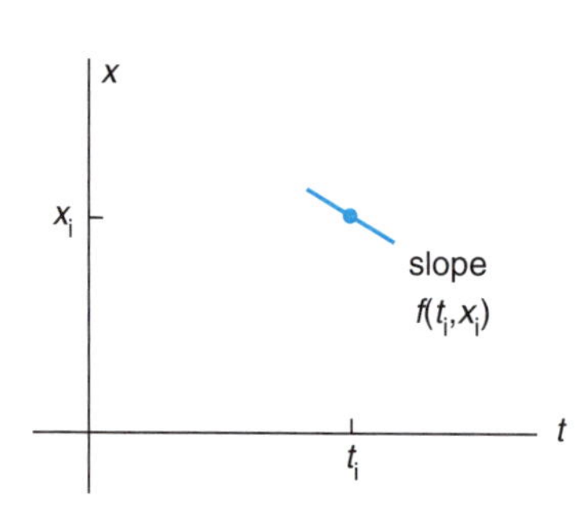

Figure 1.4.1 Slope $\frac{dx}{dt} = f(t, x)$ at a point.

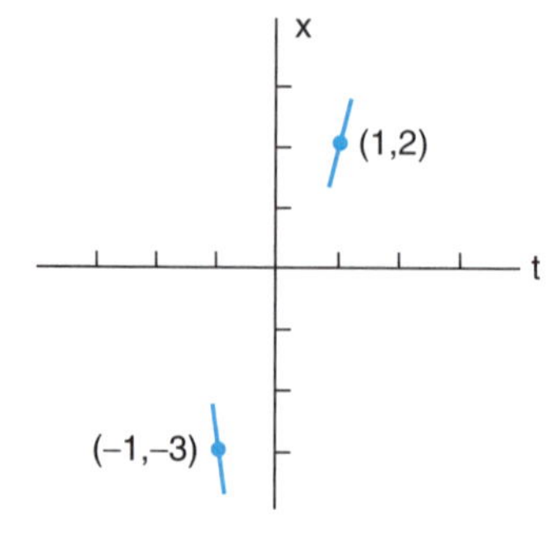

Figure 1.4.2 Slopes for Example 1.4.1.

goes through the point (t_i, x_i). This technique is most practical when the direction field is obtained on a computer. The computer can often sketch the general solution using numerical methods. The general solution is an infinite family of solutions, one corresponding to each initial condition, so that graphing all solutions is impossible. However, the computer can also be used to sketch solutions corresponding to many initial conditions. We discuss three examples.

Direction Field for Integrable Example

For the first-order differential equation $\frac{dx}{dt} = f(t, x)$, in which the $f(t, x)$ only depends on t,

$$\frac{dx}{dt} = f(t), \tag{2}$$

the direction field and its solution are simpler than the general case. The sketch of the direction field is easier to obtain because the slope of the solution $\frac{dx}{dt}$ at a given t is the same for all values of x. That is, along all vertical lines in the t, x-plane, the slope of the solution will be the same, $f(t)$. All solutions will differ only by a constant, since

$$x = \int f(t)dt + c \tag{3}$$

is the general solution. Thus, the graph of the solutions (obtained either exactly or from the direction field) will be an infinite family of curves, all parallel to one another, displaced vertically from one another. This displacement corresponds to the arbitrary additive constant in the general solution.

Example 1.4.2 *Direction Field*

Consider

$$\frac{dx}{dt} = t(1 - t). \tag{4}$$

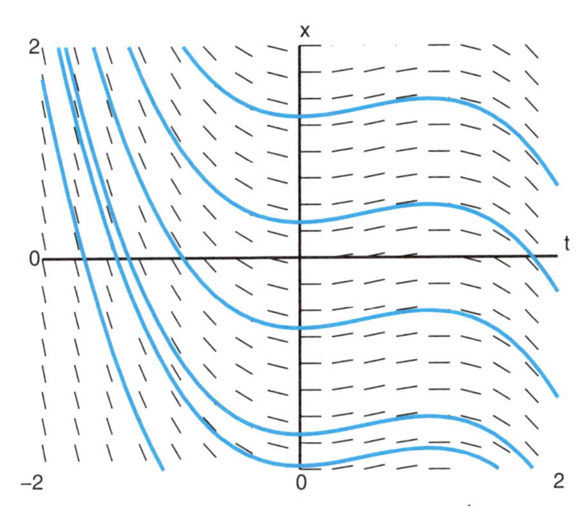

Figure 1.4.3 Direction field and solutions of (4), $\frac{dx}{dt} = t(1-t)$.

The direction field is graphed in figure 1.4.3, along with some of the solutions. The exact general solution is

$$x = \frac{1}{2}t^2 - \frac{1}{3}t^3 + c. \tag{5}$$

In figure 1.4.3, all solution curves are displaced vertically from one another. Note that the local extrema of the solutions are at $t = 0$ and $t = 1$, since from (4) that is where $\frac{dx}{dt} = 0$. ♦

Direction Field for Autonomous Example

We next determine the direction field for the important class of first-order equations $\frac{dx}{dt} = f(t, x)$, in which $f(t, x)$ does not depend on t:

$$\frac{dx}{dt} = g(x). \tag{6}$$

Such an equation is called **autonomous**. These equations are discussed extensively (including a brief explanation of the name) in Chapter 5. The direction field is simpler to obtain, since the slope of the solution $\frac{dx}{dt}$ is the same value $g(x)$ along horizontal lines. By separation, the solution of (6) is

$$\int \frac{dx}{g(x)} = t + c. \tag{7}$$

Hence all solutions are parallel to one another and are shifted horizontally from one another by a constant.

Example 1.4.3 *Direction Field*

Consider

$$\frac{dx}{dt} = x(1-x). \tag{8}$$

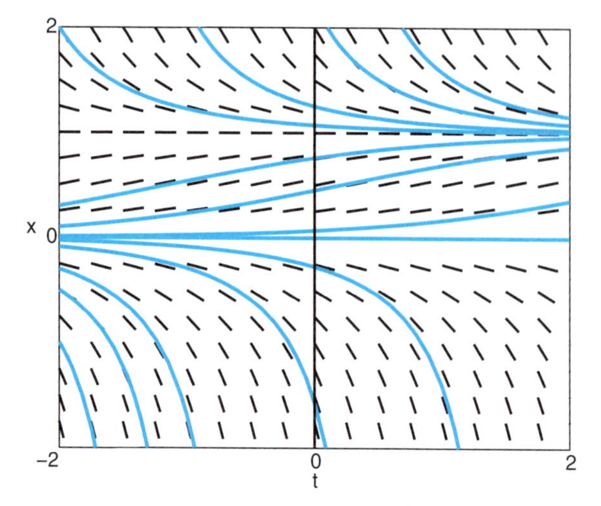

Figure 1.4.4 Direction field for (8), $\frac{dx}{dt} = x(1-x)$.

The direction field and solutions are sketched in figure 1.4.4. Note that all solutions are shifted horizontally from one another. The general solution, which expresses the horizontal shifts, is given in (12) of Section 1.3. This equation is an important model in population dynamics and is discussed in Chapter 5. ♦

As a rule, the general solution of a first-order differential equation represents an infinite family of curves that are not just horizontal or vertical shifts of one another.

Example 1.4.4 *Direction Field*

Consider

$$\frac{dx}{dt} = x^2 - t. \tag{9}$$

A sketch of the direction field obtained by a computer is shown in figure 1.4.5. From figure 1.4.5, we see that the differential equation has many solutions. However, there is a unique solution (one curve) going through each point. To find the solution of the initial value problem corresponding to the initial condition $x\left(\frac{1}{2}\right) = 1$, we locate the specific curve that goes through the point $\left(\frac{1}{2}, 1\right)$. Unlike in the previous examples, the solution curves are not shifts of one another. ♦

From a plot such as that in figure 1.4.5, we can often get a good idea of what the graph of the solutions of a first-order differential equation looks like. This procedure is not much different from determining the graph of a function by plotting a large number of points. How do we know how fine a grid to choose? How do we know that the solutions do not change their behavior at a distance from the grid? What values of t or x are important? What is the behavior of the differential equation near these important points? Why do the solutions of the differential equations behave the way they are observed numerically to act? The direction field method is basically a numerical method and, while useful, it is far from the best numerical

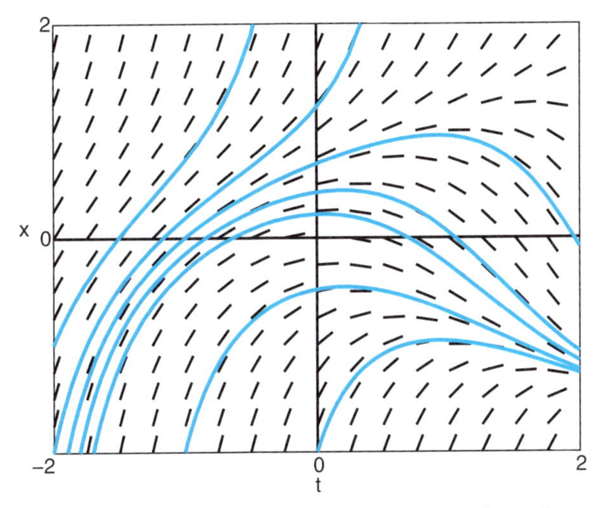

Figure 1.4.5 Direction field and solutions of (9), $\frac{dx}{dt}=x^2-t$.

method. Section 1.5 will discuss how to view it as a numerical method. We often need to get a better understanding of the solutions of differential equations than can be obtained this way. That is one reason we wish to study differential equations further.

Exercises

In Exercises 1–6, the direction field of the given differential equations has been plotted by a computer. Sketch some solutions of the differential equation. (*Hint*: First copy or trace the figure.)

1. $\dfrac{dx}{dt}=x(1-x^2)$.

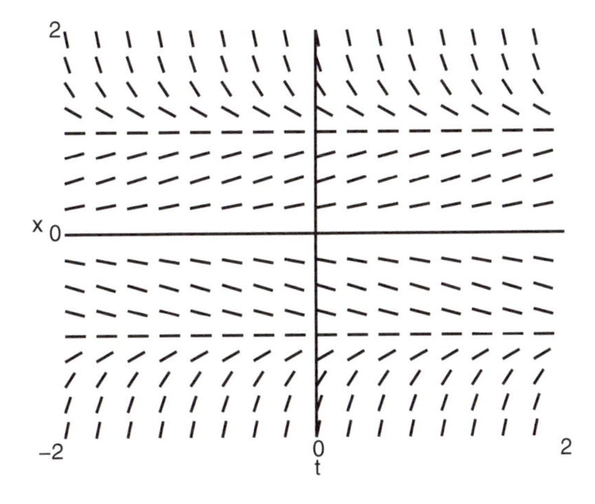

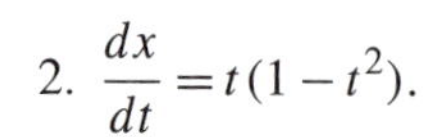

2. $\dfrac{dx}{dt} = t(1 - t^2)$.

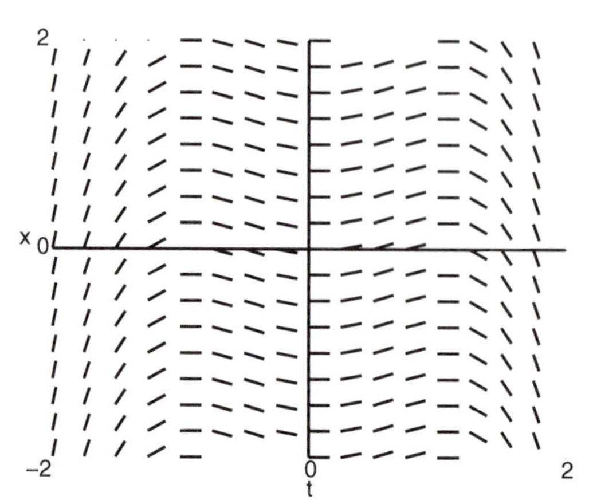

3. $\dfrac{dx}{dt} = x(t - x)$.

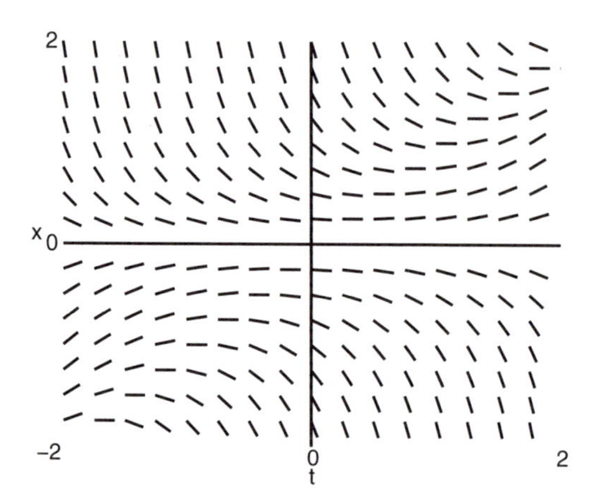

4. $\dfrac{dx}{dt} = t(x - t)$.

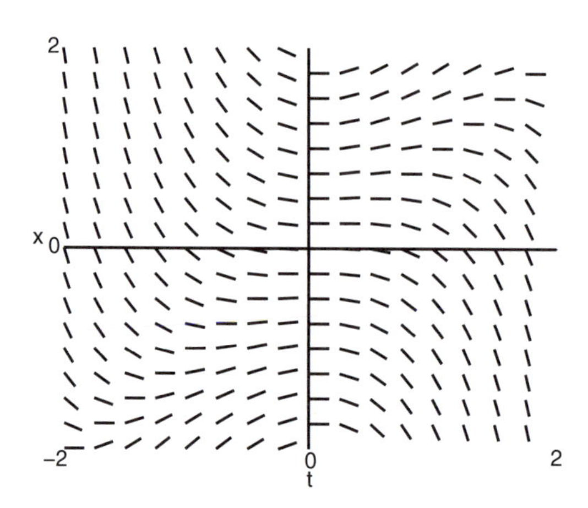

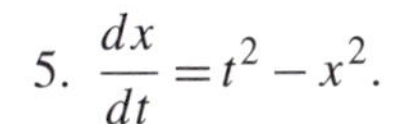

5. $\dfrac{dx}{dt} = t^2 - x^2.$

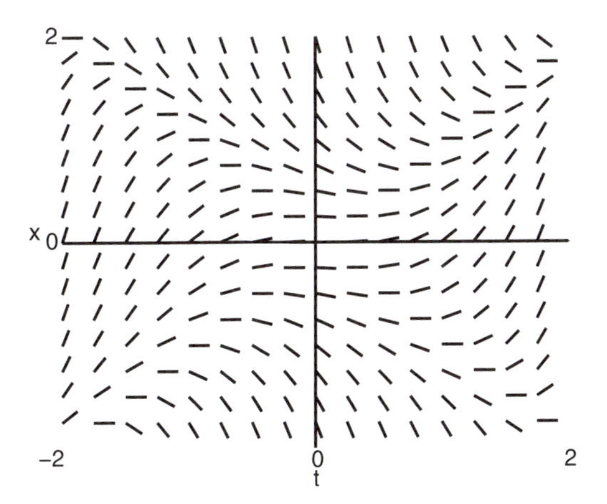

6. $\dfrac{dx}{dt} = t^2 - x.$

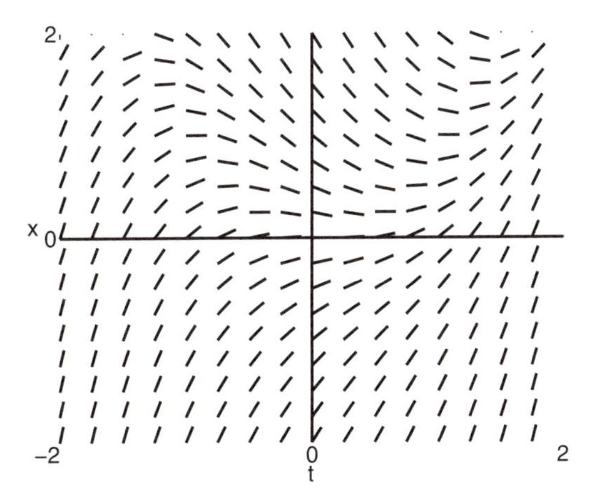

1.4.1 Existence and Uniqueness

Often the first-order differential equation $\frac{dx}{dt} = f(t, x)$ cannot be solved by simple integration. It is then important to know when there are solutions. As with the previous elementary examples, the differential equation usually has many solutions, of which only one will satisfy given initial conditions. A more precise statement of the mathematical result which guarantees this is the following.

THEOREM 1.4.1 **Basic existence and uniqueness theorem** *for the initial value problem for first-order differential equations. There exists a unique solution to the differential equation*

$$\frac{dx}{dt} = f(t, x), \tag{10}$$

which satisfies given initial conditions

$$x(t_0) = x_0, \tag{11}$$

if both the function $f(t, x)$ and its partial derivative $\frac{\partial f}{\partial x}$ are continuous functions of t and x at and near the initial point $t = t_0$, $x = x_0$.

In most cases of practical interest, the continuity conditions are satisfied for most values of (t_0, x_0), so that there exists a unique solution to the initial value problem for most initial conditions.

If the conditions of the basic existence and uniqueness theorem are not met at a point (t_0, x_0), then solutions may not exist at (t_0, x_0), or there may be more than one solution passing through the same point (t_0, x_0), or the solution may not be differentiable at t_0. Such behavior often has a physical interpretation. For example, certain types of friction and the switching behavior of some controllers can both lead to places where solutions are not differentiable.

Example 1.4.5 *Local Existence of a Solution*

The differential equation

$$\frac{dx}{dt} = tx^2 \tag{12}$$

satisfies the conditions of the uniqueness theorem for all (t, x) since $f = tx^2$ and $\frac{\partial f}{\partial x} = 2tx$ are continuous functions of t and x. Thus, there should be a unique solution of the initial value problem for any (t_0, x_0). For example, there should be a solution at the initial condition $(0, 2)$. Solving the differential equation (12) by separation gives

$$\int \frac{dx}{x^2} = \int t\, dt.$$

After integration and applying the initial condition, we obtain

$$-\frac{1}{x} + \frac{1}{2} = \frac{1}{2}t^2.$$

Solving for x yields the promised unique solution

$$x = \frac{2}{1 - t^2}. \tag{13}$$

However, this solution (13) is not defined for all t. It becomes infinite at $t = \pm 1$. While the basic existence and uniqueness theorem guarantees a solution of the differential equation which satisfies the initial condition, it does not say that the solution is defined for all t. It is a local existence theorem, only guaranteeing that the solution exists for some t near the initial t_0. In this example, the solution "explodes" at $t = 1$. ♦

The direction field for (12) is graphed in figure 1.4.6. The specific initial condition $x(0) = 2$ and others are shown. One inadequacy of the direction field method is that it may be impossible to tell whether solutions become infinite or not. From the figure it

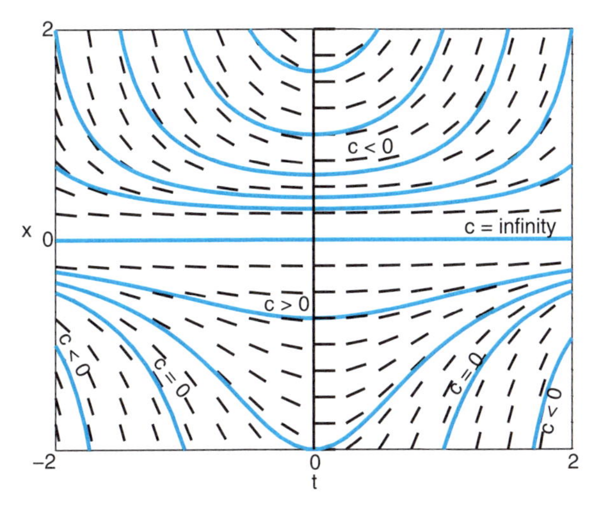

Figure 1.4.6 Direction field and some solutions for (12), $\frac{dx}{dt} = tx^2$.

is impossible to tell what occurs outside the plotted region. However, by separation, it can be shown that all solutions of (12) are described by

$$x = \frac{-2}{t^2 + c} \quad \text{and} \quad x = 0.$$

Note that $x = 0$ is a solution which corresponds to $c = \infty$. If $c > 0$, solutions exist for all t, while solutions have vertical asymptotes if $c \leq 0$. The case $c = 0$, that is, $x = \frac{-2}{t^2}$ separates solutions which become infinite from those that do not. This fact cannot be determined from the direction field.

Example 1.4.6 *Initial Condition with no Solution*

The differential equation

$$\frac{dx}{dt} = -\frac{1}{t^2} \tag{14}$$

satisfies the basic existence and uniqueness theorem everywhere $t_0 \neq 0$, since $f(t, x) = -\frac{1}{t^2}$ is continuous everywhere $t \neq 0$. By antidifferentiation, the general solution of (14) is

$$x = \frac{1}{t} + c.$$

If the initial condition is given at $t_0 \neq 0$, then the constant c can be determined uniquely by the initial condition. There is a unique solution to the initial value problem in this case, and the solution exists locally as in the previous example up to the singularity at $t = 0$. However, if the initial condition is given at $t_0 = 0$, then there are no solutions, as shown in figure 1.4.7. For example, there are no solutions of (14) with $x = 1$ at $t = 0$ and we do not have existence. ♦

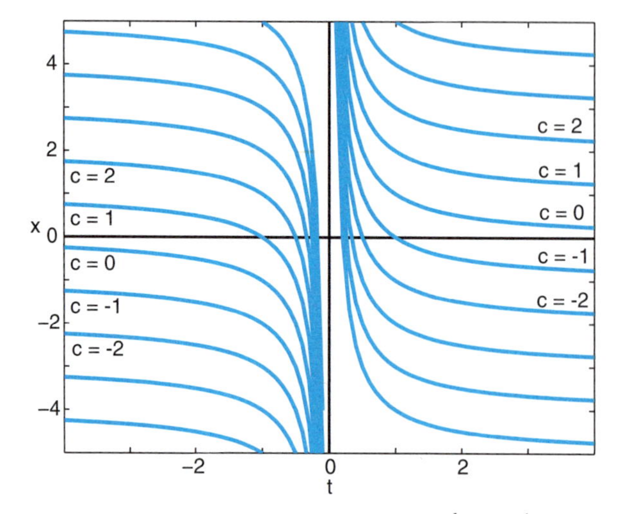

Figure 1.4.7 Several solutions of (14), $\frac{dx}{dt} = -\frac{1}{t^2}$.

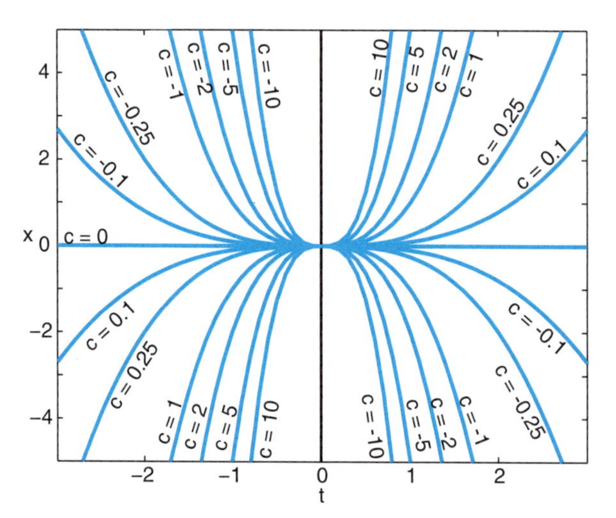

Figure 1.4.8 Several solutions $x = ct^3$ of (15), $\frac{dx}{dt} = \frac{3x}{t}$.

Example 1.4.7 *Solutions not Unique*

By separation we can show that the general solution of

$$\frac{dx}{dt} - 3\frac{x}{t} = 0 \tag{15}$$

is $x = ct^3$. Several solutions of (15) are plotted in figure 1.4.8. The function $f(t, x) = 3x/t$ is continuous everywhere that $t \neq 0$. There do not exist solutions through $(0, x)$ if $x \neq 0$. On the other hand, every solution goes through $(0, 0)$, so that at $(0, 0)$ we have existence but not uniqueness. ♦

Example 1.4.8 *Continuous but not Differentiable*

The differential equation

$$\frac{dx}{dt} = x^{1/3} \tag{16}$$

has $f(t, x) = x^{1/3}$ continuous for all t, x, but $\frac{\partial f}{\partial x}$ is not continuous if $x = 0$. Using separation of variables the solutions are

$$x = \left(\frac{3}{2}t + c\right)^{3/2} \quad \text{and} \quad x = 0.$$

If the initial condition is given as $x(t_0) = 0$, then there are two solutions, $x = 0$ and $x = \left(\frac{3}{2}t - \frac{3}{2}t_0\right)^{3/2}$. ♦

Exercises

For each of the differential equations in Exercises 1–10, give those points (t_0, x_0) in the t, x-plane for which the basic existence and uniqueness theorem guarantees that a unique solution exits.

1. $\dfrac{dx}{dt} = \dfrac{x}{1+t^2}$.
2. $\dfrac{dx}{dt} = \dfrac{t-x}{3t-7x}$.
3. $\dfrac{dx}{dt} = (1-t^2-x^2)^{7/3}$.
4. $\dfrac{dx}{dt} = (1-t^2-x^2)^{1/3}$.
5. $\dfrac{dx}{dt} = (x+t)^{1/5}$.
6. $t\dfrac{dx}{dt} = x^2 + 1$.
7. $(x-1)\dfrac{dx}{dt} = \cos t$.
8. $\dfrac{dx}{dt} = \dfrac{t^3+1}{t(x+1)}$.
9. $\dfrac{dx}{dt} = (1-x^2-2x^2)^{3/2}$.
10. $\dfrac{dx}{dt} = x^{1/3}\sqrt{x-1}$.

Exercises 11 and 12 refer to the differential equation

$$\frac{dx}{dt} = t^{-1/3}. \tag{17}$$

11. Show that the basic Existence and uniqueness theorem holds for (17) for all (t_0, x_0) such that $t_0 \neq 0$.
12. Show that, for any constant x_0, that there exists a unique solution $x(t)$ such that
   a) x satisfies (17) if $t \neq 0$.
   b) x is continuous for all t.
   c) $x(0) = x_0$.
   d) x is not differentiable at $t_0 = 0$.

In Exercises 13–19, you are given a family of curves $g(t, x) = c$. For each exercise,

a) Verify the curves implicitly define solutions of the given differential equation.
b) Graph the curves on the same axis for the indicated values of c.
c) Determine for which points (t_0, x_0) the assumptions of the basic existence and uniqueness theorem fail to hold, and graph those points on the same graph.

13. $t^2 - x^2 = c, \ \dfrac{dx}{dt} = \dfrac{t}{x}, \ c = 0, \pm 1, \pm 2.$

14. $t^2 + x^2 = c, \ \dfrac{dx}{dt} = -\dfrac{t}{x}, \ c = 1, 2, 3.$

15. $t + x^2 = c, \ 2x\dfrac{dx}{dt} + 1 = 0, \ c = 0, \pm 1.$

16. $x = ct, \ \dfrac{dx}{dt} = \dfrac{x}{t}, \ c = 0, \pm 1, \pm 2.$

17. $x = c \sin t, \ \dfrac{dx}{dt} = x \cot t, \ c = 0, \pm 1, \pm 2.$

18. $x = \tan(t + c), \ \dfrac{dx}{dt} = 1 + x^2, \ c = 0, \pm 1.$

19. $x = \dfrac{1}{t + c}, \ \dfrac{dx}{dt} = -x^2, \ c = 0, \pm 1.$

20.
   a) Show that $x = 0$ and $x = \left(\frac{2}{3}t + c\right)^{3/2}$ are solutions of

$$\frac{dx}{dt} = x^{1/3}. \tag{18}$$

   b) Show that there are at least two solutions of (18) through every point (t_0, x_0) with $x_0 = 0$.
   c) Sketch several solutions of (18), including $x = 0$ on the same graph.
   d) Note that $f(t, x) = x^{1/3}$ is continuous everywhere. Why aren't this fact and part b) a contradiction of the basic existence and uniqueness theorem?

21.
   a) Show that $x = 1$ and $x = \left(\frac{4}{5}t + c\right)^{5/4} + 1$ are solutions of

$$\frac{dx}{dt} = (x - 1)^{1/5}. \tag{19}$$

   b) Show that there are at least two solutions of (19) through every point (t_0, x_0) with $x_0 = 1$.
   c) Sketch several solutions of (19), including $x = 1$ on the same graph.

d) Note that $f(t,x)=(x-1)^{1/5}$ is continuous everywhere. Why aren't this fact and part b) a contradiction of the basic existence and uniqueness theorems?

1.5 Euler's Numerical Method (optional)

We have given examples where the solution of a differential equation is given by a formula, and we have seen in Section 1.4 how the solutions can sometimes be sketched and analyzed. But many of the differential equations that arise in applications are too complicated to have solutions that can be given by formulas, and they must be solved numerically. In this section we will give one basic method and use it to make some general comments about numerical methods.

Suppose that we wish to numerically solve

$$\frac{dx}{dt}=f(t,x),\quad x(a)=x_0 \tag{1}$$

and obtain an estimate for $x(b)$, where $a<b$. We assume that f, f_x are continuous, so that the initial value problem (1) has a unique solution. Subdivide the interval $[a, b]$ with $N+1$ mesh points $t_0,\ldots,t_N$, so that $t_0=a$, $t_N=b$, and $t_n=a+hn$, where $h=(b-a)/N$ is called the step size. The numerical approximation of the solution will begin at t_0 and successively estimate the solution at each succeeding t_n.

Suppose that $x(t)$ is the solution of (1) and that we have its value at time t_n. Then we can approximate its value at time t_n+h using the tangent line approximation (first-order Taylor polynomial) by

$$x(t_n+h)\approx x(t_n)+h\frac{dx}{dt}(t_n) \tag{2}$$

Pictorially we have figure 1.5.1 But from (1) we have that $\frac{dx}{dt}(t_n)=f(t_n,x(t_n))$. Thus (2) becomes

$$x(t_n+h)\approx x(t_n)+hf(t_n,x(t_n)). \tag{3}$$

This formula gives a way of estimating x at time t_{n+1} given an estimate x_n of x at time t_n. The resulting numerical method is called **Euler's method.**

Euler's Method for the solution of $\frac{dx}{dt}=f(t,x),\ x(t_0)=x_0$

1. Let $t_0=a$.
2. Recursively compute $x_1,\ldots,x_N$ by

$$x_{n+1}=x_n+hf(t_n,x_n), \tag{4}$$

where

$$t_n=a+nh,\qquad h=\frac{b-a}{N}.$$

3. x_N is an estimate of $x(b)$ and x_n is an estimate of $x(t_n)$.

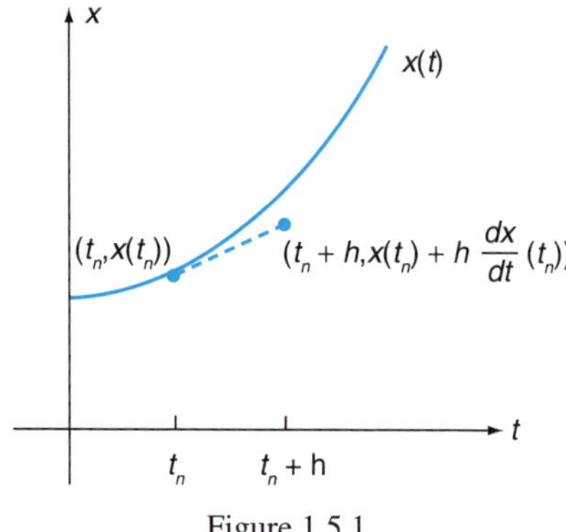

Figure 1.5.1

Example 1.5.1 *Euler's Method*

Let x be the solution of the differential equation

$$\frac{dx}{dt} - tx = t, \quad x(0) = 1. \tag{5}$$

Estimate $x(1)$ with Euler's method using a step size $h = 0.25$.

● SOLUTION. The differential equation is

$$\frac{dx}{dt} = t + tx, \quad x(0) = 1, \tag{6}$$

so that $f(t,x) = t + tx$. We have from (5) that

$$t_0 = 0, \quad x_0 = x(0) = 1.$$

The recursive relationship (4) is then

$$x_{n+1} = x_n + hf(t_n, x_n) = x_n + h(t_n + t_n x_n). \tag{7}$$

Thus we get

$$\begin{aligned}
& & x_1 &= x_0 + h(t_0 + t_0 x_0) = 1 + 0.25(0 + 0\cdot 1) = 1,\\
t_1 &= t_0 + h = 0.25, & x_2 &= x_1 + h(t_1 + t_1 x_1)\\
& & &= 1 + 0.25(0.25 + 0.25\cdot 1) = 1.125,\\
t_2 &= t_1 + h = 0.5, & x_3 &= x_2 + h(t_2 + t_2 x_2)\\
& & &= 1.125 + 0.25(0.5 + 0.5\cdot 1.125) = 1.391,
\end{aligned}$$

$$t_3 = t_2 + h = 0.75, \qquad x_4 = x_3 + h(t_3 + t_3 x_3)$$
$$= 1.391 + 0.25(0.75 + 0.75 \cdot 1.391) = 1.839.$$

Then $t_4 = t_3 + h = 0.75 + 0.25 = 1$ and the estimate for $x(1)$ is $x_4 = 1.839$ (to three places). ♦

Equation (6) can be solved using separation of variables to get the actual solution

$$x = 2e^{t^2} - 1. \tag{8}$$

It is perhaps easier just to note that Euler's method solves differential equation (5) and the condition at $t = 1$ for the exact solution satisfies

$$x(1) = 2e^{0.5} - 1 = 2.2974425. \tag{9}$$

The error in our estimate of $x(1)$ is then

$$e_4 = x(1) - x_4 = 0.458. \tag{10}$$

These calculations are represented in figure 1.5.2, which shows both the solution (8) and the numerical estimates.

It seems natural to try to get a smaller error by taking smaller steps.

Example 1.5.2 *Euler's Method with Smaller Steps*

Again estimate $x(1)$ using Euler's method, where x is the solution of (5), but this time with a step size $h = 0.125$.

- SOLUTION. We have $a = 0$, $b = 1$, $h = 0.125$, and $N = (b - a)/h = 8$. Thus

$$t_0 = 0, \quad t_n = nh, \quad t_8 = 1.$$

Again use the difference relationship (2) and compute (t_n, x_n) for $n = 1, \ldots, 8$. The computed x_n and the actual solution are graphed in figure 1.5.3. The estimate for $x(1)$ is

$$x(1) = x(t_8) \approx x_8 = 2.048. \tag{11}$$

♦

Note that the error in the estimate (11) is

$$e_8 = x(1) - x_8 = 0.249. \tag{12}$$

Thus reducing the step size by a factor of one-half has reduced the error (10) by about one-half also. This, in fact, turns out to be a general property of Euler's method.

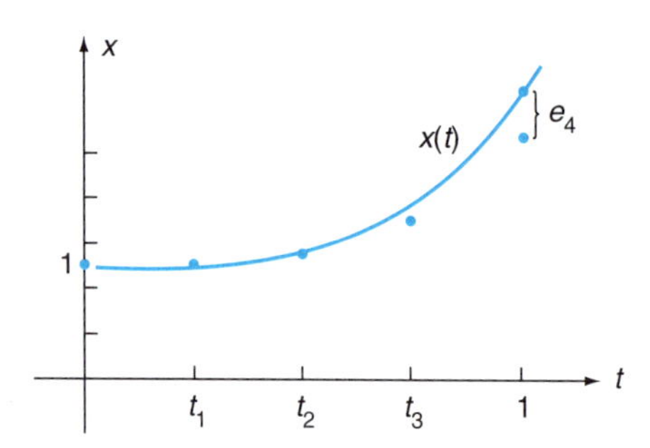

Figure 1.5.2 Euler estimate with $N = 4$ and exact solution (8).

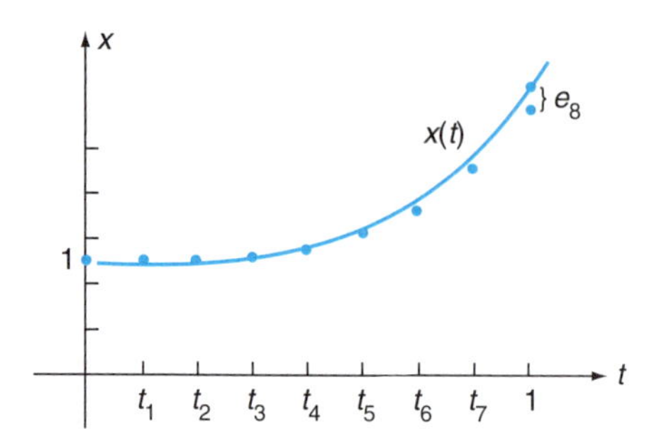

Figure 1.5.3 Euler estimates with $N = 8$ and exact solution (8).

A method for solving (1) is said to be of **order** r (written $O(h^r)$) on the interval $[a, b]$ if there is a constant M depending on a, b, f, but not on the step size h or the point $t_n \in [a, b]$, such that

$$|e_n| = |x(t_n) - x_n| \leq Mh^r. \tag{13}$$

THEOREM 1.5.1 **Order of Euler's Method** *Euler's method is a first-order method. That is, for $t_n \in [a, b]$ and a given step size h,*

$$|e_n| = |x(t_n) - x_n| \leq Mh$$

for some constant M independent of n, h (provided h is sufficiently small).

Proof of this theorem requires some technical assumptions on $f(t, x)$ and a careful consideration of different types of error. Note that (13) says that, by taking h to be half as large, we should expect about half as much error, and that is what we observed in (10), (12). Theorem 1.5.1 also seems to suggest that to get answers accurate to several significant figures using Euler's method we would have to take very small steps, and hence a large number of steps. Table 1.5.1 gives the result of solving (5) for several step sizes.

General Comments on Numerical Methods

Euler's method has provided an introduction to the idea of a numerical method. However, in practice for the more complex problems that arise in applications, the actual numerical solution procedure is much more complicated. Modern methods vary the step size and the order to get both efficiency and accuracy. They have the ability to output estimates at times other than t_i. Also, different types of problems require different methods depending on whether there are rapidly decaying or highly oscillatory solutions or events which alter the equations. Fortunately, numerical methods are included in symbolic software packages such as MAPLE and Mathematica, in general purpose commercial numerical packages such as MATLAB and SIMULINK,

TABLE 1.5.1
Euler's method for (5)

Step size h	*Steps N*	*Estimate x_N for $x(1)$*	*Error at $b=1$ $e_N = x(1) - x_N$*
0.1	10	2.09422	0.20322
0.01	100	2.27564	0.02180
0.001	1000	2.29525	0.00220
0.0001	10000	2.29722	0.00022

in open source general purpose packages such as Scilab and Scicos, and in specialized modeling software such as ACSL or SPICE. There are also programs such as XPP.

Comment on Significant Digits

When comparing your answer to those in the text, keep in mind that any of the following may affect the number of significant digits in your answer: precision of arithmetic on your machine, how round-off and floating-point operations are carried out, order in which computation is done, the accuracy of built-in functions such as sin, cos, and how underflow and overflow are handled.

Exercises

(Exercises marked with an asterisk are suggested only for progammable calculators and computers.) For Exercises 1–11, compute an estimate for the value of $x(b)$, using Euler's method, given the indicated step size.

1. $\dfrac{dx}{dt} = x - t,\ x(0) = 1, b = 2, h = 0.5.$

2. $\dfrac{dx}{dt} = tx,\ x(0) = 1, b = 1, h = 0.2.$

3. $\dfrac{dx}{dt} = -tx^2,\ x(0) = 1, b = 2, h = 1.$

4. $\dfrac{dx}{dt} = 3x - 2t,\ x(0) = 0, b = 1, h = 0.5.$

5. $\dfrac{dx}{dt} = 2x - 4t,\ x(0) = 1, b = 2, h = 0.5.$

6. $\dfrac{dx}{dt} = tx - t,\ x(0) = 1, b = 4, h = 0.2.$

7. $\dfrac{dx}{dt} = \sin x,\ x(0) = 0, b = 4, h = 0.5.$

8. $\dfrac{dx}{dt} = \sin x,\ x(0) = 1, b = 4, h = 0.2.$

Exercises 9, 10, 11 show that for some equations, Euler's method produces numerical solutions that resemble the actual solution only for small step sizes. In this example, the behavior is related to a property called **stiffness**.

9.

a) $\dfrac{dx}{dt} = -20x,\ x(0) = 1,\ b = 2, h = 0.2.$

b) On the same t, x-axis plot the points (t_n, x_n) from part (a) and the true solution $x = e^{-20t}$.

10. (Continuation of Exercise 9).

a) $\dfrac{dx}{dt} = -20x,\ x(0) = 1,\ b = 2, h = 0.1.$

b)* On the same t, x-axis plot the points (t_n, x_n) from part (a) and the true solution $x = e^{-20t}$.

11. (Continuation of Exercises 9, 10)

a) $\dfrac{dx}{dt} = -20x,\ x(0) = 1,\ b = 2, h = 0.01.$

b)* On the same t, x-axis plot the points (t_n, x_n) from part (a) and the true solution $x = e^{-20t}$.

12. Estimate $x(1)$ for the solution of $\frac{dx}{dt} = x^2$, $x(0) = 1$, using Euler's method with a step size of $h = 0.5$.

13. (Continuation of Exercise 12) Estimate $x(1)$ for the solution of $\frac{dx}{dt} = x^2$, $x(0) = 1$, using Euler's method with step sizes $h = 0.2$, $h = 0.1$, $h = 0.01*$, $h = 0.001*$.

14. (Continuation of Exercises 12, 13) Explain what you observed about the estimates for $x(1)$ by solving the differential equation $\frac{dx}{dt} = x^2$, $x(0) = 1$, by separation of variables.

Exercises 15–17 illustrate the point that, while analytically obtained solutions cannot cross equilibria if $f(t, x)$, $f_x(t, x)$ are continuous, the numerical solution can "jump" over an equilibirum.

15. Using Euler's method, calculate estimates for the solution of the differential equation

$$\frac{dx}{dt} = 1 - 2x + x^2,\ x(0) = -5, 0 \le t \le 2, \tag{14}$$

at points (t_n, x_n) where $h = 0.2$. Graph these points (t_n, x_n) and sketch what you think the solution would look like.

16. (Continuation of Exercise 15) Sketch the solutions of $\frac{dx}{dt} = 1 - 2x + x^2$ using the techniques of Section 1.4. Observe that $x = 1$ is an equilibrium solution. Compare this picture with the graph of Exericse 15.

17. (Continuation of Exercises 15, 16). Solve (14) using Euler's method with step of $h = 0.1$. Plot all values of (t_n, x_n) and compare to the analysis of Exericse 16.

1.6 First-Order Linear Differential Equations

Linear first-order differential equations,

$$\frac{dx}{dt} + p(t)x = f(t), \tag{1}$$

are comparatively easy to solve, and applications frequently are described by such equations. The function $f(t)$ is called the **forcing function**, the **input**, or the **nonhomogeneous term**, depending upon the application. When the differential equation is written in the form (1), then f is also sometimes called the **right-hand side**. When $f(t)$ is called the input, the solution of the differential equation $x(t)$ is usually called the **output**. It is important to understand which differential equations are linear and which equations are not linear. A differential equation that is not linear is called a **nonlinear** differential equation. A first-order differential equation $\frac{dx}{dt} = g(t, x)$ is **linear** if $g(t, x)$ is a linear function of x, as written in (1).

Most phenomena in nature do not actually satisfy linear differential equations. However, in many applications, the solutions of interest do not differ too greatly from the solutions of a linear differential equation. In this case, the differential equation may often be approximated by a linear differential equation. As a consequence, linear differential equations play a fundamental role in many parts of science and engineering.

1.6.1 Form of the General Solution

A **particular** solution x_p of the linear differential equation (1) is any function $x_p(t)$ satisfying the differential equation (1). Thus,

$$\frac{dx_p}{dt} + p(t)x_p = f(t). \tag{2}$$

To simplify our notation we omit the (t) from $\frac{dx}{dt}$ and x. If $f(t) = 0$, then the differential equation is called **homogeneous**. If $f(t)$ is not identically zero, then the **associated homogeneous equation** is

$$\frac{dx}{dt}(t) + p(t)x(t) = 0. \tag{3}$$

Thus a solution x_h of the associated homogeneous equation satisfies

$$\frac{dx_h}{dt} + p(t)x_h = 0. \tag{4}$$

We have introduced the ideas of a particular solution and solutions of the associated homogeneous equation because we will see that they play important roles physically and mathematically. We shall show later in this section that

The general solution of any linear first-order differential equation (2) can always be put in the form

$$x = x_p + x_h, \tag{5}$$

where x_p is a particular solution of (2) and x_h is the general solution of the associated homogeneous equation (3).

Furthermore, we will show that

The general solution of a homogeneous first-order linear differential equation is always in the form

$$x_h = cx_1, \tag{6}$$

where x_1 is a nonzero solution of the associated homogeneous equation and c is an arbitrary constant.

Thus the key result is that

For linear first-order equations (2), all the solutions (the general solution) are in the form of a particular solution x_p plus an arbitrary multiple c times one nonzero solution x_1 of the associated homogeneous equations

$$x = x_p + x_h = x_p + cx_1. \tag{7}$$

Equation (7) is fundamentally important. We use the notation x_1 here since in the next chapter, on second-order equations, we will see that the solution of the associated homogeneous equation is made up of two functions x_1 and x_2.

A complete verification of the facts (5) and (6) comes at the end of this section. Here, we will merely verify that $x = x_p + cx_1$ satisfies the first-order linear differential equation

$$\frac{dx}{dt} + p(t)x = f(t). \tag{8}$$

We substitute $x_p + cx_1$ into (8) for x and collect those terms that have c as a multiple and those that do not:

$$\begin{aligned}\frac{dx}{dt} + p(t)x &= \frac{d}{dt}(x_p + cx_1) + p(t)(x_p + cx_1)\\ &= \frac{dx_p}{dt} + p(t)x_p + c\left(\frac{dx_1}{dt} + p(t)x_1\right).\end{aligned} \tag{9}$$

However, x_p satisfies (2) and x_1 satisfies (4), and thus (9) becomes

$$\frac{dx}{dt} + p(t)x = f(t) + c \cdot 0 = f(t).$$

In this way we have verified that $x = x_p + cx_1$ satisfies (8).

We recommend that you memorize that the general solution of linear first-order differential equations (8) is in the form (7), where a particular solution x_p satisfies (2) and x_1 satisfies the associated homogeneous equation (4).

1.6.2 Solutions of Homogeneous First-Order Linear Differential Equations

The general solution of the first-order linear differential equation (8), $\frac{dx}{dt} + p(t)x = f(t)$, is always in the form $x = x_p + cx_1$. However, we have not discussed any methods to obtain the solution x_1 of the associated homogeneous equation or the particular solution x_p. In this subsection we will show that homogeneous solutions of first-order linear differential equations can always be obtained by separation.

Solutions of the related homogeneous equation satisfy

$$\frac{dx}{dt} + p(t)x = 0. \tag{10}$$

Note that $x = 0$ is always a solution of (10). It is called the **trivial solution** of the homogeneous equation. When we say that the general solution of $\frac{dx}{dt} + p(t)x = f(t)$ is always in the form $x = x_p + cx_1$, we mean that x_1 is a nontrivial solution and do not allow $x_1 = 0$.

Nontrivial solutions of the homogeneous equation can always be obtained by separation, since (10) can be written

$$\int \frac{dx}{x} = -\int p(t)dt. \tag{11}$$

Indefinite integration yields

$$\ln |x| = -\int p(t)dt + c_0,$$

where c_0 is an arbitrary constant of integration. Exponentiating gives

$$|x| = e^{-\int p(t)dt + c_0} = e^{c_0} e^{-\int p(t)dt}. \tag{12}$$

Note that c_0 is an arbitrary constant, so that e^{c_0} is an arbitrary positive constant. Solving (12) for x gives $x = [\pm e^{c_0}]e^{-\int p(t)dt}$. Let $c = \pm e^{c_0}$. Then c is an arbitrary constant, and we have

$$x = ce^{-\int p(t)dt}. \tag{13}$$

By separation, we have shown in (13) that the form cx_1 for the solution of a homogeneous first-order linear equation is correct and that we can use

$$x_1 = e^{-\int p(t)dt}. \tag{14}$$

If we instead used a definite integral, which is sometimes preferable if the function $p(t)$ is not simple, then we would obtain an equivalent expression:

$$x_1 = e^{-\int_a^t p(\bar{t})d\bar{t}}.$$

The lower limit a can be chosen as any constant. It usually is the initial value t_0 of t.

Example 1.6.1 *Homogeneous Linear Differential Equation*

Find the general solution of the homogeneous first-order linear differential equation

$$\frac{dx}{dt} + t^2 x = 0. \tag{15}$$

• SOLUTION. Since the equation is homogeneous, by separation we obtain

$$\int \frac{dx}{x} = -\int t^2 dt.$$

Indefinite integration yields

$$\ln|x| = -\frac{1}{3}t^3 + c_0.$$

Exponentiation first yields $|x| = e^{c_0} e^{-\frac{t^3}{3}}$, which becomes the general solution

$$x = ce^{-\frac{t^3}{3}}, \tag{16}$$

where c is an arbitrary constant. ♦

Example 1.6.2 *Homogeneous Equation Requiring a Definite Integral*

Find the general solution of the homogeneous first-order linear differential equation

$$\frac{dx}{dt} + \sin(t^2)x = 0. \tag{17}$$

• SOLUTION. By separation, we obtain

$$\int \frac{dx}{x} = -\int \sin(t^2)dt.$$

Integration yields

$$\ln|x| = -\int_0^t \sin(\bar{t}^2)d\bar{t} + c_0,$$

where we have chosen the lower limit to be zero and the dummy variable of integration to be $\bar{t}$. Exponentiation first yields $|x| = e^{c_0} e^{-\int_0^t \sin(\bar{t}^2)d\bar{t}}$, which becomes the general solution

$$x = ce^{-\int_0^t \sin(\bar{t}^2)d\bar{t}}, \tag{18}$$

where c is an arbitrary constant. ♦

Exercises

In the following exercises, problems where a definite integral is needed or recommended are indicated by an asterisk (*). In Exercises 1–10, find the general solution using separation.

1. $\frac{dx}{dt} = 3x.$
2. $\frac{dx}{dt} = -8x.$
3. $\frac{dx}{dt} = 2tx.$
4. $\frac{dx}{dt} = -t^3 x.$
5. $2t\frac{dx}{dt} + x = 0.$
6. $2t\frac{dx}{dt} - x = 0.$
7. $\frac{dx}{dt} + (\cos t)x = 0.$
8. $\frac{dx}{dt} + (3 + t^5)x = 0.$

9.* $\frac{dx}{dt} + \cos(t^{-1/2})x = 0.$

10.* $\frac{dx}{dt} + \frac{\ln t}{1 + e^t}x = 0$

In Exercises 11–17, solve the initial value problem using separation.

11. $\frac{dx}{dt} = -5x$ with $x(0) = 9.$
12. $\frac{dx}{dt} = 2x$ with $x(0) = 5.$
13. $\frac{dx}{dt} = 9x$ with $x(3) = 7.$
14. $\frac{dx}{dt} = -6x$ with $x(2) = 4.$

15.* $\frac{dx}{dt} + \frac{\sin t}{4 + e^t}x = 0$ with $x(5) = 10.$

16.* $\frac{dx}{dt} + \ln(7 + te^t)x = 0$ with $x(0) = 3.$

17. $\frac{dx}{dt} + t^{-2}x = 0$ with $x(1) = 3.$

18. Reconsider $\frac{dx}{dt} + p(t)x = f(t)$. Suppose x_p is a particular solution and $\overline{x}$ is any other particular solution.

(a) Show that $x = \overline{x} - x_p$ must be a solution of the associated homogeneous equation $\frac{dx}{dt} + p(t)x = 0.$

(b) Using $x_h = cx_1$, the result of this section on solutions of homogeneous equations, show that $x = x_p + cx_1$, as claimed in this section.

1.6.3 Integrating Factors for First-Order Linear Differential Equations

Now that we have a method to obtain the solution for homogeneous first-order linear differential equations, we need a method to obtain a particular solution. We will discuss two methods for obtaining particular solutions. One method uses an integrating factor. This method always works for first-order linear differential equations, but the algebraic steps necessary for a satisfactory answer can be unwieldy. We will describe another much easier method, in the next section. It is called the method of undetermined coefficients. The same method is discussed for second-order linear differential equations in Section 2.6. Unfortunately, the method works only on linear differential equations with constant coefficients and only on some such problems. However, the technique is very useful in applications.

Although there are methods for deriving this result (see Exercise 55 in this section), it is often easiest for students just to memorize that a particular solution of

$$\frac{dx}{dt} + p(t)x = f(t) \tag{19}$$

may be obtained using the **integrating factor**

$$u(t) = e^{\int p(t)dt}. \tag{20}$$

The expression $u(t)$ is an integrating factor, since we will show that the differential equation may be solved by an integration if we first multiply both sides of the differential equation (19) by the integrating factor (20). Multiplying (19) by (20) gives

$$e^{\int p(t)dt}\left(\frac{dx}{dt} + p(t)x\right) = f(t)e^{\int p(t)dt}. \tag{21}$$

The left-hand side is an exact derivative of $u(t)x(t)$ (as can be verified with care by using the product rule). Thus, (21) is equivalent to

$$\frac{d}{dt}(xe^{\int p(t)dt}) = f(t)e^{\int p(t)dt}.$$

Indefinite integration yields

$$xe^{\int p(t)dt} = \int f(t)e^{\int p(t)dt}dt + c, \tag{22}$$

where c is an arbitrary constant of integration. By multiplying both sides of (22) by $e^{-\int p(t)dt}$, we obtain the **general solution of any first-order linear differential equation:**

$$x = e^{-\int p(t)dt}\int f(t)e^{\int p(t)dt}dt + ce^{-\int p(t)dt}. \tag{23}$$

Although our goal was just to determine a particular solution, using an integrating factor gives the general solution. Formula (23) proves the facts (5), (6), and (7) that the general solution of any first-order linear differential equation is in the form $x = x_p + cx_1$, where c is an arbitrary constant. A solution x_1 of the associated homogeneous equation is seen to be $x_1 = e^{-\int p(t)dt}$, whereas a particular solution is possibly more complicated:

$$x_p = e^{-\int p(t)dt} \int f(t) e^{\int p(t)dt} dt.$$

This is an important formula that reappears frequently in applications. However, when solving problems, it is not a good idea just to memorize this result. Instead, in each example, it is better to repeat the following steps.

Summary of the Integrating Factor Method for Solving $\frac{dx}{dt} + p(t)x = f(t)$

1. Compute the integrating factor $u = e^{\int p(t)dt}$ and simplify.
2. Multiply both sides of the first-order linear differential equation (19) by u to get

$$\frac{d}{dt}(ux) = uf.$$

3. Integrate both sides with respect to t to get $u(t)x(t) = \int u(t)f(t)dt + c$.
4. Divide by $u(t)$ to get x.

If the integrals can be evaluated explicitly, then this solution is straightforward. However, often one or more of these integrals cannot be done explicitly, in which case definite integrals should be used. Also note that the differential equation must be put in the form (19) before beginning this procedure.

The integrating factor can be used even if the linear differential equation is homogeneous. Thus, we have two methods to obtain solutions of homogeneous differential equations: separation and the integrating factor.

Example 1.6.3 *Homogeneous*

Find the general solution of

$$\frac{dx}{dt} + t^2 x = 0 \tag{24}$$

using an integrating factor.

• SOLUTION. From the differential equation we have $p(t) = t^2$. To obtain the solution using an integrating factor, we multiply the linear differential equation (24) by the integrating factor $e^{\int p(t)dt} = e^{\frac{t^3}{3}}$. This gives

$$e^{\frac{t^3}{3}} \left(\frac{dx}{dt} + t^2 x \right) = 0. \tag{25}$$

The left-hand side is always the derivative of x times the integrating factor using the product rule. Thus, (25) is equivalent to

$$\frac{d}{dt}(e^{\frac{t^3}{3}}x)=0.$$

Integration gives $e^{\frac{t^3}{3}}x=c$, and hence

$$x=ce^{-\frac{t^3}{3}}.$$

This is the same general solution we obtained by separation (see (16)). ♦

Example 1.6.4 *Nonhomogeneous*

The following first-order linear differential equation is not homogeneous:

$$t\frac{dx}{dt}-3x=t^5.$$

Find the general solution for $t>0$.

• SOLUTION. In order to find the integrating factor, $e^{\int p(t)dt}$, the differential equation must first be put in the form $\frac{dx}{dt}+p(t)x=f(t)$. This requires dividing the differential equation by t to get

$$\frac{dx}{dt}-\frac{3}{t}x=t^4. \tag{26}$$

Thus, $p(t)=-\frac{3}{t}$, and hence the integrating factor is $e^{\int p(t)dt}=e^{-3\ln t}=t^{-3}$ (since we assume $t>0$). Multiplying both sides of (26) by the integrating factor gives

$$\frac{1}{t^3}\left(\frac{dx}{dt}-\frac{3}{t}x\right)=\frac{1}{t^3}t^4=t.$$

The left-hand side is always an exact derivative of x times the integrating factor:

$$\frac{d}{dt}\left(\frac{1}{t^3}x\right)=t.$$

By indefinite integration, we then obtain

$$\frac{1}{t^3}x=\frac{t^2}{2}+c.$$

Solving for x, we see that the general solution is in the proper form, a particular solution plus an arbitrary multiple of a solution of the associated homogeneous equation:

$$x=\frac{t^5}{2}+ct^3.$$

♦

Example 1.6.5 *One Definite Integration*

Solving the following first-order linear differential equation will require a definite integral:

$$\frac{dx}{dt} + t^2 x = \cos t.$$

The integrating factor $e^{\int p(t)dt} = e^{\frac{t^3}{3}}$ is straightforward and does not require a definite integral, since $p(t) = t^2$. Multiplying the differential equation by this integrating factor yields

$$e^{\frac{t^3}{3}} \left(\frac{dx}{dt} + t^2 x \right) = e^{\frac{t^3}{3}} \cos t.$$

Again, the left-hand side is an exact derivative, so that we get

$$\frac{d}{dt}\left(e^{\frac{t^3}{3}} x\right) = e^{\frac{t^3}{3}} \cos t.$$

The right-hand side does not have an elementary integral. Even if it did have one, we could still use a definite integral. We integrate both sides. For convenience we choose the lower limit to be 0:

$$e^{\frac{t^3}{3}} x = \int_0^t e^{\frac{\bar{t}^3}{3}} \cos \bar{t}\, d\bar{t} + c.$$

From this, we obtain the general solution as

$$x = e^{-\frac{t^3}{3}} \int_0^t e^{\frac{\bar{t}^3}{3}} \cos \bar{t} d\bar{t} + c e^{-\frac{t^3}{3}},$$

a particular solution plus an arbitrary multiple of a solution of the associated homogeneous equation. It is very important to use a dummy variable of integration different from t so that there is no temptation to have $e^{-\frac{t^3}{3}}$ incorrectly cancel $e^{\frac{\bar{t}^3}{3}}$. In this example, the integrating factor did not require definite integration. ♦

It is quite remarkable that we may obtain the general solution of any first-order linear differential equation. Examples that require two definite integrals are more complicated. Even a relatively easy-looking linear differential equation may still have a complicated solution.

Exercises

In Exercises 19–30, solve the differential equation. If there is no initial condition, give the general solution.

19. $t\frac{dx}{dt} + x = e^t$, $x(1) = 1$.
20. $\frac{dx}{dt} + 2tx = t$.

21. $\dfrac{dx}{dt} = 3e^t$.

22. $\dfrac{dx}{dt} = \dfrac{x+t}{t}$.

23. $(t^2+1)\dfrac{dx}{dt} + 2tx = 1$.

24. $\dfrac{du}{dt} = 3(u-1)$.

25. $\dfrac{dx}{dt} + 4x = t,\ x(0) = 0$.

26. $(t+1)\dfrac{dx}{dt} + x = t$.

27. $t\dfrac{dx}{dt} = 2x,\ x(1) = 4$.

28. $t\dfrac{dx}{dt} = -3x + \dfrac{\sin t}{t^2}$.

29. $t^2\dfrac{dx}{dt} + tx = 1$.

30. $t^3\dfrac{dx}{dt} + 4t^4 = t^7$.

For each of the linear differential equations in Exercises 31–33, find the general solution of the differential equation, sketch several solutions ($c = 0$, $c = \pm 1$, or $c = \pm 3$ would suffice), and describe the behavior of the solutions near $t = 0$.

31. $t\dfrac{dx}{dt} + 3x = t$.

32. $t\dfrac{dx}{dt} + 2x = t^{-1}$.

33. $t\dfrac{dx}{dt} - x = t^2$.

In Exercises 34–45, find the solution of the initial value problem that requires definite integrals. If there are no conditions given, find the general solution.

34. $\dfrac{dx}{dt} + t^4x = 1$.

35. $\dfrac{dx}{dt} + 2tx = 1$.

36. $\dfrac{dx}{dt} + (\sin t)x = t$.

37. $\dfrac{dx}{dt} + t^2x = t$.

38. $\dfrac{dx}{dt} - 2tx = e^t$.

39. $\dfrac{dx}{dt} + e^t x = 3$.

40. $\dfrac{dx}{dt} - x = e^{t^2}$.

41. $\dfrac{dx}{dt} + x = \dfrac{1}{t+1}$ with $x(2) = 1$.

42. $\dfrac{dx}{dt} + \dfrac{3}{t}x = \ln(3+\cos^2 t)$ with $x(4)=6$.
43. $3t\dfrac{dx}{dt} - x = t\sin t$ with $x(5)=0$.
44. $\dfrac{dx}{dt} + x = \sin(e^{4t})$.
45. $7t\dfrac{dx}{dt} + x = e^t$.

From calculus we know that

$$\frac{dt}{dx} = \frac{1}{dx/dt}.$$

That is, we can think of x as the independent variable and t as the dependent variable provided $\frac{dx}{dt} \neq 0$. In Exercises 46–48, the differential equation is nonlinear in the form $\frac{dx}{dt} = f(t,x)$, but linear (after some algebraic manipulations) if written as $\frac{dt}{dx} = \frac{1}{f(t,x)}$. Find the general solution of the differential equation.

46. $\dfrac{dx}{dt} = \dfrac{1}{x}$.
47. $\dfrac{dx}{dt} = \dfrac{1}{x+t}$.
48. $\dfrac{dx}{dt} = \dfrac{x}{t+x}$.
49. Verify that, if an arbitrary constant c_1 is introduced when $\int p(t)dt$ is computed, then the form of the general solution (23) is unchanged. (*Note*: If c is an arbitrary constant, then ce^{-c_1} is still an arbitrary constant.)
50. Show that if $u(t)$ is a function of t, and u satisfies the property

$$u\left(\frac{dx}{dt} + p(t)x\right) = \frac{d(ux)}{dt},$$

then $u = ke^{\int p(t)dt}$, k a constant.
51. Show that if u is an integrating factor for (19), then so is ku for any constant k.
52. Consider the differential equation $\frac{dx}{dt} - tx = 2t - t^3$.
   a) Show that $x = t^2$ is a particular solution. Solve the initial value problem $x(0)=2$.
   b) What is the integrating factor? Why is this method difficult?
   c) Find the general solution with an integrating factor. Compare to part (a).
53. Consider $\frac{dx}{dt} + [(\sin t)/t]x = f(t)$. Show that the particular solution with $x_p(0) = 0$ can be put into the form $x_p = \int_0^t G(t,\bar{t})q(\bar{t})d\bar{t}$ for some function $G(t,\bar{t})$. The function $G(t,\bar{t})$ is called the **influence function** or **Green's function** for the equation.
54. Consider $\frac{dx}{dt} + p(t)x = f(t)$. Show that the particular solution with $x_p(0)=0$ can be put into the form $x_p = \int_0^t G(t,\bar{t})f(\bar{t})d\bar{t}$ for some function $G(t,\bar{t})$. The function $G(t,\bar{t})$ is called the **influence function** or **Green's function** for the equation.

55. We briefly present an equivalent method (know as the **method of variation of parameters**) for solving the linear differential equation

$$\frac{dx}{dt} + p(t)x = f(t).$$

Make the change of variables $x = u(t)x_1(t)$, where $x_1(t) = e^{-\int p(t)dt}$ is the solution of the homogeneous equation that was determined previously. Find and solve the simpler first-order linear differential equation for $u(t)$. Compare your solution to the answer obtained by using an integrating factor. (A modification of this method is useful for second-order linear differential equations, since the integrating factor method cannot be applied to such equations.)

1.7 Linear First-Order Differential Equations with Constant Coefficients and Constant Input

Solutions of first-order linear differential equations can always be obtained by the integrating factor method of Section 1.6.3. In this section we will describe an easier method for a particular class of first-order linear differential equations. These equations occur quite frequently in applications.

The linear first-order differential equation.

$$\frac{dx}{dt} + p(t)x = f(t).$$

is said to have **constant coefficients** if $p(t)$ is a constant, $p(t) = k$:

$$\frac{dx}{dt} + kx = f(t). \tag{1}$$

We say that the differential equation has constant coefficients even if the forcing function or input $f(t)$ is not constant. It will be easy to obtain a solution x_1 of the associated homogeneous equation for (1). Sometimes a method of undetermined coefficients can be used to obtain a particular solution x_p of a linear differential equation with constant coefficients. Then $x_p + cx_1$ will give the general solution of (1). This section introduces this method by considering the important case when the input $f(t)$ is constant and doing one example in which the input is not constant. Other cases are considered in the exercises and in Chapter 2.

1.7.1 Homogeneous Linear Differential Equations with Constant Coefficients

The associated homogeneous equation for the linear differential equation with constant coefficients (1) is

$$\frac{dx}{dt} + kx = 0. \tag{2}$$

It is easy to verify that $x = e^{-kt}$ satisfies (2) by substituting this exponential into (2). This solution of the homogeneous equation can be also derived, since (2) is separable (Section 1.3), or by the integrating factor method (Subsection 1.6.3). However, it is easier to obtain this solution by just remembering the following simple fact. This fact will be useful in this and later chapters:

A homogeneous linear constant coefficient differential equation always has at least one solution of the form

$$x = e^{rt}. \tag{3}$$

For first-order linear equations, r will turn out to be a real number. Statement (3) is still true of second-order equations, but we will see in the next chapter that r can then be a complex number.

We can determine what specific exponential satisfies the homogeneous differential equation with constant coefficients (2) by substituting the unknown exponential $x = e^{rt}$ into the differential equation (2):

$$\frac{d}{dt}e^{rt} + ke^{rt} = re^{rt} + ke^{rt} = (r+k)e^{rt} = 0.$$

Dividing by e^{rt} then gives that $r = -k$. Since $x_1(t) = e^{-kt}$ is one solution of the homogeneous (2), our theory tells us that the general solution of the homogeneous equation will be $x = ce^{-kt}$.

The general solution of

$$\frac{dx}{dt} + kx = 0$$

with k constant is

$$x = ce^{-kt}.$$

This method is illustrated in the next example.

Example 1.7.1 *Homogeneous Linear Constant Coefficient*

Determine the general solution of

$$\frac{dx}{dt} + 4x = 0. \tag{4}$$

- SOLUTION. If we substitute $x = e^{rt}$ into (4) for x, we get

$$\frac{d}{dt}(e^{rt}) + 4e^{rt} = 0.$$

Since $\frac{d}{dt}(e^{rt}) = re^{rt}$, the differential equation (4) is satisfied if

$$re^{rt} + 4e^{rt} = (r+4)e^{rt} = 0.$$

This equation is to hold for all t. If we cancel the e^{rt} since it never equals zero (or just divide by e^{rt}), we obtain

$$r+4=0 \text{ or equivalently, } r=-4.$$

Thus, e^{-4t} solves the constant coefficient differential equation (4). Since the differential equation (4) is linear, the general solution is in the form of a particular solution plus an arbitrary multiple of the solution e^{-4t} of the associated homogeneous equation. Equation (4) is homogeneous, so that we can take $x_p=0$. Thus, the general solution of (4) is

$$x=ce^{-4t}.$$

♦

1.7.2 Constant Coefficient Linear Differential Equations with Constant Input

Particular solutions of first-order linear differential equations with constant coefficients,

$$\frac{dx}{dt}+kx=f(t), \tag{5}$$

can always be obtained by using the integrating factor method discussed in Section 1.6.3. However, sometimes it is easier to obtain a particular solution by the method of undetermined coefficients. In this subsection, we will discuss only one elementary class of examples. These ideas will be addressed in greater depth in the section on second-order linear differential equations with constant coefficients (Section 2.6).

If the right-hand side of (5) is a constant, $f(t)=I_0$, we refer to the differential equation as having a constant input:

$$\frac{dx}{dt}+kx=I_0.$$

Example 1.7.2 *Constant Coefficient Differential Equation with Constant Input*

Find the general solution of

$$\frac{dx}{dt}+8x=9. \tag{6}$$

• SOLUTION. Since the right-hand side is a constant, we might guess that a particular solution is a constant, $x_p=A$. The constant A is called an **undetermined coefficient**, and this method is called the method of undetermined coefficients. To determine the specific constant A, we substitute $x_p=A$ into (6) for x to obtain $\frac{dA}{dt}+8A=9$. But $\frac{dA}{dt}=0$, since A is a constant, so that

$$8A=9 \text{ or equivalently, } A=\frac{9}{8}.$$

Thus, a particular solution of (6) is

$$x_p = A = \frac{9}{8}.$$

The general solution of any linear first-order equation is in the form of a particular solution plus an arbitrary multiple of a solution x_1 of the associated homogeneous equation. Using the techniques of Example 1.7.1, we find that $x_1 = e^{-8t}$. Thus the general solution of (6) is

$$x = \frac{9}{8} + ce^{-8t},$$

where c is an arbitrary constant. ♦

All cases of constant coefficient differential equations with a constant input can be described by

$$\frac{dx}{dt} + kx = I_0, \tag{7}$$

where I_0 is a constant. In the same manner as Example 1.7.2, the general solution of (7) can be obtained as

$$x = \frac{I_0}{k} + ce^{-kt}, \quad k \neq 0. \tag{8}$$

A particular solution is $x_p = \frac{I_0}{k}$ if $k \neq 0$. More examples of this type of differential equation will appear in the sections on applications of first-order differential equations.

However, if $k = 0$ in (7), then the method of undetermined coefficients as described must be modified. If $k = 0$, (7) is

$$\frac{dx}{dt} = I_0. \tag{9}$$

Any constant is a solution of the homogeneous equation $\left(\frac{dx_h}{dt} = 0\right)$, and hence a constant cannot be a particular solution. Thus, $x_p \neq A$, contrary to what was assumed earlier. In this case, it is easy to see, by integration of (9), that the guess was wrong and a particular solution is $x_p = I_0 t$. This particular solution is not a constant, but rather is proportional to t. The general solution of (9) by integration is $x = I_0 t + c$. This is again a particular solution plus an arbitrary multiple of a solution, $x_1 = 1$, of the associated homogeneous equation. More will be said about this in our discussion of particular solutions for second-order linear differential equations with constant coefficients. In summary,

If $k \neq 0$, then $\frac{dx}{dt} + kx = I_0$ has a constant particular solution.

In applications the constant solutions that result from constant input are variously called **equilibriums**, **steady-state solutions**, and **operating points**.

1.7.3 Constant Coefficient Differential Equations with Exponential Input

Example 1.7.3 *Constant Coefficient Differential Equation with an Exponential Input*

Find the general solution of

$$\frac{dx}{dt} + 8x = 4e^{7t}. \tag{10}$$

• SOLUTION. Since the right-hand side is a constant times an exponential e^{7t}, we might guess that a particular solution is an unknown constant times the same exponential, $x_p = Ae^{7t}$. The constant A is called an **undetermined coefficient**, and this method is called the method of undetermined coefficients. To determine the specific constant A, we substitute $x_p = Ae^{7t}$ into (10) for x to obtain $7Ae^{7t} + 8Ae^{7t} = 4e^{7t}$, so that

$$15Ae^{7t} = 4e^{7t} \text{ or equivalently, } A = \frac{4}{15}.$$

Thus, a particular solution of (10) is

$$x_p = Ae^{7t} = \frac{4}{15}e^{7t}.$$

The general solution of any linear first-order equation is in the form of a particular solution plus an arbitrary multiple of a solution x_1 of the associated homogeneous equation. Using the techniques of Example 1.7.2, we find that $x_1 = e^{-8t}$. Thus the general solution of (10) is

$$x = \frac{4}{15}e^{7t} + ce^{-8t},$$

where c is an arbitrary constant. ♦

1.7.4 Constant Coefficient Differential Equations with Discontinuous Input

Many practical problems involve discontinuous inputs. The discontinuities arise from such physical phenomena as switches, sparks, or digital devices. The method of undetermined coefficients can also be used if the input is a discontinuous input composed of relatively simple functions.

Example 1.7.4 *Discontinuous Input*

Solve the initial value problem

$$\frac{dx}{dt} + 2x = g(t) \quad \text{with } x(0) = 0, \tag{11}$$

where the input $g(t)$ is the following piecewise constant function:

$$g(t)=\begin{cases} 1, & 0 \le t < 2, \\ 0, & 2 \le t. \end{cases}$$

• SOLUTION. We actually have a two-part problem here. A different differential equation is valid in each region.

For $0 \le t < 2$, the differential equation (11) is

$$\frac{dx}{dt}+2x=1, \tag{12}$$

and must satisfy the given initial condition $x(0)=0$. The differential equation (12) is linear with constant coefficients and a constant input. A particular solution is a constant, which is easily seen to equal $\frac{1}{2}$. A solution of the associated homogeneous equation is e^{-2t}, so that the general solution is $x=\frac{1}{2}+ce^{-2t}$. The initial condition $x(0)=0$ implies that $c=-\frac{1}{2}$, so that the solution of the initial value problem for $0 \le t < 2$ is

$$x(t)=\frac{1}{2}-\frac{1}{2}e^{-2t}. \tag{13}$$

However, the differential equation (11) for $t \ge 2$ is

$$\frac{dx}{dt}+2x=0. \tag{14}$$

The general solution of (14) will have one arbitrary constant introduced. This constant is determined not from the condition at $t=0$, but instead from a condition of (14)'s starting value $t=2$. From physical considerations, we often know that the solution x must be continuous, and thus the initial condition for the differential equation (14) for $t \ge 2$ is

$$x(2)=\lim_{x\to 2^-} x(t). \tag{15}$$

That is, the solutions for $t<2$ and $t>2$ should match up, or "meet," when $t=2$, so that x can be continuous everywhere, including $t=2$. Since the solution has been determined for $0<t<2$, we compute its limit at $t=2$ from (13) to be

$$x(2)=\lim_{x\to 2^-} x(t)=\frac{1}{2}-\frac{1}{2}e^{-4}. \tag{16}$$

Since (14) is linear and homogeneous (with constant coefficients), its general solution is seen to be

$$x(t)=c_1e^{-2t}. \tag{17}$$

A different notation for the constant has been introduced, so that the constant associated with the differential equation for $t>2$ is not confused with the constant associated

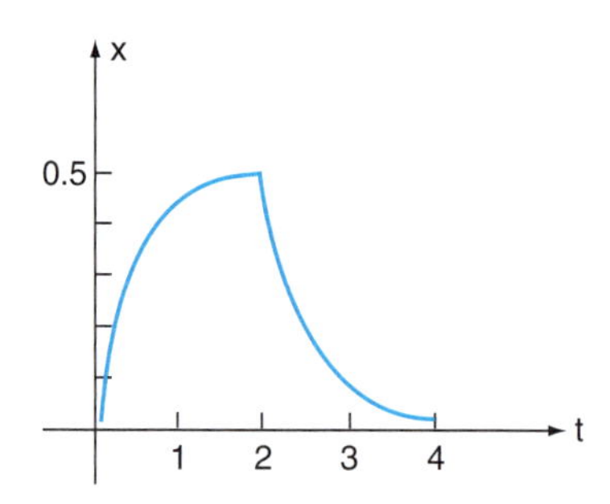

Figure 1.7.1 Graph of (19).

with the differential equation for $t < 2$. The constant c_1 is determined from the initial condition valid at $t = 2$. Letting $t = 2$ in (17) and setting it equal to (16), we get

$$\frac{1}{2} - \frac{1}{2}e^{-4} = c_1 e^{-4}.$$

Thus, $c_1 = \frac{1}{2}(e^4 - 1)$. Hence for $t > 2$ we have

$$x(t) = \frac{1}{2}(e^4 - 1)e^{-2t}. \tag{18}$$

Combining the solutions, (13) and (18), valid in the two regions, we obtain

$$x(t) = \begin{cases} \frac{1}{2}(1 - e^{-2t}), & 0 \le t < 2, \\ \frac{1}{2}(e^4 - 1)e^{-2t}, & 2 \le t. \end{cases} \tag{19}$$

The graph of $x(t)$ is given in figure 1.7.1. Some students may recognize this graph is the charge-discharge curve of a capacitor in an RC circuit. This will be discussed more carefully in Section 1.10. ♦

Exercises

In Exercises 1–10, find the general solution.

1. $\dfrac{dx}{dt} = 8x.$

2. $\dfrac{dx}{dt} = 3x.$

3. $\dfrac{dx}{dt} = -2x.$

4. $\dfrac{dx}{dt} = -\dfrac{1}{3}x.$

5. $\dfrac{dx}{dt} + 7x = 0.$

6. $\dfrac{dx}{dt} - x = 0.$

7. $\dfrac{dx}{dt} + x = 0.$

8. $\dfrac{dx}{dt} - \dfrac{3}{4}x = 0.$

9. $\dfrac{dx}{dt} = 5x.$

10. $\dfrac{dx}{dt} = -3x.$

In Exercises 11–22, find the general solution.

11. $\dfrac{dx}{dt} + 3x = 8.$

12. $\dfrac{dx}{dt} + 6x = 5.$

13. $\dfrac{dx}{dt} - 4x = -9.$

14. $\dfrac{dx}{dt} - x = -\dfrac{1}{3}.$

15. $\dfrac{dx}{dt} - \dfrac{4}{5}x = 3.$

16. $\dfrac{dx}{dt} - 8x = 7.$

17. $\dfrac{dx}{dt} + \dfrac{2}{3}x = -\dfrac{4}{3}.$

18. $\dfrac{dx}{dt} + \dfrac{3}{8}x = -4.$

19. $\dfrac{dx}{dt} = 2x + 18.$

20. $\dfrac{dx}{dt} = -8x + 9.$

21. $\dfrac{dx}{dt} = -x - \dfrac{17}{3}.$

22. $\dfrac{dx}{dt} = 6x - 15.$

In Exercises 23–30, the input is a constant multiple of an exponential e^{at}. Suppose that there is a particular solution of the form Ae^{at} for some constant A. Find the constant A and determine the particular and general solution.

23. $\dfrac{dx}{dt} + 7x = 8e^{-4t}.$

24. $\dfrac{dx}{dt} - x = 4e^{3t}.$

25. $\dfrac{dx}{dt} - 2x = -3e^{-5t}$.

26. $\dfrac{dx}{dt} + x = -2e^{8t}$.

27. $\dfrac{dx}{dt} + 4x = 3e^{4t}$.

28. $\dfrac{dx}{dt} - x = 5e^{-t}$.

29. $\dfrac{dx}{dt} + 3x = e^{-2t}$.

30. $\dfrac{dx}{dt} - 5x = 6e^{7t}$.

In Exercises 31–40, obtain just a particular solution (the general solution could always be obtained easily by adding an arbitrary multiple of a solution of the associated homogeneous equation). The forcing function is a polynomial. Guess that a **polynomial of the same degree** is the particular solution and use the **method of undetermined coefficients** to find a particular solution. For example, if you would guess a polynomial of degree 2, substitute $x = At^2 + Bt + C$ into the differential equation and determine the constants so that $x = At^2 + Bt + C$ satisfies the differential equation.

31. $\dfrac{dx}{dt} + 7x = 3 + 5t$.

32. $\dfrac{dx}{dt} - 4x = -6 + 9t$.

33. $\dfrac{dx}{dt} + 2x = 14t$.

34. $\dfrac{dx}{dt} - \dfrac{1}{2}x = -3t$.

35. $\dfrac{dx}{dt} + x = 2t^2 + 5t - 8$.

36. $\dfrac{dx}{dt} - 9x = -t^2 + 7t + 2$.

37. $\dfrac{dx}{dt} + 3x = t^3$.

38. $\dfrac{dx}{dt} - 6x = 4t^2 + 3$.

39. $\dfrac{dx}{dt} + 5x = t$.

40. $\dfrac{dx}{dt} + 8x = 5t + 11$.

In Exercises 41–52, obtain just a particular solution (the general solution can always be obtained easily by adding an arbitrary multiple of a solution of the associated homogeneous equation). In these exercises, the forcing function is an elementary sinusoidal function. If the forcing function is $\cos 5t$, then a particular solution of the form $A \cos 5t$ will not work. However, a linear combination of $\cos 5t$ and $\sin 5t$

will work. Thus, the form for x_p is $A\cos 5t + B\sin 5t$, where the constants A and B are determined by substituting the assumed form of the particular solution into the differential equation for x. In Exercises 47–52, the forcing function has several different terms. Use a form for x_p consisting of the sum of the forms for each term.

41. $\dfrac{dx}{dt} + 2x = 3\sin 6t$.

42. $\dfrac{dx}{dt} - 3x = 4\cos 3t$.

43. $\dfrac{dx}{dt} - 5x = 2\cos t$.

44. $\dfrac{dx}{dt} + 8x = \sin t$.

45. $\dfrac{dx}{dt} + 4x = 3\cos 2t + 5\sin 2t$.

46. $\dfrac{dx}{dt} - x = 2\cos 4t + \sin 4t$.

47. $\dfrac{dx}{dt} + 6x = \cos t + \sin 5t$.

48. $\dfrac{dx}{dt} + x = \cos 2t + \cos 4t$.

49. $\dfrac{dx}{dt} - 9x = 5 + 2\sin 3t$.

50. $\dfrac{dx}{dt} - 2x = -3 + 4\cos t$.

51. $\dfrac{dx}{dt} + x = 2e^{3t} + \sin t$.

52. $\dfrac{dx}{dt} + 9x = 4e^{-5t} + \cos 2t$.

53. Consider the differential equation $\frac{dx}{dt} + 3x = 8e^{-3t}$. Show that a simple exponential Ae^{-3t} does not work as a particular solution because e^{-3t} is a solution of the associated homogeneous equation. Instead, solve the differential equation by the integrating factor method. Make a conjecture as to how to modify the method of undetermined coefficients when the simple exponential forcing is a solution of the associated homogeneous equation.
54. Consider the differential equation $\frac{dx}{dt} - kx = P(t)e^{kt}$, where $P(t)$ is a polynomial of degree n. Solve this differential equation by making the change of variables $x = ve^{kt}$. Find and solve the differential equation for v. Show that v is a polynomial of degree $n+1$. This proves that a particular solution is in the form of an exponential e^{kt} times a polynomial of one higher degree, if the exponential part of the forcing function is a solution of the associated homogeneous equation. A slight generalization of this will be important in Chapter 2.
55. Consider the differential equation $\frac{dx}{dt} - 2x = 7e^{2t}$. Show that a simple exponential Ae^{2t} does not work as a particular solution because e^{2t} is a solution of the associated homogeneous equation. Instead, solve the differential equation by making

a change of variables $x = ve^{2t}$. Find and solve the differential equation for v. Make a conjecture as to how to modify the method of undetermined coefficients when the simple exponential forcing is a solution of the associated homogeneous equation.

In Exercises 56–62, a simple exponential will not work as a particular solution, and x_p must have the form Ate^{at}. Find the general solution.

56. $\dfrac{dx}{dt} - 3x = 8e^{3t}$.

57. $\dfrac{dx}{dt} - x = 4e^{t}$.

58. $\dfrac{dx}{dt} + 4x = 3e^{-4t}$.

59. $\dfrac{dx}{dt} + x = 5e^{-t}$.

60. $\dfrac{dx}{dt} - 9x = -2e^{9t}$.

61. $\dfrac{dx}{dt} - 7x = 8e^{7t}$.

62. $\dfrac{dx}{dt} + 5x = -3e^{5t}$.

In Exercises 63–68, find the continuous solution of

$$\frac{dx}{dt} + p(t)x = f(t), \quad 0 \le t \le 2,$$

and sketch the solution.

63.

$$p(t) = \begin{cases} 2, & 0 \le t < 1, \\ 1, & 1 \le t \le 2, \end{cases} \quad f(t) = 0, \; x(0) = 2.$$

64.

$$p(t) = 1, \quad f(t) = \begin{cases} 1, & 0 \le t < 1, \\ 0, & 1 \le t \le 2, \end{cases} \quad x(0) = 1.$$

65.

$$p(t) = 0, \quad f(t) = \begin{cases} 1, & 0 \le t < 1, \\ -1, & 1 \le t \le 2, \end{cases} \quad x(0) = 0.$$

66.

$$p(t) = \begin{cases} 0, & 0 \le t < 1, \\ 1, & 1 \le t \le 2, \end{cases} \quad f(t) = 1, \; x(0) = 0.$$

67.

$$p(t) = \begin{cases} 1, & 0 \le t < 1, \\ 0, & 1 \le t \le 2, \end{cases} \quad f(t) = \begin{cases} 0, & 0 \le t < 1, \\ 1, & 1 \le t \le 2, \end{cases} \quad x(0) = 2.$$

68.

$$p(t)=\begin{cases} 1, & 0\le t<1, \\ -1, & 1\le t\le 2, \end{cases} \quad f(t)=2,\ x(0)=1.$$

1.8 Growth and Decay Problems

Differential equations are studied by scientists and engineers because they describe physical problems. Solving a differential equation is only part of applying differential equations. Equally important is showing how the differential equation describes a real-world situation. Experts from a given field will sometimes argue about whether one differential equation or another applies to a given situation. Knowledge about both mathematics and the field of application is required.

This section introduces several applications of first-order equations. In these applications, the reader should pay close attention to how the equations are derived from the given physical problem and should not seek to merely memorize formulas.

1.8.1 A First Model of Population Growth

The first application that we discuss will describe a very idealized situation. More realistic situations can be described, but more complicated mathematical models are required. In this section, one of the simplest models of population growth of a single species will be developed. Let

$$P(t)= \text{ population } P \text{ as a function of time } t. \tag{1}$$

The population could be that of a country or the world. Other examples are the number of bacteria in a laboratory experiment or the population of a particularly insidious insect attacking a crop. The derivative of the population with respect to time is important:

$$\frac{dP}{dt}= \text{rate of change of the population.} \tag{2}$$

The derivative is measured in units of number per unit time. For example, in 2005 the population of the United States was increasing at the rate of 2.7 million per year. However, the **growth rate** is often a more significant measure of growth:

$$\frac{1}{P}\frac{dP}{dt}= \text{growth rate.} \tag{3}$$

The growth rate is the rate of change of the population per individual. Since the population of the United States was 296 million in 2005, the growth rate was $\frac{2.7}{296}=0.0092$ per year or about 0.9%, a little less than 1% per year. It is common to measure the growth rate of human populations in percent per year. For other organisms, other time units, such as hours or days, may be used.

To illustrate how such a differential equation is used to predict the future, we will make the simplest assumption about growth. We assume the growth rate is a constant k. Then

$$\frac{1}{P}\frac{dP}{dt} = k,$$

or equivalently,

$$\frac{dP}{dt} = kP. \tag{4}$$

Our assumption that the growth rate is a constant is equivalent to assuming that the rate of change of the population is proportional to the population. The differential equation (4) by itself is not enough to predict the future. In addition, we need to know the initial population $P(0)$. The first-order differential equation (4) is linear with constant coefficients. It is also homogeneous, since it has a zero right-hand side $\left(\frac{dP}{dt} - kP = 0\right)$. We could solve (4) by separation or integrating factors. We shall use the approach of Section 1.7. From Section 1.7 we know that the general solution of (4) will be an arbitrary multiple of a solution of the homogeneous equation. For constant coefficient equations, solutions of homogeneous equations are in the form e^{rt}. If $P = e^{rt}$ is substituted into the differential equation (4), we obtain $r = k$, so that e^{kt} is a solution of the homogeneous equation. Thus, the general solution of (4) is

$$P(t) = ce^{kt}. \tag{5}$$

The arbitrary constant is determined by evaluating the solution (5) at the initial time $t = 0$ using the initial population $P(0)$:

$$P(0) = c.$$

The solution of the initial value problem for the differential equation (4) is then

$$P(t) = P(0)e^{kt}. \tag{6}$$

The solution (6) is an elementary exponential and is sketched for one initial condition in figure 1.8.1 for the exponential growth case, $k > 0$. Solutions corresponding to other initial conditions are graphed in figure 1.8.2. Solution curves for $t > 0$ and $t < 0$, $P(0)$ positive and negative, are included in figure 1.8.3, which we earlier called the direction field. If $k < 0$, the growth rate is negative and the population exponentially decays, as sketched in figures 1.8.4 and 1.8.5. One reason that exponentials are studied so much in high school mathematics and in calculus is that the solution of the simplest differential equation models involves the exponential. Other continuous growth problems, such as the increase in the amount of money in a bank due to interest and the increase in the cost of living due to inflation, satisfy differential equation (6), and so they also involve exponentials. Examples from physics and chemistry involving exponentials are given in the next sections.

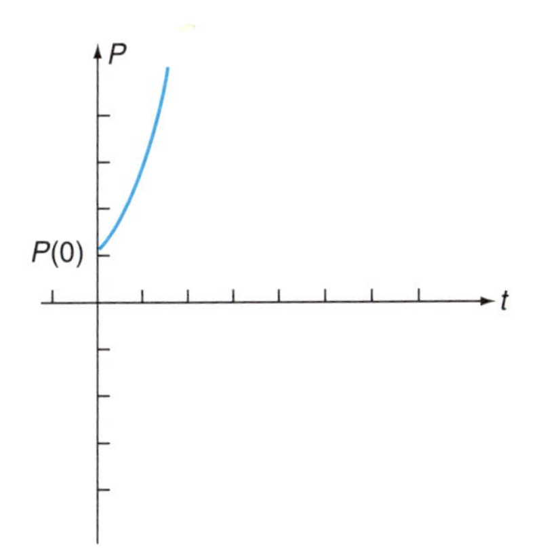

Figure 1.8.1 $P(t) = P(0)e^{kt}$ for $k > 0$.

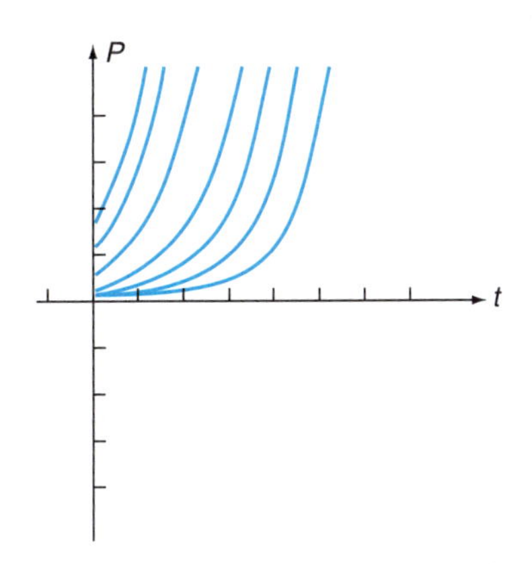

Figure 1.8.2 Graphs of $P(t) = P(0)e^{kt}$, $k > 0$, for several $P(0) > 0$.

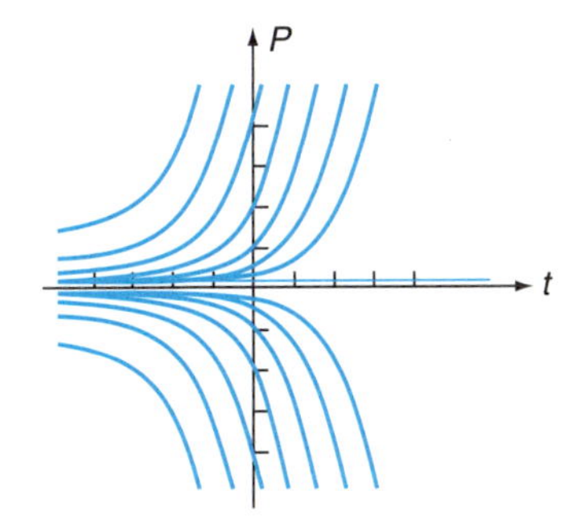

Figure 1.8.3 Graphs of $P(t) = P(0)e^{kt}$ for $k > 0$.

Doubling Time

An additional way to understand the implications of exponential growth is to investigate how long it takes a population to double, assuming it grows at a constant rate k. We are asking at what time t the population will be twice its initial population $P(0)$. That is, we wish to find t such that $2P(0) = P(t)$. Using (6), the requirement is

$$2P(0) = P(0)e^{kt}. \tag{7}$$

In order to solve (7) for the doubling time t, we first observe that $P(0)$ cancels, so that (7) becomes

$$2 = e^{kt}. \tag{8}$$

Thus, the doubling time does not depend on the initial population. For example, with continuous compounding, your money will double in the same time whether you deposit \$100 or \$1000 (this assumes that the bank doesn't offer a higher rate to the larger depositor and that you do not make additional deposits). To determine the doubling time, we take the natural logarithm of both sides of (8):

$$\ln 2 = kt.$$

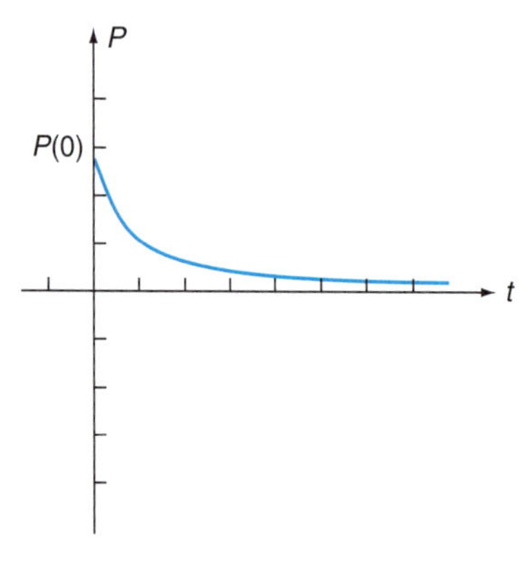

Figure 1.8.4 $P(t) = P(0)e^{kt}$ if $k < 0$.

Figure 1.8.5 Solutions of $\frac{dP}{dt} = kP$ with $k < 0$.

Thus, the doubling time is given by

$$t_{\text{doubling time}} = t_d = \frac{\ln 2}{k}, \tag{9}$$

where k is the growth rate. Using a calculator or tables, $\ln 2 \approx 0.69315$. Since $0.7 = 70/100$, this yields a useful observation. From (9), the doubling time for exponential growth is approximately $0.69315/k$. This motivates the **Rule of 70** (some prefer a rule of 72 since 72 is divisible by 3, 6, 9, and 12) when the time unit is years. As an approximation,

$$t_{\text{doubling time (in years)}} \approx \frac{70}{\text{growth rate (measured in \% per year)}}. \tag{10}$$

This is a great formula to memorize, since it communicates so well the implications of the compounding effect of continuous growth.

Example 1.8.1 *Using (10)*

If the population of the world is growing at 2% per year, then the world's population is expected to double in approximately $\frac{70}{2} = 35$ years (not 50 years, as you get without compounding). If inflation is 5% per year, the cost of living is expected to double in approximately $\frac{70}{5} = 14$ years (not 20 years!). ♦

Sometimes it is advantageous to utilize the doubling time more explicitly in the solution of the differential equation. From (9), the growth rate k may be replaced by the doubling time t_d:

$$k = \frac{\ln 2}{t_d}.$$

The solution (6) of the differential equation (4) becomes

$$P(t) = P(0)e^{kt} = P(0)(e^{\ln 2})^{\frac{t}{t_d}} = P(0)2^{\frac{t}{t_d}}, \tag{11}$$

thus expressing the solution in terms of its doubling time. This shows clearly that the solution doubles after every time t_d.

Example 1.8.2 *Using (11)*

Suppose that it is given that some population triples every 7 years. Find the population after 10 years.

- SOLUTION. Using the ideas in (11), the solution of the differential equation (14) is

$$P(t) = P(0)3^{\frac{t}{7}}.$$

Thus the population could be predicted at $t = 10$ to be $P(0)3^{\frac{10}{7}}$. ♦

Yield

We would like to examine more carefully the effect of compounding when a quantity grows continually at a constant growth rate k:

$$P(t) = P(0)e^{kt}.$$

When a quantity is growing continuously at the rate of 100 $k\%$ per year, the actual amount at the end of one year ($t = 1$) is

$$P(1) = P(0)e^{k}.$$

In banking, $P(t)$ is interpreted as the amount of money in a fund at time t. The initial amount $P(0)$ is called the **principal**. Exponential growth is referred to as **continuous compounding**. The interest earned after one year is $P(1) - P(0) = P(0)(e^k - 1)$. The **yield** (**effective interest rate**) is defined to be the actual interest earned over one year divided by the principal:

$$\text{Yield} = \frac{P(1) - P(0)}{P(0)} = e^k - 1. \tag{12}$$

In this analysis, money is being compounded continuously, so that growth can be described by a differential equation.

Example 1.8.3 *Yield*

Suppose a bank has an interest rate of 5% per year that is compounded continually. What will the yield be?

- SOLUTION. Using (12), we have

$$\text{Yield} = e^{0.05} - 1 \approx 1.0513 - 1 = 0.0513.$$

This corresponds to the 5.13% yield a bank might advertise when its interest rate is 5% per year. (The real world can be more complicated than this. Banks have in the past, from the days of hand computation of interest, defined a year to be only 360 days. The consumer would actually benefit by receiving 5 extra days of interest beyond the 5% promised.) ♦

The differential equation $\frac{dP}{dt} = kP$ models the continuous compounding of an initial deposit, where $P(t)$ is the amount of money and k is the growth rate. This model is often sufficiently accurate if money is compounded daily. However, there may be additional deposits or inputs. This leads to a differential equation that is not homogeneous.

Example 1.8.4 *Constant Input*

Suppose the interest rate on a $1,000 deposit is 7% per year. This means $k = 0.07$. Suppose that instead of leaving the money to grow, we deposit an additional $10 per day. Determine the amount of money in the account as a function of time. Assume continuous compounding and continuous deposit.

• SOLUTION. We must first determine the differential equation that applies. Here, the amount of money increases for two reasons: interest and deposits. In general, we can write a word equation for the rate of change of the amount:

$$\begin{pmatrix} \text{Rate of} \\ \text{change of} \\ \text{money} \end{pmatrix} = \begin{pmatrix} \text{rate of} \\ \text{increase due} \\ \text{to interest} \end{pmatrix} + \begin{pmatrix} \text{rate of} \\ \text{increase due} \\ \text{to deposit} \end{pmatrix}. \tag{13}$$

All three terms in (13) must have the same units, which we take as dollars per year. As discussed before, $\frac{dP}{dt}$ is the rate of change of the amount of money and $kP = 0.07P$ is the amount of interest per year. The deposit rate is a constant, $10 per day, equivalent to $3,650 per year. Thus, the differential equation (13) is

$$\frac{dP}{dt} = 0.07P + 3650.$$

This is a first-order linear differential equation with constant coefficients with a constant input of 3650:

$$\frac{dP}{dt} - 0.07P = 3650. \tag{14}$$

The general solution will be a particular solution plus an arbitrary constant times the solution of the associated homogeneous equation:

$$P = P_p + cP_1.$$

Equation (14) could be solved by separation or integration factors. We shall use the approach of Section 1.7. Since the input is constant, we shall look for a particular solution that is a constant $P_p = A$. Substituting $P = A$ into the differential equation, we find that $P_p = a = -\frac{3650}{0.07}$. A solution of the associated homogeneous equation grows exponentially, $P_1 = e^{kt} = e^{0.07t}$. Thus, the general solution of (14) is

$$P(t) = -\frac{3650}{0.07} + ce^{0.07t}. \tag{15}$$

The initial deposit was \$1000, so the initial condition for the differential equation is $P(0) = 1000$. Thus, the arbitrary constant satisfies $1000 = -\frac{3650}{0.07} + c$. Solving for c, we get the solution to the initial value problem:

$$P(t) = -\frac{3650}{0.07} + \left(1000 + \frac{3650}{0.07}\right) e^{0.07t}. \tag{16}$$

Having solved the differential equation, we can now analyze the effect of making continual deposits at the rate of \$3650 per year. Suppose we were interested in the amount of money that would be accumulated after 10 and 20 years. Without interest, the amount of money at the end of 10 years is \$37,500, and at the end of 20 years is \$74,000. With interest, the amount of money is determined from (16) and is considerably larger:

$$\begin{aligned} P(10) &= -\frac{3650}{0.07} + \left(1000 + \frac{3650}{0.07}\right) e^{0.7} \approx \$54,876, \\ P(20) &= -\frac{3650}{0.07} + \left(1000 + \frac{3650}{0.07}\right) e^{1.4} \approx \$163,372. \end{aligned}$$

We could also make the necessary modifications to compare the effects of various interest rates on the amount of money. ♦

1.8.2 Radioactive Decay

First-order differential equations are very important in understanding radioactive decay. For a given atom of radioactive material, not acted on by radiation or other particles, there is a fixed probability that it will decay in a given time period. For example, suppose the probability is 0.001 that the atom will decay in one year. Now suppose that we have $x(t)$ of these atoms at time t. We would expect at that time to have atoms decaying at a rate of $(0.001)x(t)$ atoms per year. This leads to the following law of radioactive decay:

> The rate of decay for a radioactive material is proportional to the number of atoms present. (17)

Note that the law of decay is not really a law. However, if we have a reasonably large number of atoms and the material is not bombarded by other radiation or particles, then the law of decay is often sufficiently accurate. We do not intend to belabor the point, but it is important to realize that the suitability of a law like (17) depends on the problem and the intended applications of the answer.

Since $x(t)$ is the amount of radioactive material at time t, (17) may be written as a first-order linear differential equation:

$$\frac{dx}{dt} = -kx. \tag{18}$$

We introduce the minus sign, since in radioactive decay the number of atoms diminishes. Here the proportionality constant k, the **decay rate**, will be positive since there is a loss of material.

Equation (18) is homogeneous (this is more easily seen if (18) is rewritten $\frac{dx}{dt} + kx = 0$). Using the technique of Section 1.7, we note that a solution of the homogeneous differential equation with constant coefficients is in the form $x = e^{rt}$, which when substituted into (18) yields $r = -k$. Thus, e^{-kt} is a solution of the homogeneous equation, and so the general solution of (18) will be an arbitrary multiple of the solution of the homogeneous equation:

$$x(t) = ce^{-kt}.$$

(The differential equation (18) can also be solved by the integrating factor technique or by separation of variables.) The constant c is determined by satisfying the initial condition $x(0) = c$. Thus, we finally arrive at

$$x(t) = x(0)e^{-kt}. \tag{19}$$

Thus, radioactive decay is exponential decay, since $k > 0$ implies that $x(t) \to 0$ at $t \to \infty$, as sketched in figure 1.8.5.

Half-life

The **half-life** of a radioactive substance is the length of time it takes the material to decay to half its original amount. If T is the half-life, we have by definition

$$x(T) = \frac{1}{2}x(0). \tag{20}$$

To determine the half-life, we substitute the formula (19) for $x(t)$ into (20) to get

$$x(0)e^{-kT} = \frac{1}{2}x(0)$$

or

$$e^{-kT} = \frac{1}{2}.$$

Taking the natural logarithm of both sides and solving for T gives the half-life as

$$T_{\text{half-life}} = T = \frac{\ln 2}{k}. \tag{21}$$

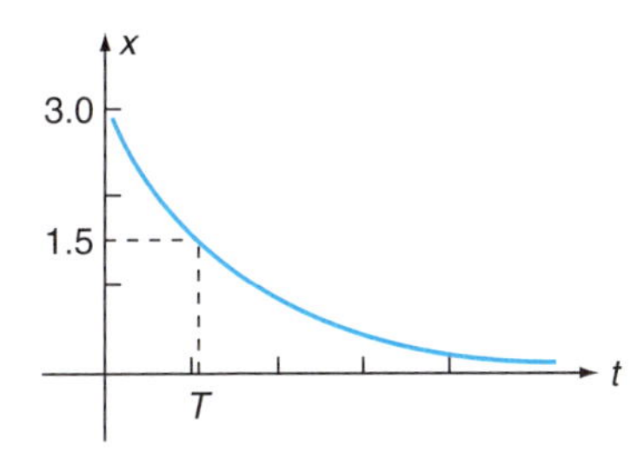

Figure 1.8.6 Graph of solution of Example 1.8.2.

The half-life and the decay rate k are closely related. Determining the half-life for radioactive decay is mathematically analogous to finding the doubling time for population growth.

Example 1.8.5 *Radioactive Decay*

In two years, 3 g of a radioisotope decay to 0.9 g. Determine both the half-life T and the decay rate k.

• SOLUTION. Let $x(t)$ be the amount of the radioisotope at time t. Measure x in grams and t in years. The basic law is given by (18): $\frac{dx}{dt} = -kx$. Solving the equation gives (19):

$$x(t) = x(0)e^{-kt}. \tag{22}$$

This equation holds for any radioisotope. For the isotope of this example, we are given the initial amount and the amount at $t = 2$:

$$x(0) = 3, \quad x(2) = 0.9. \tag{23}$$

Substituting (23) into the general formula (22) with $t = 2$ yields

$$0.9 = 3e^{-2k}.$$

Thus,

$$k = \frac{\ln\left(\frac{1}{0.3}\right)}{2} \approx 0.6.$$

From (21), the half-life is

$$T = \frac{2\ln 2}{-\ln 0.3} \approx 1.2 \text{ years}.$$

See figure 1.8.6. ♦

1.8.3 Thermal Cooling

Differential equations govern thermal cooling. Ignoring circulation and other effects, **Newton's law of cooling** states that

> The rate of change of the surface temperature of an object is proportional to the difference between the temperature of the object and the temperature of its surroundings (also called the ambient temperature) at that time. (24)

If $T(t)$ is the surface temperature of the object at time t and Q_0 is the ambient temperature at time t, then Newton's law of cooling becomes

$$\frac{dT}{dt} = -k(T - Q_0). \tag{25}$$

It is important to understand why we have introduced a minus sign on the right-hand side of (25). If $T > Q_0$, the body's surface is hotter than its surrounding environment, and hence there is a loss of surface temperature. It follows that in this situation, $\frac{dT}{dt}$ must be negative, and consequently k is positive. We have introduced the minus sign so that the proportionality constant k is positive, which is more convenient. On the other hand, if $T < Q_0$, then the surface temperature increases and $\frac{dT}{dt} > 0$. Again the constant k is positive. Note that (24) does not require that the outside temperature Q_0 be constant over time. In using (25) in what follows, we shall assume uniform cooling. That is, we consider the temperature of an object to be the same as its surface temperature.

Equation (25) is a first-order linear differential equation with constant coefficients. It is often written in the form

$$\frac{dT}{dt} + kT = kQ_0. \tag{26}$$

Equation (26) can always be solved by the integrating factor technique even if Q_0 depends on t. (Examples with nonconstant Q_0 are in the exercises.)

We will analyze only the simplest case, in which the outside (ambient) temperature Q_0 is constant. In this case, the constant coefficient differential equation has a constant input. Using the constant input technique, it can be shown that

$$T(t) = Q_0 + ce^{-kt}. \tag{27}$$

We could also solve (26) by separation or by using an integrating factor. The integrating factor is $u = e^{\int k\,dt} = e^{kt}$. Multiplying both sides of (26) gives

$$\frac{d}{dt}(e^{kT}T) = kQ_0e^{kt}.$$

Integrating both sides with regard to t, we have

$$e^{kt}T = Q_0e^{kt} + c.$$

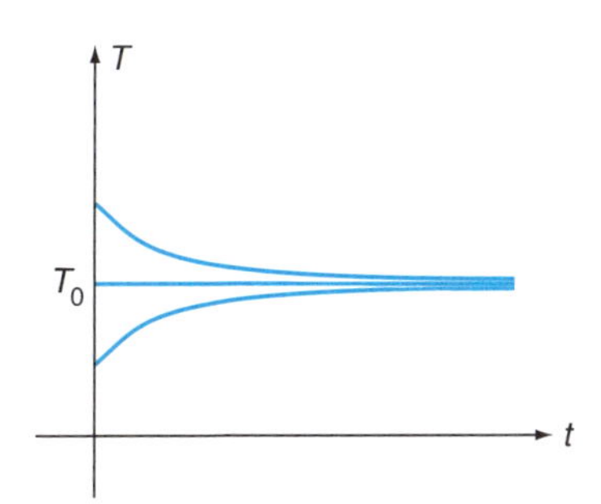

Figure 1.8.7 Graph of three solutions to (26).

Dividing by e^{kt}, we get that the solution is

$$T(t) = Q_0 + ce^{-kt},$$

which is (27).

No matter what the initial condition, we observe that as $t \to \infty$,

$$T(t) \to Q_0.$$

That is, the surface temperature approaches the ambient temperature as time increases. The manner in which the ambient temperature is approached is described by (27). Usually we wish to express the arbitrary constant c in terms of the initial condition $T(0)$ for the temperature. Letting $t = 0$ in (27), we get $T(0) = Q_0 + c$, so that $c = T(0) - Q_0$. Thus, (27) becomes

$$T(t) = Q_0 + [T(0) - Q_0]e^{-kt}. \tag{28}$$

If the initial temperature $T(0)$ equals the ambient temperature Q_0, then (28) shows that the temperature stays constant, equaling the ambient temperature. Graphs of the temperature as a function of time for various initial conditions are given in figure 1.8.7. Note that in all cases the solution approaches the ambient temperature Q_0 as $t \to \infty$.

Equilibrium

An **equilibrium solution** of a differential equation is a solution of the differential equation that is constant in time. For the differential equation (25), we see that $T(t) = Q_0$ is an equilibrium solution. For Newton's law of cooling, we have shown that the temperature approaches this equilibrium as $t \to \infty$ independent of the initial conditions. In Chapters 4, 5, and 6, we will discuss equilibrium solutions more fully and introduce the notion of whether an equilibrium solution is stable or unstable.

Example 1.8.6 *Cooling*

Suppose the room temperature in your office is 70°F. Experience has taught you that the temperature of a cup of coffee brought to your office will drop from 120°F to 100°F in 10 min. What should be the temperature of your cup of coffee when it is brought into the room if you want it to take 20 minutes for the temperature to drop to 100°F?

• SOLUTION. First, we define our notation, set up the basic equations, and express the given data in terms of these. Let t be the time in minutes and $T(t)$ the temperature in degrees Fahrenheit at time t. The governing differential equation follows from the cooling law (24):

$$\frac{dT}{dt} = -k(T - Q_0). \tag{29}$$

From the problem description, the ambient temperature is 70°F:

$$Q_0 = 70. \tag{30}$$

Not only is the initial temperature prescribed, but we are given the temperature at 10 min:

$$T(10) = 100 \quad \text{when } T(0) = 120. \tag{31}$$

We shall break this problem into two subproblems. The first is to determine k for the flow of thermal energy from our coffee cup into the room. Equation (28) is the solution of the initial value problem for (29):

$$T(t) = Q_0 + [T(0) - Q_0]e^{-kt} = 70 + 50e^{-kt}. \tag{32}$$

We are not given the proportionality constant k and must determine it from (32) using the given temperature at 10 min:

$$T(10) = 100 = 70 + 50e^{-10k}.$$

Solving this equation, we obtain

$$e^{-10k} = \frac{3}{5} \quad \text{so that } k = \frac{\ln\left(\frac{3}{5}\right)}{-10} \approx 0.05. \tag{33}$$

Now we can solve the second subproblem, which is to determine the initial temperature if $T(20) = 100$. From (28) again,

$$T(t) = Q_0 + [T(0) - Q_0]e^{-kt} = 70 + [T(0) - 70]e^{-kt}.$$

However, k is determined in (33). Thus,

$$100 = T(20) = 70 + e^{2\ln 3/5}[T(0) - 70]$$

or

$$30 = \frac{9}{25}[T(0) - 70].$$

Finally, solving this equation for the desired initial temperature $T(0)$, we obtain

$$T(0) = 70 + \frac{30 \cdot 25}{9} \approx 153.33°\text{F}.$$

In this example, the coffee cup surface temperature dropped from 120°F to 100°F in 10 min and from $153\frac{1}{3}$°F to 100°F in 20 min. This shows the impossibility of guessing the answer without using the differential equation and its solution.

Exercises

In all the following problems, assume exponential growth from continuous compounding.

1. One estimate of the growth rate of the United States is 1.5% per year. How many years will it take for the population to double?
2. A crystal grows by 5% in one day. When would you expect the crystal to be twice its original size?
3. A certain bacteria population is observed to double in 8 h. What is its growth rate?
4. The population of the world is expected to double in the next 30 yr. What is its growth rate?
5. The growth rate of a certain strain of bacteria is unknown, but assumed to be constant. When an experiment started, it was estimated that there were about 1500 bacteria, and an hour later there were 2000. How many bacteria would you predict there are 4 h after the experiment started?
6. A population of bacteria is initially given and grows at a constant rate k_1. Suppose τ hours later the bacteria are put into a different culture such that the population now grows at the constant rate k_2. Determine the population of bacteria for all time.
7. The doubling time for a certain virus is 3 yr. How long will it take for the virus to increase to 10 times its current population level?
8. Initially you have 0.1 g of a bacteria in a large container; 2 hr later you have 0.15 g. What is the doubling time for these bacteria?
9. An organism living in a pond reproduces at a rate proportional to the population size. Organisms also die off at a rate proportional to the population size. In addition, organisms are continuously added at a rate of k g/yr. Give the differential equation that models this situation.
10. A bacteria population is reproducing in a large vat of nutrients according to an exponential growth law that would cause the population to double in 0.5 h. However, bacteria are continuously siphoned off at a rate of 5 g/h. Initially, there are 10 g of bacteria. How many bacteria are there after 2 h?
11. The rate of interest at one bank is 3% per year, whereas the yield at another bank is 3% per year. Both offer continuous compounding. Find the two doubling times.
12. The cost of a liter bottle of water was 85 cents two years ago, but it now costs 95 cents. If this rate of increase continues, approximately when will the bottle cost $1.50?
13. The GDP (gross domestic product) of a certain country increased by 6.4% during the last year. If it continued to increase at that rate, approximately how many years would it take for the GDP to double?
14. During one year, food prices increase by 15%. At that rate, in approximately how many years would food prices triple?

15. **Cost of living** is the amount required to purchase a certain fixed list of goods and services in a single year. We assume that it is subject to exponential growth. The growth rate is known as the **rate of inflation**. If the cost of living rose from \$10,000 to \$11,000 in one year (a 10% net increase), what is the instantaneous rate of increase in the cost of living that year? Equivalently, what is the rate of inflation?
16. Over a 3-year period, housing prices increased 15%. At that rate, how many years would it take for housing prices to increase 50%?
17. A radioactive isotope has a half-life of 16 days. You wish to have 30 g at the end of 30 days. How much radioisotope should you start with?
18. A radioisotope is going to be used in an experiment. At the end of 10 days, only 5% is to be left. What should the half-life be?
19. A radioactive isotope sits unused in your laboratory for 10 yr, at which time it is found to contain only 80% of the original radioactive material.
   a) What is the half-life of this isotope?
   b) How many *additional* years will it take until only 15% of the original amount is left?
20. At the time an item was produced, 0.01 of the carbon it contained was carbon-14, a radioisotope with a half-life of about 5745 yr.
   a) You examine the item and discover that only 0.0001 of the carbon is carbon-14. How old is the object? (This process of determining the age of an object from the amount of carbon-14 it contains is known as carbon-14 dating.)
   b) Derive a formula that gives the age A of the object in terms of the fraction of carbon that is carbon-14 at the present time, T.
21. The temperature of an engine at the time it is shut off is 200°C. The surrounding air temperature is 30°C. After 10 min have elapsed, the surface temperature of the engine is 180°C.
   a) How long will it take for the surface temperature of the engine to cool to 40°C?
   b) For a given temperature T between 200°C and 30°C, let $t(T)$ be the time it takes to cool the engine from 200°C to T. (For example, $t(200) = 0$ and $t(40)$ is the answer to part (a).) Find the formula for $t(T)$ in terms of T and graph the function. (The ambient temperature is still 30°C.)
22. Earlier experiments have shown that a certain component cools in air according to the cooling law (25) with constant of proportionality 0.2. At the end of the first processing stage, the temperature of the component is 120°C. The component remains for 10 min in a large room and then enters the next processing stage. At that time the surface temperature is supposed to be 60°C.
   a) What must the room temperature be for the desired cooling to take place?
   b) Suppose that the entrance and exit temperatures are still set at 120°C and 60°C, respectively, but the length of the wait in the room is w, a constant. Find the desired room temperature as a function of w and graph it.
23. An object at 100°C is to be placed in a 40°C room. What should the constant of proportionality be in (25) in order that the object be at 60°C after 10 min?
24. The air in a room is cooling. At time t (in hrs) the air temperature is $Q_0(t) = 70 + 20e^{-t/2}$. An object is placed in the room at time $t = 0$. The object is initially at 50°C and changes temperature according to (25) with $k = \frac{1}{2}$.
   a) Find the temperature, $T(t)$, of the object for $0 \le t \le 5$.
   b) Graph both Q_0 and T on the same axes.

25. An instrument at an initial temperature of 40°C is placed in a room whose temperature is 20°C. For the next 5 h the room temperature $Q_0(t)$ gradually rises and is given by $Q_0(t) = 20 + 10t$, t in hours.
  a) Give the form the cooling law (25) takes for the instrument.
  b) From prior experience you know that your instrument cools according to (25) with $k = 1$ if t is measured in hours. If $T(t)$ is the surface temperature at time t, solve the equation in part (a) for $T(t)$.
  c) Graph $Q_0(t)$ and $T(t)$ on the same axes for $0 \le t \le 5$.

Most banks compound daily. Continuous compounding gives answers that are very close to those given by daily compounding. In Exercises 26–31, approximate the effect of daily compounding by assuming continuous compounding.

26. You have \$10,000 and intend to invest it for 5 years in a bank that offers continuous compounding. If you want to have \$15,000 in your account at the end of these 5 years, what annual interest rate do you have to get?
27. You invest \$2,000 in an account paying 6% annually, compounded daily, on the amount in excess of \$500.
  a) Express this as a differential equation by assuming continuous compounding.
  b) Solve the differential equation and determine the amount in the account after 10 yr.
  c) How much more would the account have after 10 yr if the full amount earned interest?
28. An amount of \$10,000 is deposited in a bank that pays 9% annual interest compounded daily.
  a) If you withdraw \$10 a day, how much money do you have after 3 yr?
  b) How much can you withdraw each day if the account is to be depleted in exactly 10 yr?
29. An amount of \$1000 is deposited in a bank that pays 8% annual interest compounded daily. A deposit of B dollars is made daily.
  a) What should B be in order to have \$10,000 after 5 yr?
  b) Determine the function $B(t)$ that gives the daily deposit needed to have \$10,000 after t years ($B(5)$ is computed in part (a)).
30. An amount of \$10,000 is invested at 12% annual interest compounded daily. An additional investment of B \$ is made daily. What should B be in order for the investment account to be \$100,000 after 10 yr?
31. An amount of \$100 is deposited in a foreign bank that pays 20% annual interest compounded daily. Each day you make a transaction of amount $f(t)$ dollars. Part of the year you are able to deposit money [$f(t) > 0$], and part of the year you make withdrawals [$f(t) < 0$]. If t is in years, these transactions occur in the following cyclical yearly pattern:

$$\begin{aligned} f(t) &= \frac{400}{365} \cos(2\pi t) \ \$/\text{day} \\ &= 400 \cos(2\pi t) \ \$/\text{year}. \end{aligned}$$

  a) Find the amount of money in the account as a function of t.
  b) Graph your answer to (a) for 10 yr (a sketch will do if no computer is available).

1.9 Mixture Problems

In this section, a quantity $Q(t)$, such as the amount of some pollutant in a water tank, varies with time. Further amounts of this quantity are being added. The addition will be called inflow. Simultaneously, some of this quantity is being lost. The quantity lost will be called the outflow. In the case of a water tank, the loss could be due to evaporation, overflow, an open valve, or all three. In all cases we assume that the tank is very well mixed (through fast stirring if necessary), so that the concentration of the pollutant will be assumed to be the same throughout all portions of the tank. The beginning idea for analyzing these problems is the fundamental physical principle of conservation of the quantity Q:

$$\left[\begin{array}{c}\text{rate of change}\\ \text{of } Q\end{array}\right] = \frac{dQ}{dt} = \left[\begin{array}{c}\text{inflow rate}\\ \text{of } Q\end{array}\right] - \left[\begin{array}{c}\text{outflow rate}\\ \text{of } Q\end{array}\right], \tag{1}$$

where we have noted that $\frac{dQ}{dt}$, the time derivative of Q, is the rate of change of Q with regard to time t.

Procedure for Flow Problems

1. It is helpful to draw a rough sketch of a tank illustrating the inflow and outflow with pipes (see figure 1.9.1).
2. Label quantities and note the given data.
3. Express inflow rates and outflow rates in terms of the given variables and substitute them into (1).
4. Solve the resulting differential equation.
5. Answer any questions such as "how long?"

We will discuss problems of two different kinds. In one, the volume of water in the tank is fixed, and the resulting differential equation will be easier to solve. In the other, the volume of water is changing in time, and the resulting differential equation will be harder to solve. In some of the exercises, the process of setting up the differential equations will be emphasized, and you will not be asked to solve the differential equations.

1.9.1 Mixture Problems with a Fixed Volume

We begin with a fixed volume example.

Example 1.9.1 *Fixed Volume*

Consider a 100-m^3 tank full of water. The water contains a pollutant at a concentration of 0.6 g/m^3. Cleaner water, with a pollutant concentration of 0.15 g/m^3, is pumped into the well-mixed tank at a rate of 5 m^3/s. Water flows out of the tank through an overflow valve at the same rate as it is pumped in.

a) Determine the amount and concentration of the pollutant in the tank as a function of time. Graph the result.
b) At what time will the concentration be 0.3 g/m^3?

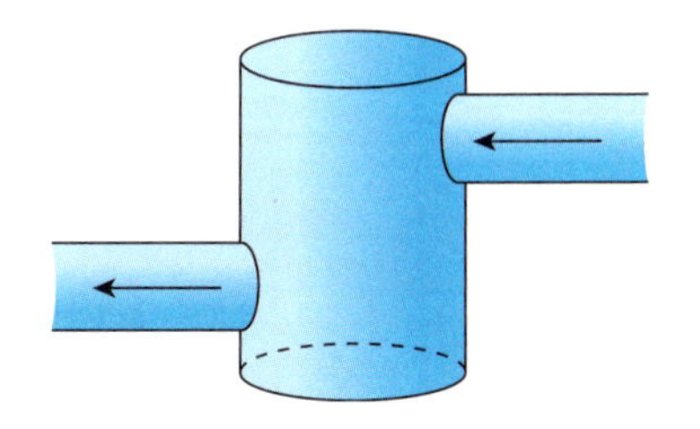

Figure 1.9.1 Tank with inflow and outflow.

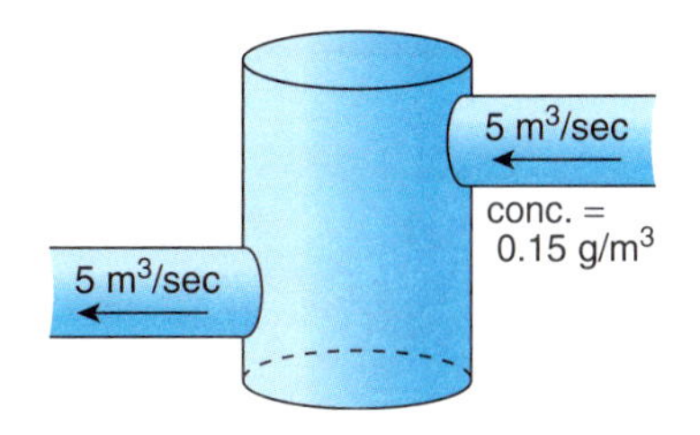

Figure 1.9.2 Picture for Example 1.9.1.

• SOLUTION. In order to illustrate the general principles, and since this is our first mixing problem, we shall include a few more steps than are necessary to solve the particular problem.

In mixture problems, it is best to first draw a rough diagram of a tank indicating the inflow and the outflow (see figure 1.9.2). Water flows in at the rate of 5 m^3/s, with concentration 0.15 g/m^3, and the mixture flows out at the same rate of 5 m^3/s. Thus, the volume of water in the tank stays the same, 100 m^3. Usually it is easier to formulate a differential equation for the amount of the pollutant. We let $Q(t)$ be the amount in grams of pollutant in the tank. Q depends on the time t, which we measure in seconds. The amount of pollutant in the tank changes in time as a result of inflow and outflow, so that the rate of change of the amount of pollutant satisfies (1):

$$\frac{dQ}{dt} = \begin{bmatrix} \text{inflow rate} \\ \text{of pollutant} \end{bmatrix} - \begin{bmatrix} \text{outflow rate} \\ \text{of pollutant} \end{bmatrix}. \tag{2}$$

It is given that 5 m^3 of water per second flows in, with the concentration of pollutant being 0.15g/m^3. Thus, $5 \cdot (0.15)$ g of pollutant per second flows in, since each cubic meter has 0.15 g of pollutant:

$$\begin{bmatrix} \text{Inflow rate} \\ \text{of pollutant} \end{bmatrix} = \begin{bmatrix} \text{flow of volume} \\ \text{of water in} \end{bmatrix} \cdot \begin{bmatrix} \text{concentration of} \\ \text{pollutant flowing in} \end{bmatrix}.$$

We now compute the outflow of pollutant. Again 5 m^3per second flows out:

$$\begin{aligned} \begin{bmatrix} \text{Outflow rate} \\ \text{of pollutant} \end{bmatrix} &= \begin{bmatrix} \text{flow of volume} \\ \text{of water out} \end{bmatrix} \cdot \begin{bmatrix} \text{concentration of} \\ \text{pollutant flowing out} \end{bmatrix} \\ &= 5 \cdot \begin{bmatrix} \text{concentration of} \\ \text{pollutant flowing out} \end{bmatrix}. \end{aligned}$$

But we are not given the concentration of the pollutant that flows out. We must compute this concentration. The water in the tank is assumed to be well mixed, so the concentration that flows out is assumed to be the same as the overall concentration of the pollutant in the tank. To compute the concentration of a pollutant in a tank, we simply divide the total amount of pollutant by the total volume:

$$\begin{bmatrix} \text{Concentration of pollutant} \\ \text{in the tank} \end{bmatrix} = \frac{\text{amount of pollutant}}{\text{volume}} = \frac{Q(t)}{100}.$$

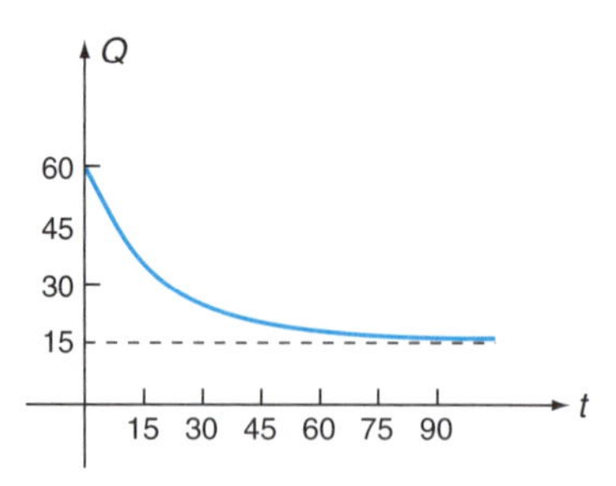

Figure 1.9.3 Graph of $15+45e^{-t/20}$.

In this case, the volume of water is fixed at 100 m^3. The amount of pollutant in the tank is unknown, but we called it $Q(t)$. The differential equation for the amount of pollutant follows from (2):

$$\frac{dQ}{dt}=5\cdot 0.15-5\cdot\frac{Q}{100}=0.75-\frac{1}{20}Q. \tag{3}$$

The differential equation (3) is linear, and we often rewrite it as

$$\frac{dQ}{dt}+\frac{1}{20}Q=0.75. \tag{4}$$

Since (4) is a linear differential equation with constant coefficients and the input 0.75 is a constant, we can use the method of undetermined coefficients of Section 1.7. (Integrating factors from Section 1.6 also work well.) A particular solution will be a constant, $Q_p=A$. Substituting A for Q in (4), we compute that $A=\frac{3}{4}\cdot 20=15$. A solution of the associated homogeneous equation is easily seen to be $Q_1=e^{rt}=e^{-\frac{t}{20}}$. Thus the general solution of (4) is

$$Q(t)=15+ce^{-\frac{t}{20}}. \tag{5}$$

The arbitrary constant c is determined from the initial conditions. The word problem states that the 100-m^{3}tank initially contains the pollutant at a concentration of 0.6 g/m^3. Thus, initially the amount of pollutant is $Q(0)=0.6(100)=60$ g. By substituting this initial condition into the general solution (5), we obtain $60=15+c$ or $c=45$. Thus, the solution of the initial value problem is

$$Q(t)=15+45e^{-\frac{t}{20}}, \tag{6}$$

which is graphed in figure 1.9.3. The amount of pollutant is initially 60, but exponentially decays to 15 as time increases. As $t\to\infty$, $Q(t)\to 15$. It should be physically obvious that the amount of pollutant approaches 15 g as time approaches infinity, because the water entering the tank has a concentration of 0.15 g/m^3. Eventually the concentration of the pollutant in the tank will approach a level equal to the inflow concentration. Since the tank is 100 m^3, the amount of pollutant in the tank corresponding to the inflow would approach $100\cdot 0.15=15$ g of pollutant.

Next we calculate the concentration $c(t)$:

$$c(t) = \frac{Q(t)}{\text{volume}} = \frac{15 + 45e^{-\frac{t}{20}}}{100} = 0.15 + 0.45e^{-\frac{t}{20}}. \tag{7}$$

This also shows that the concentration approaches the concentration of the inflow, 0.15, as time increases. We are asked to determine when the concentration equals 0.3, $c(t) = 0.3$. Using (7), we obtain

$$0.3 = 0.15 + 0.45e^{-\frac{t}{20}}.$$

Solving for t, we find $t = -20 \ln\left(\frac{1}{3}\right) \approx 21.97$ s. ♦

For this particular problem, it would have been just as easy to set up the equation for the concentration instead of the amount of the pollutant. However, if the volume varies, then it is usually easier to proceed as we did in this example.

1.9.2 Mixture Problems with Variable Volumes

Let us reconsider mixture problems in a tank. If the rate of water entering the tank is different from that of water leaving the tank, the volume of water in the tank will not be constant as in the previous example. We will show that the resulting differential equation for the amount of a pollutant will still often be linear, but that the coefficients will not be constant. Before we do an example, let us discuss a somewhat general problem.

Suppose we are given the amount of salt initially dissolved in a tank of water. Let us assume that salt water with a concentration of c_i pounds per gallon flows into the tank at the rate of r_i gallons per minute. Here the subscript i stands for the flow in. The tank contents are well mixed, and the salt water mixture is pumped out at the rate of r_o gallons per minute (subscript o for out). This problem is shown diagrammatically in figure 1.9.4. The volume $V(t)$ of salt water in the tank may vary. Since $V'(t) = r_i - r_o$, we have

$$V(t) = V(0) + (r_i - r_o)t. \tag{8}$$

Actually this is the somewhat obvious solution of the initial value problem for the differential equation $\frac{dV}{dt} = r_i - r_o$, representing the rate of change of the volume of water in the tank. If the inflow rate of water differs from the outflow rate of water, then the volume varies.

We let $Q(t)$ be the amount of salt in the tank. The inflow rate of salt is $r_i c_i$ (the inflow rate of water r_i times the given concentration of salt c_i in the inflow). The outflow rate of salt is the outflow rate of water r_o times the concentration of salt c_o in the outflow. As before, the concentration of salt in the outflow is the same as the concentration of salt $c(t)$ in the tank:

$$c_o = c(t) = \frac{Q(t)}{\text{volume}} = \frac{Q(t)}{V(0) + (r_i - r_0)t}. \tag{9}$$

The rate of change of the amount of salt is given by the basic relationship (1):

$$\frac{dQ}{dt} = \begin{bmatrix} \text{inflow rate} \\ \text{of salt} \end{bmatrix} - \begin{bmatrix} \text{outflow rate} \\ \text{of salt} \end{bmatrix}.$$

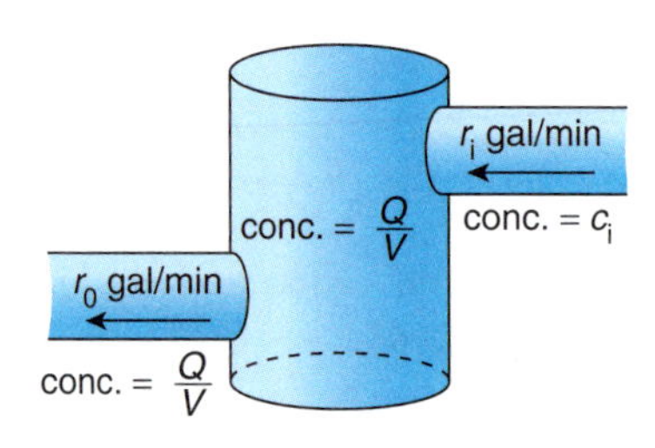

Figure 1.9.4

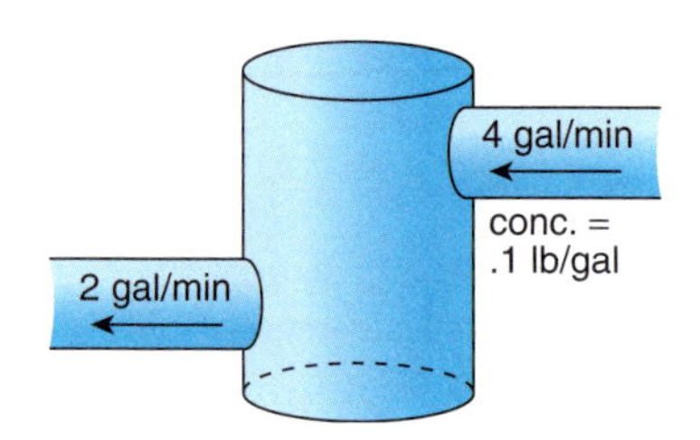

Figure 1.9.5 Example 1.9.2.

Using (9), we have

$$\frac{dQ}{dt} = r_i c_i - r_o c_o = r_i c_i - r_o \frac{Q}{V(0) + (r_i - r_o)t}. \tag{10}$$

Equation (10) is a linear differential equation for $Q(t)$, the amount of salt. The coefficients are not constant if $r_i \neq r_o$. If the coefficients are not constant, (10) may be solved by the integrating factor method.

Example 1.9.2 *Variable Volume*

A 100-gal tank is initially half full of pure water. Then water containing 0.1 lb/gal of salt is added at a rate of 4 gal/min. The well-mixed contents of the tank flow out a pipe at the rate of 2 gal/min. When the tank is full, it overflows. Find the amount and concentration of salt.

• SOLUTION. This problem is sketched in figure 1.9.5. The volume of salt water is increasing and is

$$V(t) = 50 + 2t, \tag{11}$$

since $r_i - r_o = 4 - 2 = 2$ and $V(0) = 50$. The 100-gal tank will fill up at $t = 25$ min. First, we solve the problem before the tank is filled.

THE FIRST 25 MINUTES. We let $Q(t)$ be the amount of salt in the tank, so that the concentration of salt is $c(t) = Q(t)/(50 + 2t)$. The rate of change of the amount of salt satisfies (10):

$$\frac{dQ}{dt} = [4] \cdot [0.1] - [2] \cdot \frac{Q}{50 + 2t}.$$

This can be rewritten as

$$\frac{dQ}{dt} + \frac{2Q}{50 + 2t} = 0.4. \tag{12}$$

We solve this linear differential equation by the integrating factor method. The integrating factor is

$$\exp\left(\int \frac{2}{50 + 2t} dt\right) = \exp\left(\ln(50 + 2t)\right) = 50 + 2t.$$

We next multiply (12) by this integrating factor to get

$$\frac{d}{dt}[(50+2t)Q] = (50+2t)0.4.$$

Antidifferentiating yields

$$(50+2t)Q = 0.1(50+2t)^2 + c.$$

Solving for Q yields the general solution of the linear differential equation:

$$Q(t) = 0.1(50+2t) + c(50+2t)^{-1}, \quad t \leq 25. \tag{13}$$

In this form a particular solution and a solution of the associated homogeneous equation are evident. The concentration can now be determined easily from (12) and (13) as

$$c(t) = \frac{Q(t)}{V(t)} = 0.1 + c(50+2t)^{-2}.$$

These formulas are valid until the tank fills, that is, for $t < 25$ min. The constant c can be determined from the initial condition:

$$Q(0) = 0.1 \cdot 50 + c(50)^{-1}, \quad \text{so that} \quad c = 50Q(0) - 0.1(50)^2$$

In our example, the water is initially pure, so that $Q(0) = 0$. Thus,

$$c = -0.1(50)^2 = -250, \tag{14}$$

which can be substituted into the general solution (13).

AFTER 25 MINUTES. After the tank fills up ($t > 25$), the inflow and outflow of salt water are the same, 4 gal/min. Thus, the volume is now a constant, 100 gal, and the method of solution is similar to that for the earlier example with constant volume. The differential equation will be

$$\frac{dQ}{dt} = (4) \cdot (0.1) - (4)\frac{Q}{100} = 0.4 - \frac{Q}{25}. \tag{15}$$

A particular solution is easily seen to be $Q_p = A = 0.4 \cdot 25 = 10$, and a solution of the associated homogeneous equation is $Q_1 = e^{rt} = e^{-t/25}$. Thus, a general solution is

$$Q(t) = 10 + c_1 e^{-t/25}, \tag{16}$$

where we have used a different arbitrary constant. This solution (16) is valid for $t > 25$. As $t \to \infty$, the amount of salt $Q(t)$ approaches 10 lb, as expected. The initial condition to determine the constant c_1 occurs at $t = 25$, by assuming the amount of

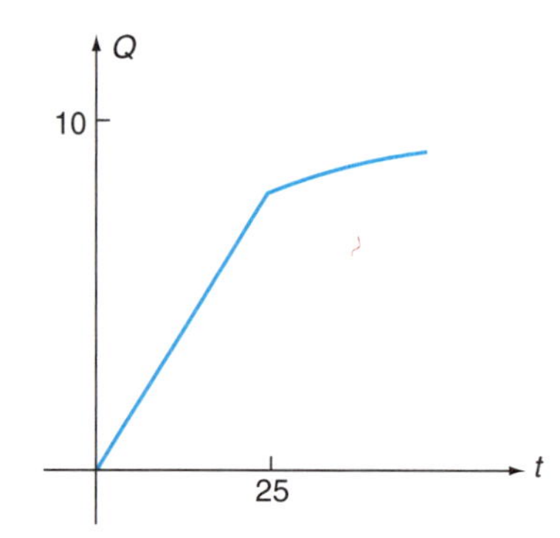

Figure 1.9.6 Graph of solutions of Example 1.9.2.

salt Q to vary continuously at $t=25$. That is, $Q(25)=\lim_{t\to 25^-} Q(t)$. From (13) and (14), we have

$$\lim_{t\to 25^-} Q(t) = 10 + \frac{c}{100} = 10 + \frac{-250}{100} = 7.5. \tag{17}$$

This condition (17) determines the constant c_1 in (16):

$$7.5 = Q(25) = 10 + c_1 e^{-1}, \quad \text{so that} \quad c_1 = -2.5e.$$

Thus, the solution of the initial value problem (valid for $t > 25$) follows from (16):

$$Q(t) = 10 - (2.5e)e^{-t/25} = 10 - 2.5e^{(25-t)/25}, \quad t > 25. \tag{18}$$

The solution, given by (13), (14), and (18) is shown in figure 1.9.6. ♦

Exercises

1. A well-mixed tank contains 300 gal of water with a salt concentration of 0.2 lb/gal. Water containing salt at a concentration of 0.4 lb/gal enters at a rate of 2 gal/min. An open valve allows water to leave the tank at the same rate.
a) Determine the amount and the concentration of salt in the tank as functions of time.
b) How long will it take for the concentration to increase to 0.3 lb/gal?
2. A room has a volume of 800 ft^3. The air in the room contains chlorine at a concentration of 0.1 g/ft^3. Fresh air enters at a rate of 8 ft^3/min. The air in the room is well mixed and flows out of a door at the same rate as the fresh air comes in.
a) Find the chlorine concentration in the room as a function of t.
b) Suppose that the flow rate of fresh air is adjustable. Determine the flow rate required to reduce the chlorine concentration to 0.001 g/ft^3 within 20 min.
3. A well-mixed tank contains 100 L of water with a salt concentration of 0.1 kg/L. Water containing salt at a concentration of 0.2 kg/L enters at a rate of 5 L/h. An open valve allows water to leave at 4 L/h. Water evaporates from the tank at 1 L/h.

a) Determine the amount and concentration of salt as a function of time.
b) Is the limiting concentration the same as that of the inflow?

4. A well-circulated pond contains 1 million L of water that contains a pollutant at a concentration of 0.01 kg/L. Pure water enters from a stream at 100 L/h. Water evaporates from the pond (leaving the pollutant behind) at 50 L/h. How many days will it take for the concentration of pollutant to drop to 0.001 kg/L?
5. A 1000-gal tank is initially half full of water and contains 10 lb of iodine in solution. Pure water enters the tank at a rate of 6 gal/min. An open valve allows water to leave at a rate of 1 gal/min. When full, the tank overflows.

a) Find the amount and concentration of iodine in the tank for the first 100 min.
b) Find the amount and concentration of iodine for the next 100 min.
c) Graph both concentration and amount of iodine.

6. A 100-gal tank is initially full of water containing 10 lb of salt in solution. The tank will overflow whenever additional water is added. A pump attached to a sensor pumps fresh water into the tank at a rate proportional to the concentration of salt in the tank. The constant of proportionality is 10 $(\text{gal})^2/\text{lb} \cdot \text{min}$. Find the amount and concentration of salt in the tank as functions of t.
7. A well-circulated lake contains 1000 kL of water that is polluted at a concentration of 2 kg/kL. Water from the effluent of a factory enters the lake at a rate of 5 kL/h with a concentration of 7 kg/kL of the pollutant. Water flows out of the lake through an outlet at the rate of 2 kL/h. Determine the amount and concentration of the pollutant as a function of time.

In Exercises 8*–14*, formulate the differential equations (before any tanks empty or overflow) that would be used to solve the problems, but do *not* solve the differential equations. Since Exercises 9*–13* involve more than one tank, they require one equation for each tank.

8.* A 600-gal tank initially contains 200 gal of brine (salt water) with 25 lb of salt. Brine containing 2 lb of salt per gallon enters the tank at the rate of 13 gal/s. The mixed brine in the tank flows out at the rate of 8 gal/s. How much salt is there in each tank as a function of time?

9.* Consider two tanks. Initially, tank 1 contains 150 gal of brine (salt water) with 8 lb of salt, and tank 2 contains 250 gal of brine with 14 lb of salt. Brine containing 3 lb of salt per gallon enters tank 1 at the rate of 13 gal/s. Mixed brine in tank 1 flows out into tank 2 at the rate of 7 gal/s. The mixture in tank 2 flows away at the rate of 28 gal/s. How much salt is there in each tank as a function of time?

10.* Consider two tanks. Initially, tank 1 contains 100 gal of brine (salt water) with 35 lb of salt, and tank 2 contains 400 gal of brine with 15 lb of salt. Suppose the mixture flows out of tank 1 into tank 2 at the rate of 17 gal/min and the mixture flows out of tank 2 into tank 1 at the rate of 6 gal/min. How much salt is there in each tank as a function of time?

11.* Consider two tanks. Initially, tank 1 contains 230 gal of brine (salt water) with 28 lb of salt, and tank 2 contains 275 gal of brine with 7 lb of salt. Brine containing 5 lb of salt per gallon enters tank 1 at the rate of 21 gal/s. Mixed brine in tank 1 flows out at the rate of 18 gal/s, half flowing into tank 2. The mixture in tank 2 flows away at the rate of 4 gal/s. How much salt is there in each tank as a function of time?

12.* Consider three tanks. Initially, tank 1 contains 200 gal of brine (salt water) with 55 lb of salt, tank 2 contains 500 gal of brine with 35 lb of salt, and tank 3 is empty. Suppose the mixture flows out of tank 1 into tank 2 at the rate of 8 gal/min and the mixture flows out tank 2 into tank 3 at the rate of 22 gal/min. How much salt is there in each tank as a function of time?

13.* Consider three tanks. Initially, tank 1 contains 100 gal of brine (salt water) with 17 lb of salt, tank 2 contains 200 gal of brine with 19 lb of salt, and tank 3 contains 300 gal of brine with 21 lb of salt. Brine containing 5 lb of salt per gallon enters tank 1 at the rate of 11 gal/s. Mixed brine in tank 1 flows out into tank 2 at the rate of 18 gal/s, and the mixture also flows out of tank 2 into tank 3 at the rate of 18 gal/s. How much salt is there in each tank as a function of time?

14.* A large tank contains 7 gal of pure water. Polluted water containing 7 g of bacteria per gallon enters at the rate of 14 gal/h. A well-mixed mixture is removed at the rate of 4 gal/h. However, it is also known that the bacteria multiply inside the tank at a growth rate of 2% per hour. Determine the amount of bacteria in the tank as a function of time.

1.10 Electronic Circuits

One of the applications of differential equations that will frequently recur throughout this book is the theory of electronic circuits. There are several reasons for this, among them the importance of circuit theory and the pervasiveness of differential equations in circuit theory. (One of the authors received his first exposure to circuit theory from a highly mathematical electrical engineering professor by the name of Amar Bose. You would be correct if you recognized the significance of his last name.) Also, circuits are one example of what could be called network models. Network models are widely used, for example, in manufacturing and other economic systems.

Circuits will be covered again in greater detail later. In this section we shall introduce the basic circuit concepts we shall use and give some simple examples of circuits that are described by first-order differential equations.

We consider only lumped-parameter circuits. In circuit theory, the word **circuit** means that quantities such as current are determined solely by position along the path. A wire has a finite thickness, and interesting electrical behavior occurs across the wire. We ignore such field effects. **Lumped parameter** means that the effects of the various electrical components may be considered to be concentrated at one point. The converse of a lumped-parameter circuit is a **distributed-parameter** circuit. The analysis of distributed-parameter circuits often involves partial differential equations which we do not discuss in this text. Antennas are an example of distributed-parameter systems.

At each point in the circuit there are two quantities of interest: **voltage** (or potential) and **current** (or flow of charge). Current is, by convention, the net flow of positive charge. A **branch** is part of a circuit with two terminals to which connections can be made, a **node** is a point where two or more branches come together (a node is denoted in our sketches as a large dot), and a **loop** is a closed path formed by connecting

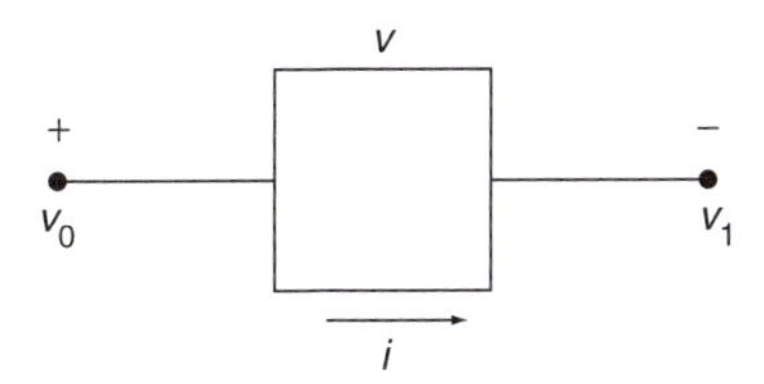

Figure 1.10.1 A two-terminal device. The nodal voltages are denoted v_0 and v_1, the voltage drop $v = v_0 - v_1$, and the current is denoted by i.

branches. Our basic modeling laws are Kirchhoff's circuit laws, the circuit and voltage laws:

Current law:	The algebraic sum of the currents entering a node at any instant is zero.	(1)
Voltage law:	the algebraic sum of the voltage drops around a loop at any instant is zero.	

The voltage law is equivalent to saying that the voltage drop from one point to another is the same in any direction along the circuit. We will discuss these laws shortly.

To set up the circuit equations, a current variable is assigned to each branch. One can talk about either the potentials (voltages) at the nodes or the potential drops across the branches. Kirchoff's current law may then be applied to each node, and the voltage law to each loop. This procedure exhibits a certain amount of arbitrariness. There is usually some redundancy among the equations, and the determination of a minimal number of equations is generally computationally nontrivial.

In this section we discuss only single- or double-loop circuits. Let us consider the branch containing a two-terminal device shown in figure 1.10.1. The current is denoted by i. The voltages at the two nodes are denoted v_0 and v_1. The voltage drop is defined to be the difference, $v = v_0 - v_1$. For our purposes, the behavior of the device is completely determined if we know v and i at any time t. The relationship between v and i is called the $v - i$ **characteristic** of the particular device.

We shall consider only the following five basic types of devices.

RESISTOR. If the voltage drop v (measured in volts) is uniquely determined by the current i (measured in amperes) and the time,

$$v = f(i, t),$$

the device is called a current-controlled resistor. If the voltage is proportional to the current,

$$v = iR, \tag{2}$$

and R depends only on t, then the device is a linear resistor and R is called the resistance (usually measured in ohms; one ohm is the resistance that would give a voltage drop of 1 volt if the current were 1 ampere). In many applications R may be approximated by a constant. Resistors will be denoted -⩘⩘-. The coefficient R measures the resistance of the device to the flow of electricity. For a given voltage v, $i = v/R$. Large values of R correspond to small currents, as the device resists the flow of electricity.

TABLE 1.10.1
Electrical Units

Device	*Symbol*	*Unit*	*$v - I$ characteristic*
Resistor	R	ohm (Ω)	$v = IR$
Capacitor	C	farad (F)	$i = Cdv/dt$
Inductor	L	henry (H)	$v = Ldi/dt$
Variable			
Current	i	ampere (A)	I

CAPACITOR. A capacitor stores energy in the form of a charge q (measured in coulombs). The charge q and the voltage drop v across the capacitor are proportional,

$$q = Cv, \tag{3}$$

where C is the capacitance (measured in farads). One farad is the capacitance when 1 coulomb of electricity raises the potential by 1 volt. Charge builds up across a capacitor. The current is due to the flow of electrons. If the charge on a capacitor is constant in time, there is no flow of electrons. The rate of change of the charge is the flow of electrons or the current:

$$i = \frac{dq}{dt}. \tag{4}$$

If the capacitance C is constant, we may differentiate (3), using (4), to obtain the $v - i$ characteristic for a capacitor:

$$i = C\frac{dv}{dt}. \tag{5}$$

The symbol for a capacitor is ·⊣ ⊢·. We define only such linear capacitors in this text.

INDUCTOR. An inductor stores energy in a magnetic field. The voltage-current relationship for a (linear) inductor is

$$v = L\frac{di}{dt}, \tag{6}$$

where L is called the inductance (and measured in henrys). One henry is the inductance in which one volt is induced by a current varying at the rate of one ampere per second. The symbol for an inductor is -ʘʘʘ-.

For many devices, such as transistors, we design models by considering them to be made up of linear capacitors, inductors, and current-controlled resistors. Linear resistors will not suffice to fully model a transistor.

No device, of course, is only a resistor or an inductor. However, we can analyze many circuits by considering them to be made up of resistors, inductors, and capacitors. Also, no device is truly linear. However, if restrictions are put on the allowable current and voltage, we can often make the assumption that the device is linear. As with all physical problems, some assumptions are necessary in order for a mathematical model to approximate a physical situation.

Many students know the fundamental relations for capacitors, inductors, and resistors. For others, we provide the short table, table 1.10.1.

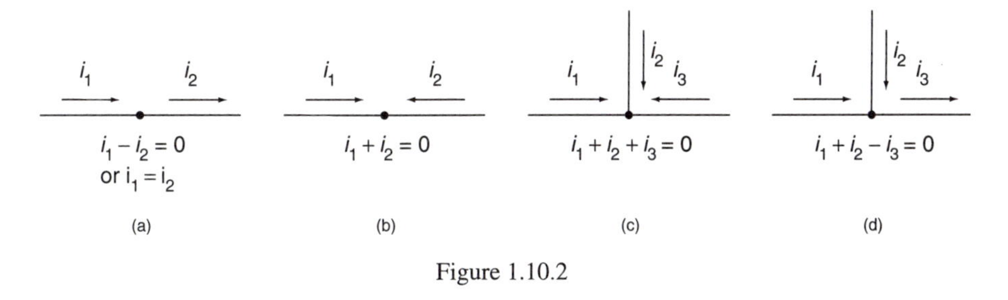

Figure 1.10.2

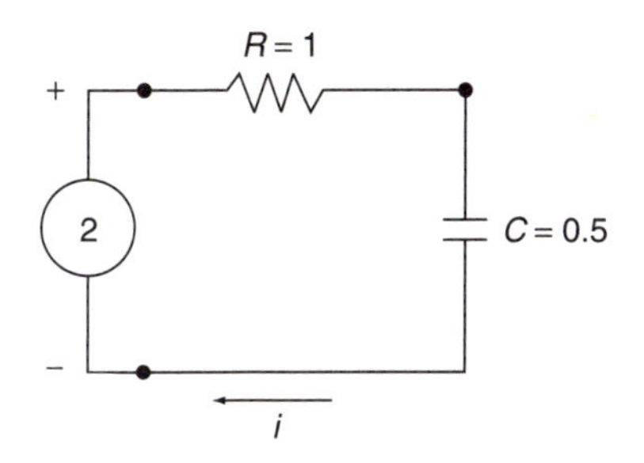

Figure 1.10.3

VOLTAGE SOURCE. Some circuits we will consider will also contain a voltage source, which is denoted by $-$ •—ⓔ—• $+$. The voltage is higher by e on the $+$ side, so that we say the voltage drop is $-e$ for any current. Batteries are an example of a voltage source.

CURRENT SOURCE. Similarly, a current source is denoted by •—ⓘ—•. In a current source, the current is i for any voltage. Some solar cells are current sources.

Applying Kirchhoff's Laws

In applications of the loop and node laws, it is important to use the correct signs. When we apply the current law at a node, currents entering the node are given the opposite sign to that given the currents leaving the node. Figure 1.10.2 shows several nodes and the corresponding current equations. The voltage drops for resistors, capacitors, and inductors are added if the current variable in that branch is in the same direction as we are moving around the loop. The voltage drops are subtracted if the current variable is in the direction opposite to that in which we are moving around the loop. The converse holds for a voltage source, since in that case we have a voltage gain (negative drop).

Example 1.10.1 *Circuit with a Resistor, Capacitor, and Voltage Source*

For the linear circuit (with resistance 1 Ω, capacitance 0.5 F, and a 2-V battery) shown in figure 1.10.3, find the current through the loop and the charge $q(t)$ on the capacitor, given that initially there is no charge on the capacitor, $q(0) = 0$.

• SOLUTION. Taking the current in each branch to move in a clockwise direction, we see, by the current law (figure 1.10.2a), that the current is the same in all three branches. We call the current i. By the voltage law, the sum of the voltage drops, starting at the voltage source, is (note the different sign for the voltage gain at the source)

$$-2 + iR + \frac{q}{C} = 0 \quad \text{or} \quad -2 + 1i + 2q = 0. \tag{7}$$

The voltage law can be written as a differential equation for the charge or the current by recalling that $i = \frac{dq}{dt}$. In terms of q, we obtain the linear first-order differential equation

$$-2 + \frac{dq}{dt} + 2q = 0 \quad \text{or} \quad \frac{dq}{dt} + 2q = 2. \tag{8}$$

Since (8) has constant coefficients and the input (right-hand side) is a constant, we know that a particular solution will be a constant, $q_p = A = 1$. A solution of the associated homogeneous equation is easily seen to be $q = e^{rt} = e^{-2t}$. Thus, the general solution of (8) is

$$q(t) = 1 + ce^{-2t}. \tag{9}$$

In this example, the capacitor is not charged initially, $q(0) = 0$, so that $0 = 1 + c$. Thus, the solution of the initial value problem is

$$q(t) = 1 - e^{-2t}. \tag{10}$$

We see that the capacitor charges and the charge on the capacitor approaches 1 as $t \to \infty$. From the solution (10) we can obtain the current, $i = \frac{dq}{dt}$, by differentiating

$$i(t) = 2e^{-2t}. \tag{11}$$

The current approaches 0 as $t \to \infty$. As $t \to \infty$, there is no voltage drop across the resistor. Thus, as $t \to \infty$, the voltage drop across the capacitor must equal the 2 V of the battery, corresponding to the charge across the capacitor approaching 1 C. ♦

An alternative way to solve this problem is to consider the first-order linear differential equation for the current obtained by differentiating (8) with regard to t:

$$\frac{di}{dt} + 2i = 0. \tag{12}$$

This differential equation is easy to solve, but the initial conditions for the current were not given. Instead we must determine the initial conditions for the current by evaluating the voltage law (7) at $t = 0$. Since $q(0) = 0$, it follows that $i(0) = 2$. In this manner (11) can be obtained from (12). In this alternative method, the charge would

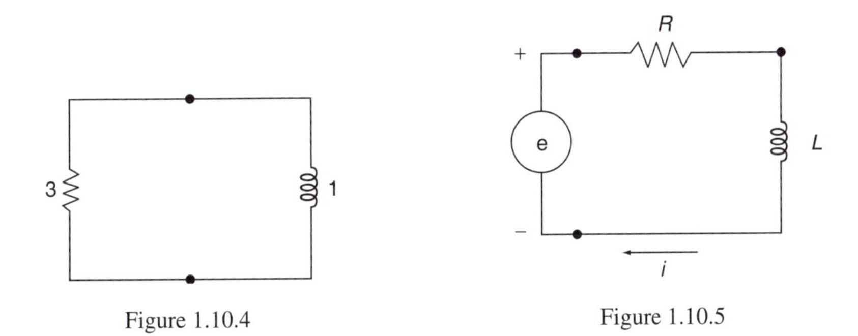

Figure 1.10.4

Figure 1.10.5

be determined by integration, since $\frac{dq}{dt} = i$. Using the initial condition for the charge would give (10).

Example 1.10.2 *Circuit with a Resistor and an Inductor*

Consider the circuit in figure 1.10.4 with a 3-Ω resistor and a 1-H inductor. Determine the current as a function of time in this circuit given that its initial value is 6 A (defined clockwise).

• SOLUTION. The current i is the same in each branch. In this case, from the voltage law, we directly obtain a differential equation for the current:

$$L\frac{di}{dt} + iR = 0 \quad \text{or} \quad \frac{di}{dt} + 3i = 0. \tag{13}$$

This linear differential equation is homogeneous (there is no input), so that zero is a particular solution, $i_p = 0$. A solution of the homogeneous equation is $i_1 = e^{rt} = e^{-3t}$, so that the general solution of (13) is

$$i(t) = ce^{-3t}. \tag{14}$$

The current is given initially to be 6 A, $i(0) = 6$, so that $c = 6$. The solution of the initial value problem is

$$i(t) = ce^{-3t}.$$

♦

Exercises

Exercises 1–6 refer to the circuit given in figure 1.10.5. The inductor is linear with inductance L; the voltage source is $e(t)$.

1. The voltage source is constant, $e = 1$, the resistor is linear with v-i characteristic $v = 2i$, and $L = 1$. Set up the differential equation for the current. Determine the current as a function of time for any initial current $i(0)$. Find any steady-state (equilibrium) solutions.
2. The voltage source is constant, $e = 4$, the resistor is linear with v-i characteristic $v = 6i$, and $L = 2$. Set up the differential equation for the current. Determine the

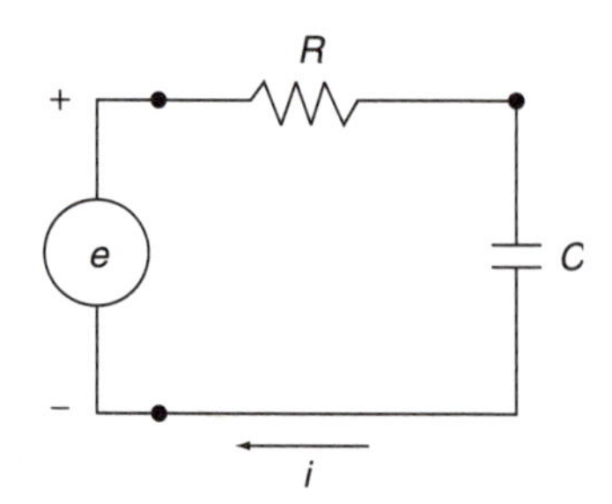

Figure 1.10.6

current as a function of time for any initial current $i(0)$. Find any steady-state (equilibrium) solutions.

3. The voltage source is constant, $e(t) = \sin t$, the resistor is linear with v-i characteristic $v = i$, and $L = 1$. Set up the differential equation for the current. Determine the current as a function of time for any initial current $i(0)$.
4. The voltage source is a constant e, the resistor is linear with v-i characteristic $v = iR$, $R > 0$, and the inductance is $L > 0$.

a) Show that $\lim_{t\to\infty} i(t)$ exists and is the same for any initial current.

b) If $e = 8.5$, for what values of R and L will $\lim_{t\to\infty} i(t) = 4.2$?

5. The resistor has v-i characteristic $v = i$ and $L = 1$. The voltage source is a 9-V battery that is shorted out after 10 s. That is,

$$e(t) = \begin{cases} 9, & 0 \le t < 10, \\ 0, & t \ge 10. \end{cases}$$

The current is initially zero. Find $i(t)$ for $0 \le t \le 20$ and graph the result.

6. The resistor has v-i characteristic $v = 2i$ and $L = 1$. The voltage source is a 1.5-V battery that is shorted out for 10 s and then unshorted (by a switch) so that

$$e(t) = \begin{cases} 0, & 0 \le t < 10, \\ 1.5, & t \ge 10. \end{cases}$$

The current is initially zero. Find $i(t)$ for $0 \le t \le 20$ and graph the result.

7. For the circuit in figure 1.10.6, the capacitance is 0.5, the resistor is linear with v-i characteristic $v = 2i$, and the sinusoidal voltage source is given by $e(t) = 6 \sin t$. Find the charge on the capacitor as a function of t given that it was initially 1.

1.11 Mechanics II: Including Air Resistance

The mechanics problems to be considered here are those of straight-line motion of a constant mass m with a resistive force. As stated in Section 1.2.3, Newton's second law of motion implies that

$$m\frac{d^2x}{dt^2} = F\left(x, \frac{dx}{dt}, t\right), \tag{1}$$

where F is the sum of the applied forces. If F depends on x, $\frac{dx}{dt}$, and t, then (1) is a second-order differential equation for the position $x(t)$. We have already solved some very elementary problems of this type in Section 1.2.3.

If the force does not depend on the position x,

$$m\frac{d^2x}{dt^2} = F\left(\frac{dx}{dt}, t\right), \tag{2}$$

then this second-order differential equation can be reduced to a first-order equation by solving for the velocity instead,

$$v = \frac{dx}{dt}. \tag{3}$$

In this case, (2) becomes a first-order equation

$$m\frac{dv}{dt} = F(v, t). \tag{4}$$

A common example in which an applied force depends on velocity but not position is air resistance. In general, if an object moves through a fluid (liquid or gas), the fluid exerts a force called **resistance** on the body. If the fluid has approximately uniform density and the velocity is not too large, then resistance in the real world may often be approximated by the following law:

> Resistance is proportional to the magnitude of the velocity and acts in a direction opposite to that of velocity. (5)

This law is called **linear damping,** and it states that the resistive force satisfies

$$\text{Resistive force } = -kv \tag{6}$$

with $k \geq 0$. The constant of proportionality depends on the shape of the body and the nature of the medium the object travels through. The dimensions of k are force/velocity = mass $\times$ acceleration/velocity = mass/time.

In the real world, determining resistance can be quite complicated. A significant amount of effort goes into designing automobiles and airplanes in order to minimize drag (air resistance). There can be other considerations such as lift (for aircraft and race cars) and cost. Much of the design process in the past relied on experimental wind tunnels, but now mathematical models of the fluid dynamic processes of flow around complicated three-dimensional objects can be analyzed on supercomputers at a tremendous savings in cost over building realistic models and testing them in wind tunnels (although wind tunnels are still very important). At high speeds, experiments show that linear damping is not valid, and in some cases the resistive force is instead proportional to the velocity squared.

Here we will analyze problems in which a mass is acted upon only by gravity and a linear resistive force. In this case, the first-order differential equation for the velocity becomes

$$m\frac{dv}{dt} = -mg - kv. \tag{7}$$

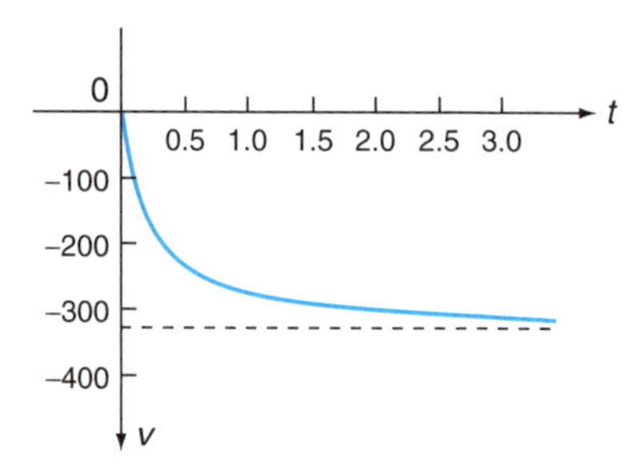

Figure 1.11.1 Graph of (9).

We note that (7) is a first-order linear differential equation with constant coefficients with a constant input, $-mg$. The coordinate system has been chosen so that positive x is upward.

Example 1.11.1 *Air Resistance*

Consider a 4-g mass dropped from a height of 6 m. Assume that air resistance acts on the mass with a constant of proportionality 12 g/s. Determine the velocity as a function of time.

• SOLUTION. The differential equation (7) is

$$4\frac{dv}{dt} = -4 \cdot 980 - 12v \quad \text{or} \quad \frac{dv}{dt} + 3v = -980. \tag{8}$$

Since this is a linear differential equation, the general solution is in the form of a particular solution plus an arbitrary multiple of a solution of the associated homogeneous equation. Since the input $-mg$ is a constant, we can guess that a particular solution is a constant A. Substituting A into the differential equation (8) for v, we find that $v_p = A = \frac{-980}{3}$. A solution of the associated homogeneous equation is seen to be $v_1 = e^{rt} = e^{-3t}$. Thus, the general solution of (8) is

$$v(t) = \frac{-980}{3} + ce^{-3t}. \tag{9}$$

Formula (9) can also be found by using an integrating factor of $u(t) = e^{3t}$ for (8).

The phrase "dropping an object" means that the initial velocity is zero, $v(0) = 0$. The constant c is then easily determined to be $c = \frac{980}{3}$. Thus,

$$v(t) = \frac{-980}{3} + \frac{980}{3}e^{-3t}. \tag{10}$$

♦

The velocity (10) is sketched in figure 1.11.1. The figure shows that the speed continues to increase as time increases. Note that as $t \to \infty$, $e^{-3t} \to 0$, so that $v(t) \to \frac{-980}{3}$. This limit is referred to as the **terminal velocity**. This is the fastest a body will travel (from rest) if there is a linear resistive force. From (9), we see that the terminal velocity does not depend on the initial condition.

With gravity and no resistance, an object keeps accelerating, and its velocity gets larger and larger. There is no limit to how fast an object will move under gravitational acceleration without air resistance. However, with air resistance, an object has the terminal velocity

$$v_{\text{terminal}} = \frac{-mg}{k} \tag{11}$$

derived from (7). Heavier objects have a larger terminal velocity. If the resistance were reduced, the terminal velocity would of course be larger. Employing a parachute increases the resistance and hence lowers the terminal velocity.

If the object above was really dropped from 6 m, the question could be how fast the object was going when it hit the ground and how close it was to its terminal velocity when it hit. We leave to the exercises the investigation of these questions.

Exercises

1. A mass of 20 g is dropped from an airplane flying horizontally. Air resistance acts according to (6) with a constant of proportionality of 10 g/s. Considering only vertical motion:

a) Find the velocity as a function of time.

b) Find the velocity after 10 s, assuming that the body has not hit the ground (or a bird).

c) If the gravitational force is assumed constant, what is the limiting velocity?

2. A weight of 64 lb is flung vertically up into the air from the earth's surface. At the instant it leaves the launcher, it has a velocity of 192 ft/s.

a) Ignoring air resistance, determine how long it takes for the object to reach its maximum altitude.

b) If air resistance acts according to (6) with a constant of proportionality of 128 lb/s, how long does it take for the object to reach its maximum altitude?

3. A mass of 70 g is to be ejected downward from a stationary helicopter. Air resistance acts according to (6) with the constant of proportionality of 7 g/s. At what velocity should the mass be ejected if it is to have a velocity of 12,600 cm/s after 5 s?

4. Suppose that an object of mass m is ejected downward with velocity v_0 over the surface of a planet whose gravitational constant is G. The atmosphere exerts resistance on the body according to (6) with constant of proportionality of r. (All parameters are in cgs units.)

a) Find the formula for $v(t)$.

b) Find the formula for the limiting velocity.

Assume that the object consists of a payload and a parachute. The payload is 80% of the total mass, and the parachute takes up the balance of the total mass. Assume also that the parachute accounts for essentially all the resistive force.

c) How much would the payload have to be reduced to cut the limiting velocity in half?

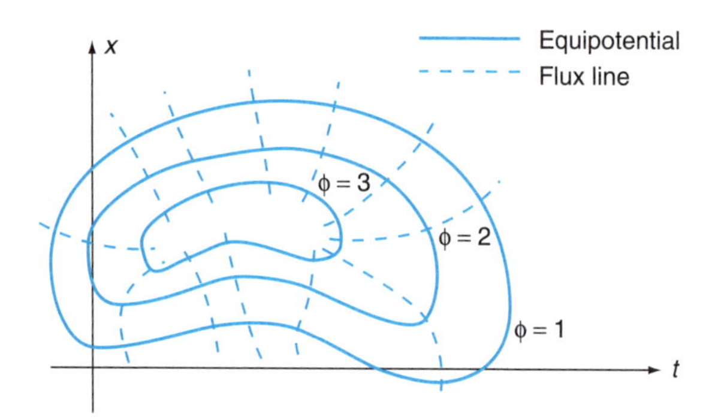

Figure 1.12.1 Equipotentials and flux lines.

5. A 32-lb weight is dropping through a gas near the earth's surface. The resistance is proportional to the square of the velocity, with constant of proportionality 1. At time $t = 0$, the velocity is 1000 ft/s.

a) Find the velocity as a function of time.

b) Is there a limiting velocity? If there is, find it.

1.12 Orthogonal Trajectories (optional)

Many physical problems are described using a function $\phi(t, x)$ called a **potential** that represents a quantity such as height, temperature, or pressure. The curves $\phi(t, x) = c$ are called **equipotentials** or **level curves** because the potential is constant along these curves. In the case of height, the equipotentials are the familiar lines connecting points of equal altitude on a topographic map. For temperature, the equipotentials are usually called **isotherms**, and for pressure, they are called **isobars** (see figure 1.12.1).

A potential often causes an action perpendicular to the equipotentials. These curves perpendicular to the equipotentials are usually called **flux** (or **stream**) **lines**. On a topographic map, the flux lines show the direction in which an object will roll downhill (at least initially). In the case of isotherms, the flux lines show the direction of heat flow.

Given the equipotentials, we can often find the flux lines. Since the flux lines are everywhere orthogonal (perpendicular) to the equipotentials, they are sometimes also called an **orthogonal family** of curves.

Procedure for Calculating Flux Lines from Equipotentials

1. First write the family of curves (equipotentials) in the form

$$\phi(t, x) = c,$$

where c is a constant and $x = x(t)$.

2. Differentiate with regard to t to get

$$\phi_t(t, x) + \phi_x(t, x)\frac{dx}{dt} = 0$$

($\phi_t = \frac{\partial \phi}{\partial t}$ and $\phi_x = \frac{\partial \phi}{\partial x}$).

3. Solve for $\frac{dx}{dt}$:

$$\frac{dx}{dt} = -\frac{\phi_t(t, x)}{\phi_x(t, x)}. \tag{1}$$

Equation (1) gives the slope of the equipotentials at (t, x). Since the flux lines are to be orthogonal to the equipotentials, the flux slopes will be the negative reciprocals of the equipotential slopes. Thus the flux lines satisfy the differential equation

$$\frac{dx}{dt} = \frac{-1}{-\phi_t(t, x)/\phi_x(t, x)} = \frac{\phi_x(t, x)}{\phi_t(t, x)}. \tag{2}$$

4. Solve (2).

Example 1.12.1 *Orthogonal Family for $\phi(t, x) = c$*

The expression $t^2 = c - 2x^2$ defines a family of curves (equipotentials) which are ellipses. Find the orthogonal family (flux lines).

• SOLUTION. First let us write the family of curves in the form $t^2 + 2x^2 = c$. We next differentiate $t^2 + 2x^2 = c$ with regard to t:

$$2t + 4x\frac{dx}{dt} = 0.$$

Solving for $\frac{dx}{dt}$ yields

$$\frac{dx}{dt} = -\frac{t}{2x}.$$

The slope of the orthogonal family at (t, x) is thus given by

$$\frac{dx}{dt} = -\frac{1}{-t/2x} = \frac{2x}{t}.$$

Solving this differential equation by separation of variables, we get

$$\int \frac{1}{x} dx = \int \frac{2}{t} dt,$$

so that integrating gives

$$\ln |x| = 2 \ln |t| + c_1.$$

We now exponentiate both sides to arrive at the orthogonal family of curves

$$x = kt^2, \quad k = \pm e^{c_1}.$$

Note that $k = 0$ also gives a solution which is $x = 0$. Figure 1.12.2 shows both families of curves. ♦

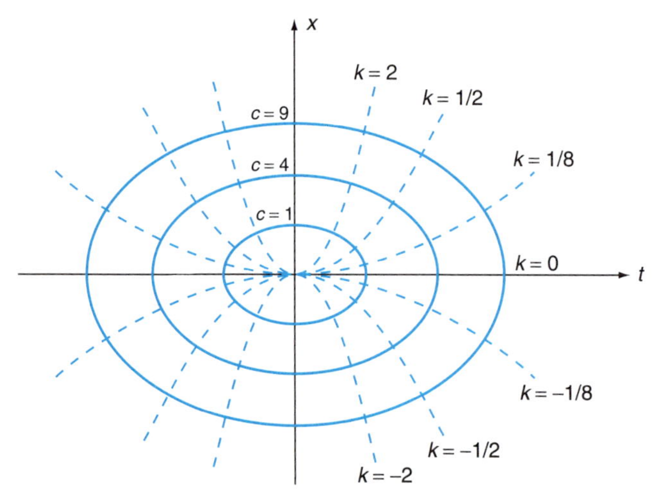

Figure 1.12.2 The families of curves $t^2 + 2x^2 = c$ and $x = kt^2$.

Sometimes the original formula is not of the form $\phi(t, x) = c$ but rather $F(t, x, c) = 0$. That is, the parameter c is not solved for. We still need to rewrite the family as $\phi(t, x) = c$. However, if ϕ is complicated to differentiate, we may differentiate $F(t, x, c) = 0$ with regard to t to get

$$F_t(t, x, c) + F_x(t, x, c)\frac{dx}{dt} = 0,$$

and then substitute ϕ for c.

Example 1.12.2 *Orthogonal Family for $F(t, x, c) = 0$*

Find the differential equation for the orthogonal family to $t^3 + 3x^2 = t^2c$.

• SOLUTION. Differentiating with respect to t gives $3t^2 + 6x\frac{dx}{dt} = 2tc$. But

$$c = \frac{t^3 + 3x^2}{t^2}.$$

Thus

$$\frac{dx}{dt} = \frac{1}{6x}(2tc - 3t^2) = \frac{6x^2 - t^3}{6tx},$$

and the orthogonal family satisfies

$$\frac{dx}{dt} = -\frac{6tx}{6x^2 - t^3}.$$

♦

Exercises

For each family of curves (equipotentials) in Exercises 1–18, compute the orthogonal family (flux lines). Sketch both families for Exercises 1–3.

1. $x = t + c$.
2. $x = ct$.
3. $xt = c$.
4. $x^2 = c(t + x)$.
5. $x = t^c$.
6. $x = (t + c)^3$.
7. $x = t^2 + c$.
8. $x^3(t + 1) = c$.
9. $t = (x - c)^2$.
10. $x = e^{ct}$.
11. $x = \tan(t + c)$.
12. $t^{1/5} + x^{1/5} = c$.
13. $x = c \cos t$.
14. $x = \cos t + c$.
15. $x = ct^2$.
16. $e^x - e^t = c$.
17. $x^3 + t^2 = c$.
18. $\tan x + \tan t = c$.

CHAPTER 2

Linear Second- and Higher-Order Differential Equations

2.1 General Solution of Second-Order Linear Differential Equations

In Section 1.6, we analyzed first-order linear differential equations

$$\frac{dx}{dt} + p(t)x = f(t). \tag{1}$$

We showed that the general solution of (1) is always in the form

$$x = x_p + cx_1, \tag{2}$$

where x_p is a particular solution of (1), x_1 is a nonzero solution of the associated homogeneous equation $\frac{dx}{dt} + p(t)x = 0$, and c is an arbitrary constant.

A similar property holds for second-order linear differential equations:

$$\frac{d^2x}{dt^2} + p(t)\frac{dx}{dt} + q(t)x = f(t). \tag{3}$$

If the input $f(t) = 0$, then the linear differential equation is called **homogeneous**. We will show:

Form of Solution of Second-Order Linear Differential Equation

The general solution of

$$\frac{d^2x}{dt^2} + p(t)\frac{dx}{dt} + q(t)x = f(t) \tag{4}$$

is always in the form

$$x = x_p + c_1x_1 + c_2x_2, \tag{5}$$

where

- x_p is a particular solution of (4) and
- x_1 and x_2 are solutions for the associated homogeneous equation

$$\frac{d^2x}{dt^2} + p(t)\frac{dx}{dt} + q(t)x = 0. \tag{6}$$

The solutions x_1, x_2 cannot be a constant multiple of each other. This is sometimes expressed as saying x_1 and x_2 are linearly independent. Such a pair of solutions of the associated homogeneous equation is called a **fundamental set** of solutions.

The constants c_1 and c_2 are arbitrary constants. The expression $c_1x_1 + c_2x_2$ is called a **linear combination** of x_1 and x_2.

For a second-order linear differential equation, there are two solutions of the associated homogeneous differential equation in the fundamental set (for a first-order linear equation, there is only one solution of the associated homogeneous equation in the general solution; see (2)). The general solution for a second-order linear differential equation has two arbitrary constants (first-order linear differential equations have one arbitrary constant). For a second-order linear differential equation, the two arbitrary constants will be determined from the two initial conditions (for first-order linear differential equations, the one arbitrary constant is determined from the one initial condition).

Note that $x = 0$ always satisfies the homogeneous differential equation

$$\frac{d^2x}{dt^2} + p(t)\frac{dx}{dt} + q(t)x = 0. \tag{7}$$

Thus, if we are solving (7), we choose $x_p = 0$. It then follows from (5) that

The general solution of a homogeneous differential equation (7) is a linear combination

$$x = c_1x_1 + c_2x_2 \tag{8}$$

of linearly independent solutions x_1, x_2 of the homogeneous equation.

Procedures for determining the x_1, x_2, and x_p will be presented later in this chapter. After doing an example, we will try to understand why (5) is valid.

Example 2.1.1 *Initial Value Problem*

a) Verify that $\sin t$, $\cos t$ are solutions of $x'' + x = 0$ ($x'' = \frac{d^2x}{dt^2}$).

b) Verify that e^{-3t} is a solution of $x'' + x = 10e^{-3t}$.

c) Give the general solution of

$$x'' + x = 10e^{-3t}. \tag{9}$$

d) Solve the initial value problem

$$x'' + x = 10e^{-3t}, \quad x(0) = 0, \quad x'(0) = 1. \tag{10}$$

- SOLUTION

a) The verification that $\cos t$ is a solution of the homogeneous equation $x'' + x = 0$ is straightforward but provides good practice. $x = \cos t$, $x' = -\sin t$, $x'' = -\cos t$, and thus $x'' + x = -\cos t + \cos t = 0$. Similarly, it can be shown that $x = \sin t$ is a solution. Also, $x'' + x = 0$ is of second order. Thus, by (8), $\{\sin t, \cos t\}$ is a set of solutions and $c_1 \sin t + c_2 \cos t$ gives all solutions of $x'' + x = 0$.

b) It is straightforward to verify that $x_p = e^{-3t}$ satisfies (9). Note that $x_p' = -3e^{-3t}$ and $x_p'' = 9e^{-3t}$, so that $x_p'' + x_p = 9e^{-3t} + e^{-3t} = 10e^{-3t}$, as claimed. Thus our particular solution of the nonhomogeneous equation is $x_p = e^{-3t}$. Later we will have methods to find the particular solution.

c) Thus, by (5), the general solution of $x'' + x = 10e^{-3t}$ is

$$x = e^{-3t} + c_1 \sin t + c_2 \cos t. \tag{11}$$

d) To solve (10), we apply the initial conditions in (10) to the general solution (11). Evaluating (11) at zero and using the initial condition in (10) gives

$$0 = x(0) = 1 + c_2 \quad \text{so that} \quad c_2 = -1.$$

Differentiate (11) to get

$$x' = -3e^{-3t} + c_1 \cos t - c_2 \sin t.$$

Evaluate x' at $t = 0$ and use the given initial condition $x'(0) = 1$ to get

$$1 = x'(0) = -3 + c_1$$

so that $c_1 = 4$. Thus the solution of the initial value problem (10) is

$$x = e^{-3t} + 4 \sin t - \cos t.$$

♦

Form of General Solution

We have claimed that the form of the general solutions of

$$\frac{d^2x}{dt^2} + p(t)\frac{dx}{dt} + q(t)x = f(t) \tag{12}$$

is

$$x = x_p + c_1 x_1 + c_2 x_2. \tag{13}$$

We show first that $x = x_p + x_h$, and then that $x_h = c_1 x_1 + c_2 x_2$ satisfies the homogeneous equation, so that $x_p + c_1 x_1 + c_2 x_2$ solves (12). In the next section, we will discuss the more difficult question of showing that all solutions must be in this form.

Difference Between Particular Solutions Must Be a Solution of the Associated Homogeneous Equation

We consider

$$\frac{d^2x}{dt^2} + p(t)\frac{dx}{dt} + q(t)x = f(t). \tag{14}$$

We first show that

$$x = x_p + x_h, \tag{15}$$

where x_p is a particular solution and x_h is a solution of the associated homogeneous equation

$$\frac{d^2x}{dt^2} + p(t)\frac{dx}{dt} + q(t)x = 0. \tag{16}$$

Suppose that x and x_p are both solutions of (14). Thus

$$x'' + px' + qx = f(t) \quad \text{and} \quad x_p'' + px_p' + qx_p = f(t). \tag{17}$$

Introduce $x_h = x - x_p$, the difference between x and x_p. Clearly $x = x_p + x_h$. To determine the differential equation satisfied by x_h, we calculate $x_h'' + px_h' + qx_h$ as follows:

$$\begin{aligned} x_h'' + px_h' + qx_h &= (x - x_p)'' + p(x - x_p)' + q(x - x_p) \\ &= x'' - x_p'' + px - px_p' + qx - qx_p \\ &= (x'' + px' + qx) - (x_p'' + px_p' + qx_p) \\ &= f - f = 0, \end{aligned}$$

where we have used (17). Thus x_h is a solution of $x_h'' + px_h' + qx_h = 0$, the associated homogeneous equation, as claimed in (15).

Solutions of Homogeneous Equations

A linear combination

$$c_1x_1 + c_2x_2$$

of two solutions $x_1,\ x_2$ of a linear homogeneous differential equation

$$L(x) = \frac{d^2x}{dt^2} + p(t)\frac{dx}{dt} + q(t)x = 0 \tag{18}$$

is also a solution of the homogeneous differential equation.

Suppose that x_1 and x_2 are solutions of the homogeneous equation (18). Thus

$$x_1'' + px_1' + qx_1 = 0 \text{ and } x_2'' + px_2' + qx_2 = 0. \tag{19}$$

We substitute $c_1x_1 + c_2x_2$, where c_1 and c_2 are constants, for x in $x'' + px' + qx = 0$ to see if it solves that equation:

$$\begin{aligned} &(c_1x_1 + c_2x_2)'' + p(c_1x_1 + c_2x_2)' + q(c_1x_1 + c_2x_2) \\ &= c_1x_1'' + c_2x_2'' + pc_1x_1' + pc_2x_2' + qc_1x_1 + qc_2x_2 \\ &= c_1(x_1'' + px_1' + qx_1) + c_2(x_2'' + px_2' + qx_2) \\ &= c_1 0 + c_2 0 = 0, \end{aligned}$$

using (19). Thus, $c_1x_1 + c_2x_2$ is also a homogeneous solution of $x'' + px' + qx = 0$, as claimed in (18).

Exercises

For Exercises 1–8, verify that the given set is a set of solutions for the associated homogeneous equation, and verify that x_p is a particular solution. Then give the general solution of the differential equation, and solve the initial value problem.

1. $x''+x=1$, $x(0)=0$, $x'(0)=0$, $\{\sin t, \cos t\}$, $x_p=1$.
2. $x''-x=e^{3t}$, $x(0)=0$, $x'(0)=1$, $\{e^t, e^{-t}\}$, $x_p=\frac{1}{8}e^{3t}$.
3. $x''-3x'+2x=2t$, $x(0)=1$, $x'(0)=0$, $\{e^t, e^{2t}\}$, $x_p=t+\frac{3}{2}$.
4. $x''-2x'+x=4e^{2t}$, $x(0)=0$, $x'(0)=0$, $\{e^t, te^t\}$, $x_p=4e^{2t}$.
5. $x''+2x'+2x=6$, $x(0)=1$, $x'(0)=1$, $\{e^{-t}\cos t, e^{-t}\sin t\}$, $x_p=3$.
6. $t^2x''-2tx'+2x=2t^3$, $x(1)=2$, $x'(1)=3$, $\{t, t^2\}$, $x_p=t^3$, $t>0$.
7. $x''+x=2\cos t$, $x(0)=1$, $x'(0)=-1$, $\{\sin t, \cos t\}$, $x_p=t\sin t$.
8. $t^2x''+4tx'+2t=2\ln t+3$, $x(1)=1$, $x'(1)=2$, $\{t^{-1}, t^{-2}\}$, $x_p=\ln t$, $t>0$.
9.
   a) Verify that both
   $$x=c_1e^t+c_2e^{2t}+2\cosh t \tag{20}$$
   and
   $$\tilde{x}=\tilde{c}_1e^t+\tilde{c}_2e^{2t}+e^{-t} \tag{21}$$
   are a general solution of $x''-3x'+2x=6e^{-t}$.

   b) Verify that, if the initial conditions $x(0)=4$, $x'(0)=3$ are applied to (20) and (21), both give the same solution.
10.
   a) Verify that both
   $$x=c_1+c_2t+t^2 \tag{22}$$
   and
   $$\tilde{x}=\tilde{c}_1(1+t)+\tilde{c}_2(1-t)+t^2+3t+1 \tag{23}$$
   are general solutions of $x''=2$.

   b) Verify that, if the initial conditions $x(0)=0$, $x'(0)=1$ are applied to (22) and (23), then both give the same solution.

2.2 Initial Value Problem (for Homogeneous Equations)

We continue to study second-order linear differential equations

$$\frac{d^2x}{dt^2}+p(t)\frac{dx}{dt}+q(t)x=f(t). \tag{1}$$

We have shown that

$$x=x_p+x_h,$$

where x_p is a particular solution of (1) and x_h solves the associated homogeneous equation

$$\frac{d^2x}{dt^2} + p(t)\frac{dx}{dt} + q(t)x = 0. \tag{2}$$

We have claimed that

$$x_h = c_1 x_1 + c_2 x_2, \tag{3}$$

where x_1, x_2 are a set of solutions of (2). We will now introduce several important concepts that will be used several times in our study of differential equations.

For *any* numbers x_0, v_0, we first consider the very practical question of solving the **initial value problem**. In general we need to find the two constants c_1 and c_2 in order to satisfy the two given initial conditions

$$\begin{aligned} x(t_0) &= x_0, \\ x'(t_0) &= v_0. \end{aligned} \tag{4}$$

We assume that

$$x(t) = c_1 x_1(t) + c_2 x_2(t). \tag{5}$$

The initial conditions are satisfied if

$$x_0 = c_1 x_1(t_0) + c_2 x_2(t_0) \tag{6}$$

and

$$v_0 = c_1 x_1'(t_0) + c_2 x_2'(t_0). \tag{7}$$

The mathematical problem of solving for the constants consists of solving two linear equations (6) and (7) for the two unknowns c_1 and c_2. Systems of linear equations may have a unique solution, no solution, or an infinite number of solutions, depending on the coefficients of the linear equations. Since the linear equations (6) and (7) each define a line in the (c_1, c_2)-plane, these three conditions correspond to intersecting nonparallel lines, parallel lines that do not coincide, and parallel lines that do coincide. The result of the analysis to follow is that if the two solutions of a homogeneous linear second-order differential equation are chosen not to be multiples of each other, then no difficulties will arise in solving these linear equations, and a unique solution for c_1 and c_2 will occur. This is graphed in figure 2.2.1.

We may solve the linear system (6) and (7) for c_1 and c_2 in several ways, although all are mathematically equivalent. For example, we can eliminate c_2 by multiplying (6) by $x_2'(t_0)$, multiplying (7) by $x_2(t_0)$, and then subtracting. This calculation gives

$$c_1[x_1(t_0)x_2'(t_0) - x_2(t_0)x_1'(t_0)] = x_2'(t_0)x_0 - x_2(t_0)v_0. \tag{8}$$

If $[x_1(t_0)x_2'(t_0) - x_2(t_0)x_1'(t_0)] = 0$, the coefficient multiplying c_1 in (8) is zero, and it turns out that either there will not be a solution for the constants or the solution will not be unique. If $[x_1(t_0)x_2'(t_0) - x_2(t_0)x_1'(t_0)] \neq 0$, we can solve (8) for c_1 to get

$$c_1 = \frac{x_2'(t_0)x_0 - x_2(t_0)v_0}{[x_1(t_0)x_2'(t_0) - x_2(t_0)x_1'(t_0)]}.$$

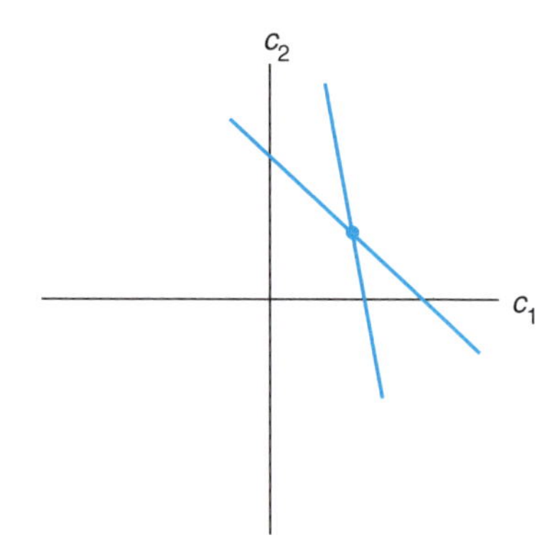

Figure 2.2.1

A similar calculation works for c_2. We see that

> There is a unique solution of the intial value problem (6) and (7) precisely when $x_1(t_0)x_2'(t_0) - x_2(t_0)x_1'(t_0) \neq 0$.

We have just rediscovered the fact from the theory of linear equations that there is a unique solution of (6) and (7) for c_1 and c_2 if and only if the determinant of the coefficients is nonzero at t_0, that is,

$$\det\begin{bmatrix} x_1 & x_2 \\ x_1' & x_2' \end{bmatrix} = x_1x_2' - x_2x_1' \neq 0. \tag{9}$$

Wronskian

We define the **Wronskian** W of the two functions x_1, x_2 to be

$$W[x_1, x_2] = \det\begin{bmatrix} x_1 & x_2 \\ x_1' & x_2' \end{bmatrix} = x_1(t)x_2'(t) - x_2(t)x_1'(t). \tag{10}$$

Here x_1, x_2 are two solutions of the homogeneous second-order linear differential equation, each satisfying

$$\frac{d^2x}{dt^2} + p(t)\frac{dx}{dt} + q(t)x = 0. \tag{11}$$

We now show that the Wronskian of two solutions of x_1, x_2 of (11) itself satisfies a linear first-order differential equation. From that we will analyze the meaning of the Wronskian being zero. We begin by calculating the first derivative with respect to t of the Wronskian of two solutions x_1, x_2 of (11):

$$\begin{aligned}
\frac{dW}{dt} &= [x_1x_2' - x_2x_1']' \\
&= x_1x_2'' + x_1'x_2' - x_2x_1'' - x_2'x_1' \quad \text{(product rule)} \\
&= x_1x_2'' - x_2x_1'' \quad \text{(cancellation of two terms)} \\
&= x_1(-px_2' - qx_2) - x_2(-px_1' - qx_1) \quad (x_1 \text{ and } x_2 \text{ are solutions of (11)}) \\
&= -p[x_1x_2' - x_2x_1'] \quad \text{(cancellation and rearrangement)} \\
&= -pW \quad \text{(definition of the Wronskian (10))}.
\end{aligned}$$

Thus, the Wronskian satisfies the first-order linear differential equation

$$\frac{dW}{dt} = -p(t)W. \tag{12}$$

Here $p(t)$ is the same p that appears as a coefficient in the second-order linear differential equation (11).

The first-order linear differential equation for the Wronskian (12) is separable. It can also be solved using integrating factors. From Section 1.3 or 1.6, the solution of (12) is

$$W(t) = W(t_0)e^{-\int_{t_0}^{t} p(t)dt}. \tag{13}$$

The exponential in (13) is never zero. $W(t_0)$ is a constant, the Wronskian evaluated at the initial time. Thus, the Wronskian $W(t)$ is either always zero (if $W(t_0) = 0$) or never zero (if $W(t_0) \neq 0$). In all problems, the two homogeneous solutions are chosen such that $W(t_0) \neq 0$, and hence in this case $W(t) \neq 0$. The Wronskian arises in many places in the study of linear differential equations. Thus (13) will be quite useful.

Fundamental Solution

We have shown in the previous section that $c_1x_1 + c_2x_2$ solves the homogeneous differential equation. However, it is possible that we cannot solve the initial value problem with this solution. In order to solve this initial value problem the initial Wronskian of x_1, x_2 must be nonzero. Thus two solutions x_1, x_2 of a homogeneous second-order linear differential equation form a **fundamental set** and can be shown to be independent if $W(t_0) \neq 0$. This is an easy test to determine if a set of solutions is fundamental. It follows that any homogeneous solutions can be written as a **linear combination** of that fundamental set

$$x_h = c_1x_1 + c_2x_2,$$

and that the general solution of the nonhomogeneous differential equation (1) is in the form

$$x = x_p + c_1x_1 + c_2x_2.$$

Example 2.2.1 *Fundamental Set and Initial Value Problem*

a) Show that $\{\sin t, \cos t\}$ forms a fundamental set of solutions of the homogeneous linear differential equation

$$x'' + x = 0. \tag{14}$$

b) Find the general solution.
c) Find the solution of the initial value problem $x(0) = 6$, $x'(0) = 3$.

- SOLUTION

a) First we should verify that $x_1 = \sin t$ and $x_2 = \cos t$ each satisfy (14). We omit this step since it was done in Example 2.1.1. To show that $\{\sin t, \cos t\}$ is a fundamental set of solutions, we verify that $W[x_1, x_2] \neq 0$ at $t = 0$. Note that $x_1' = \cos t$ and $x_2' = -\sin t$. Thus

$$W[x_1, x_2] = \det \begin{bmatrix} x_1(0) & x_2(0) \\ x_1'(0) & x_2'(0) \end{bmatrix} = \det \begin{bmatrix} 0 & 1 \\ 1 & 0 \end{bmatrix} = -1 \neq 0.$$

Hence $\{\sin t, \cos t\}$ forms a fundamental set.

b) Since (14) is homogeneous, its general solution is a linear combination of $\{\sin t \cos t\}$,

$$x = c_1 \sin t + c_2 \cos t. \tag{15}$$

c) To satisfy the initial conditions, we first calculate the derivative of (15):

$$x' = c_1 \cos t - c_2 \sin t.$$

The initial conditions are satisfied if

$$x(0) = 6 = c_2,$$
$$x'(0) = 3 = c_1.$$

The solution of the initial value problem is thus

$$x = 3 \sin t + 6 \cos t.$$

♦

nth-Order Linear Differential Equations

We now present the form of the general solution for nth-order linear differential equations

$$a_n(t)\frac{d^n x}{dt^n} + a_{n-1}(t)\frac{d^{n-1}x}{dt^{n-1}} + \cdots + a_1(t)\frac{dx}{dt} + a_0(t)x = f(t), \tag{16}$$

which is very similar to the theory for second-order equations developed in Sections 2.1–2.3. For first-order linear differential equations, we had one arbitrary constant in the general solution, and for second-order equations there were two. Thus we would expect to have n arbitrary constants for an nth-order differential equation. The solutions of (16) may be broken into a particular solution and a homogeneous solution, just as for second-order linear differential equations. We have

The general solution (16) is always in the form

$$x = x_p + c_1 x_1 + c_2 x_2 + \cdots + c_n x_n, \tag{17}$$

where

1. x_p is a particular solution of the original equation (16).
2. $\{x_1, x_2, \ldots, x_n\}$ is a fundamental set of n solutions of the associated homogeneous equation (16) with $f(t)$=0. That is,

there are n solutions of the homogeneous equation and they must be linearly independent. A more technical definition (which we omit) of independent is needed than the one we have given for the $n = 2$ case. There is also a Wronskian which involves the determinant of an $n \times n$ matrix.

The general solution of a homogeneous linear differential equation is always in the form

$$x = c_1x_1 + c_2x_2 + \cdots + c_nx_n, \tag{18}$$

where $\{x_1, x_2, \ldots, x_n\}$ is a fundamental set of solutions of the associated homogeneous equation (16) with $f(t) = 0$.

Exercises

In Exercises 1–7, use that $W \neq 0$ is equivalent to the solutions of the homogeneous equation forming a fundamental set.

1. Verify that $\sin t$, $\cos t$ are solutions of $x'' + x = 0$. Determine whether they are a fundamental set of solutions.
2. Find all solutions of the form t^r for $t^2x'' - 6x = 0$ on $(0, \infty)$ and determine whether they form a fundamental set of solutions.
3. Find all solutions of the form t^r for $t^2x'' - tx' + x = 0$ on $(0, \infty)$ and determine whether they form a fundamental set of solutions.
4. Find all solutions of the form e^{rt} for $x'' - 4x' + 4x = 0$ and determine whether they form a fundamental set of solutions.
5. Let x_1 be the solution on $(0, \infty)$ of

$$t^2x'' + x' + tx = 0, \quad x(1) = 1, \quad x'(1) = 1,$$

and x_2 the solution on $(0, \infty)$ of

$$t^2x'' + x' + tx = 0, \quad x(1) = 0, \quad x'(1) = -1.$$

a) Verify that $\{x_1, x_2\}$ is a fundamental set of solutions of $t^2x'' + x' + tx = 0$.
b) Let x_3 be the solution of

$$t^2x'' + x' + tx = 0, \quad x(1) = 2, \quad x'(1) = 0.$$

Find constants c_1, c_2 such that

$$x_3 = c_1x_1 + c_2x_2.$$

6. Let x_1 be the solution of

$$x'' + tx' + x = 0, \quad x(0) = 1, \quad x'(0) = 2,$$

and x_2 the solution of

$$x'' + tx' + x = 0, \quad x(0) = 1, \quad x'(0) = -1.$$

a) Verify that $\{x_1, x_2\}$ is a fundamental set of solutions of $x'' + tx' + x = 0$.
b) Let x_3 be the solution of

$$x'' + tx' + x = 0, \quad x(1) = 2, \quad x'(1) = 0.$$

Find constants c_1, c_2 such that

$$x_3 = c_1 x_1 + c_2 x_2.$$

7. Let x_1 be the solution of

$$x'' + px' + qx = 0, \quad x(t_0) = 1, \quad x'(t_0) = 0,$$

and x_2 the solution of

$$x'' + px' + qx = 0, \quad x(t_0) = 0, \quad x'(t_0) = 1.$$

Verify that $\{x_1, x_2\}$ is a fundamental set of solutions and that the solution x_3 of

$$x'' + px' + qx = 0, \quad x(t_0) = \alpha, \quad x'(t_0) = \beta$$

is $x_3 = \alpha x_1 + \beta x_2$. (The significance of this exercise is that it proves that a fundamental set always exists.)

8. In Section 2.6 we will derive the important **Euler's formula**

$$e^{it} = \cos t + i \sin t, \tag{19}$$

where $i = \sqrt{-1}$. In this exercise, we outline a different derivation of (19). Consider the differential equation

$$x'' + x = 0. \tag{20}$$

a) Show that $\cos t$ and $\sin t$ are solutions of (20).
b) Show that e^{it} is another solution of (20).
c) Determine the specific linear combination of $\cos t$ and $\sin t$ that will equal e^{it}. (*Hint*: Use the initial conditions that e^{it} satisfies.)

9. Show that the Wronskian of two solutions of Airy's differential equation $\frac{d^2x}{dt^2} + tx = 0$ is a constant.
10. Show that the Wronskian of two solutions for $\frac{d^2x}{dt^2} + q(t)x = 0$ is a constant.

In Exercises 11–15, verify that the given functions are solutions of the given equation, compute the Wronskian, and compare the Wronskian you obtain to (13).

11. $\cos t$ and $\sin t$ for $x'' + x = 0$.
12. e^{2t} and e^{-2t} for $x'' - 4x = 0$.
13. e^t and e^{2t} for $x'' - 3x' + 2x = 0$.
14. t and t^2 for $t^2x'' - 2tx' + 2x = 0$.
15. t^{-1} and t^{-2} for $t^2x'' + 4tx' + 2x = 0$.
16. Verify that $\{\sin 2t, \cos 2t\}$ form a fundamental set of solutions of $x'' + 4x = 0$. Verify that $x_3(t) = \sin(2t + \pi/4)$ is another solution of $x'' + 4x = 0$. Find c_1, c_2 so that

$$\sin\left(2t + \frac{\pi}{4}\right) = c_1 \sin 2t + c_2 \cos 2t.$$

2.3 Reduction of Order

In the previous sections we have begun to study the second-order linear differential equation

$$x'' + p(t)x' + q(t)x = f(t).$$

We have shown that the general solution is in the form

$$x = x_p + c_1x_1 + c_2x_2.$$

At this point we have not described any methods for obtaining the solutions, x_1, x_2, of the associated homogeneous equation, or a particular solution, x_p. In fact, there are no general methods for obtaining x_1, x_2, x_p for all linear second-order equations (other than numerical methods). This contrasts with our study of linear first-order equations, for which the general solutions can always be obtained by an integrating factor. Integrating factors do not exist for general second-order linear differential equations.

The **method of reduction of order** shows that if we have one solution x_1 of a homogeneous differential equation, then we can always obtain the second solution x_2 of the homogeneous equation. (However, there is no general method for obtaining one homogeneous solution in the first place.) We shall discuss the method, work three examples, and then summarize the method.

Suppose we wish to solve the homogeneous equation

$$x'' + p(t)x' + q(t)x = 0, \tag{1}$$

and that we are lucky enough to know one solution $x_1(t)$ of (1). We now show how to reduce (1) to a first-order differential equation. The key is to look for solutions of (1) in the form

$$x = vx_1, \tag{2}$$

where v is an unknown function of t and x_1 is our known solution of the homogeneous equation (1). Substituting $x = vx_1$ into (1) for x gives

$$(vx_1)'' + p(vx_1)' + q(vx_1) = 0.$$

Perform the differentiations using the product rule and note that $(vx_1)''$ is the derivative of $(vx_1)'$:

$$(v''x_1 + 2v'x_1' + vx_1'') + p(v'x_1 + vx_1') + qvx_1 = 0.$$

Now regroup by derivatives of v:

$$x_1v'' + (2x_1' + px_1)v' + (x_1'' + px_1' + qx_1)v = 0. \tag{3}$$

But by assumption, x_1 is a solution of the equation $x_1'' + px_1' + qx_1 = 0$. Thus the v terms in (3) always vanish, leaving

$$x_1v'' + (2x_1' + px_1)v' = 0. \tag{4}$$

Now x_1 and $2x_1' + px_1$ are known functions of t. Letting

$$w = v' \tag{5}$$

in (4) gives a first-order homogeneous linear equation in w:

$$x_1 w' + (2x_1' + px_1)w = 0. \tag{6}$$

Equation (6) may be solved by separation (Section 1.3) or by integrating factors (Section 1.6). Then from (5) we must integrate w to get v. Then vx_1 gives us the second solution for our fundamental set of solutions.

Instead of continuing with the general case, we now illustrate the method of reduction of order by several examples. This method works whether or not the coefficient in front of the second derivative in (1) is 1.

Example 2.3.1 *Reduction of Order*

First verify that $x_1 = t$ is a solution of

$$t^2 x'' + 3tx' - 3x = 0, \quad t > 0, \tag{7}$$

where $x_1' = 1$ and $x_1'' = 0$ so that $3t - 3t = 0$. Find a second solution of (7) and give the general solution.

• SOLUTION. Let

$$x = vx_1 = vt, \tag{8}$$

and substitute for x in (7):

$$t^2(vt)'' + 3t(vt)' - 3(vt) = 0.$$

Performing the differentiations, using the product rule,

$$t^2(v''t + 2v') + 3t(v't + v) - 3vt = 0.$$

Regrouping the terms, we see that the v terms cancel as promised, to leave

$$t^3 v'' + 5t^2 v' = 0.$$

Let $w = v'$ and divide by t^3 to get

$$\frac{dw}{dt} + \frac{5}{t} w = 0.$$

This is a first-order homogeneous linear differential equation. It can be solved by separation or integrating factors. By separation,

$$\int \frac{dw}{w} = -\int \frac{5}{t} dt.$$

Integrating (assuming $t > 0$) gives

$$\ln |w| = -5 \ln t + c_0.$$

Solving for w yields

$$w = c_1 t^{-5}.$$

But $w = v'$, so that

$$v' = c_1 t^{-5}.$$

Now integrate to find v:

$$v = \frac{c_1}{-4} t^{-4} + c_2 = \tilde{c}_1 t^{-4} + c_2,$$

where $\tilde{c}_1$ is a new arbitrary constant. (We could just as well have used c_1 again instead of $\tilde{c}_1$.) Finally,

$$x = v x_1 = vt = \tilde{c}_1 t^{-3} + c_2 t \tag{9}$$

is the general solution. The previously unknown second solution is t^{-3}. When correctly applied, reduction of order will always give us the general solution, a linear combination of two solutions. ♦

Example 2.3.2 *Solution with Definite Integral*

First verify that $x_1 = t$ is a solution of

$$x'' - tx' + x = 0. \tag{10}$$

since $x_1' = 1$ and $x_1'' = 0$ so that $-t + t = 0$. Find a second solution of (10) and give the general solution.

• SOLUTION. Let

$$x = v x_1 = vt, \tag{11}$$

and substitute for x in (10):

$$(vt)'' - t(vt)' + (vt) = 0.$$

Perform the differentiation, using the product rule,

$$(v''t + 2v') - t(v't + v) + vt = 0.$$

Again the v terms cancel, as can be shown to happen in general, to give

$$tv'' + (2 - t^2)v' = 0.$$

Let $w = v'$ and divide by t to get the first-order linear equation

$$\frac{dw}{dt} + \left(\frac{2}{t} - t\right) w = 0,$$

which can be solved by separation or by integrating factors. By integrating factors, we have

$$w = c_1 e^{-\int (2/t - t)dt} = c_1 e^{-2\ln t + \frac{1}{2}t^2} = c_1 t^{-2} e^{\frac{t^2}{2}}.$$

Since $w = v'$,

$$v' = c_1 t^{-2} e^{\frac{t^2}{2}}.$$

Antidifferentiation now requires a definite integral to find v:

$$v = c_1 \int_1^t \bar{t}^{-2} e^{\frac{\bar{t}^2}{2}} \, d\bar{t} + c_2.$$

Finally,

$$x = vx_1 = vt = c_1 t \int_1^t \bar{t}^{-2} e^{\frac{\bar{t}^2}{2}} \, d\bar{t} + c_2 t \tag{12}$$

is the general solution. It is a linear combination of two solutions of the homogeneous equation (10). The method of reduction of order has been used to obtain the second solution,

$$x_2 = t \int_1^t \bar{t}^{-2} e^{\frac{\bar{t}^2}{2}} \, d\bar{t}.$$

♦

The next example will be important in Section 2.4.

Example 2.3.3 *Reduction of Order*

a) Find a solution of $x'' - 2x' + x = 0$ of the form e^{rt}.
b) Use the solution from part (a) to find a fundamental set of solutions for $x'' - 2x' + x = 0$, using reduction of order.

• SOLUTION

a) Let $x = e^{rt}$ and substitute into

$$x'' - 2x' + x = 0 \tag{13}$$

to get

$$r^2 e^{rt} - 2re^{rt} + e^{rt} = 0$$

or

$$r^2 - 2r + 1 = 0.$$

Thus $(r-1)^2 = 0$ and $r = 1$, so e^t is a solution of (13).

b) We have one solution $x_1 = e^t$ of (13). We shall use reduction of order to find a second solution and the general solution. Let $x = vx_1 = ve^t$, so that (13) becomes

$$(ve^t)'' - 2(ve^t)' + ve^t = 0.$$

Differentiate using the product rule,

$$(v''e^t + 2v'e^t + ve^t) - 2(v'e^t + ve^t) + ve^t = 0.$$

Note that the v terms cancel as promised. In addition, in this example (but not in general), the v_1' also vanish, so that

$$v''e^t = 0, \quad \text{or} \quad v'' = 0.$$

Antidifferentiate twice:

$$v' = c_1, \tag{14}$$

$$v = c_1 t + c_2. \tag{15}$$

Thus,

$$x = vx_1 = (c_1 t + c_2)e^t = c_1 te^t + c_2 e^t$$

is the general solution of (13). A fundamental set of solutions would be $\{e^t, te^t\}$. A second solution te^t is obtained by taking $c_1 = 1$ in (14) and $c_2 = 0$ in (15). ♦

Summary of Reduction of Order

Reduction of order can be used to find a fundamental set of two solutions of the homogeneous equation $x'' + px' + qx = 0$, given one homogeneous solution x_1, as follows:

1. Let $x = vx_1$ and substitute into $x'' + px' + qx = 0$.
2. This leads to a second-order linear equation in v with only v'' and v' terms. Let $w = v'$ to get a first-order linear differential equation (6) in w.
3. Find a nonzero solution for w by either separation or integrating factors.
4. Let v be an antiderivative of w.
5. Let $x_2 = vx_1$. Then $\{x_1, x_2\}$ is a fundamental set of solutions for $x'' + px' + qx = 0$.

Exercises

In Exercises 1–8, verify that the given function x_1 is a solution of the given homogeneous differential equation. Then using reduction of order find a fundamental set of solutions and the general solution for the differential equation. Exercises 7 and 8 require definite integrals.

1. $t^2 x + 3tx' + x = 0$, $x_1 = t^{-1}$.
2. $t^2 x'' + 5tx' + 3x = 0$, $x_1 = t^{-1}$.
3. $x'' + 10x' + 25x = 0$, $x_1 = e^{-5t}$.
4. $tx'' - (1 + t)x' + x = 0$, $x_1 = e^t$.
5. $tx'' + (t - 1)x' - x = 0$, $x_1 = e^{-t}$.
6. $x'' + 6x' + 9x = 0$, $x_1 = e^{-3t}$.
7. $x'' + tx' - x = 0$, $x_1 = t$.
8. $t^2 x'' + t^2 x' - 2(1 + t)x = 0$, $x_1 = t^2$.

In Exercises 9–12, find a solution x_1 of the given differential equations in the form t^r for some constant r. Then find a second solution by reduction of order.

9. $t^2x'' - 3tx' + 4x = 0$.
10. $t^2x'' + 5tx' + 4x = 0$.
11. $t^2x'' + 7tx' + 9x = 0$.
12. $t^2x'' - 5tx' + 9x = 0$.
13. Find a solution of $x'' + 4x = 0$ of the form $\sin rt$ for some constant r. Then find the general solution of $x'' + 4x = 0$ by reduction of order.
14. Find a solution of $x'' + 16x = 0$ of the form $\sin rt$ for some constant r. Then find the general solution of $x'' + 16x = 0$ by reduction of order.

In Exercises 15–19, find a solution x_1 of the given differential equation in the form e^{rt} for some constant r. Then find a second solution and the general solution by reduction of order.

15. $x'' - 4x' + 4x = 0$.
16. $x'' - 6x' + 9x = 0$.
17. $x'' + 2x' + x = 0$.
18. $x'' - 5x' + 6x = 0$.
19. $x'' - 4x = 0$.
20. Verify that $x_1(t) = t$ is a solution of (**Legendre equation of order one**) $(1 - t^2)x'' - 2tx' + 2x = 0$ on the interval $(-1, 1)$. Find the general solution on $(-1, 1)$. (*Note*: The integrations are a little more difficult, but can be worked using our techniques.)
21. Verify the statement in the summary of reduction of order that $\{x_1, x_2\}$ form a fundamental set of solutions.
22. Using reduction of order, find a general formula for the second solution of $x'' + p(t)x' + q(t)x = 0$ if x_1 is one solution.

2.4 Homogeneous Linear Constant Coefficient Differential Equations (Second Order)

Linear constant coefficient differential equations form an important class of differential equations that appear both in physical models and as approximations for more complicated equations. Applications to electric circuits and mechanical systems will be given in Sections 2.5, 2.7, and 2.8.

This section will consider the general linear, second-order homogeneous, **constant coefficient** differential equation,

$$ax'' + bx' + cx = 0, \tag{1}$$

where the coefficients a, b, and c are real constants and $a \neq 0$. From Section 2.1, we know that the general solution of $ax'' + bx' + cx = 0$ will be in the form

$$x = c_1x_1 + c_2x_2, \tag{2}$$

where $\{x_1, x_2\}$ is a fundamental set of solutions for $ax'' + bx' + cx = 0$.

From linear first-order homogeneous constant coefficient differential equations,

$$\frac{dx}{dt} + kx = 0,$$

we stated in Section 1.6 that

> A homogeneous linear constant coefficient differential equation always has at least one solution of the form $x = e^{rt}$.

This also applies to second- and higher-order homogeneous linear constant coefficient differential equations.

The key to finding two homogeneous solutions $\{x_1, x_2\}$ for linear differential equations with constant coefficients is to look for a solution of the form $x = e^{rt}$, where r is a constant. Substituting $x = e^{rt}$ into $ax'' + bx' + cx = 0$ gives

$$a(e^{rt})'' + b(e^{rt})' + ce^{rt} = 0,$$

and, upon differentiation,

$$ar^2 e^{rt} + bre^{rt} + ce^{rt} = 0.$$

Finally, divide by e^{rt}, which is always nonzero:

$$ar^2 + br + c = 0. \tag{3}$$

The equation $ar^2 + br + c = 0$ is the **characteristic equation** of

$$ax'' + bx' + cx = 0.$$

The polynomial $ar^2 + br + c$ is called the **characteristic polynomial** of

$$ax'' + bx' + cx = 0.$$

The above sequence of calculations can be reversed, so we have shown that

> If r satisfies the characteristic equation $ar^2 + br + c = 0$, then e^{rt} is a solution of $ax'' + bx' + cx = 0$. (4)

Every second-degree polynomial has two roots. For $ar^2 + br + c = 0$, the roots are given by $r = (-b \pm \sqrt{b^2 - 4ac})/2a$. There are three cases, depending on whether $b^2 - 4ac > 0$, $b^2 - 4ac = 0$, or $b^2 - 4ac < 0$.

Case 1: Characteristic Equation Has Distinct Real Roots ($b^2 - 4ac > 0$)

Suppose the characteristic equation $ar^2 + br + c = 0$ has two distinct real roots r_1, r_2. Then by (4), $e^{r_1 t}, e^{r_2 t}$ are two solutions of $ax'' + bx' + c = 0$. These solutions form a fundamental set (see Exercise 36). Thus

$$x = c_1 e^{r_1 t} + c_2 e^{r_2 t},$$

where c_1 and c_2 are arbitrary constants would be the general solution of $ax'' + bx' + c = 0$.

Example 2.4.1 *Distinct Real Roots*

Find the general solution of $x'' - x' - 20x = 0$.

• SOLUTION. By substituting $x = e^{rt}$ into the differential equation, we derive the characteristic equation $r^2 - r - 20 = 0$. Factoring gives $r^2 - r - 20 = (r - 5) \times (r + 4) = 0$. There are two distinct roots $r = 5$, $r = -4$. Thus e^{5t}, e^{-4t} are solutions, and the general solution is

$$x = c_1 e^{5t} + c_2 e^{-4t},$$

where c_1 and c_2 are arbitrary constants. ♦

Example 2.4.2 *Distinct Real Roots*

Find the general solution of $x'' + 4x' = 0$.

• SOLUTION. By substituting $x = e^{rt}$, we find that the characteristic equation is $r^2 + 4r = r(r + 4) = 0$. There are two distinct roots $r = 0$, $r = -4$. A fundamental set of solutions would be $\{1, e^{-4t}\}$ since $e^{0t} = 1$. The general solution is

$$x = c_1 \cdot 1 + c_2 e^{-4t} = c_1 + c_2 e^{-4t},$$

where c_1 and c_2 are arbitrary constants ♦

Case 2: Characteristic Equation Has Repeated Real Root ($b^2 - 4ac = 0$)

Suppose that the characteristic equation has a single repeated root r_1. In this case, there is only one solution of the differential equation of the form $e^{rt} = e^{r_1 t}$. The second solution can always be found by the reduction of order (Section 2.3). We shall give an example to illustrate and motivate the general case. However, we shall see that the reduction always produces the same type of second solution, so that it will not be necessary to carry the reduction out each time a differential equation is solved. Students who have omitted reduction of order may skip to the summary of repeated roots.

Example 2.4.3 *Reduction of Order*

Find the general solution of $x'' - 6x' + 9x = 0$.

• SOLUTION. Substituting $x = e^{rt}$ into $x'' - 6x' + 9x = 0$, we find that the characteristic equation is $r^2 - 6r + 9 = (r - 3)^2 = 0$. Thus $r = 3$ is a repeated root. One solution is e^{3t}. To get the second solution, we will use reduction of order. Let $x = ve^{3t}$ and substitute into the differential equation to get

$$(ve^{3t})'' - 6(ve^{3t})' + 9(ve^{3t}) = 0,$$

or, upon differentiation, using the product rule,

$$(v''e^{3t} + 6v'e^{3t} + 9ve^{3t}) - 6(v'e^{3t} + 3ve^{3t}) + 9ve^{3t} = 0.$$

Note the v terms cancel as claimed in Section 2.3 by reduction of order. Here the v' terms also cancel, so this simplifies to

$$v''e^{3t} = 0 \quad \text{or} \quad v'' = 0.$$

Antidifferentiating twice yields

$$v = c_1 t + c_2.$$

Thus

$$x = ve^{3t} = c_1 te^{3t} + c_2 e^{3t}$$

is the general solution of $x'' - 6x' + 9x = 0$. Note that e^{3t} was one solution, and we have found that a second solution is te^{3t}. We claim that this is a general result. ♦

Summary of Repeated Roots

When$x_1 = e^{r_1 t}$ is a solution of (1) corresponding to repeated roots, a second solution is always of the form

$$x_2 = te^{r_1 t}. \tag{5}$$

In the examples, exercises, and text that follow, do not use reduction of order to obtain the second solution unless asked to. Use the result (5) instead.

Example 2.4.4 *Repeated Real Roots*

Find the general solution of $x'' + 2x' + x = 0$.

• SOLUTION. By substituting $x = e^{rt}$, we find that the characteristic equation is $r^2 + 2r + 1 = (r+1)^2 = 0$. Thus $r = -1$ is a repeated root. One homogeneous solution is e^{-t}. Since the root is repeated, the second solution is te^{-t} according to (5). Thus the general solution of the homogeneous equation is

$$x = c_1 e^{-t} + c_2 te^{-t},$$

where c_1 and c_2 are arbitrary constants. ♦

Example 2.4.5 *Repeated Real Roots for Higher Order Differential Equations*

Find the general solution of

$$x^{(5)} - 3x^{(4)} + 3y^{(3)} - x^{(2)} = 0.$$

Here $x^{(n)} = \frac{d^n x}{dt^n}$.

• SOLUTION. The characteristic equation is

$$r^5 - 3r^4 + 3r^3 - r^2 = r^2(r^3 - 3r^2 + 3r - 1)$$
$$= r^2(r-1)^3 = 0.$$

There are two distinct roots $r = 0$, 1, of multiplicities 2 and 3. Using (5), the root $r = 0$ of multiplicity 2 means that we include

$$\{e^{0t}, te^{0t}\} = \{1, t\}$$

in the fundamental set of solutions. The root $r = 1$ of multiplicity 3 means that we include $\{e^t, te^t, t^2e^t\}$ in the fundamental set of solutions. The general solution is thus

$$x = c_1 + c_2t + c_3e^t + c_4te^t + c_5t^2e^t,$$

where c_1, c_2, c_3, c_4, and c_5 are arbitrary constants. ♦

Case 3: Characteristic Equation Has Complex Roots ($b^2 - 4ac < 0$)

Since the differential equation

$$ax'' + bx' + cx = 0 \tag{6}$$

has constant coefficients, solutions exist in the form $x = e^{rt}$ if r satisfies the characteristic equation

$$ar^2 + br + c = 0. \tag{7}$$

If $b^2 - 4ac < 0$, then the roots are complex. Suppose that

$$r_1 = \alpha + i\beta \quad (i^2 = -1),$$

with α, β real numbers, is a complex root of (7). Since a, b, c are real, the other root r_2 must be the **complex conjugate** of r_2. That is,

$$r_2 = \alpha - i\beta.$$

The general solution of (6) can be represented by a linear combination of these two complex exponentials:

$$x = \tilde{c}_1e^{r_1t} + \tilde{c}_2e^{r_2t}. \tag{8}$$

However, (8) is not particularly useful in this form for some physical problems, since both e^{r_1t} and e^{r_2t} involve complex numbers. We shall replace $\{e^{r_1t}, e^{r_2t}\}$ by a different fundamental set of solutions. The arbitrary constants in (8) are denoted by $\tilde{c}_1$, $\tilde{c}_2$ to distinguish them from the c_1, c_2 we use in the other set of solutions.

Complex exponentials satisfy the usual algebraic properties of exponentials, $e^{r_1t} = e^{(\alpha+i\beta)t} = e^{\alpha t}e^{i\beta t}$. Thus, the general solution (8) can be written

$$x = \tilde{c}_1e^{\alpha t}e^{i\beta t} + \tilde{c}_2e^{\alpha t}e^{-i\beta t}$$

or

$$x = e^{\alpha t}(\tilde{c}_1e^{i\beta t} + \tilde{c}_2e^{-i\beta t}). \tag{9}$$

This solution is still not real. However, because of the Euler formula, $e^{i\beta t} = \cos\beta t + i\sin\beta t$, we will shortly show that an arbitrary linear combination of $e^{\pm i\beta t}$ is equivalent to an arbitrary linear combination of $\cos\beta t$ and $\sin\beta t$. Thus, (9) can be written $x=e^{\alpha t}(c_1\cos\beta t+c_2\sin\beta t)$. To summarize:

In the case of complex roots, $\alpha\pm\beta i$, the general solution of $ax''+bx'+cx=0$ is given by

$$\begin{aligned} x &= e^{\alpha t}(c_1\cos\beta t+c_2\sin\beta t)\\ &= c_1e^{\alpha t}\cos\beta t+c_2e^{\alpha t}\sin\beta t. \end{aligned} \tag{10}$$

To understand this result better requires some further discussion of complex numbers.

Euler's Formula

The relationship between complex exponentials and sines and cosines follows from **Euler's formula**,

$$e^{i\theta}=\cos\theta+i\sin\theta. \tag{11}$$

An additional important relationship follows from replacing θ by $-\theta$ in (11) and using the evenness of cosine, $\cos(-\theta)=\cos\theta$, and the oddness of sine, $\sin(-\theta)=-\sin\theta$, to get

$$e^{-i\theta}=\cos\theta-i\sin\theta. \tag{12}$$

To explain Euler's formula (11), we assume that the Taylor series for the real exponential,

$$e^t=1+t+\frac{t^2}{2!}+\frac{t^3}{3!}+\frac{t^4}{4!}+\cdots, \tag{13}$$

is valid for the complex exponential. That is, we can let $t=i\theta$ in (13) to get

$$e^{i\theta}=1+i\theta-\frac{\theta^2}{2!}-i\frac{\theta^3}{3!}+\frac{\theta^4}{4!}+\cdots.$$

By collecting the real and imaginary parts, we have

$$e^{i\theta}=\left(1-\frac{\theta^2}{2!}+\frac{\theta^4}{4!}+\cdots\right)+i\left(\theta-\frac{\theta^3}{3!}+\cdots\right).$$

Euler's formula (11) now follows from the Taylor series for $\cos\theta$ and $\sin\theta$:

$$\begin{aligned} \cos\theta &= 1-\frac{\theta^2}{2!}+\frac{\theta^4}{4!}+\cdots,\\ \sin\theta &= \theta-\frac{\theta^3}{3!}+\cdots. \end{aligned}$$

In our applications to differential equations, $\theta=\beta t$.

Sines and Cosines are Equivalent to the Complex Exponentials

We have shown that the general solution (9) of the linear differential equation is

$$x = e^{\alpha t}(\tilde{c}_1 e^{i\beta t} + \tilde{c}_2 e^{-i\beta t}). \tag{14}$$

The linear combination of complex exponentials that appears in (14) may be related to $\cos \beta t$ and $\sin \beta t$ using Euler's formulas (11) and (12), as follows:

$$\tilde{c}_1 e^{i\beta t} + \tilde{c}_2 e^{-i\beta t} = \tilde{c}_1(\cos \beta t + i \sin \beta t) + \tilde{c}_2(\cos \beta t - i \sin \beta t).$$

By collecting the $\cos \beta t$ and $\sin \beta t$, we obtain

$$\tilde{c}_1 e^{i\beta t} + \tilde{c}_2 e^{-i\beta t} = c_1 \cos \beta t + c_2 \sin \beta t, \tag{15}$$

where $c_1 = \tilde{c}_1 + \tilde{c}_2$ and $c_2 = i(\tilde{c}_1 - \tilde{c}_2)$. Equation (15) shows that

An arbitrary linear combination of $e^{i\beta t}$ and $e^{-i\beta t}$,

$$\tilde{c}_1 c^{i\beta t} + \tilde{c}_2 e^{-i\beta t}, \tag{16}$$

is equivalent to an arbitrary linear combination of $\cos \beta t$ and $\sin \beta t$.

$$c_1 \cos \beta t + c_2 \sin \beta t.$$

It is helpful to memorize (16).

Alternative Derivation

By adding and subtracting Euler's formula ((11) and (12)), we derive fundamental relationships for $\cos \theta$ and $\sin \theta$:

$$\begin{aligned} \cos \theta &= \frac{1}{2} e^{i\theta} + \frac{1}{2} e^{-i\theta}, \\ \sin \theta &= \frac{1}{2i} e^{i\theta} - \frac{1}{2i} e^{-i\theta}. \end{aligned} \tag{17}$$

For example, $\cos \theta$ is a specific linear combination of $e^{i\theta}$ and $e^{-i\theta}$. By letting $\theta = \beta t$ in (17) and multiplying by $e^{\alpha t}$, we obtain

$$\begin{aligned} e^{\alpha t} \cos \beta t &= e^{\alpha t}\left(\frac{1}{2} e^{i\beta t} + \frac{1}{2} e^{-i\beta t}\right), \\ e^{\alpha t} \sin \beta t &= e^{\alpha t}\left(\frac{1}{2i} e^{i\beta t} - \frac{1}{2i} e^{-i\beta t}\right). \end{aligned} \tag{18}$$

Since the right-hand sides of (18) are linear combinations of homogeneous solutions, then $e^{\alpha t} \cos \beta t$ and $e^{\alpha t} \sin \beta t$ are also solutions of $ax'' + bx' + cx = 0$.

Example 2.4.6 *Complex Roots*

Find the general solution of

$$x'' + x' + x = 0.$$

• SOLUTION. By substituting $x = e^{rt}$, we see the characteristic equation is $r^2 + r + 1 = 0$. By the quadratic formula, the roots are

$$r = \frac{-1 \pm \sqrt{1-4}}{2} = -\frac{1}{2} \pm i\frac{\sqrt{3}}{2}.$$

Since $e^{\left(-\frac{1}{2} \pm i\frac{\sqrt{3}}{2}\right)t} = e^{-\frac{t}{2}} e^{\pm i\left(\frac{\sqrt{3}}{2}\right)t}$, we have

$$x = c_1 e^{-\frac{t}{2}} \cos\left(\frac{\sqrt{3}}{2}t\right) + c_2 e^{-\frac{t}{2}} \sin\left(\frac{\sqrt{3}}{2}t\right)$$

is the general solution of $x'' + x' + x = 0$.

Note that the root $-\frac{1}{2} + i\left(\frac{\sqrt{3}}{2}\right)$ does not give the solution $e^{-t/2}\cos\left[\left(\frac{\sqrt{3}}{2}\right)t\right]$. Rather, the *pair* of solutions $e^{-t/2}\cos\left[\left(\frac{\sqrt{3}}{2}\right)t\right]$, $e^{-t/2}\sin\left[\left(\frac{\sqrt{3}}{2}\right)t\right]$ comes from the *pair* of roots

$$-\frac{1}{2} + i\frac{\sqrt{3}}{2}, \quad -\frac{1}{2} - i\frac{\sqrt{3}}{2}.$$ ♦

Example 2.4.7 *Pure Imaginary Roots*

Find the general solution of $x'' + 4x = 0$.

• SOLUTION. By inspection, or by substituting $x = e^{rt}$, we see the characteristic equation is $r^2 + 4 = 0$, so that the roots are $r = \alpha \pm i\beta = \pm 2i$. Thus $\alpha = 0$, $\beta = 2$, and a fundamental set of solutions is

$$e^{0t} \cos 2t = \cos 2t \quad \text{and} \quad e^{0t} \sin 2t = \sin 2t.$$

The general solution is

$$x = c_1 \cos 2t + c_2 \sin 2t.$$ ♦

For convenience, we summarize the three cases.

Solution of $ax'' + bx' + cx = 0$ with a, b, c Real Constants

First find the roots r_1, r_2 of the characteristic equation $ar^2 + br + c = 0$. There are three cases:

1. If r_1, r_2 are distinct real roots, then the general solution is $x = c_1 e^{r_1 t} + c_2 e^{r_2 t}$.
2. If $r_1 = r_2$ is a repeated real root, the general solution is $x = c_1 e^{r_1 t} + c_2 t e^{r_1 t}$.
3. If r_1, r_2 are complex roots, they are a conjugate pair:

$$r_1 = \alpha + i\beta, \quad r_2 = \alpha - i\beta.$$

The general solution is

$$x = c_1 e^{\alpha t} \cos \beta t + c_2 e^{\alpha t} \sin \beta t.$$

As will be shown in Sections 2.5 and 2.7, there is a close relationship between electrical circuits and mechanical systems and linear differential equations with constant coefficients. In these problems, one often starts knowing the desired response (solution) and wants to design the device (make up the differential equation).

Example 2.4.8 *Obtaining a Differential Equation from the Solution*

Find a second-order linear homogeneous constant coefficient differential equation that has $c_1e^{-3t}+c_2e^{-2t}$ as its general solution.

• SOLUTION. $c_1e^{-3t}+c_2e^{-2t}$ will be the general solution if $r=-3$ and $r=2$ are roots of the characteristic equation. One such characteristic equation would be

$$[r-(-3)][r-(-2)]=(r+3)(r+2)=r^2+5r+6=0.$$

A corresponding differential equation is

$$x''+5x'+6x=0.$$

Note that $2x''+10x'+12x=0$ would be another correct answer, since the roots determine the polynomial only up to a constant factor. ♦

Exercises

In Exercises 1–33, solve the differential equation. Determine the general solution if no initial conditions are given.

1. $x''+x'-6x=0.$
2. $x''-x=0.$
3. $x''+x=0.$
4. $x''+4x'+4x=0.$
5. $x''+4x'+5x=0.$
6. $x''-2x'+x=0.$
7. $x''-3x'+2x=0.$
8. $2x''-2x'+x=0.$
9. $x''-x'=0.$
10. $4x''+8x'+3x=0.$
11. $3x''=0.$
12. $x''-2x'+2x=0.$
13. $3x''+2x'-x=0.$
14. $x''+9x=0,\ x(0)=1,\ x'(0)=1.$
15. $x''+x'-2x=0,\ x(0)=0,\ x'(0)=1.$
16. $2x''+12x'+18x=0,\ x(0)=1,\ x'(0)=0.$
17. $2x''+4x=0.$

18. $3x'' - 24x' + 45x = 0.$
19. $2x'' + 8x' + 6x = 0, x(0) = 2, x'(0) = 0.$
20. $x'' - 16x = 0.$
21. $x'' + 10x' + 25x = 0.$
22. $2x'' + 3x' = 0.$
23. $x'' - 14x' + 49x = 0.$
24. $x'' + 4x' + 20x = 0.$
25. $x'' - 6x' + 25x = 0.$
26. $x'' + x' + x = 0.$
27. $x'' - 12x = 0.$
28. $x'' + 2x' + 8x = 0.$
29. $x'' + 4x' + 8x = 0.$
30. $x'' + 10x = 0.$
31. $x'' + 8x = 0.$
32. $x'' + 6x' + 11x = 0.$
33. $x'' + 6x' + 7x = 0.$
34. Suppose that r is a real number. Verify that e^{rt}, te^{rt} form a fundamental set of solutions.
35. Suppose α, β are real numbers and $\beta \neq 0$. Verify that $e^{\alpha t}\cos\beta t$, $e^{\alpha t}\sin\beta t$ form a fundamental set of solutions.
36. Suppose that $r_1 \neq r_2$. Verify that $\{e^{r_1 t}, e^{r_2 t}\}$ form a fundamental set of solutions.
37. Show that if $ar^2 + br + c = 0$ has the repeated root r_1, then $ar^2 + br + c = a(r - r_1)^2$.
38. Using reduction of order and the previous exercise, show that if r_1 is a repeated root of $ar^2 + br + c = 0$, then $te^{r_1 t}$ is a solution of $ax'' + bx' + cx = 0$.
39. From the solutions of Exercises 1–33, find an example with repeated roots and *derive* the second solution using reduction of order.

In Exercises 40–50, determine a homogeneous second-order linear constant coefficient differential equation with the given expression as its general solution.

40. $x = c_1e^{3t} + c_2e^{-4t}.$
41. $x = c_1e^{-t} + c_2e^{-2t}.$
42. $x = c_1e^{2t} + c_2te^{2t}.$
43. $x = c_1e^{3t} + c_2te^{3t}.$
44. $x = c_1 + c_2e^{-5t}.$
45. $x = c_1\sin 4t + c_2\cos 4t.$
46. $x = c_1e^{-t}\sin 2t + c_2e^{-t}\cos 2t.$
47. $x = c_1e^{t}\sin t + c_2e^{t}\cos t.$
48. $x = c_1e^{2t}\sin 3t + c_2e^{2t}\cos 3t.$
49. $x = c_1 + c_2t.$
50. $x = c_1\sin 2t + c_2\cos 2t.$

2.4.1 Homogeneous Linear Constant Coefficient Differential Equations (nth-Order)

With very small changes we can use the previous ideas to solve homogeneous linear constant coefficient differential equations

$$a_n x^{(n)} + a_{n-1} x^{(n-1)} + \cdots + a_1 x' + a_0 x = 0,$$

where $x^{(i)} = \frac{d^i x}{dt^i}$. The procedure for the solution of (1) is summarized in the following algorithm.

Procedure for the Solution of $a_n x^{(n)} + a_{n-1} x^{(n-1)} + \cdots + a_1 x' + a_0 x = 0$ When $a_n, a_{n-1}, \ldots, a_0$ are Constants

1. Form the characteristic equation $a_n r^n + a_{n-1} r^{n-1} + \cdots + a_1 r + a_0 = 0$ and determine its roots and their multiplicities. The characteristic equation can be found by substituting $x = e^{rt}$.
2. A fundamental set of solutions for (1) is determined as follows:
   a) If $r = r_1$ is a real root of multiplicity m, then include the m functions
   $$\{e^{r_1 t}, te^{r_1 t}, \ldots, t^{m-1} e^{r_1 t}\} \tag{19}$$
   in the fundamental set of solutions.
   b) If $r = \alpha \pm \beta i$ is a pair of complex conjugate roots and they each have multiplicity m, then include the $2m$ functions
   $$\begin{aligned}\{e^{\alpha t} \cos \beta t, e^{\alpha t} \sin \beta t, te^{\alpha t} \cos \beta t, te^{\alpha t} \sin \beta t, \ldots, \\ t^{m-1} e^{\alpha t} \cos \beta t, t^{m-1} e^{\alpha t} \sin \beta t\}\end{aligned} \tag{20}$$
   in the fundamental set of solutions. (If $\alpha + \beta i$ is a root of multiplicity m, then $\alpha - \beta i$ will be a root of multiplicity m also, since $a_n, \ldots, a_0$ are real.)

Example 2.4.9 *Repeated Imaginary Roots*

Find the general solution of $x'''' + 2x'' + x = 0$.

• SOLUTION. The characteristic equation is $r^4 + 2r^2 + 1 = (r^2 + 1)^2 = 0$, as can be found by substituting $x = e^{rt}$. Thus $r = \pm i$ are complex conjugate roots of multiplicity 2. In the notation of (6):

$$i = 0 + 1 \cdot i, \quad \text{so that } \alpha = 0 \quad \text{and} \quad \beta = 1.$$

Two solutions are $\cos t$, $\sin t$ corresponding to the roots $r = \pm i$. Since the complex roots have multiplicity 2, two other linearly independent solutions are $t \sin t$, $t \cos t$. Thus, a fundamental set of solutions is

$$\{\cos t, \sin t, t \cos t, t \sin t\}.$$

The general solution is

$$x = c_1 \cos t + c_2 \sin t + c_3 t \cos t + c_4 t \sin t.$$ ♦

Example 2.4.10 *Repeated Roots*

Find the general solution of the linear homogeneous constant coefficient differential equation if the roots of the characteristic equation are known to be $r = 7 \pm 3i$ repeated twice, $r = \pm 4i$, and $r = 5$.

• SOLUTION. Since there are 7 roots, the general solution is a linear combination of 7 homogeneous solutions using 7 arbitrary constants.

$$x = c_1 e^{7t} \cos 3t + c_2 e^{7t} \sin 3t + c_3 t e^{7t} \cos 3t + c_4 t e^{7t} \sin 3t$$
$$+ c_5 \cos 4t + c_6 \sin 4t + c_7 e^{5t}.$$ ♦

In actual applications the roots of the characteristic equation will usually not be integers. They are often found (estimated) using a numerical procedure. In some cases this could be Newton's method from calculus. In this case, care must be taken when determining whether values such as, say, 1.123, 1.124 represent distinct real roots that are close together, or one root of multiplicity two whose computation has been influenced by roundoff error or the numerical method.

Exercises

For Exercises 51–73, find the general solution. Here $x^{(n)}$ denotes the nth derivative. For example, $x^{(4)}$ is the fourth derivative.

51. $x''' - 6x'' + 12x' - 8x = 0.$
52. $x''' + 5x'' + 4x = 0.$
53. $x''' - 5x'' + 4x = 0.$
54. $x''' + 8x'' + 16x = 0.$
55. $x''' + x'' - 2x = 0.$
56. $x'''' - 2x''' = 0.$
57. $x'''' + 4x''' + 6x'' + 4x' + x = 0.$
58. $x''' - 2x'' - x' + 2x = 0.$
59. $x' - 3x = 0.$
60. $x' + 4x = 0.$
61. $x''' + x'' - 2x = 0.$
62. $x'''' + 4x'' + 4x = 0.$
63. $x'''' + 4x''' + 8x'' + 8x' + 4x = 0.$
64. $x''' + 3x'' + 3x' + x = 0.$
65. $x^{(4)} - 4x^{(3)} + 6x^{(2)} - 4x^{(1)} + x = 0.$
66. $x^{(6)} + 3x^{(4)} + 3x^{(2)} + x = 0.$
67. $x^{(6)} - 3x^{(4)} + 3x^{(2)} - x = 0.$
68. $x^{(4)} - 16x = 0.$

69. $x^{(4)} - x = 0$.
70. $x'''' + 2x''' + 2x'' = 0$.
71. $x'''' + 50x'' + 625x = 0$.
72. $3x' + 4x = 0$.
73. $x''' + 3x'' + x' - 5x = 0$.

For Exercises 74–83, you are given the general solution. Write a homogeneous linear constant coefficient differential equation that has that general solution.

74. $x = c_1 + c_2 t + c_3 t^2 + c_4 t^3$.
75. $x = c_1 e^{2t} + c_2 t e^{2t} + c_3$.
76. $x = c_1 e^t \sin 2t + c_2 e^t \cos 2t + c_3 e^{-t} + c_4 t e^{-t}$.
77. $x = c_1 \sin 5t + c_2 \cos 5t + c_3 t \sin 5t + c_4 t \cos 5t$.
78. $x = c_1 e^t + c_2 e^{2t} + c_3 e^{-t}$.
79. $x = c_1 e^{-2t} + c_2 t e^{-2t} + c_3 t^2 e^{-2t}$.
80. $x = c_1 e^t + c_2 t e^t + c_3 e^{-t} + c_4 t e^{-t}$.
81. $x = c_1 e^t \sin t + c_2 e^t \cos t + c_3 t e^t \sin t + c_4 t e^t \cos t$.
82. $x = c_1 e^t + c_2 e^{2t} + c_3 e^{3t} + c_4 e^{4t}$.
83. $x = c_1 \sin t + c_2 \cos t + c_3 t \sin t + c_4 t \cos t$.
84. Find all solutions of $x''' - 3x'' + 3x' - x = 0$ of the form e^{rt}. Then find the general solution by reduction of order (Section 2.3). Compare you answer to that obtained by using (19).

In Exercises 85–92, you are given the roots of the characteristic equation of a linear constant coefficient homogeneous differential equation. Find the general solution of the differential equation.

85. 8 repeated four times and $8 \pm 7i$.
86. $3 \pm 8i$ repeated two times.
87. $4 \pm 5i$, $\pm 5i$ repeated two times, and 0 repeated three times.
88. $\pm 5i$ repeated three times, $\pm i$, and 0.
89. $2 \pm i$ repeated three times.
90. $7 \pm 3i$, 7, -7, $\pm 3i$.
91. $2 \pm 6i$, $-2 \pm 6i$, and $\pm 6i$ repeated twice.
92. $3 \pm 5i$, $-3 \pm 5i$, and 0 repeated four times.

2.5 Mechanical Vibrations I: Formulation and Free Response

2.5.1 Formulation of equations

This section, like Section 1.11, will study a problem involving the linear motion of a rigid body governed by **Newton's law**,

$$\frac{d(mv)}{dt} = F_T, \tag{1}$$

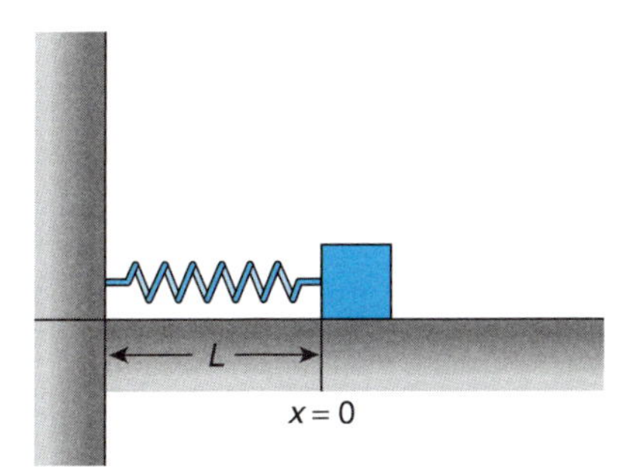

Figure 2.5.1

where m is mass, v is velocity, mv is momentum, t is time, and F_T is the total force acting on the body. The resulting differential equations will be analyzed using the preceding sections on linear constant coefficient differential equations. We shall again use the centimeter-gram-second (cgs) and foot-pound-second (fps) systems of measurement (see Table 2.5.1 at the end of this section).

Horizontal Spring-Mass System

Suppose, then, that we have a horizontal spring connected to a wall with a constant mass m attached (see figure 2.5.1). We shall assume that the total force F_T is made up of three forces:

F_s = force exerted by the spring,

F_r = a friction or resistive force acing on the mass (air resistance, for example),

F_e = any other external forces (such as magnetism).

Then Newton's law (1) takes the form

$$m\frac{d^2x}{dt^2} = F_T = F_s + F_r + F_e. \tag{2}$$

Let L be the length of the spring with no mass attached. We introduce a coordinate system in which $x = 0$ is located a distance L from the wall, as illustrated in figure 2.5.1. Then $x(t)$ is the distance of the mass from this position. If $x > 0$ (as illustrated in figure 2.5.2), the spring is stretched from its natural length a distance x, and it is known that the spring exerts a force to the left. In figure 2.5.3, $x < 0$, in which case the spring is compressed and the spring exerts a force to the right.

We can now express the different forces in (2) in terms of x. To begin with, assume that the spring satisfies **Hooke's law**,

> The force exerted by the spring is proportional to the distance the spring is stretched (or compressed).

This is an assumption concerning not only the type of spring but also the mass, external forces, and initial conditions. In particular, we are assuming that the resulting motion

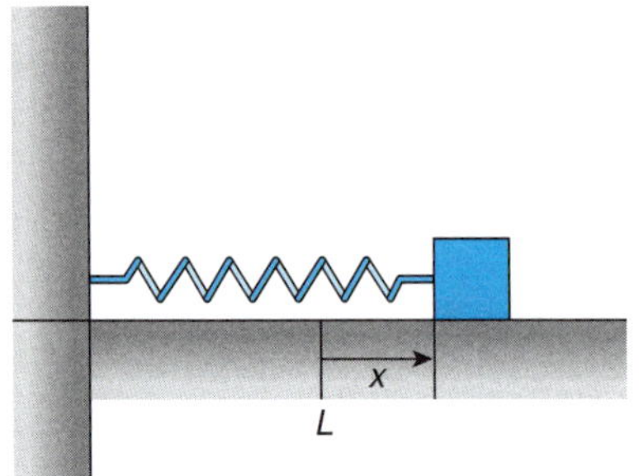

Figure 2.5.2 $x > 0$.

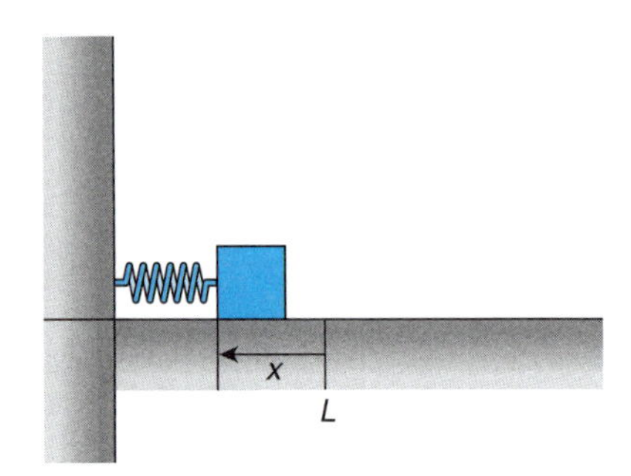

Figure 2.5.3 $x < 0$.

is not of too great an amplitude, so that linear equations may be used. Because of the coordinate system we have chosen, the distance the spring is stretched is x, and Hooke's law becomes

$$F_s = -kx, \tag{3}$$

where k is a positive constant, called the **spring constant**. k is positive because the spring force acts opposite to the sign of x. That is, if the spring is compressed, $x < 0$, then the spring force is positive.

Many types of resistive forces at low velocities are (approximately) proportional to the velocity. That is,

$$F_r = -\delta v = -\delta \frac{dx}{dt} \quad \text{with } \delta > 0. \tag{4}$$

The constant δ is called the **damping constant**. The constant of proportionality, $-\delta$, is negative, since the force of friction acts in the opposite direction to the motion and we want $\delta \geq 0$. Note that (4), like Hooke's law, represents a restriction on the motion to be studied and not just an assumption concerning the type of spring and the type of resistance. Finally, we assume the external forces F_e depend only on time t and not on position x or velocity $\frac{dx}{dt}$.

Then substituting (3) and (4) into Newton's law (2), we get the linear differential equation

$$m\frac{d^2x}{dt^2} = -kx - \delta\frac{dx}{dt} + f(t),$$

where $f(t) = F_e(t)$. This differential equation is usually written in the form

$$m\frac{d^2x}{dt^2} + \delta\frac{dx}{dt} + kx = f(t). \tag{5}$$

Note that m, δ, and k are all nonnegative constants. The initial conditions usually used with (5) are the initial position $x(0)$ and the initial velocity $v(0) = x'(0)$.

Sometimes it is convenient to think of the resistive force as being due to an attached device, such as a piston moving in a fluid. One such device is called a **dashpot**.

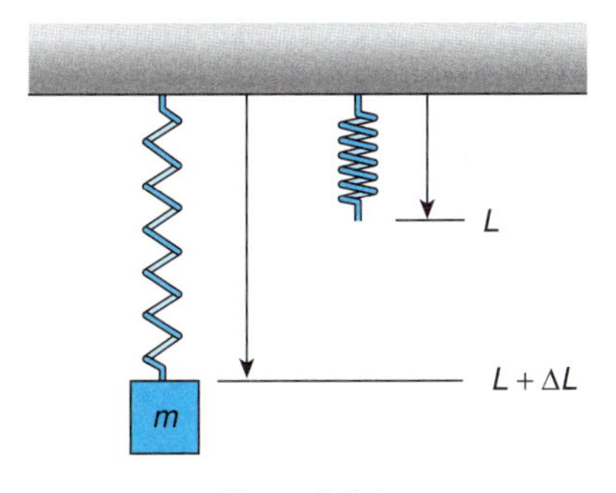

Figure 2.5.4

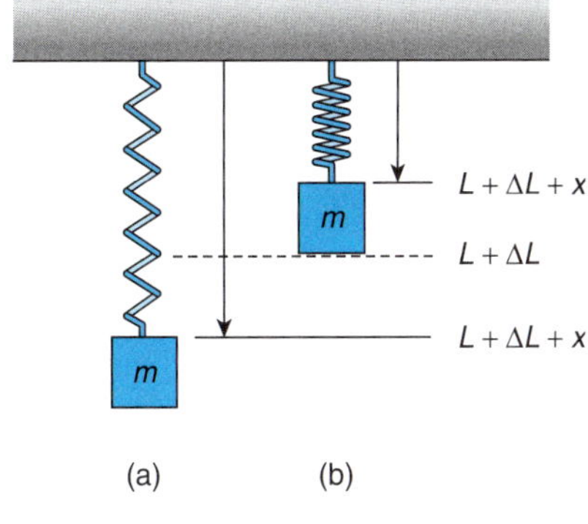

Figure 2.5.5

Conceptually, this approach is similar to modeling the resistance in wire by the inclusion of a small resistor in the circuit equations instead of thinking of the resistance as distributed throughout the wire.

Spring-Mass System with Gravity

A few additional complications occur for a vertically vibrating spring-mass system in which the force due to gravity must be included. After some effort in choosing an appropriate coordinate system, we will show that (5) is still valid.

Suppose that a spring-mass system is hanging from a fixed support. The force of gravity acting downward on the mass (the weight) is

$$F_g = mg, \tag{6}$$

where $g = 32$ ft/s^2 (fps) or 980 cm/s^2 (cgs). Again let L be the length of the hanging spring with no mass attached. Let y measure a change from this rest length. Thus $y = 0$ is a distance L from the support. We measure y downward so that positive y corresponds to additional spring length. Then Newton's law with the gravitational force (6) becomes

$$m\frac{d^2y}{dt^2} + \delta\frac{dy}{dt} + ky = mg + f(t). \tag{7}$$

First we analyze what happens without an external force, $f(t) = 0$. From our experience, we know that with gravity the mass will sag (see figure 2.5.4) and be at rest at $y = \Delta L$ (which we will determine). If the mass is at rest ($y = \Delta L$), then the force of gravity must balance out the spring forces, so that

$$k\Delta L = mg. \tag{8}$$

The stretching ΔL of the spring with gravitational force is given by (8). Often in experiments the weight of the object mg is known and ΔL is easily measured. Thus, the spring constant k can be determined from (8).

If we choose (even with an external force) a coordinate system x centered at the sagged rest system (see figure 2.5.5), then

$$y = x + \Delta L. \tag{9}$$

Note that $x=0$ corresponds to $y=\Delta L$. If (9) is substituted into (7), we obtain

$$m\frac{d^2x}{dt^2}+\delta\frac{dx}{dt}+k(x+\Delta L)=mg+f(t).$$

But $x\Delta L=mg$ from (8), and so we have finally

$$m\frac{d^2x}{dt^2}+\delta\frac{dx}{dt}+kx=f(t). \tag{10}$$

Equation (10) is the same as (5), showing that a vertically vibrating spring-mass system with a gravitational force moves the same way as a horizontally vibrating spring-mass system if a coordinate system is chosen that takes into account the sag due to gravity.

Free Response

Suppose that there is no external force other than gravity, so that the external force is $f(t)=0$. The solutions of the resulting homogeneous differential equation,

$$m\frac{d^2x}{dt^2}+\delta\frac{dx}{dt}+kx=0, \tag{11}$$

are called the **free** (or **natural**) **response** of the spring-mass system. This is how the system reacts for given initial conditions if it is allowed to proceed without external interference. Note that the free response is the same thing as the solution of the associated homogeneous equation. The resulting behavior is quite different, depending on whether or not there is any damping present.

2.5.2 Simple Harmonic Motion (No Damping, $\delta=0$)

If there is no damping ($\delta=0$), the free response of a spring-mass system is given by the differential equation

$$m\frac{d^2x}{dt^2}+kx=0. \tag{12}$$

We can solve (12) using the techniques of Section 2.4. Since the differential equation has constant coefficients (and is homogeneous), we substitute $x=e^{rt}$. The characteristic equation is $mr^2+k=0$, with roots $r=\pm i\sqrt{k/m}$, since $m>0$, $k>0$. Thus, the general solution of (12) is

$$x=c_1\cos\sqrt{\frac{k}{m}}t+c_2\sin\sqrt{\frac{k}{m}}t, \tag{13}$$

where c_1 and c_2 are arbitrary constants.

It is sometimes convenient to introduce the notation

$$\omega_0=\sqrt{\frac{k}{m}}. \tag{14}$$

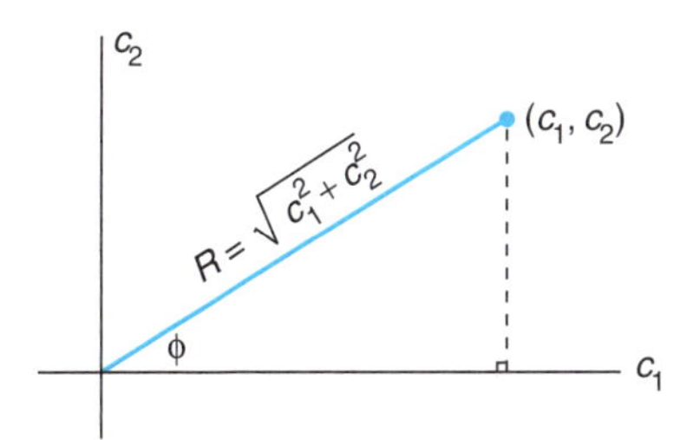

Figure 2.5.6 Amplitude R and phase ϕ.

This will be explained shortly. With this notation, the differential equation (12) becomes, after dividing by m,

$$\frac{d^2x}{dt^2} + \omega_0^2 x = 0, \tag{15}$$

and its general solution (13) is written more easily:

$$x = c_1 \cos \omega_0 t + c_2 \sin \omega_0 t. \tag{16}$$

Note that the number ω_0 that appears in the solution (16) is the square root of the positive constant coefficient in the differential equation (15).

Amplitude-Phase Form

The general solution (13) of the homogeneous equation is a linear combination of $\cos \omega_0 t$ and $\sin \omega_0 t$. We will show, in general, that the addition of these two trigonometric functions yields a simple trigonometric function because

$$c_1 \cos \omega_0 t + c_2 \sin \omega_0 t = R \cos(\omega_0 t - \phi). \tag{17}$$

To show (17), we use the trigonometric addition formula, $\cos(a - b) = \cos a \cos b + \sin a \sin b$, to rewrite the right-hand side of (17) to get

$$c_1 \cos \omega_0 t + c_2 \sin \omega_0 t = R(\cos \omega_0 t \cos \phi + \sin \omega_0 t \sin \phi). \tag{18}$$

Equation (18) holds if

$$\begin{aligned} c_1 &= R \cos \phi, \\ c_2 &= R \sin \phi. \end{aligned} \tag{19}$$

It is seen in figure 2.5.6 from (19) that R and ϕ are polar coordinates for the point (c_1, c_2). From the Pythagorean theorem,

$$R = \sqrt{c_1^2 + c_2^2}. \tag{20}$$

The variable ϕ is called the **phase angle**, and $R = \sqrt{c_1^2 + c_2^2}$ the **amplitude** (or **amplitude of oscillation**). Drawing the triangle and specifying the quadrant are helpful in determining the phase angle. The formula

$$\tan \phi = \frac{c_2}{c_1} \tag{21}$$

TABLE 2.5.1
Units

	cgs	*mks*	*fps*
Length	centimeter (cm)	meter (m)	foot (ft)
Mass	gram (g)	kilogram (kg)	slug = 1 lb $\cdot$s^2/ft
Time	second (s)	second (s)	second (s)
Velocity	cm/s	m/s	ft/s
Acceleration	cm/s^2	m/s^2	ft/s^2
Force ($F = ma$)	dyne = 1 g $\cdot$cm/s^2	newton (N) = 1 kg $\cdot$m/s^2	pound (lb)
Gravity g	980 cm/s^2	9.8 m/s^2	32 ft/s^2

is sometimes convenient. However, $\phi = \tan^{-1}(c_2/c_1)$ is correct only if ϕ is in the first or fourth quadrant. If ϕ is in the second or third quadrant, then $\phi = \pi + \tan^{-1}(c_2/c_1)$. Fewer errors are made if the quadrant of the phase angle is stated. In summary:

Amplitude-Phase Form:

$$c_1 \cos \omega_0 t + c_2 \sin \omega_0 t = R \cos(\omega_0 t - \phi),$$
$$R = \sqrt{c_1^2 + c_2^2},$$
$$\tan \phi = \frac{c_2}{c_1},$$
$$\phi = \tan^{-1}(c_2/c_1) \quad \text{if } c_1 \geq 0,$$
$$\phi = \tan^{-1}(c_2/c_1) + \pi \quad \text{if } c_1 < 0,$$

One of the significances of the amplitude and phase form is that it allows a relatively simple graphing of the solution

$$x(t) = R \cos(\omega_0 t - \phi). \tag{22}$$

R is the amplitude, and the solution is a shifted cosine:

$$x(t) = R \cos \left[\omega_0 \left(t - \frac{\phi}{\omega_0} \right) \right].$$

One of the times at which the solution is a maximum is $t = \phi/\omega_0$. Formula (22) and figure 2.5.7 show that the resulting free response, in the absence of friction, is **simple harmonic motion** (given by a cosine function). The mass oscillates sinusoidally around $x = 0$. The solution is periodic in time. Since the trigonometric functions have period 2π, the **period** of motion T is

$$\text{period} = T = \frac{2\pi}{\omega_0}.$$

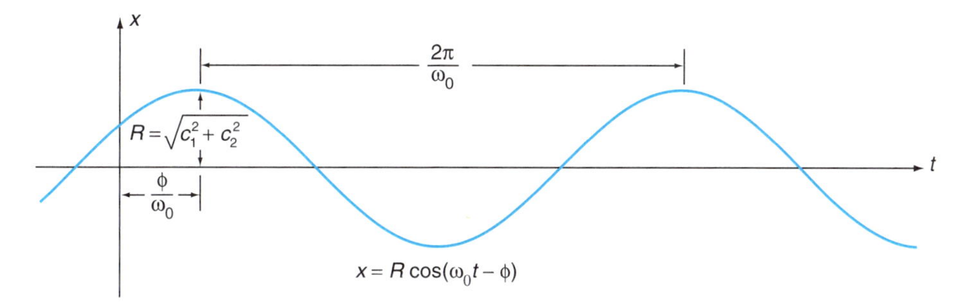

Figure 2.5.7 Simple harmonic motion.

The **frequency** is the number of periods (cycles) per unit time. If the units of time were seconds and the period was $\frac{1}{5}$ s, then the frequency would be 5 cycles per second (or 5 **hertz** (Hz)). In general, the frequency satisfies

$$\text{frequency} = \frac{1}{T} = \frac{\omega_0}{2\pi}.$$

This also gives an interpretation of the important parameter ω_0:

$$\omega_0 = 2\pi \cdot \text{ frequency}.$$

Since in one period the angle of the cosine changes by 2π radians, ω_0 is sometimes called the **circular frequency**, the number of radians per unit of time.

For the spring-mass system, (12) or (15), the frequency $(1/2\pi)\sqrt{k/m}$ is called the **natural frequency** since it is the frequency that arises naturally without any forcing to the system. Stiffer springs (larger k) have larger frequencies.

Table 2.5.1 reviews various units commonly used.

Example 2.5.1 *Resulting Motion of Spring-Mass System*

The spring constant of a 24-in. steel spring is measured by hanging a 1-slug (32-lb) mass from the spring and observing that the spring stretches 3 in. Now a $\frac{1}{2}$-slug (16-lb) mass is attached to the spring. The mass is pulled down 3 in. and released with a velocity of 1 ft/s downward. Determine the resulting motion.

● SOLUTION. First we need to determine the spring constant. From (8),

$$k\Delta L = mg.$$

In fps units this is

$$k\left(\frac{1}{4}\text{ft}\right) = 1\frac{\text{lb} \cdot \text{s}^2}{\text{ft}} \cdot \left(32\frac{\text{ft}}{\text{s}^2}\right),$$

or $k = 128$ lb/ft. Since $m = \frac{1}{2}$, the equation of motion is (12),

$$\frac{1}{2}\frac{d^2x}{dt^2} + 128x = 0,$$

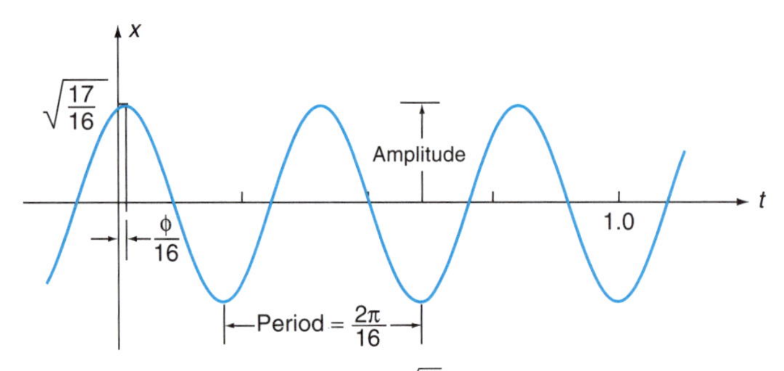

Figure 2.5.8 Graph of $t = \frac{\sqrt{17}}{16} \times \cos(16t - 0.2498)$.

or $d^2x/dt^2 + 256x = 0$. The characteristic equation is $r^2 + 256 = 0$, which has roots $r = \pm 16i$. Thus the general solution of the differential equation is

$$x = c_1 \cos 16t + c_2 \sin 16t. \tag{23}$$

The initial conditions are

$$x(0) = \frac{1}{4}\text{ft}, \quad x'(0) = 1\text{ft/s}.$$

Thus

$$\frac{1}{4} = x(0) = c_1, \quad 1 = x'(0) = 16c_2,$$

and the resulting motion is

$$x = \frac{1}{4} \cos 16t + \frac{1}{16} \sin 16t. \tag{24}$$

In order to visualize this motion more easily, we rewrite the solution in the form $x = R\cos(16t - \phi)$. The amplitude is

$$R = \sqrt{c_1^2 + c_2^2} = \sqrt{\left(\frac{1}{4}\right)^2 + \left(\frac{1}{16}\right)^2} = \frac{\sqrt{17}}{16}.$$

The phase angle ϕ is in the first quadrant with

$$\tan\phi = \frac{c_2}{c_1} = \frac{1/16}{1/4} = \frac{1}{4},$$

so that

$$\phi = \tan^{-1}\left(\frac{1}{4}\right) = 0.2498 \text{ radian}.$$

Thus the motion (24) may be rewritten as

$$x = \frac{\sqrt{17}}{16} \cos(16t - 0.2498).$$

The period is $\frac{2\pi}{16}$ and the frequency is $\frac{(16/2\pi)}{s}$ (see figure 2.5.8). ♦

Exercises

In Exercises 1–10, put $x(t)$ into the amplitude-phase form $x(t) = R\cos(\omega_0 t - \phi)$:

1. $x(t) = 3\cos 5t - 7\sin 5t$.
2. $x(t) = -5\cos 7t + 3\sin 7t$.
3. $x(t) = \sqrt{3}\cos 14t + \sin 14t$.
4. $x(t) = 5\cos 4t + 5\sqrt{3}\sin 4t$.
5. $x(t) = -6\cos 5t + 6\sin 5t$.
6. $x(t) = 4\cos 7t - 4\sin 7t$.
7. $x(t) = \sqrt{3}\cos 6t - \sin 6t$.
8. $x(t) = -\sqrt{3}\cos 9t - \sin 9t$.
9. $x(t) = -4\cos 2t + 4\sqrt{3}\sin 2t$.
10. $x(t) = 2\cos 2t - 2\sqrt{3}\sin 2t$.

Free Response. In Exercises 11–24, we assume that Hooke's law is applicable, resistance is negligible, and the mass of the spring is negligible.

11. A mass of 30 g is attached to a spring. At equilibrium the spring has stretched 20 cm. The spring is pulled down another 10 cm and released. Set up the differential equation for the motion, and solve it to determine the resulting motion, ignoring friction.
12. A mass of 400 g is attached to a spring. At equilibrium the point has stretched 245 cm. The spring is pulled down and released such that the mass is 10 cm below equilibrium and traveling upward at $\sqrt{84}$ cm/s. Set up the differential equation for the motion. Solve the differential equation and express the solution in the amplitude-phase form.
13. A mass of 8 slugs ($8 \cdot 32$ lb) is attached to a long spring. The spring stretches 2 ft before coming to rest. The 8-slug mass is removed and a 2-slug mass (64 lb) is attached and placed at equilibrium. The mass is pushed down and released. At the time it is released, the mass is 2 ft below equilibrium and traveling downward at 1 ft/s. Derive the differential equation for the motion of the spring-mass system. Solve the differential equation.
14. A spring-mass system has a spring constant of $k = 5$ g/s^2. What mass should be attached to make the resulting motion have a frequency of 30 Hz?
15. A spring is to be attached to a mass of 10 slugs (320 lb). What should the spring constant be to make the resulting motion have a frequency of 5 Hz?
16. A mass of 6 g is attached to a spring-mass system with spring constant 30 g/s^2. What should the initial conditions be to give a response with amplitude 3 and phase angle $\pi/4$?
17. A mass of 16 g is attached to a spring-mass system with spring constant 64 g/s^2. What should the initial conditions be to give a response with amplitude 2 and phase angle $\pi/3$?
18. Suppose you have a spring-mass system as in figure 2.5.5.
    a) What is the effect on the period and frequency of doubling the mass?
    b) What is the effect on the period and frequency of doubling the spring constant?
    c) What is the effect on the period and frequency of doubling both the spring constant and the mass?

19. A spring-mass system with mass m and spring constant k is subjected to a sudden impulse. The result is that, at time $t = 0$, the mass is at the equilibrium position but has a velocity of 10 cm/s downward.
   a) Determine the subsequent motion.
   b) Determine the amplitude of the resulting motion as a function of m and k.
   c) What is the effect on amplitude of increasing k?
   d) What is the effect on amplitude of increasing m?
20. (Same situation as Exercise 19) If the mass is 50 g, what should the spring constant be to give an amplitude of 20 cm?
21. At time $t = 0$ the mass in a spring-mass system with mass m and spring constant k is observed to be 1 ft below equilibrium and traveling downward at 1 ft/s.
   a) Determine the subsequent motion.
   b) Determine the amplitude as a function of m and k.
   c) What is the effect on amplitude of increasing m or k?
22. (Same situation as Exercise 21) If the spring constant is $k = 8$ lb/ft, what should the mass be to give an amplitude of 4 ft? At what time will this amplitude be first obtained?
23. (**Conservation of mechanical energy**) Suppose that a spring-mass system is modeled by $mx'' + kx = 0$. The quantities $\frac{1}{2}mv^2$ and $\frac{1}{2}kx^2$ are the **kinetic energy** and the **elastic potential energy**. Their sum is the total **mechanical energy**. Multiply $mx'' + kx = 0$ by x' and integrate to show that $\frac{1}{2}mv^2 + \frac{1}{2}kx^2 = \text{constant}$.
24. Let $E = \frac{1}{2}mv^2 + \frac{1}{2}kx^2 = \frac{1}{2}m(x')^2 + \frac{1}{2}kx^2$. Show that if x is a solution of $mx'' + \delta x' + kx = 0$ with $\delta > 0$, then $\frac{dE}{dt} < 0$. Thus E is monotonically decreasing, so that mechanical energy decreases in the presence of resistance.

In Exercises 25–30, find the general solution. (If the initial conditions are given, put the solution in amplitude and phase form if possible.)

25. $\dfrac{d^2x}{dt^2} + 7x = 0.$

26. $\dfrac{d^2x}{dt^2} + 23x = 0.$

27. $\dfrac{d^2x}{dt^2} - 7x = 0.$

28. $\dfrac{d^2x}{dt^2} - 23x = 0.$

29. $\dfrac{d^2x}{dt^2} + 5x = 0,\ x(0) = 2,\ \dfrac{dx}{dt}(0) = 3.$

30. $\dfrac{d^2x}{dt^2} + 9x = 0,\ x(0) = 7,\ \dfrac{dx}{dt}(0) = 5.$

31. Consider an object moving in a circle of radius r, such that the angle θ is steadily increasing, $\frac{d\theta}{dt} = \text{constant}$. How many cycles per second does the object make? Show that the x component of the object satisfies simple harmonic motion (22). Show that $\frac{d\theta}{dt}$ equals the circular frequency.
32. Consider a mass m, constrained to move only horizontally, located between two fixed walls a distance D apart and connected to two springs. The left spring has length L_1 and spring constant k_1 while the right spring has length L_2 and spring

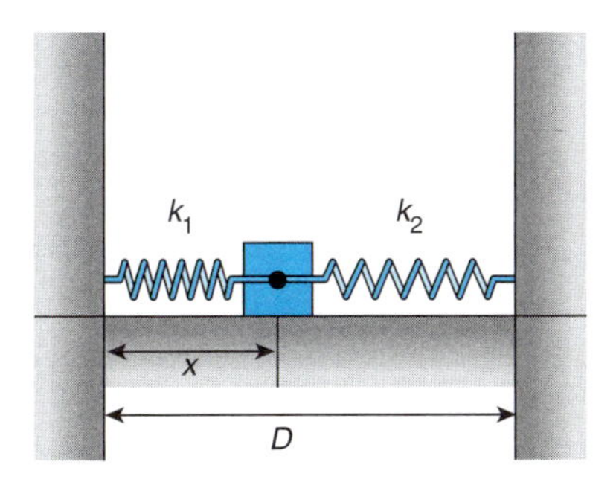

Figure 2.5.9

constant k_2. Let the position of the mass from the left wall be x, as illustrated in figure 2.5.9. Set up one differential equation for the one mass (acted upon by two forces).

2.5.3 Free Response with Friction ($\delta > 0$)

Suppose now that friction is not negligible, so that the dynamics of the spring-mass system in figure 2.5.4 are described by the differential equation

$$m\frac{d^2x}{dt^2} + \delta\frac{dx}{dt} + kx = 0. \tag{25}$$

Since the differential equation has constant coefficients, we may use the characteristic equation of (25), $mr^2 + \delta r + k = 0$, which has the roots

$$r = -\frac{\delta}{2m} \pm \frac{\sqrt{\delta^2 - 4mk}}{2m}. \tag{26}$$

There are three cases to consider, depending on whether $\delta^2 - 4mk$ is negative, zero, or positive. Since we have just discussed the $\delta = 0$ (no friction) case, it is convenient to think of the mass m and spring constant k as fixed, and explain what happens as the friction coefficient δ increases.

Case 1: Underdamping $(0 < \delta < \sqrt{4mk})$

In this case, $\delta^2 - 4mk < 0$. There is a pair of complex conjugate roots

$$r = -\alpha \pm \beta i, \quad \alpha = \frac{\delta}{2m}, \quad \beta = \frac{\sqrt{4mk - \delta^2}}{2m}, \tag{27}$$

so that the solutions of the differential equation $mx'' + \delta x' + kx = 0$ are

$$x = e^{-\alpha t}(c_1 \cos \beta t + c_2 \sin \beta t), \tag{28}$$

or, using (22),

$$x = e^{-\alpha t}\sqrt{c_1^2 + c_2^2}\cos(\beta t - \phi). \tag{29}$$

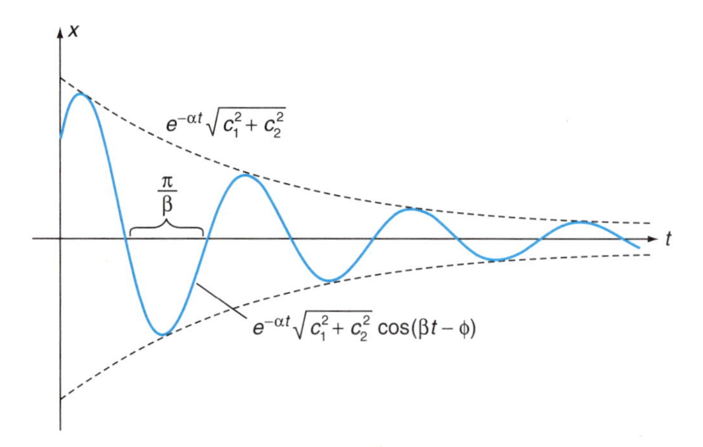

Figure 2.5.10 Underdamped oscillation.

This is a **damped oscillation**, known as **underdamped** ($\delta < \sqrt{4mk}$). The solution is graphed in figure 2.5.10. The sinusoidal function $\cos(\beta t - \phi)$ varies between 1 and -1. Thus the solution stays between $\sqrt{c_1^2+c_2^2}e^{-\alpha t}$ and $-\sqrt{c_1^2+c_2^2}e^{-\alpha t}$. To qualitatively sketch (29), first sketch $\sqrt{c_1^2+c_2^2}e^{-\alpha t}$ and its negative $-\sqrt{c_1^2+c_2^2}e^{-\alpha t}$. The solution oscillates between these two curves, as shown in figure 2.5.10. For underdamped oscillations, the term

$$\sqrt{c_1^2+c_2^2}e^{-\alpha t}$$

is called the **time-damped amplitude** of the oscillation.

With friction the amplitude oscillates with a **pseudo-(circular) frequency**

$$\beta = \frac{\sqrt{4mk-\delta^2}}{2m}.$$

This frequency can be related to the (circular) frequency ($\omega_0 = \sqrt{k/m}$) at which the spring-mass system would oscillate without damping:

$$\beta = \sqrt{\frac{k}{m} - \frac{\delta^2}{4m^2}} = \sqrt{\omega_0^2 - \frac{\delta^2}{4m^2}}.$$

The pseudofrequency of an underdamped oscillator is less than the natural frequency. As the resistance δ is increased, the pseudofrequency decreases (the period increases).

One experimental method of determining the damping constant δ is by observing the exponential decay of the amplitude of the oscillation. Another method is to compare the frequencies with and without damping.

For a given mass m and spring constant k, increasing the resistance δ (but keeping $\delta < \sqrt{4mk}$) has two effects. First, the oscillation is damped faster; that is, the time-varying amplitude $e^{-\alpha t}\sqrt{c_1^2+c_2^2}$ decreases faster. Second, the period of the oscillation being damped, $2\pi/\beta$, increases (figure 2.5.10).

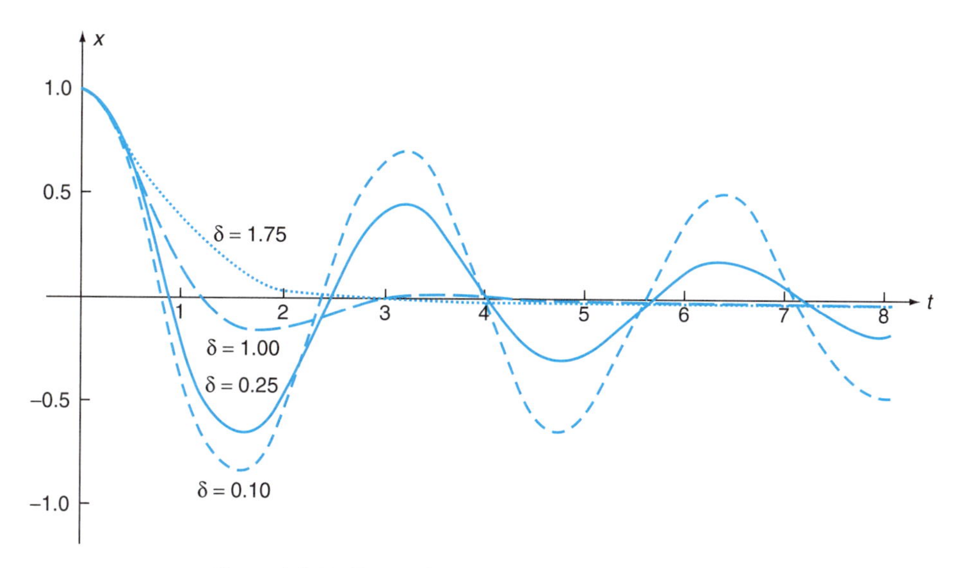

Figure 2.5.11 Graph of (33) for $\delta = 0.1$, 0.25, 1, and 1.75.

Example 2.5.2 *Underdamped Motion*

A mass of 0.5 slug (weight of 16 lb) is suspended on a spring with a spring constant of $k = 2$ lb/ft. The damping is δ lb · s/ft. The mass is pulled down 1 ft. and released. Describe and graph the motion for $\delta = 0.1$, 0.25, 1, and 1.75.

• SOLUTION. The differential equation describing the motion is

$$\frac{1}{2}\frac{d^2x}{dt^2} + \delta\frac{dx}{dt} + 2x = 0, \quad x(0) = 1, \quad x'(0) = 0. \tag{30}$$

The characteristic equation, which may be found by inspection or by substituting in $x = e^{rt}$, of (30) is $\frac{1}{2}r^2 + \delta r + 2 = 0$, which has roots $r = -\delta \pm i\sqrt{4-\delta^2}$, so that the solution of the differential equation (30) is

$$x = e^{-\alpha t}\left[c_1 \cos(\sqrt{4-\delta^2}t) + c_2 \sin(\sqrt{4-\delta^2}t)\right], \tag{31}$$

which is underdamped if $\delta < 2$. The initial conditions imply that

$$1 = x(0) = c_1,$$
$$0 = x'(0) = -\delta c_1 + \sqrt{4-\delta^2}c_2.$$

Thus $c_1 = 1$, $c_2 = \delta(4-\delta^2)^{-1/2}$, and

$$x = e^{-\delta t}\left[\cos(\sqrt{4-\delta^2}t) + \frac{\delta}{(4-\delta^2)^{1/2}} \sin(\sqrt{4-\delta^2}t)\right]. \tag{32}$$

We shall rewrite (32) using (22). Now

$$1+\left[\frac{\delta}{(4-\delta^2)^{1/2}}\right]^2=\frac{4}{4-\delta^2}.$$

Thus (32) is

$$x=e^{-\delta t}\frac{2}{\sqrt{4-\delta^2}}\cos(\sqrt{4-\delta^2}t-\phi), \tag{33}$$

where

$$\tan\phi=\delta(4-\delta^2)^{-1/2}.$$

The graphs of (33) for $\delta=0.1$, 0.25, 1, and 1.75 are shown in figure 2.5.11. ♦

Case 2: Overdamping ($\delta>\sqrt{4mk}$)

In this case, called **overdamping**, the characteristic equation $mr^2+\delta r+k=0$ for the differential equation $mx''+\delta x'+kx=0$ has two distinct negative roots:

$$r_1=-\frac{\delta}{2m}+\frac{\sqrt{\delta^2-4mk}}{2m}, \tag{34a}$$

$$r_1=-\frac{\delta}{2m}-\frac{\sqrt{\delta^2-4mk}}{2m}. \tag{34b}$$

(*Note*: $r_1<0$ since $\sqrt{\delta^2-4mk}<\sqrt{\delta^2}=\delta$.) Thus the solution is

$$x=c_1e^{r_1t}+c_2e^{r_2t}. \tag{35}$$

Note that $x\to 0$ as $t\to\infty$, since $r_1<0$ and $r_2<0$.

In general, for overdamped systems, depending on the initial conditions, the solution for x takes one of the three forms shown in figure 2.5.12, or their negative (see Exercises 41, 42). Figure 2.5.12(a) results if the mass is initially moving away from equilibrium ($x=0$). The mass slows down and then returns toward the equilibrium. If the mass is initially moving toward equilibrium, it either approaches the equilibrium position [figure 2.5.12(b)] or, if the initial velocity is great enough, overshoots the equilibrium and then returns toward the equilibrium position, as in figure 2.5.12(c).

Case 3: Critical Damping ($\delta=\sqrt{4mk}$)

In this case, $\delta^2-4mk=0$ and $x=e^{rt}$ with (26) shows that there is a repeated real root $r=-\delta/(2m)$. The general solution of the differential equation $mx''+\delta x'+kx=0$ is then

$$x=c_1e^{-(\delta/2m)t}+c_2te^{-(\delta/2m)t}. \tag{36}$$

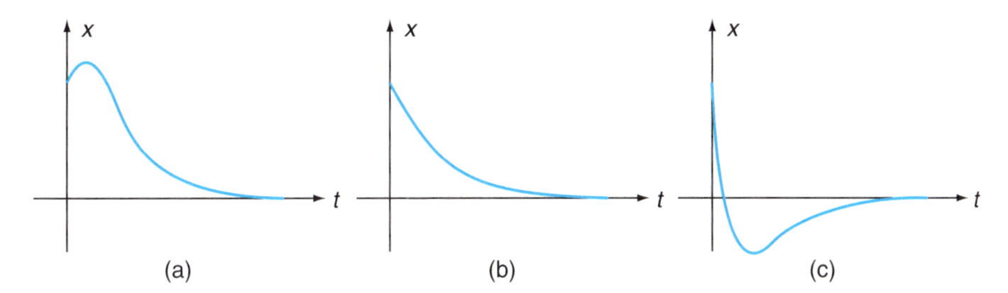

Figure 2.5.12 Critically damped or overdamped.

The solution (36) for the **critically damped** case is different from that for the overdamped case (35). However, depending on the values of c_1, c_2, the critically damped solution (36) has a graph that has the same general shape as one of those given for the overdamped case (see figure 2.5.12).

Example 2.5.3 *Critical Damping*

A device is being designed that can be modeled as a spring-mass system. The spring constant is $k = 10$ g/s^2, and the damping constant is $\delta = 20$ g/s.

a) Determine the mass so that the resulting spring-mass system will be critically damped.
b) The mass is pulled down 5 cm from the rest position and released with a downward velocity of 10 cm/s. Determine and solve the equations of motion. Graph the resulting motion.

• SOLUTION

a) Critical damping occurs if $\delta^2 - rmk = 0$. That is, $400 - 40m = 0$. Thus, the desired mass is 10 g.
b) The equation of motion is

$$10x'' + 20x' + 10x = 0, \quad x(0) = 5, \quad x'(0) = 10. \tag{37}$$

Substituting $x = e^{rt}$, we see the characteristic equation for (37) is $10r^2 + 20r + 10 = 0$, or $10(r+1)^2 = 0$, which has a repeated real root of $r = -1$. The general solution of $10x'' + 20x' + 10x = 0$ is thus

$$x = c_1e^{-t} + c_2te^{-t}.$$

Applying the initial conditions in (37) to this general solution gives

$$5 = x(0) = c_1,$$
$$10 = x'(0) = -c_1 + c_2.$$

Solve for c_1, c_2 to get $c_1 = 5$, $c_2 = 15$. Thus

$$x = 5e^{-t} + 15te^{-t},$$

which has a graph like that in figure 2.5.12(a). ♦

It is important to note that for all three cases in which there is damping ($\delta > 0$), we have

$$\lim_{t \to \infty} x(t) = 0.$$

The importance of this is discussed in Section 2.7.

Exercises

In Exercises 33–40, set up the differential equation that describes the motion under the assumptions of this section. Solve the differential equation. State whether the motion of the spring-mass system is harmonic, damped oscillation, critically damped, or overdamped. If the motion is a damped oscillation, rewrite in the form (22).

33. The spring-mass system has an attached mass of 10 g. The spring constant is 30 g/s^2. A dashpot mechanism is attached, which has a damping coefficient of 40 g/s. The mass is pulled down and released. At time $t = 0$, the mass is 3 cm below the rest position and moving upward at 5 cm/s.
34. A long spring has a mass of 1 slug attached to it. The spring stretches 16/13 ft and comes to rest. The damping coefficient is 2 slug/s. The mass is subjected to an impulsive force at time $t = 0$, which imparts a velocity of 5 ft/s downward.
35. A mass of 1 g is attached to a spring-mass system for which friction is negligible. The spring stretches 20 cm and comes to rest. The mass is pulled down 1 cm from rest and released with a velocity of 7 cm/s downward.
36. A spring system has an attached mass of 1 g, a spring coefficient of 5 g/s^2, and a damping coefficient of 4 g/s. The mass is pushed upward 1 cm from the rest position and released with a velocity of 3 cm/s downward.
37. A spring with spring constant $k = 12$ slug/s^2 has a mass attached that stretches the spring $2\frac{2}{3}$ ft. The damping coefficient is 7 slug/s. The mass is pushed 1 ft above the rest position and then released with a velocity of 1 ft/s downward.
38. A long spring with spring constant $k = 8$ g/s^2 has a mass attached that stretches the spring 245 cm. The damping coefficient is $\delta = 8$ g/s. At time $t = 0$, the mass is at the equilibrium position and has a velocity of 3 cm/s downward.
39. A spring-mass system has a spring with spring constant $k = 5$ g/s^2, attached mass of $m = 1$ g, and a friction coefficient of $\delta = 4$ g/s. The mass is pulled 2 cm below the equilibrium position and released.
40. A spring-mass system has a spring constant of $k = 1$ lb/ft, a damping coefficient of 2 slug/s, and an attached mass of 1 slug. The mass is pushed upward and released 1 ft above the equilibrium position with a velocity of 1 ft/s upward.
41. Show that in the case of critical damping, $\delta = \sqrt{4mk}$, the mass may change direction at most once (the general solution (36) has at most one horizontal tangent).
42. Show that in the case of overdamping, the mass can change direction at most once.
43. Suppose $\delta > 0$, $k > 0$ are fixed. Describe how varying the mass affects whether the motion is a damped oscillation, overdamped, or critically damped.

44. Suppose a 1-slug mass is attached to a spring with spring constant 400 lb/ft. Determine the damping coefficient if underdamped oscillations are observed with frequenzy 3 Hz.
45. Suppose a spring-mass system with a 1-slug mass is known to vibrate at 5 Hz without damping but only at 4 Hz when underdamped. Determine the damping coefficient.

For Exercises 46–55, solve the differential equation and state whether the corresponding motion is underdamped, overdamped, or critically damped.

46. $2\frac{d^2x}{dt^2}+5\frac{dx}{dt}+3x=0.$
47. $2\frac{d^2x}{dt^2}+4\frac{dx}{dt}+3x=0.$
48. $2\frac{d^2x}{dt^2}+\sqrt{48}\frac{dx}{dt}+3x=0.$
49. $3\frac{d^2x}{dt^2}+5\frac{dx}{dt}+4x=0.$
50. $3\frac{d^2x}{dt^2}+7\frac{dx}{dt}+4x=0.$
51. $3\frac{d^2x}{dt^2}+8\frac{dx}{dt}+4x=0.$
52. $9\frac{d^2x}{dt^2}+10\frac{dx}{dt}+4x=0.$
53. $9\frac{d^2x}{dt^2}+12\frac{dx}{dt}+4x=0.$
54. $9\frac{d^2x}{dt^2}+15\frac{dx}{dt}+4x=0.$
55. $3\frac{d^2x}{dt^2}+4\frac{dx}{dt}+x=0.$
56. Consider a spring-mass system with critical damping. Suppose that $x(0)=1$, $x'(0)=v_0$. For what values of v_0 will the solution look like (*a*), (*b*), (*c*) of Figure 2.5.12? (*Hint*: The time of a local maximum or minimum must be positive.)
57. Consider an overdamped spring-mass system. Suppose that $x(0)=1$, $x'(0)=v_0$. Find a formula for the time of a local maximum or minimum, assuming that one exists.

Exercises 58–61 illustrate a general fact but are given only for $m=1$, $k=1$. For Exercises 58 and 59, let x_0 be the solution of

$$x''+\delta x'+x=0, \quad x(0)=1, x'(0)=-1 \tag{38}$$

for a given δ. Note that $\sqrt{4mk}=\sqrt{4}=2$.

58. Solve (38) for $0<\delta<2$ and $\delta=2$. Show that

$$\lim_{\delta\to 2^-} x_\delta(t)=x_2(t) \quad \text{for all } t\geq 0.$$

This shows that as the damping coefficient approaches critical damping, the damped oscillation approaches the critical damping solution.

59. Solve (38) for $2 < \delta$ and $\delta = 2$. Show that

$$\lim_{\delta \to 2^+} x_\delta(t) = x_2(t) \quad \text{for all } t \geq 0.$$

This shows that the overdamped solution approaches the critical damping solution.

For Exercises 60 and 61, let x_δ be the solution of

$$x'' + \delta x' + x = 0, \quad x(0) = 0, x'(0) = 1. \tag{39}$$

60. Solve (39) for $0 < \delta < 2$ and $\delta = 2$. Show that

$$\lim_{\delta \to 2^-} x_\delta(t) = x_2(t) \quad \text{for all } t \geq 0.$$

61. Solve (39) for $\delta > 2$ and $\delta = 2$. Show that

$$\lim_{\delta \to 2^+} x_\delta(t) = x_2(t) \quad \text{for all } t \geq 0.$$

62. (Requires computer with fairly good graphic capabilities) Illustrate the limits in Exercises 58–61 by graphing x_δ for several values of δ on the same graph. For example, in Exercise 58 you might take $\delta = 2 - \frac{1}{n}$ or $\delta = 2 - \frac{1}{n^2}$ for $n = 1, 2, 3, \ldots$, and plot on an interval of, say, $0 \leq t \leq 5$ (see figure 2.5.11).

2.6 The Method of Undetermined Coefficients

As pointed out in Section 2.1, the general solution of

$$ax''(t) + bx'(t) + cx(t) = f(t),$$

where a, b, c are constants and $a \neq 0$, may be written in the form $x = x_p + c_1x_1 + c_2x_2$, where x_1, x_2 are solutions of the associated homogeneous equation $ax'' + bx' + cx = 0$ and x_p is a particular solution of $ax'' + bx' + cx = f$. The homogeneous equation was solved in Section 2.4. Two methods for finding the particular solution x_p will be given in this chapter. The method of undetermined coefficients will be developed in this section. The method of variation of parameters will be discussed Sections 2.10 and 2.11.

Method of Undetermined Coefficients

The method of undetermined coefficients is used to find a particular solution for a linear differential equation with constant coefficients,

$$ax'' + bx' + cx = f(t), \tag{1}$$

where a, b, c are constant coefficients and $f(t)$ is one of the following:

1. $f(t) = \text{polynomial}$;
2. $f(t) = \text{exponential} = e^{\alpha t}$;
3. $f(t) = \text{sinusoid} = \sin \beta t$ or $\cos \beta t$;
4. $f(t) = \text{products of } 1, 2, 3$;
5. $f(t) = \text{linear combinations of } 1, 2, 3, 4$ (using superposition, see (23)).

Slightly different procedures are used in the method of undetermined coefficients to find a particular solution depending on the type of forcing function. We begin by considering when $f(t)$ involves polynomials or exponentials. Proofs will not be given in this section. The basic understanding of this technique will be gained from examples.

FIRST STEP. The first step using undetermined coefficients is always the same: analyze solutions of the associated homogeneous equation. Determine the characteristic equation and all roots (including the multiplicity of roots).

SECOND STEP. The second step depends on $f(t)$. For example,

> If $f(t)$ in (1) is an exponential, $e^{\alpha t}$, times an m th-degree polynomial
> $f(t) = e^{\alpha t}$ (polynomial of degree m), then x_p is of the **form**
> $$x_p = e^{\alpha t} t^k (A_0 + A_1 t + \cdots + A_m t^m), \tag{2}$$
> for appropriate constants $A_0, \ldots, A_m$, where k is the multiplicity of $r = \alpha$ as a root of the characteristic equation of the homogeneous differential equation.

The degree of the polynomial in the particular solution is the same as the degree of the polynomial in the forcing function $f(t)$ if $e^{\alpha t}$ is not a homogeneous solution (that is, $r = \alpha$ is not a root of the characteristic equation). The degree of the polynomial in the particular solution is increased by the number of times $r = \alpha$ is a root of the characteristic equation.

The equation (2) gives the form of the particular solution. The A_i are called **undetermined coefficients** and may be determined by substituting the form (2) into the differential equation (1). This method as described by (2), is valid for first-order equations, and it generalizes the simple examples discussed in Section 1.6. The method of undetermined coefficients is also valid with no alterations for finding a particular solution of higher-order constant coefficient linear differential equations. With minor modifications (2) can be used for all seven cases we now consider.

Case 1: Polynomial Forcing

A special case of (2) occurs if the forcing equation $f(t)$ is just a polynomial. This corresponds to $\alpha = 0$. If the forcing function is a polynomial, then a particular solution is also a polynomial (of possibly larger degree). The degree of the particular solution is the same as the degree of the polynomial forcing if $r = 0$ is not a root of the characteristic equation. The degree of the particular solution is increased by the number of times $r = 0$ is a root of the characteristic equation. To be more precise

> If $f(t)$ in (1) is an mth-degree polynomial, then x_p is of the form
> $$x_p = t^k (A_0 + A_1 t + \cdots + A_m t^m), \tag{3}$$
> where k is the multiplicity of $r = 0$ as a root of the characteristic equation.

Example 2.6.1 *Polynomial Forcing ($r = 0$ is not a root)*

Find the general solution of

$$x'' - 3x' + 2x = 2t^2 + 1. \tag{4}$$

• SOLUTION. The characteristic equation $r^2 - 3r + 2 = (r-2)(r-1) = 0$ has roots $r = 1$ and $r = 2$. Thus $x_h = c_1 e^t + c_2 e^{2t}$. Here $f(t) = 2t^2 + 1$ is a second-degree polynomial. Thus x_p has the form $t^k(A_0 + A_1 t + A_2 t^2)$. $r = 0$ is not a root of the characteristic equation, so $k = 0$ and we have

$$x_p = A_0 + A_1 t + A_2 t^2. \tag{5}$$

Substituting this expression for x in the differential equation (4), we get

$$(A_0 + A_1 t + A_2 t^2)'' - 3(A_0 + A_1 t + A_2 t^2)' + 2(A_0 + A_1 t + A_2 t^2) = 2t^2 + 1$$

or

$$2A_2 - 3A_1 - 6A_2 t + 2A_0 + 2A_1 t + 2A_2 t^2 = 2t^2 + 1.$$

Equating coefficients of like powers of t gives

$$\begin{aligned} 1: &\quad 2A_2 - 3A_1 + 2A_0 = 1, \\ t: &\quad -6A_2 + 2A_1 = 0, \\ t^2: &\quad 2A_2 = 2, \end{aligned}$$

which has the solution $A_0 = 4$, $A_1 = 3$, $A_2 = 1$. Thus the particular solution is $x_p = 4 + 3t + t^2$, and the general solution is

$$x = x_p + x_h = 4 + 3t + t^2 + c_1 e^t + c_2 e^{2t}.$$

♦

Example 2.6.2 *Polynomial Forcing ($r = 0$ is a root)*

Suppose that you were solving the differential equation

$$x'' - 3x' = 2t^2 + 1 \tag{6}$$

using the method of undetermined coefficients. Give the form of the particular solution that you would use.

• SOLUTION. The characteristic equation $r^2 - 3r = (r-3)r = 0$ has roots $r = 0$ and $r = 3$. Thus $x_h = c_1 + c_2 e^{3t}$. Here $f(t) = 2t^2 + 1$ is a second-degree polynomial. Thus x_p has the form $t^k(A_0 + A_1 t + A_2 t^2)$. Here $r = 0$ is a root of the characteristic equation of multiplicity 1, so $k = 1$ and we have

$$x_p = t(A_0 + A_1 t + A_2 t^2) = A_0 t + A_1 t^2 + A_2 t^3. \tag{7}$$

♦

It is important to note that in (4) and (6) we had identical right-hand sides $f(t) = 2t^2 + 1$, but different forms (5) and (7) for x_p. To get the correct form for x_p, one has to look at not only the forcing term but also the roots of the characteristic equation. We shall give a physical application of this later in this chapter.

Case 2: Exponential Forcing

A special case of (2) occurs if the forcing function $f(t)$ is just a constant times $e^{\alpha t}$. The constant corresponds to a zero-degree polynomial, $m=0$. If the forcing function is this type of exponential, then the form of a partcular solution is a similar exponential $x_p = Ae^{\alpha t}$ if $e^{\alpha t}$ is not a homogeneous solution (i.e., if $r=\alpha$ is not a root of the characteristic equation). The form of a particular solution is the same exponential multiplied by a polynomial whose degree is increased (from $m=0$) by the number of times $r=\alpha$ is a root of the characteristic equation. To be more precise,

If $f(t)$ in (1) is a constant times $e^{\alpha t}$, then x_p has the form

$$x_p = At^k e^{\alpha t}, \tag{8}$$

where k is the multiplicity of $r=\alpha$ as a root of the characteristic equation.

Example 2.6.3 *Exponential Forcing ($r=\alpha$ is not a root)*

Find a particular solution of

$$x'' + x = e^{-t}. \tag{9}$$

• SOLUTION. By substituting $x = e^{rt}$ into the homogeneous differential equation or by inspection, we see the characteristic equation is $r^2+1=0$, which has roots $r=i$ and $r=-i$. Here $f(t)=e^{-t}$ is an exponential. By (8), $x_p = At^k e^{-t}$, where k is the number of times $r=\alpha=-1$ is a root of the characteristic equation $r^2+1=0$. Since $r=-1$ is not a root, we have $k=0$ and

$$x_p = Ae^{-t}.$$

Substituting this form for x_p into (9) gives

$$(Ae^{-t})'' + Ae^{-t} = 3e^{-t}.$$

Differentiating and simplifying gives $2A=3$, so that $x_p = \frac{3}{2}e^{-t}$ is a particular solution of (9). ♦

Example 2.6.4 *Exponential Forcing ($r=\alpha$ is a root)*

Find a particular solution of

$$x'' - x = 3e^{-t}. \tag{10}$$

• SOLUTION. The characteristic equation is $r^2-1=0$, which has roots $r=1,-1$. Here $f(t)=3e^{-t}$ is an exponential. By (8), $x_p = At^k e^{-t}$, where k is the number of times $\alpha=-1$ is a root of the characteristic equation r^2-1, Since $r=-1$ is a root once, we have $k=1$ and

$$x_p = Ate^{-t}.$$

Substituting this form for x_p into (10) gives

$$(Ate^{-t})'' - Ate^{-t} = 3e^{-t}.$$

Differentiating once,

$$A(e^{-t} - te^{-t})' - Ate^{-t} = 3e^{-t},$$

and a second time gives

$$A(-2e^{-t} + te^{-t}) - Ate^{-t} = 3e^{-t}.$$

The te^{-t} terms cancel. Simplifying gives $-2A = 3$, so that $x_p = -\frac{3}{2}te^{-t}$ is a particular solution of (10). ♦

Again note that both (9) and (10) had the same forcing function $f(t)$, but different forms for x_p because of the different homogeneous equations. This points out the important fact:

> In general, the form for x_p cannot be determined by looking only at the forcing function f. The solution of the homogeneous equation (roots of the characteristic equation) must also be considered.

In the examples to follow, we shall make frequent use of the **principle of superposition** (see (23))

> If $f(t) = f_1(t) + \cdots + f_m(t)$, then there is a particular solution of the form obtained by adding up the forms of the particular solutions for each $f_i(t)$. (11)

Also, in many of the examples we shall merely determine the form for x_p and not actually find the constants in the form.

Case 3: Exponential times Polynomial Forcing

We give examples of the general case of (2).

Example 2.6.5 *Superposition*

Give the form for x_p if

$$x'' - x' = t^3 + t + e^{2t} - 2te^{2t}$$

is to be solved by the method of undetermined coefficients.

• SOLUTION. The characteristic equation is $r^2 - r = r(r-1) = 0$, which has the roots $r = 0, 1$, so that $x_h = c_1 + c_2e^t$. The forcing term is

$$f = \underbrace{(t^3 + t)} + \underbrace{(1 - 2t)e^{2t}}.$$

The first term is a third-degree polynomial. Since $r = 0$ is a root of multiplicity 1 of the characteristic equation (Case 1), x_p must include a term of the form $t^k(A_0 + A_1t +$

$a_2t^2 + A_3t^3)$ with $k = 1$. The second term is of the form $p(t)e^{\alpha t}$, where $p(t) = 1 - 2t$ is a first-degree polynomial and $\alpha = 1$. Since $r = 1$ is not a root of the characteristic equation, by (2) x_p must include a term of the form $x^k(A_4 + A_5t)e^{2t}$ with $k = 0$. Thus x_p has the form

$$x_p = t(A_0 + A_1t + A_2t^2 + A_3t^3) + (A_4 + A_5t)e^{2t}.$$ ♦

Example 2.6.6 *Exponential times Polynomial Forcing ($r = \alpha$ is a root)*

Give the form for x_p if

$$x'' - 2x' + x = 7te^t$$

is to be solved by the method of undetermined coefficients.

• SOLUTION. The characteristic equation is $r^2 - 2r + 1 = (r - 1)^2 = 0$ has a root $r = 1$ of multiplicity 2. Thus $x_h = c_1e^t + c_2te^t$. The forcing term is of the form $p(t)e^{\alpha t}$, where $p(t) = 7t$ is a first-degree polynomial and $r = \alpha = 1$. Since $r = 1$ is a root of multiplicity 2, by (2) with $k = 2$, $m = 1$, the form for x_p is

$$x_p = t^2(A_0 + A_1t)e^t.$$ ♦

The method of undetermined coefficients works for any order linear constant coefficient differential equation as long as the forcing term has the correct types of functions. That is, the forcing terms are all of the form described in this Section.

Example 2.6.7 *Third-Order Example*

Find the form of the general solution of

$$x''' - 6x'' + 12x' - 8x = te^{2t}. \tag{12}$$

• SOLUTION. First we need to solve the associated homogeneous equation

$$x''' - 6x'' + 12x' - 8x = 0.$$

The characteristic equation is $r^3 - 6r^2 + 12r - 8 = (r - 2)^3 = 0$, so that $r = 2$ is a root of multiplicity 3. A fundamental set of solutions for the associated homogeneous equation is $\{e^{2t}, te^{2t}, t^2e^{2t}\}$, so that $x_h = c_1e^{2t} + c_2te^{2t} + c_3t^2e^{2t}$. The forcing term is $f(t) = te^{2t}$. By (2) with $\alpha = 2$, $m = 1$, x_p must include $t^k[A_0 + A_1t]e^{2t}$, where k is the multiplicity of 2 as a root of the characteristic equation. Thus $k = 3$ and x_p is in the form

$$x_p = t^3[A_0 + A_1t]e^{2t} = A_0t^3e^{2t} + A_1t^4e^{2t}.$$

The general solution is $x = x_p + x_h$. ♦

Case 4: Sinusoidal Forcing

If the forcing figure is sinusoidal, $f(t) = E_1 \cos \beta t + E_2 \sin \beta t$, we think of the forcing being a linear combination of the complex exponentials. Then, according to (8) (which is a special case of (2)), the particular solution should involve the same exponentials $e^{\pm i\beta t}$, which are equivalent to a linear combination of $\cos \beta t$ and $\sin \beta t$. Of course, this should be multiplied by a power of t if $r = \pm \beta i$ are roots of the characteristic equation. To be more precise,

If $f(t) = E_1 \cos \beta t + E_2 \sin \beta t$ in (1), then x_p has the **form**

$$x_p = t^k (A \cos \beta t + B \sin \beta t), \quad (13)$$

where k is the multiplicity of $r = \beta i$ as a root of the characteristic equation.

Example 2.6.8 *Sinusoidal Forcing ($r = i\beta$ is not a root)*

Give the form for x_p if

$$x'' + 2x' + 2x = 3e^{-t} + 4 \cos t. \quad (14)$$

is to be solved by the method of undetermined coefficients.

• SOLUTION. The characteristic equation is $r^2 + 2r + 2 = 0$. Its roots are $r = -1 + i$ and $r = -1 - i$. Thus $x_h = c_1 e^{-t} \cos t + c_2 e^{-t} \sin t$. Now we will determine x_p. Note that f is a sum of two terms, $3e^{-t}$ and $4 \cos t$. Since -1 is not a root of the characteristic equation, Case 2 says that x_p includes a term of the form $A_1 e^{-t}$. Now consider $4 \cos t$. Since i is not a root of the characteristic equation, Case 4 with $\beta = 1$ says that x_p includes terms of the form $A_2 \cos t + B_2 \sin t$. Thus

$$x_p = A_1 e^{-t} + A_2 \cos t + B_2 \sin t$$

for some constants A_1, A_2, B_2. ♦

This example emphasizes another common source of errors:

Even though only $\cos \beta t$ appeared in the forcing term f, the form for x_p may require both $t^k A \cos \beta t$ and $t^k B \sin \beta t$.

It can be proved, using properties of linear independence and the Wronskian, that

If a, b, c are constants and f is the type of function described in the method of undetermined coefficients, then the method always works (perhaps messily). In particular, if the equations for the undetermined constants are not consistent (don't have a solution), then an error has been made.

Example 2.6.9 *Sinusoidal Forcing ($r = i\beta$ is a root)*

Give the form for x_p if

$$x'' + 4x = \sin 2t$$

is to be solved by the method of undetermined coefficients.

• SOLUTION. The characteristic equation is $r^2 + 4 = 0$, whose roots are $r = \beta i = \pm 2i$. Thus $x_h = c_1 \cos 2t + c_2 \sin 2t$. The forcing term $\sin 2t$ is $\sin \beta t$, where $\beta = 2$.

Since $r = \beta i$ is a root of the characteristic equation of multiplicity 1, we have $k = 1$, and by Case 4 the form of the particular solution is

$$x_p = t[A\cos 2t + B\sin 2t]. \qquad \blacklozenge$$

Case 5: Polynomial times Sinusoidal Forcing

If the forcing function is a polynomial times a sinusoidal forcing function, the sinusoidal function is treated as a complex exponential as in the previous case. To be more precise,

> If $f(t) = p(t)\cos\beta t + q(t)\sin\beta t$ in (1), where $p(t)$ is an mth-degree polynomial and $q(t)$ an nth-degree polynomial, then x_p has the **form**
>
> $$\begin{aligned} x_p = t^k[&(A_0 + A_1 t + \cdots + A_s t^s)\cos\beta t \\ &+ (B_0 + B_1 t + \cdots + B_s t^s)\sin\beta t], \end{aligned} \tag{15}$$
>
> where k is the multiplicity of $r = \beta i$ as a root of the characteristic equation and s is the larger of m, n.

Example 2.6.10 *Polynomial times Sinusoidal Forcing ($r = i\beta$ is a root)*

Give the form for x_p if

$$x'' + 4x = t^2\cos 2t - t\sin 2t + \sin 2t = t^2\cos 2t + (1 - t)\sin 2t$$

is to be solved by the method of undetermined coefficients.

• SOLUTION. The characteristic equation $r^2 + 4 = 0$ has roots $\pm 2i$ and thus $x_h = c_1\cos 2t + c_2\sin 2t$. The forcing term is of the form

$$p(t)\cos\beta t + q(t)\sin\beta t,$$

where $p(t) = t^2$ is a second-degree polynomial, $q(t) = 1 - t$ is a first-degree polynomial, and $\beta = 2$. Since $r = \beta i = 2i$ is a root of the characteristic equation of multiplicity 1 by Case 5, with $k = 1$, we have

$$x_p = t[(A_1 + A_1 t + A_2 t^2)\cos 2t + (B_0 + B_1 t + B_2 t^2)\sin 2t]. \qquad \blacklozenge$$

Example 2.6.11 *Fourth-Order Example*

Give the form for x_p if

$$x'''' + 8x'' + 16x = t\sin t + t^2\cos 2t \tag{16}$$

is to be solved by the method of undetermined coefficients.

• SOLUTION. First we solve the associated homogeneous equation

$$x'''' + 8x'' + 16x = 0.$$

The characteristic equation is $r^4+8r^2+16=(r^2+4)^2=0$ and has repeated complex roots $r=\pm 2i,\ \pm 2i$. A fundamental set of solutions for the associated homogeneous equation is thus

$$\{\sin 2t, \cos 2t, t\sin 2t, t\cos 2t\}.$$

Now we find the form for x_p. Consider the forcing function

$$f=t\sin t+t^2\cos 2t.$$

By (15), the $t\sin t$ term implies that x_p includes

$$t^{k_1}[(A_0+A_1t)\sin t+(B_1+B_1t)\cos t], \tag{17}$$

with $k_1=0$, since $r=i$ is not a root of the characteristic equation. The $t^2\cos 2t$ term implies that x_p includes

$$t^{k_2}[(A_2+A_3t+A_4t^2)\sin 2t+(B_2+B_3t+B_4t^2)\cos 2t], \tag{18}$$

with $k_2=2$, since $r=2i$ has multiplicity 2 as a root of the characteristic equation. In actually using (17) and (18), one may add (17) and (18) to get the form for x_p, substitute into the original differential equation (16), and solve for $A_o, \ldots, A_4, B_0, \ldots, B_4$. Alternatively, one could use (17) to find a particular solution of

$$x''''+8x''+16x=t\sin t,$$

and then use (18) to find a particular solution of

$$x''''+8x''+16x=t^2\cos 2t.$$

Adding these two particular solutions (17) and (18) gives, by the superposition principle, a solution of $x''''+8x''+16x=t\sin t+t^2\cos 2t$ as desired. ♦

Case 6: Exponential Times Sinusoidal Forcing

If the forcing function is an exponential times a sinusoid, $f(t)=E_1e^{\alpha t}\cos\beta t+E_2e^{\alpha t}\sin\beta t$, we think of the forcing as a linear combination of the complex exponentials $e^{(\alpha\pm\beta i)t}$. Then, according to (8) (which is a special case of (2)), the particular solution should involve the same exponentials $e^{(\alpha\pm\beta i)t}$, which are equivalent to a linear combination of $e^{\alpha t}\cos\beta t$ and $e^{\alpha t}\sin\beta t$. Of course, this should be multiplied by a power of t if $r=\alpha\pm i\beta$ are roots of the characteristic equation. To be more precise,

If $f(t)=E_1e^{\alpha t}\cos\beta t+E_2e^{\alpha t}\sin\beta t$ in (1), then x_p has the form

$$x_p=t^k(Ae^{\alpha t}\cos\beta t+Be^{\alpha t}\sin\beta t), \tag{19}$$

where k is the multiplicity of $r=\alpha+\beta i$ as a root of the characteristic equation.

Example 2.6.12 *Exponential times Sinusoidal Forcing ($r=\alpha+i\beta$ is a root)*

Give the form for x_p if

$$x''+2x'+2x=5e^{-t}\cos t$$

is to be solved by the method of undetermined coefficients.

• SOLUTION. The characteristic equation is $r^2+2r+2=0$, which has roots $r=-1\pm i$, so that

$$x_h=c_1e^{-t}\cos t+c_2e^{-t}\sin t.$$

The forcing term is of the form $e^{\alpha t}\cos\beta t$, where $\alpha=-1$, $\beta=1$. Since $r=-1+i$ is a root of the characteristic equation of multiplicity 1, by Case 6, with $k=1$, the form of the particular solution is

$$x_p=t(Ae^{-t}\cos t+Be^{-t}\sin t).$$ ♦

Example 2.6.13 *Superposition (see (23))*

Give the form for x_p if

$$x''+2x'+2x=e^{-t}\cos 2t+e^{-t}\sin 2t+e^{-t}-3\cos t \tag{20}$$

is to be solved by undetermined coefficients.

• SOLUTION. Equation (20) has the same characteristic equation as Example 2.6.12, so the roots are $r=-1\pm i$. The forcing term in (20) is the sum of three groups of terms:

$e^{-t}\cos 2t+e^{-t}\sin 2t$: since $r=-1+2i$ is not a root, we include $A_0e^{-t}\cos 2t+B_0e^{-t}\sin 2t$, by Case 6.

e^{-t}: since $r=-1$ is not a root, we include A_2e^{-t}, by Case 2.

$-3\cos t$: since $r=i$ is not a root, we include $A_1\cos t+B_1\sin t$, by Case 4.

Thus the form for x_p is

$$x_p=A_0e^{-t}\cos 2t+B_0e^{-t}\sin 2t+A_1\cos t+B_1\sin t+A_2e^{-t}.$$ ♦

Case 7: Polynomial times Exponential times Sinusoid Forcing

If the forcing function is a polynomial times an exponential times a sinusoidal function, then the exponential times the sinusoidal function is treated as a complex exponential as in the previous case. To be more precise,

If $f(t) = p(t)e^{\alpha t} \cos \beta t + q(t)e^{\alpha t} \sin \beta t$ is in (1), where $p(t)$ is an m th-degree polynomial and $q(t)$ is an nth-degree polynomial, then x_p has the **form**

$$x_p = t^k[(A_0 + A_1 t + \cdots + A_s t^s)e^{\alpha t} \cos \beta t + (A_0 + A_1 t + \cdots + A_s t^s)e^{\alpha t} \sin \beta t], \tag{21}$$

where k is the multiplicity of $r = \alpha + \beta i$ as a root of the characteristic equation and s is the larger of m, n.

This case is actually the general case. It includes all the previous cases for appropriate choices of α, β, m, n.

Example 2.6.14 *Polynomial times Exponential times Sinusoidal Forcing ($r = \alpha + i\beta$ is not a root)*

Give the form for x_p if

$$x'' + 3x' + 2x = te^{-t} \cos 2t$$

is to be solved by the method of undetermined coefficients.

• SOLUTION. The characteristic equation is $r^2 + 3r + 2 = (r + 1)(r + 2) = 0$, so the roots are $r = -1$ and $r = -2$. Thus, two independent homogeneous solutions are e^{-t} and e^{-2t}. The forcing term is in the form $p(t)e^{\alpha t} \cos \beta t$, where $p(t) = t$ is a first-degree polynomial and $\alpha + \beta i = -1 + 2i$. Since $-1 + 2i$ is not a root of the characteristic equation, from Case 7 (with $k = 0$), the form of x_p is

$$x_p = (A_0 + A_1 t)e^{-t} \cos 2t + (B_0 + B_1 t)e^{-t} \sin 2t.$$

♦

Example 2.6.15 *Polynomial Times Exponential Times Sinusoidal Forcing ($r = \alpha + i\beta$ is a root)*

Give the form for x_p if

$$x'' + 2x' + 5x = t^3 e^{-t} \sin 2t$$

is to be solved by the method of undetermined coefficients.

• SOLUTION. The characteristic equation is $r^2 + 2r + 5 = 0$, and its roots are $r = -1 \pm 2i$. Thus, the two independent homogeneous solutions are $e^{-t} \cos 2t$ and $e^{-t} \sin 2t$. The forcing term is in the form $p(t)e^{\alpha t} \sin \beta t$, where $p(t) = t^3$ is a third-degree polynomial and $r = \alpha + \beta i = -1 + 2i$. Since $r = -1 + 2i$ is a root of the characteristic equation of multiplicity 1, from Case 7 (with $k = 1$), the form of x_p is

$$x_p = t(A_0 + A_1 t + A_2 t^2 + A_3 t^3)e^{-t} \cos 2t + t(A_0 + A_1 t + A_2 t^2 + A_3 t^3)e^{-t} \sin 2t.$$

♦

When initial conditions are present and the forcing function is nonzero, it is important to remember to apply the initial conditions to the general solution $x_p + c_1x_1 + c_2x_2$ and not just to $x_h = c_1x_1 + c_2x_2$.

Example 2.6.16 *Initial Value Problem*

Solve

$$x'' - x = 3e^{-t}, \quad x(0) = 0, \quad x'(0) = 1. \tag{22}$$

• SOLUTION. First we must find the general solution of $x'' - x = 3e^{-t}$. In Example 2.6 we found that $x_p = -\frac{3}{2}te^{-t}$, and the characteristic equation $r^2 - 1 = 0$ has roots $r = 1, -1$. Thus the general solution is

$$x = -\frac{3}{2}te^{-t} + c_1e^t + c_2e^{-t}.$$

In order to apply the initial conditions, we compute

$$x' = -\frac{3}{2}e^{-t} + \frac{3}{2}te^{-t} + c_1e^t - c_2e^{-t}.$$

The initial conditions then give

$$0 = x(0) = c_1 + c_2,$$
$$1 = x'(0) = -\frac{3}{2} + c_1 - c_2.$$

Solving for c_1, c_2 yields $c_1 = \frac{5}{4}$, $c_2 = -\frac{5}{4}$, and the solution of (22) is

$$x = -\frac{3}{2}te^{-t} + \frac{5}{4}e^t - \frac{5}{4}e^{-t}.$$ ♦

Method of Undetermined Coefficients

In summary, the **method of undetermined coefficients** can be used on $ax'' + bx' + cx = f$ if a, b, c are constants and f is a linear combination of functions of the form

$$t^m e^{\alpha t} \cos \beta t, \quad t^m e^{\alpha t} \sin \beta t,$$

where m is a nonnegative integer and α, β are real numbers. Special cases are

$$t^m, t^m e^{\alpha t}, e^{\alpha t}, e^{\alpha t} \cos \beta t, e^{\alpha t} \sin \beta t, t^m \cos \beta t, t^m \sin \beta t.$$

The method is as follows.

1. First solve the associated homogeneous equation

$$ax'' + bx' + cx = 0.$$

2. Determine x_p as a linear combination of functions with unknown coefficients using the following rules.

TABLE 2.6.1

Case	*f includes summands of form*	x_p *then includes*	*k is the multiplicity of the root*
1.	$p(t)$, mth-degree polynomial	$t^k(A_0 + A_1 t + \cdots + A_m t^m)$	$r = 0$
2.	$Ee^{\alpha t}$	$t^k e^{\alpha t}$	$r = \alpha$
3.	$p(t)e^{\alpha t}$ $p(t)$, mth-degree polynomial	$t^k(A_0 + A_1 t + \cdots + A_m t^m)$	$r = \alpha$
4.	$E_1 \cos \beta t + E_2 \sin \beta t$ (One of E_1, E_2 may be zero)	$t^k(A \cos \beta t + B \sin \beta t)$	$r = \beta i$
5.	$p(t) \cos \beta t + q(t) \sin \beta t$ $p(t)$, mth-degree polynomial $q(t)$, nth-degree polynomial	$t^k[(A_0 + A_1 t + \cdots + A_s t^s) \cos \beta t$ $+(B_0 + B_1 t + \cdots + B_s t^s) \sin \beta t$ $s =$ larger of m, n	$r = \beta i$
6.	$E_1 e^{\alpha t} \cos\beta t + E_2 e^{\alpha t} \sin \beta t$	$t^k(Ae^{\alpha t} \cos \beta t + Be^{\alpha t} \sin \beta t)$	$r = \alpha + \beta i$
7.	$p(t)e^{\alpha t} \cos \beta t + q(t)e^{\alpha t} \sin \beta t$ $p(t)$, mth-degree polynomial $q(t)$, nth-degree polynomial	$t^k[(A_0 + A_1 t + \cdots + A_s t^s)e^{\alpha t} \cos \beta t$ $+(B_0 + B_1 t + \cdots + B_s t^s)e^{\alpha t} \sin \beta t$ $s =$ larger of m, n	$r = \alpha + \beta i$

Rule 1. If f includes a sum of terms of the form $p(t)e^{\alpha t}$, where $p(t)$ is an mth-degree polynomial, then the form for x_p should include

$$t^k[A_0 + A_1 t + \cdots + A_m t^m]e^{\alpha t},$$

where k is the multiplicity of $r = \alpha$ as a root of the characteristic equation $ar^2 + br + c = 0$.

Rule 2. If f includes a sum of terms in the form

$$p(t)e^{\alpha t} \cos \beta t + q(t)e^{\alpha t} \sin \beta t,$$

where $p(t)$ is an mth-degree polynomial and $q(t)$ is an nth-degree polynomial, then the form for x_p should include

$$t^k[A_0 + A_1 t + \cdots + A_s t^s]e^{\alpha t} \cos \beta t$$
$$+ t^k[B_0 + B_1 t + \cdots + B_s t^s]e^{\alpha t} \sin \beta t,$$

where s is the larger of m and n and k is the multiplicity of $r = \alpha + \beta i$ as a root of the characteristic equation $ar^2 + br + c = 0$.

3. Substitute the expression for x_p into the differential equation $ax'' + bx' + cx = f$ to determine the unknown coefficients A_i, B_i.
4. The general solution of $ax'' + bx' + cx = f$ is $x = x_p + x_h$.
5. Apply any conditions to $x_p + x_h$ in order to determine arbitrary constants.

Note that if $f(t)$ includes terms like $\ln t$, $t^{1/3} \sin t$, $\tan t$, then undetermined coefficients will not generally work.

In Table 2.6.1, E, E_1 and E_2 are constants.

The Superposition Principle

The *superposition principle* for linear differential equations says that

> If x_{p1} is a particular solution of $ax'' + bx' + cx = f_1$ and x_{p2} is a particular solution of $ax'' + bx' + cx = f_2$, then $x_{p1} + x_{p2}$ is a particular solution of $ax'' + bx' + cx = f_1 + f_2$. (23)

Intuitively, the superposition principle means that the response (output) that results from the sum (superposition) of two forcing terms (inputs) is the sum of the responses from each forcing term. In applications, knowing whether our device or physical problem acts in this manner is a key factor in deciding whether linear equations can be used to analyze the problem.

Exercises

In Exercises 1–12, state whether the method of undetermined coefficients can be applied to the differential equation. If it cannot, explain why not.

1. $x'' + x = t \sin t$.
2. $x'' + 3x = t^{1/2} \sin t$.
3. $x'' + x = t^2 + t + \ln |t|$.
4. $x'' + x = e^{t+1}$.
5. $x'' + x = \dfrac{\sin t}{\cos t}$.
6. $x'' + x' + x = \cosh t$.
7. $x'' + tx' = 3e^{2t}$.
8. $x'' + x = t \sinh 2t$.
9. $x'' + x = t^{-1} e^t$.
10. $x'' + xx' = e^{2t}$.
11. $x'' + 3x = e^{-2t} \cos 3t + \sinh 3t$.
12. $x'' + 3x' + 4x = \sin^2 t$.

In Exercises 13–52, solve the differential equation using the method of undetermined coefficients. If no initial conditions are given, give the general solution.

13. $x'' + 9x = t^3 + 6$.
14. $x'' - 4x = t^2 + 17t$.
15. $x'' + 8x' = 7t + 11$.
16. $x'' - 7x' = 5t - 3$.
17. $x'' - 7x' + 12x = 5e^{2t}$, $x(0) = 1$, $x'(0) = 0$.
18. $x'' + 7x' + 12x = 2e^{5t}$, $x(0) = 0$, $x'(0) = 2$.
19. $x'' - 3x' + 2x = 2e^t$.
20. $x'' + 4x = 3e^{2t}$.
21. $x'' - 2x' + 5x = 4e^t$.

22. $x'' + 2x' + 5x = 3e^{-t}$.
23. $x'' - 9x = 5e^{-3t}$.
24. $x'' - 3x' + 2x = 2e^{-t}$.
25. $x'' + 2x' + 5x = 3 \sin t$, $x(0) = 1$, $x'(0) = 1$.
26. $x'' + 9x = 5 \cos t$, $x(0) = 0$, $x'(0) = 0$.
27. $x'' + 9x = 4 \sin 3t$.
28. $x'' + x' + x = 3e^{-t}$.
29. $x'' + x = \cos 2t$, $x(0) = 0$, $x'(0) = 2$.
30. $x'' + 4x' + 8x = t^2 + 1$, $x(0) = 0$, $x'(0) = 0$.
31. $x'' - 4x = te^{3t}$.
32. $x'' - 4x' + 3x = te^{2t}$.
33. $x'' - 4x' + 3x = te^{t}$.
34. $x'' - 7x' + 12x = (t + 5)e^{4t}$.
35. $x'' + 16x = 3 \cos 4t$.
36. $x'' + x = \sin t$.
37. $x'' - 2x' + 5x = e^t \cos 3t$.
38. $x'' + 2x' + x = 3e^t$, $x(0) = 0$, $x'(0) = 2$.
39. $x'' + 4x' = 12t^2 + e^t$, $x(0) = 1$, $x'(0) = 1$.
40. $x'' + x' + x = \cos 2t$.
41. $x'' + 2x' + x = 3e^{-t}$.
42. $x'' + 2x' + x = 3te^{-t} + 2e^{-t}$.
43. $x'' - 4x' + 4x = te^t - e^t + 2e^{3t}$.
44. $x'' - 5x' + 4x = 17 \sin t + 3e^{2t}$.
45. $x'' + 5x' + 4x = 8t^2 + 3 + 2 \cos 2t$.
46. $x' + x = 2e^{-t}$.
47. $x' + 3x = t^2 + 1$.
48. $x' - x = \sin t$.
49. $x'' + 4x = \sin 2t$.
50. $2x' + 4x = t$.
51. $3x' - 2x = te^t$.
52. $x' - 3x = e^t \sin t$.

In Exercises 53–96, give the form for x_p if the method of undetermined coefficients were to be used. You need not actually compute x_p.

53. $x'' - 5x' + 6x = t^5 + 7t^3 + 4t$.
54. $x'' + 5x' + 6x = t^4 + 3t^2 + 7$.
55. $x'' + 9x' = t^3$.
56. $x'' - 9x' = 3t^3 + 2t^2 + t + 11$.
57. $x'' + 5x' + 6x = 5e^{4t}$.
58. $x'' - 5x' + 6x = 6e^{5t}$.
59. $x'' - 4x = 5e^{2t}$.

60. $x'' + 7x' + 12x = 5e^{-3t}$.
61. $x'' + 6x' + 9x = e^{-3t}$.
62. $x'' - 8x' + 16x = e^{4t}$.
63. $x'' + 2x' + x = t^3 e^{-t}$.
64. $x'' - 2x' + x = t^2 e^t$.
65. $x'' - 7x' + 12x = t^5 e^{4t}$.
66. $x'' - 16x' = (t^3 + 3t^2 + 7)e^{-4t}$.
67. $x'' - 6x' + 9x = t^4 e^{3t}$.
68. $x'' + 2x' - 3x = t^3 e^t - e^t + e^{-2t} + e^{-3t}$.
69. $x'' + 3x' - 10x = t^2 e^{2t} + e^{5t}$.
70. $x'' - 6x' + 5x = e^{5t} + t^2 e^{-5t}$.
71. $x'' + 5x' = \cos 5t$.
72. $x'' - 25x = \sin 5t$.
73. $x'' - 7x' + 12x = t^2 \sin 4t$.
74. $x'' + 3x' - 10x = t^2 \cos t$.
75. $x'' + 25x = \cos 5t$.
76. $x'' + 9x = t^2 \sin 3t + \cos 2t$.
77. $x'' - 2x' + 5x = 3e^t \sin 2t$.
78. $x'' + 2x' + 5x = 5e^{-t} \cos 3t$.
79. $x'' + 9x = te^{-t} \sin 2t$.
80. $x'' - x = t^2 e^t \cos t$.
81. $x'' + 2x' + 2x = t^2 e^t \sin t$.
82. $x'' + 2x' + 2x = 3t^2 e^{-t} \cos 2t + te^{-t} \sin 2t$.
83. $x'' + 2x' + 2x = e^{-t} \cos t + e^t \sin t$.
84. $x'' + 2x' + 2x = e^{-t} \sin t + \cos t$.
85. $x'' - x = e^{-t} - e^t + e^t \cos t$.
86. $x'' + x = e^{-t} - e^t + e^t \cos t$.
87. $x'' + 16x = t \cos 4t + e^{-t} \sin 4t + 3e^{-4t}$.
88. $x'' + 2x' + 2x = t^2 e^{-t} \cos t$.
89. $x'' - 2x' + 2x = t^3 e^{-t} \sin t + e^t \cos t$.
90. $x'' - 2x' + x = t^2 e^t \sin 3t$.
91. $x'' + 2x' + x = te^t \sin 3t + e^t \cos 3t$.
92. $x'' + 2x' + 2x = te^{-t} \sin t$.
93. $x'' + 4x' + 8x = t^2 e^{-2t} \sin 2t + te^{-2t} \cos 2t$.
94. $x'' + 4x' + 8x = te^{-2t} \cos t$.
95. $x'' + 4x' + 13x = t^3 e^{-2t} \sin 3t$.
96. $x'' + 4x' + 13x = te^t \sin 3t$.
97. Consider the differential equation $x'' = t^3 + 7t - 2$.
   a) Find the form of a particular solution using the method of undetermined coefficients.
   b) Find the particular solution by integration.

98. In this exercise, we will show that the method of undetermined coefficients is valid for forcing functions that are an exponential times a polynomial if the method has been proved for polynomials. We assume that if $p(t)$ is an mth-degree polynomial, then there is a particular solution in the form

$$x_p = q(t),$$

where $q(t)$ is a polynomial of degree m if $r = 0$ is not a root of the characteristic equation, $q(t)$ is degree $m+1$ if $r = 0$ is a root one time, and $q(t)$ has degree $m+2$ if $r = 0$ is a double root. Suppose

$$ax'' + bx' + cx = e^{\alpha t} p(t), \tag{24}$$

where $p(t)$ is an mth-degree polynomial.

a) Use the change of variables, $x = e^{\alpha t} v$, to show that

$$av'' + (2a\alpha + b)v' + (a\alpha^2 + b\alpha a + c)v = p(t).$$

b) Show that there is a particular solution for $v(t)$ that is a polynomial whose degree depends on the number of times $r = \alpha$ is a root of the characteristic equation for (24).

As noted in Examples 2.6.7 and 2.6.11, the method of undetermined coefficients can be applied to equations of order different from two with no changes. In Exercises 99–108, solve the differential equation by the method of undetermined coefficients. If no initial conditions are given, give the general solution.

99. $x''' - 3x'' + 3x' - x = e^{2t}$.

100. $x''' - 3x'' + 3x' - x = e^{t}$.

101. $x''' - x' = \sin t$.

102. $x'''' - 25x'' + 144x = t^2 - 1$.

103. $x''' + x' = 3 + 2\cos t$, $x(0) = x'(0) = x''(0) = 0$.

104. $x'''' - x = 2e^{3t} - e^{t}$.

105. $x'''' - 16x = 5te^{t}$.

106. $x'''' + 4x'' + 4x = \cos 2t$.

107. $x'''' - 5x'' + 4x = e^{2t} - e^{3t}$.

108. $x'''' + x''' - 6x'' = 72t + 24$, $x(0) = 0$, $x'(0) = 0$, $x''(0) = -6$, $x'''(0) = -57$.

In Exercises 109–116, give the form for x_p that you would use to find a particular solution by the method of undetermined coefficients. Do not actually solve for x_p.

109. $x''' - 3x'' + 3x' - x = t^2e^{t} - 3e^{t}$.

110. $x'''' + 2x'' + x = t\sin t$.

111. $x'''' - 4x''' + 6x'' - 4x' + x = t^3e^{t} + t^2e^{-t}$.

112. $x'''' + 5x'' + 4x = \sin t + \cos 2t + \sin 3t$.

113. $x''' + 2x'' + 2x' = 3e^{-t}\cos t$.

114. $x''' + 2x'' + 2x' = t^2e^{-t}\cos t - te^{-t}\sin t$.

115. $x'''' + 4x''' + 8x'' + 8x' + 4x = 7e^{-t}\cos t$.

116. $x'''' - 2x'' + x = te^{t} + t^2e^{-t} + e^{2t}$.

2.7 Mechanical Vibrations II: Forced Response

In Section 2.5 we carefully examined the free response of a spring-mass system. In this section we investigate what happens when an additional external force $f(t)$, which depends only on the time t, is applied to the mass. The equation of motion from Section 2.5 is then

$$m\frac{d^2x}{dt^2}+\delta\frac{dx}{dt}+kx=f(t). \tag{1}$$

The external force $f(t)$ is often referred to as the **forcing function** or the **input**. The solution x is then called the **response** or the **output** of the system. From our theory developed earlier, we know that the solution x of (1) has the form $x=x_p+x_h$, where x_p is a particular solution of (1) and $x_h=c_1x_1+c_2x_2$ is the general solution of the associated homogeneous equation

$$mx''+\delta x'+kx=0. \tag{2}$$

x_h is also called the free response. Note that the arbitrary constants appear only in x_h. We have then:

> The possible responses to the input $f(t)$ consist of a particular response to $f(t)$ added to the possible free responses of the system.

Our discussion will be restricted to sinusoidal forcing,

$$f(t)=F\cos\omega t,$$

where F is the **amplitude of the forcing** and ω is the **forcing frequency**. This is an important case to consider for two reasons. First, many forcing functions, such as alternating currents in circuits, are approximately in this form. Also, the theory of Fourier series, discussed in more advanced courses, may be used to express many forcing terms as a sum (series) of terms of the form $\sin\alpha nt$, $\cos\alpha nt$, α a constant.

The study of forced responses will be broken into two cases, depending on whether or not friction is present.

2.7.1 Friction is Absent ($\delta=0$)

Suppose that friction is negligible ($\delta=0$), so that the model for the spring-mass system with periodic forcing is

$$mx''+kx=F\cos\omega t. \tag{3}$$

The solution of (3) has the form

$$x=x_p+c_1\cos\left(\sqrt{\frac{k}{m}}t\right)+c_2\sin\left(\sqrt{\frac{k}{m}}t\right).$$

Here x_p is a particular solution of (3), and $\cos\omega_0t$, $\sin\omega_0t$ are a fundamental set of homogeneous solutions of $mx''+kx=0$ with $\omega_0=\sqrt{k/m}$, since $r=\pm i\sqrt{k/m}$ are roots of the characteristic equation $mr^2+k=0$.

We shall now find a particular solution for (3) by the method of undetermined coefficients (Section 2.6). The characteristic equation $mr^2 + k = 0$ has roots $r = \pm i\sqrt{k/m} = \pm i\omega_0$. For a spring-mass system (without friction) the natural circular frequency is $\omega_0 = \sqrt{k/m}$. There are two cases, depending on whether the forcing function is a solution of the associated homogeneous equation or not. This depends on whether the forcing circular frequency ω equals the natural circular frequency $\omega_0 = \sqrt{k/m}$.

Forcing Frequency Unequal to the Natural Frequency of Free Response ($\omega \neq \sqrt{k/m}$)

(Frequency of input unequal to natural frequency of free response) In this case, the form for x_p given by the method of undetermined coefficients is

$$x_p = A \cos \omega t + B \sin \omega t. \tag{4}$$

Substituting this form (4) into the differential equation (3) gives

$$m(A \cos \omega t + B \sin \omega t)'' + k(A \cos \omega t + B \sin \omega t) = F \cos \omega t$$

or

$$-mA\omega^2 \cos \omega t - mB\omega^2 \sin \omega t + kA \cos \omega t + kB \sin \omega t = F \cos \omega t.$$

Equating coefficients of like terms in order to find A, B gives

$$\begin{aligned} \cos \omega t: &\quad -mA\omega^2 + kA = F \Rightarrow A = \frac{F}{k - m\omega^2}, \\ \sin \omega t: &\quad -mB\omega^2 + kB = 0 \Rightarrow B = 0, \end{aligned}$$

so that

$$x_p = \frac{F}{k - m\omega^2} \cos \omega t. \tag{5}$$

The amplitude of this particular response is $\frac{F}{k - m\omega^2}$, where F is the amplitude of the input. It is usual to introduce the **frequency-response diagram**, in which the response (the ratio of the response amplitude to the input amplitude) is graphed (figure 2.7.1) as a function of the forcing frequency ω:

$$\frac{\text{Amplitude of the response}}{\text{Amplitude of the input}} = \frac{1}{k - m\omega^2} = \frac{1}{m(\omega_0^2 - \omega^2)}. \tag{6}$$

The response increases as the forcing frequency approaches the natural frequency. Note that the response diagram has a vertical asymptote when the forcing frequency equals the natural frequency. According to (5) or (6), the amplitude of the response is infinite (undefined) when $\omega = \omega_0 = \sqrt{\frac{k}{m}}$. However, in our analysis we have assumed $\omega \neq \omega_0 = \sqrt{\frac{k}{m}}$. This will be explained in the next subsection. Note from figure 2.7.1 that the amplitude is negative if $\omega > \omega_o$. In this case we say the response is 180° out of phase from the input.

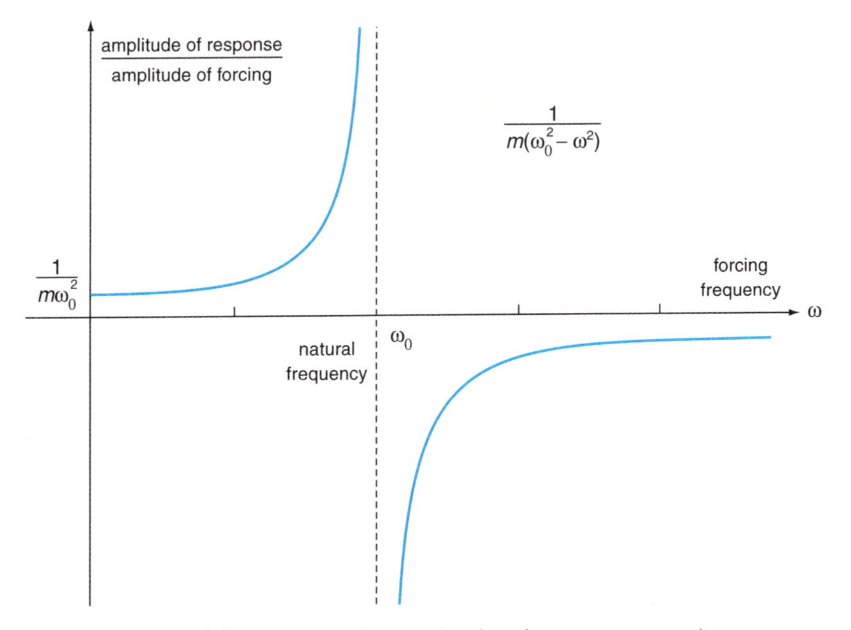

Figure 2.7.1 Response diagram (no damping-note resonance).

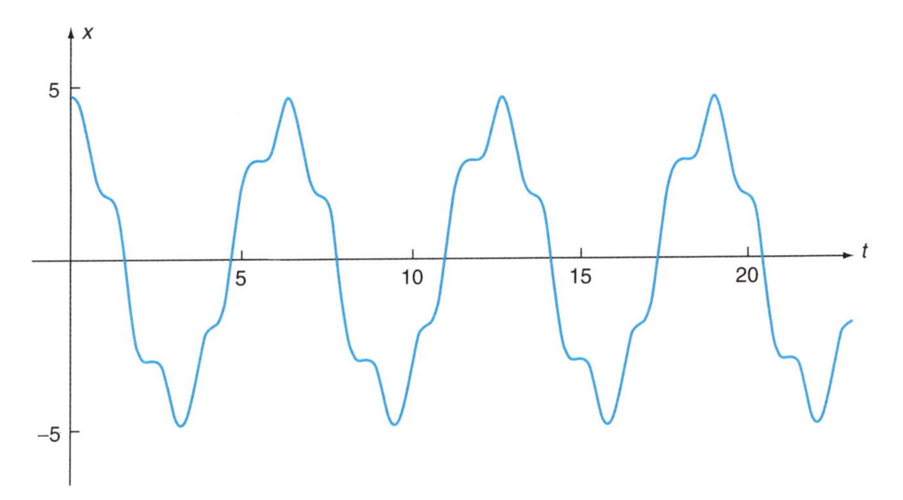

Figure 2.7.2 Graph of (7) with $k = 0.25$, $m = 0.01$, $c_1 = c_2 = 0.5$, $\omega = 1$.

The general solution of the differential equation (3) is

$$x = \frac{F}{k - m\omega^2} \cos \omega t + c_1 \cos\left(\sqrt{\frac{k}{m}} t\right) + c_2 \sin\left(\sqrt{\frac{k}{m}} t\right). \tag{7}$$

This is the superposition of two harmonic motions. One is the free response of the system and the other is a periodic forced response with the same period as the forcing function. Figure 2.7.2 gives the graph of (7) for one choice of parameters $F, m, k, \omega, c_1, c_2$.

Resonance: Forcing Frequency Equals the Natural Frequency of Free Response ($\omega = \sqrt{k/m}$)

If $\omega = \sqrt{k/m}$, then the method of undetermined coefficients says solutions must be multiplied by t, so the form of the particular solution of (3) as

$$x_p = At\cos\omega t + Bt\sin\omega t. \tag{8}$$

Substituting this form into the differential equation (3) gives

$$m(At\cos\omega t + Bt\sin\omega t)'' + k(At\cos\omega t + Bt\sin\omega t) = F\cos\omega t.$$

Using the product rule yields

$$[t(A\cos\omega t + B\sin\omega t)]' = t(-A\omega\sin\omega t + B\omega\cos\omega t) + A\cos\omega t + B\sin\omega t,$$

so that using the product rule a second time yields

$$m(-2A\omega\sin\omega t - At\omega^2\cos\omega t + 2B\omega\cos\omega t - Bt\omega^2\sin\omega t) \\ + kAt\cos\omega t + kBt\sin\omega t = F\cos\omega t.$$

Equating coefficients of like terms gives four equations in A, B:

$$\begin{aligned} \sin\omega t: &\quad -2Am\omega = 0, \\ \cos\omega t: &\quad 2Bm\omega = F, \\ t\sin\omega t: &\quad -Bm\omega^2 + kB = 0, \\ t\cos\omega t: &\quad -Am\omega^2 + kA = 0. \end{aligned}$$

Since $\omega^2 = \frac{k}{m}$ by assumption, the last $t\cos\omega t$ and $t\sin\omega t$ terms cancel. We find $A = 0$, $B = F/(2m\omega)$, from the first two equations. Thus,

$$x = \frac{F}{2m\omega}t\sin\omega t + c_1\cos\omega t + c_2\sin\omega t \tag{9}$$

is the general solution. The particular solution

$$x_p = \frac{F}{2m\omega}t\sin\omega t \tag{10}$$

is of special interest since it illustrates the phenomenon of **resonance**. The amplitude $\frac{F}{2m\omega}t$ of the oscillation grows proportionally with time. The graph of (10) is given in figure 2.7.3.

The forced response is now an unbounded oscillation. Intuitively, the phenomena of resonance may be summarized as follows:

> If the period (frequency) of the forcing function is the same as the period (frequency) of the free response of the system, then large-amplitude oscillations may result from forcing terms of small amplitude.

In structures such as bridges, resonance is to be avoided. The 1940 collapse of the Tacoma Narrows Bridge was particularly catastrophic, and a highly recommended

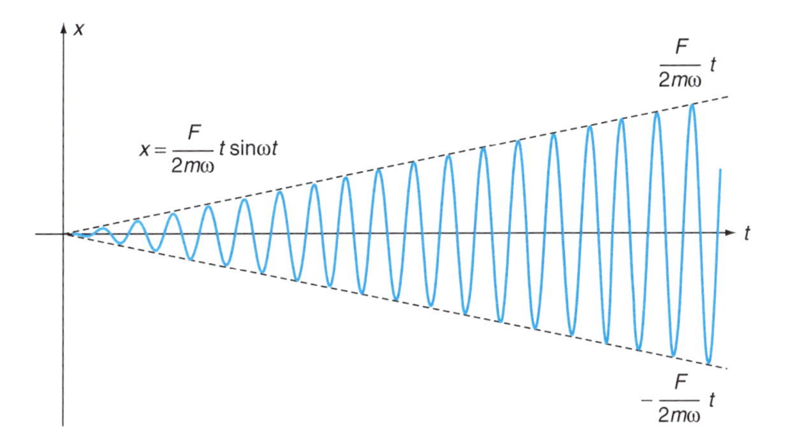

Figure 2.7.3 Resonance (forcing frequency = natural frequency).

dramatic film shows large-amplitude oscillations before its failure. Aircraft wings have also failed on occasion due to resonance. On the other hand, resonance is sometimes desirable in the manipulation of sound waves in musical instruments and detectors of various kinds. Of course, in practice the oscillations do not become arbitrarily large. Rather, either the system is changed (the spring breaks, for example) or the original linear model $mx'' + \delta x' + kx = f$ is no longer a valid model.

Also, "pure resonance" is never observed. In every system there is some friction, so δ, while perhaps very small, is greater than zero. Also, it is almost impossible to make ω *exactly* $\sqrt{\frac{k}{m}}$. However, if δ is close to zero and ω close to $\sqrt{\frac{k}{m}}$, then the forced response may exhibit a large response. The next example illustrates this.

Example 2.7.1 *Resonance and Near Resonance*

Suppose that the model for a spring-mass system with forcing term $\cos \omega t$ is

$$x'' + x = \cos \omega t, \quad x(0) = 0, \quad x'(0) = 0. \tag{11}$$

Determine the solution for all ω, and graph the solution for $\omega = 1$ and for several values of ω close to 1.

• SOLUTION. Using the method of undetermined coefficients for (7) and (9), we find that the general solution of $x'' + x = \cos \omega t$ is

$$w \neq 1: \quad x = \frac{1}{1-\omega^2} \cos \omega t + c_1 \cos t + c_2 \sin t, \tag{12}$$

$$w = 1: \quad x = \frac{1}{2} t \sin t + c_1 \cos t + c_2 \sin t. \tag{13}$$

Applying the initial condition to (12) gives

$$0 = x(0) = \frac{1}{1-\omega^2} + c_1,$$

$$0 = x'(0) = c_2,$$

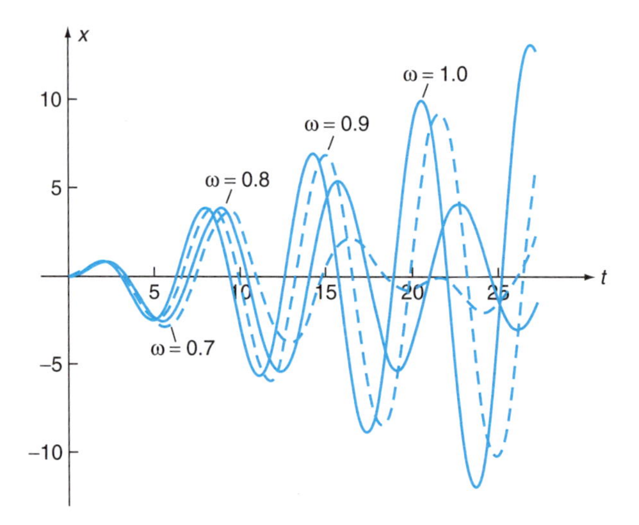

Figure 2.7.4 Graph of (14) for $\omega = 0.7, 0.8, 0.9, 1.0$.

so $c_2 = 0$, $c_1 = -\frac{1}{(1-\omega^2)}$. Thus $x = (1-\omega^2)^{-1}\cos\omega t - (1-\omega^2)^{-1}$. Applying the initial conditions to (13) gives

$$0 = x(0) = c_1, \quad 0 = x'(0) = c_2.$$

Thus the solution of (11) is

$$x(t) = \begin{cases} \dfrac{1}{1-\omega^2}(\cos\omega t - \cos t) & \text{if } \omega \neq 1, \\ \dfrac{1}{2} t \sin t & \text{if } \omega = 1. \end{cases} \tag{14}$$

The graph of (14) for several values of ω is given in figure 2.7.4. Notice that for ω close to 1, the solutions exhibit large-amplitude oscillations.

Example 2.7.2 *Resonance*

For what values of the mass m will $mx'' + 25x = 12\cos\omega t$ exhibit resonance if $12\cos\omega t$ has a frequency of 28 Hz?

• SOLUTION. The characteristic equation $mr^2 + 25 = 0$ has pure imaginary roots $r = \pm i\sqrt{\frac{25}{m}}$. Resonance occurs if

$$\omega = \sqrt{\frac{25}{m}}. \tag{15}$$

If $\cos\omega t$ has a frequency of 28 Hz, then

$$\frac{\omega}{2\pi} = 28 \quad \text{or} \quad \omega = 56\pi. \tag{16}$$

Combining (15) and (16) gives

$$56\pi = \sqrt{\frac{25}{m}}$$

or $m = \left(\frac{25}{56\pi}\right)^2$. ♦

Example 2.7.3 *Road-induced Resonance*

A car is moving along a roadway at a constant horizontal velocity v. Suppose that the vehicle can be modeled by a mass with a vertical spring between the mass and the road. Also suppose that as the car moves along the road, the only forces acting on the spring-mass system are the weight of the car and the changing length of the spring caused by the rising and falling of the road surface. Assume that the road has a surface that is approximately sinusoidal with wavelength L and that the spring-mass system has negligible friction. At what constant horizontal velocity v would resonance be observed?

• SOLUTION. Let x measure distance horizontally. Thus the varying of the road surface is given by

$$y_r(x) = a \sin \frac{2\pi x}{L}.$$

Let $y(x)$ be the vertical position of the mass at position x. To set up the coordinates, assume that when $x = 0$, the base of the spring is at the average road height, which is $y_r = 0$. Let $y = 0$ correspond to the rest position of the mass there. Thus the spring force and the vehicle weight are balanced out. We measure y in the positive direction. The rest configuration is shown in figure 2.7.5. Let m be the mass. Since we have measured from the rest position, the gravity term cancels the compression force at rest, and Hooke's law takes the form

$$m\frac{d^2y}{dt^2} = -k(\text{stretching of spring from rest length}). \tag{17}$$

The amount that the spring is stretched from its rest length at position x is $y(x) - y_r(x)$, (see figure 2.7.6). Thus

$$m\frac{d^2y}{dt^2} = -k(y - y_r) = -k\left(y - a \sin \frac{2\pi x}{L}\right). \tag{18}$$

The differential equation is usually written in the equivalent form

$$m\frac{d^2y}{dt^2} + ky = ka \sin \frac{2\pi x}{L}.$$

However, the position x is not a constant, but a function of time. If we assume that the car is moving at a constant velocity v, then (assuming $x = 0$ is defined to be where the car is at $t = 0$)

$$x = vt.$$

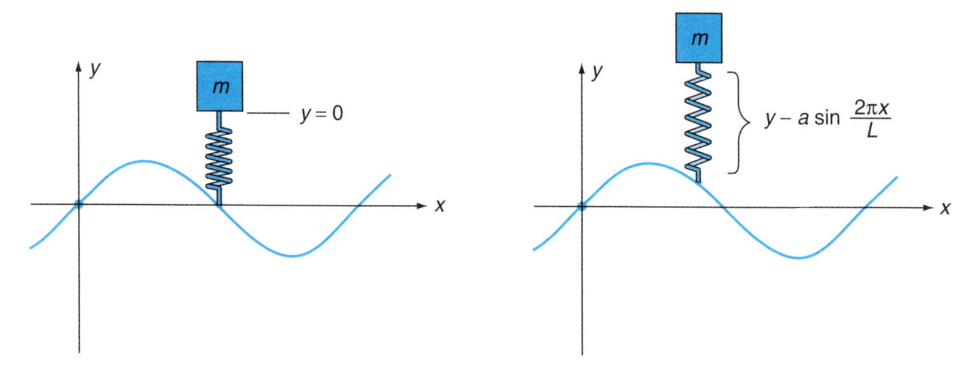

Figure 2.7.5 Rest configuration.

Figure 2.7.6 Motion.

The differential equation of the vertical motion of the car (spring-mass system) is

$$m\frac{d^2y}{dt^2} + ky = ka \sin \frac{2\pi vt}{L}.$$

The spring-mass system is forced at frequency v/L. The system resonates if the forcing frequency equals the natural frequency, that is,

$$\frac{2\pi v}{L} = \sqrt{\frac{k}{m}},$$

Thus, the velocity at which resonance occurs is $v = \frac{L}{2\pi}\sqrt{\frac{k}{m}}$. ♦

Exercises

1. For what values of m will $mx'' + 4x = 13 \cos \omega t$ exhibit resonance if $13 \cos \omega t$ has a frequency of 20 Hz?
2. For what values of m will $mx'' + 16x = 12 \cos \omega t$ exhibit resonance if $12 \cos \omega t$ has a frequency of 28 Hz?
3. For what values of k will $36x'' + kx = 4 \cos \omega t$ exhibit resonance if $4 \cos \omega t$ has a frequency of 22 Hz?
4. For what values of k will $25x'' + kx = 2 \cos \omega t$ exhibit resonance if $2 \cos \omega t$ has a frequency of 18 Hz?
5. You know that the forcing function $f(t) = F \cos \omega t$ has a frequency between 10 and 70 Hz. The spring-mass system is $mx'' + 10x = f(t)$. What values of m might lead to resonance?
6. You know that the forcing function $f(t) = F \cos \omega t$ has a frequency between 30 and 50 Hz. The spring-mass system is $mx'' + 9x = f(t)$. What values of m might lead to resonance?
7. If your spring-mass system had $m = 15$, $\delta = 0$, and $k = 8$, what forcing frequency would cause resonance?
8. If your spring-mass system had $m = 5$, $\delta = 0$, and $k = 12$, what forcing frequency would cause resonance?

9. You are building a detector that is to be sensitive (exhibit resonance) to harmonic vibrations at a frequency of 30 Hz. The detector will be a spring-mass system with spring constant $k = 15$ g/s^2. Assuming friction is negligible, what should the attached mass be?
10. A device is being built that can be modeled as a simple spring-mass system with negligible friction. The attached mass is 20 g. The spring constant is adjustable. What values of the spring constant are to be avoided if the device is subject to external harmonic forces in the *range* of 10 to 50 Hz, and your design goal is to avoid resonance?
11. (**Beats**) Using the trigonometric identities $\cos(\theta \pm \phi) = \cos\theta\cos\phi \mp \sin\phi\sin\theta$, verify that

$$\cos\omega t - \cos\beta t = 2\sin\left(\frac{(\beta-\omega)}{2}t\right)\sin\left(\frac{(\beta+\omega)}{2}t\right). \tag{19}$$

12. A spring-mass system has an attached mass of 4 g, a spring constant of 16 g/s^2 and a negligible friction. It is subject to a force of $4\cos(2.2t)$ downward, and is initially $t = 0$ at rest. Determine the subsequent motion. Using (19) from Exercise 11, rewrite the solution as the product of two sine functions, and graph the result. The resulting function has a periodic variation in amplitude, or a **beat**.
13. A rectangular closed box is floating in the coean with its top always parallel to the water's surface. The buoyancy force is proportional to the volume of water displaced, which is also proportional to the depth of the bottom of the box. Let m be the mass of the box, and let d be the depth of the bottom of the box when the box is at rest. Let $x(t)$ measure how much deeper than this rest depth the bottom of the box is at time t. Ignore resistance and assume small displacements and velocities. Explain why $mx'' + kx = 0$ is a reasonable model for the vertical movement of the box.
14. Suppose it is known that a 1000-lb weight sags a 3000-lb car by 1 in. What would be a poor wavelength for a road to have if traffic usually moves at 55 mi/h?
15. Consider a mass m attached to a spring with spring constant k without friction ($\delta = 0$). The forcing function is $F\cos\omega t$, where $\omega \neq \sqrt{\frac{k}{m}}$. Determine the solution of the initial value problem if $x(0) = x_0$ and $x'(0) = 0$.
16. Consider a mass m attached to a spring with spring constant k without friction ($\delta = 0$). The forcing function is $F\cos\omega t$, where $\omega \neq \sqrt{\frac{k}{m}}$. Determine the solution of the initial value problem if $x(0) = 0$ and $x'(0) = v_0$.
17. Consider a mass m attached to a spring with spring constant k without friction ($\delta = 0$). The forcing function is $F\sin\omega t$, where $\omega \neq \sqrt{\frac{k}{m}}$. Determine the solution of the initial value problem if $x(0) = x_0$ and $x'(0) = 0$.
18. Consider a mass m attached to a spring with spring constant k without friction ($\delta = 0$). The forcing function is $F\sin\omega t$, where $\omega \neq \sqrt{\frac{k}{m}}$. Determine the solution of the initial value problem if $x(0) = 0$ and $x'(0) = v_0$.
19. Consider a mass m attached to a spring with spring constant k without friction ($\delta = 0$). The forcing function is $F\cos\omega t$, where $\omega = \sqrt{\frac{k}{m}}$. Determine the solution of the initial value problem if $x(0) = 0$ and $x'(0) = 0$.

20. Consider a mass m attached to a spring with spring constant k without friction ($\delta = 0$). The forcing function is $F \sin \omega t$, where $\omega = \sqrt{k/m}$. Determine the solution of the initial value problem if $x(0) = 0$ and $x'(0) = 0$.

In Exercises 21–28, find the general solution.

21. $\dfrac{d^2x}{dt^2} + 4x = 8 \cos 5t.$

22. $\dfrac{d^2x}{dt^2} + 9x = -5 \sin 5t.$

23. $\dfrac{d^2x}{dt^2} - 4x = 8 \cos 5t.$

24. $\dfrac{d^2x}{dt^2} - 9x = -5 \sin 5t.$

25. $\dfrac{d^2x}{dt^2} + 4x = 3 \cos 2t.$

26. $\dfrac{d^2x}{dt^2} + 9x = -4 \sin 3t.$

27. $m\dfrac{d^2x}{dt^2} + kx = F \cos \omega t$ with $\omega = \sqrt{\dfrac{k}{m}}.$

28. $m\dfrac{d^2x}{dt^2} + kx = F \sin \omega t$ with $\omega = \sqrt{\dfrac{k}{m}}.$

2.7.2 Friction is Present ($\delta > 0$) (Damped Forced Oscillations)

If friction is present, then $mx'' + \delta x' + kx = f(t)$ takes the form

$$x = x_p + x_h = x_p + c_1 x_1 + c_2 x_2.$$

As noted earlier, there are three possibilities for the fundamental set of homogeneous solutions $\{x_1, x_2\}$:

$$\begin{aligned} \{e^{-\alpha t} \cos \beta t, e^{-\alpha t} \sin \beta t\}: &\quad \text{underdamped,} \\ \{e^{-\alpha t}, te^{-\alpha t}\}: &\quad \text{critically damped,} \\ \{e^{-\alpha_1 t}, e^{-\alpha_2 t}\}: &\quad \text{overdamped.} \end{aligned}$$

The initial conditions determine the arbitrary constants c_1, c_2, but $\lim_{t\to\infty} x_1 = 0$, $\lim_{t\to\infty} x_2 = 0$ in all three cases since they are all damped. Thus

> If friction is present ($\delta > 0$), the free response x_h is always **transient**. That is, $\lim_{t\to\infty} x_h(t) = 0$.

Another way to say the same thing is that, if friction is present, the effect of the initial conditions dies out (is transient). Thus for many important types of forcing functions, such as periodic, the response is eventually determined almost completely by the forcing function $f(t)$. In many situations, particularly the circuits discussed in the next section, this is a desirable phenomenon.

We consider the differential equation that models a spring-mass system with damping with a periodic forcing function with forcing circular frequency ω of the form $F\cos\omega t$:

$$m\frac{d^2x}{dt^2}+\delta\frac{dx}{dt}+kx=F\cos\omega t. \tag{20}$$

The forcing function $F\cos\omega t$ is not a solution of the associated homogeneous equation $mx''+\delta x'+kx=0$, since $\delta>0$. Thus, according to the method of undetermined coefficients, a particular solution is in the form of a linear combination of sines and cosines of the same frequency as the forcing:

$$x_p=A\cos\omega t+B\sin\omega t. \tag{21}$$

Substituting (21) into (20) yields

$$\begin{aligned}m\frac{d^2}{dt^2}(A\cos\omega t+B\sin\omega t)+\delta\frac{d}{dt}(A\cos\omega t+B\sin\omega t)\\+k(A\cos\omega t+B\sin\omega t)=F\cos\omega t,\end{aligned}$$

or

$$\begin{aligned}-m\omega^2(A\cos\omega t+B\sin\omega t)+\omega\delta(-A\sin\omega t+B\cos\omega t)\\k(A\cos\omega t+B\sin\omega t)=F\cos\omega t.\end{aligned}$$

Equating coefficients of like terms in order to find A and B gives

$$\cos\omega t:(k-m\omega^2)A+\delta\omega B=F, \tag{22}$$

$$\sin\omega t:-\delta\omega A+(k-m\omega^2)B=0. \tag{23}$$

The system (22)–(23) can be solved by elimination. For example, multiplying the first equation by $\delta\omega$, the second by $k-m\omega^2$, and adding yields

$$B\left[(k-m\omega^2)^2+\delta^2\omega^2\right]=F\delta\omega.$$

In this way we determine B:

$$B=\frac{F\delta\omega}{(k-m\omega^2)^2+\delta^2\omega^2}. \tag{24}$$

The equation for A is best determined using B and the second equation in (23) to get

$$A=\frac{F(k-m\omega^2)}{(k-m\omega^2)^2+\delta^2\omega^2}. \tag{25}$$

A particular solution of (19) is

$$x_p=A\cos\omega t+B\sin\omega t=R\cos(\omega t-\phi), \tag{26}$$

where A and B are given by (22)–(23) and (24)–(25). The amplitude of this particular response is

$$R = \sqrt{A^2 + B^2} = F\frac{\sqrt{(k - m\omega^2)^2 + \delta^2\omega^2}}{(k - m\omega^2)^2 + \delta^2\omega^2} = \frac{F}{\sqrt{(k - m\omega^2)^2 + \delta^2\omega^2}}. \tag{27}$$

The amplitude-response diagram from damped oscillators follows from (27). The ratio of the amplitude of the response R to the amplitude of the forcing F depends on four parameters: m, k, δ of the spring-mass system and the forcing circular frequency ω. Instead of graphing this relationship as a function of the forcing frequency, it is best to introduce a dimensionless frequency, the ratio of the forcing frequency ω to the natural circular frequency $w_0 = \sqrt{k/m}$. In this way, it follows from (27) that

$$R = \frac{F}{\sqrt{(k - m\omega^2)^2 + \delta^2\omega^2}} = \frac{F/k}{\sqrt{(1 - \omega^2/\omega_0^2)^2 + (\delta^2/km)(\omega^2/\omega_0^2)}}, \tag{28}$$

which is graphed in figure 2.7.7. We note that in this way, the graph can be made to depend on only one parameter:

$$\frac{kR}{F} = \frac{1}{\sqrt{(1 - \omega^2/\omega_0^2)^2 + 4\gamma^2\omega^2/\omega_0^2}}. \tag{29}$$

The parameter γ is the critical damping ratio

$$\gamma = \frac{\delta}{2/\sqrt{km}} = \frac{\delta/2m}{\sqrt{k/m}}. \tag{30}$$

It is also the ratio of the decay rate of underdamped (and critically damped) oscillators to the natural frequency. It can be seen that if $\gamma = 0$, the response curve is the one that we analyzed for forced vibrations without damping. If the forcing frequency equals the natural frequency, $\omega = \omega_0$, then from (29)

$$R = \frac{F}{2\gamma k}.$$

This shows that the amplitude of the response is large when the damping is small (γ small). It can be shown that the amplitude response diagram has a maximum for $\omega \neq 0$ if $\gamma < \sqrt{2}/2$. However, if $\gamma \geq \sqrt{2}/2$, then the maximum of the response diagram occurs at $\omega = 0$.

The phase ϕ is probably less important. Recall that $\tan\phi = B/A = \delta\omega/(k - m\omega^2)$.

Example 2.7.4 *Damped Forced Oscillator*

A spring-mass system, with an attached mass of $m = 1$ g, has a spring constant of 50 g/s^2 and a damping coefficient of 2 g/s. At time $t = 0$, the mass is pushed down $\frac{23}{26}$ cm and released with a velocity of $28\frac{2}{13}$ cm/s downward. A force of $41\cos 2t$ dynes acts downward on the mass for $t \geq 0$. Determine the resulting motion and graph it.

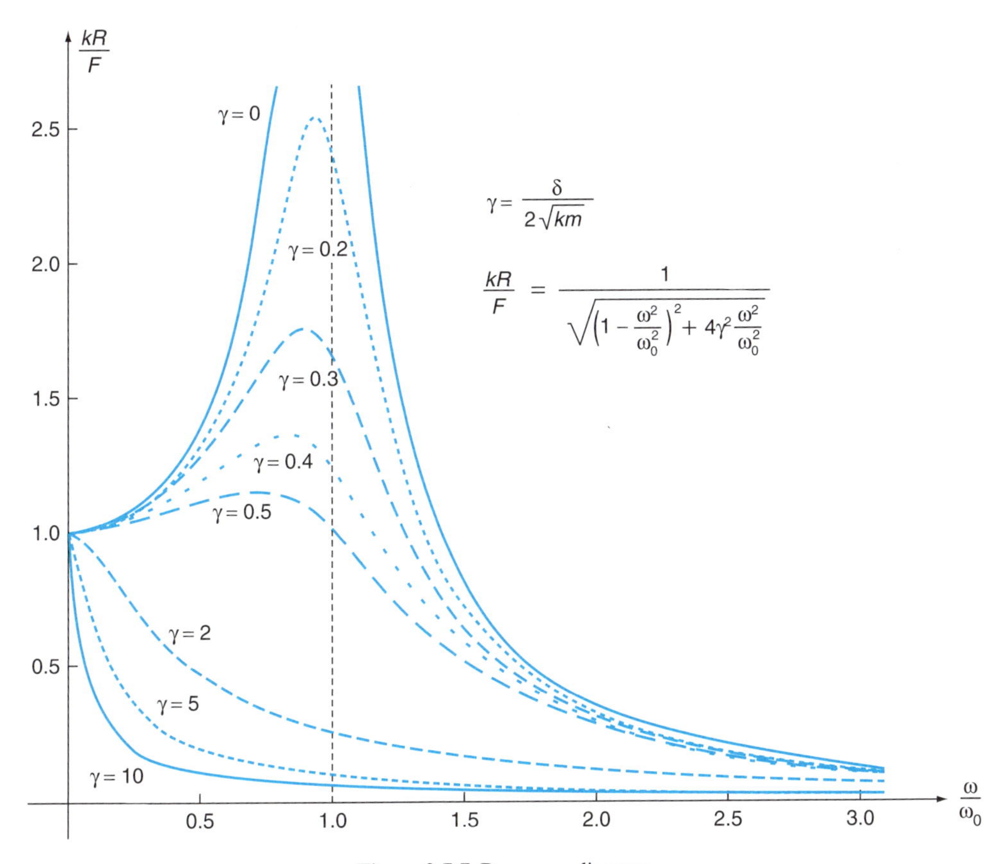

Figure 2.7.7 Response diagram.

• SOLUTION. The differential equation governing the motion is

$$x'' + 2x' + 50x = 41\cos 2t, \quad x(0) = \frac{23}{26}, \quad x'(0) = 28\frac{2}{13}. \tag{31}$$

The characteristic equation $r^2 + 2r + 50 = 0$ has roots $r = -1 \pm 7i$. Thus the solution of the associated homogeneous equation is

$$x_h = c_1 e^{-t}\cos 7t + c_2 e^{-t}\sin 7t.$$

From Section 2.6, the method of undetermined coefficients tells us that there will be a particular solution of the form

$$x_p = A\cos 2t + B\sin 2t.$$

Substituting this form into the differential equation (31) yields

$$(A\cos 2t + B\sin 2t)'' + 2(A\cos 2t + B\sin 2t)' \\ +50(A\cos 2t + B\sin 2t) = 41\cos 2t,$$

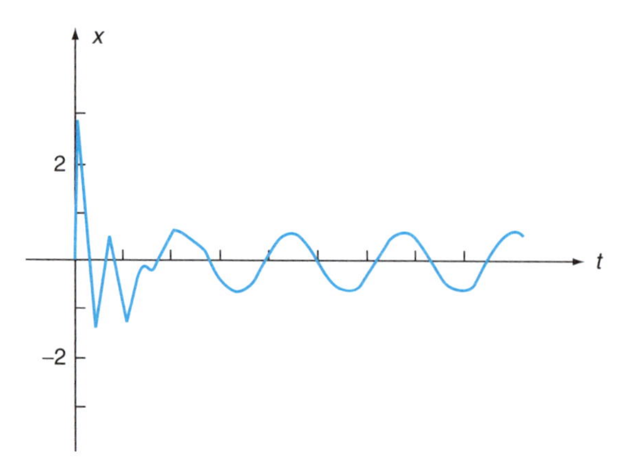

Figure 2.7.8 Graph of (32).

or

$$-4A\cos 2t - 4B\sin 2t - 4A\sin 2t + 4B\cos 2t$$
$$+50A\cos 2t + 50B\sin 2t = 41\cos 2t.$$

Equating coefficients of $\cos 2t$, $\sin 2t$ gives us equations in A, B:

$$\begin{aligned} \cos 2t: &\quad 46A + 4B = 41, \\ \sin 2t: &\quad -4A + 46B = 0, \end{aligned}$$

so that $B = \frac{1}{13}$, $A = \frac{23}{26}$, and

$$x = x_p + x_h = \frac{23}{26}\cos 2t + \frac{1}{13}\sin 2t + c_1e^{-t}\cos 7t + c_2e^{-t}\sin 7t.$$

Applying the initial conditions in (31) in order to determine c_1, c_2, we find

$$\begin{aligned} \frac{23}{26} &= x(0) = \frac{23}{26} + c_1, \\ 28\frac{2}{13} &= x'(0) = \frac{2}{13} - c_1 + 7c_2. \end{aligned}$$

Thus $c_1 = 0$, $c_2 = 4$, and

$$x = \frac{23}{26}\cos 2t + \frac{1}{13}\sin 2t + 4e^{-t}\sin 7t.$$

This solution may be simplified, using the form (26) and (27), to give

$$x = \sqrt{\frac{41}{52}}\cos(2t - \phi) + 4e^{-t}\sin 7t, \tag{32}$$

where $\tan\phi = 2/23$. Note that the long-term or steady response is periodic, with the same period as the input. It is out of phase, however (achieves its maxima at a different time determined by ϕ). As figure 2.7.8 shows, this motion may be viewed as a damped (transient) oscillation superimposed (added) onto the forced (steady-state) harmonic motion. ♦

Exercises

In Exercises 29–40, find the particular solution with no transient terms and put it into amplitude-phase form

29. $\dfrac{d^2x}{dt^2} + 6\dfrac{dx}{dt} + 25x = 3\cos 4t.$
30. $\dfrac{d^2x}{dt^2} + 6\dfrac{dx}{dt} + 25x = -5\sin 3t.$
31. $x'' + 4x' + 13x = 5\cos 2t.$
32. $x'' + 4x' + 13x = -2\sin 2t.$
33. $x'' + 8x' + 41x = 3\sin t.$
34. $x'' + 8x' + 41x = 9\cos t.$
35. $x'' + 3x' + 2x = \sin t.$
36. $x'' + 7x' + 12x = \sin t.$
37. $x'' + 2x' + 2x = \sin t.$
38. $x'' + 2x' + 5x = \sin t.$
39. $x'' + 2x' + x = \sin t.$
40. $x'' + 6x' + 9x = \sin t.$
41. Show that the amplitude response curve,
$$y = \frac{1}{\sqrt{(1-\omega^2/\omega_0^2)^2 + 4\gamma^2\omega^2/\omega_0^2}},$$
graphed in figure 2.7.7, has a maximum for $\omega \neq 0$ if $\gamma < \sqrt{2}/2$. For what value of ω does a maximum occur? Show that this value of ω approaches ω_0 as $\gamma \to 0$. However, if $\gamma \geq \sqrt{2}/2$, show that the maximum of the response diagram occurs at $\omega = 0$. *Hint*: Let $z = \omega^2/\omega_0^2$.
42. Show that γ defined by (30) is the ratio of the decay rate of underdamped oscillations and the natural frequency. Why does γ have no dimensions?
43. Show that γ defined by (30) is the ratio of the damping coefficient to the damping coefficient at critical damping. That is why it is called the critical damping ratio.
44. Let x_δ be the solution of $x'' + \delta x' + x = \sin t$, $x(0) = 0$, $x'(0) = 1$. Find the formula for x_δ when $\delta = 0$ and $0 < \delta < 2$. Show that, for all $t > 0$, $\lim_{\delta \to 0^+} x_\delta(t) = x_0(t)$. This illustrates the point made earlier in this section that if $\omega = \sqrt{\frac{k}{m}}$ and δ is small, there will be a response similar to resonance.
45. Let $x_\delta(t)$ be the function of t found in Exercise 44. Graph $x_\delta(t)$ for $0 \leq t \leq 30$ for $\delta = 0.5, 0.2, 0.1, 0.05$ on the same set of coordinates. (Requires a computer with graphics capability; illustrates the limit of Exercise 44.)
46. Suppose that the top of our vertical spring-mass system is solidly attached to a mechanism that causes the point of attachment also to move in a vertical direction. Let $h(t)$ measure the vertical displacement of the attachment point, with a downward displacement being positive. Explain why a reasonable model for small movements of the mass in the spring-mass system would be $mx'' + \delta x' + kx = kh(t)$, which is in the form (1).

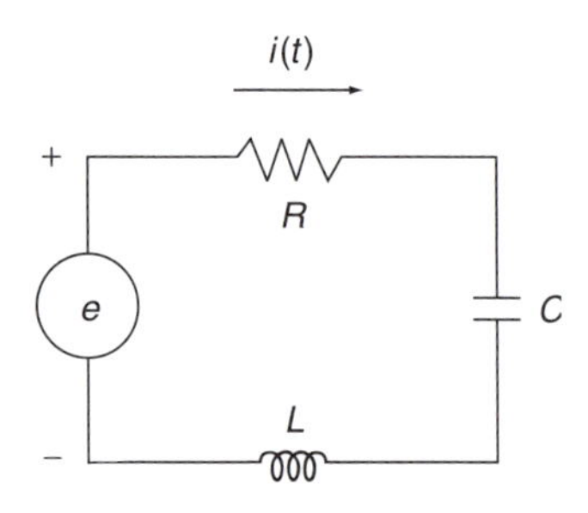

Figure 2.8.1

2.8 Linear Electric Circuits

We have already considered circuits in Section 1.10, and will use the units from that section. In this section we shall consider single-loop linear RLC circuits, as shown in figure 2.8.1. The figure shows a linear resistor of resistance R ohms, capacitor of capacitance C farads, inductor of inductance L henries, and voltage source with voltage $e(t)$ volts. Furthermore, we shall assume that R, L, C are nonnegative, which is often the case. (A negative resistance means that we have a device that puts power into the loop rather than dissipating it, as a positive resistor does.)

Recall from Section 1.10 that the **voltage law** says that

> The algebraic sum of the voltage drops around a loop at any instant is zero.

Since there is a single loop, the **current law** says that the current is the same in the resistor, inductor, and capacitor. Let this current at time t be $i(t)$ amperes. Let $q(t)$ be the charge in coulombs in the capacitor at time t. From Section 1.10 we know that the voltage drops for each device are:

$$\begin{aligned} \text{Resistor:} \quad & V_R = iR, \\ \text{Capacitor:} \quad & V_C = \frac{1}{C}q, \\ \text{Inductor:} \quad & V_I = L\frac{di}{dt}, \end{aligned}$$

where we assume R, L, C are constants. Thus the voltage law may be written as

$$L\frac{di}{dt} + Ri + \frac{1}{C}q = e.$$

However, current is the time rate of change of charge, so that $i = \frac{dq}{dt}$, and we have the second-order linear differential equation in q

$$L\frac{d^2q}{dt^2} + R\frac{dq}{dt} + \frac{1}{C}q = e(t). \tag{1}$$

or, upon differentiation,

$$L\frac{d^2i}{dt^2} + R\frac{di}{dt} + \frac{1}{C}i = e'(t), \tag{2}$$

which is a second-order linear differential equation in i. Note that this differential equation has exactly the same form as the spring-mass system and L, R, C are also nonnegative by assumption. Thus all of the analysis of the preceding section is still valid, but now mass has become inductance ($m = L$), friction is resistance ($\delta = R$), and the spring constant is the reciprocal of capacitance ($k = \frac{1}{C}$). In particular, the discussion of resonance, damping, amplitude, and phase angle is still appropriate.

That many mechanical and nonelectrical problems have the same differential equation (1) as a model is the idea behind the analog computer. To solve the differential equation (1) for a given L, R, C, e with an analog computer, one would build the circuit and then measure the resulting charge (or current) to determine the values of the solution.

Most modern circuits, of course, involve many loops and hence are modeled by systems of differential equations. As circuits have become more and more complex, it has become increasingly expensive and cumbersome to design them by building numerous prototypes. Increasingly, preliminary design work is done by computer simulations, which often involve the solution of systems of differential equations.

Typical Initial Conditions

Mathematically, the appropriate conditions for (2) are given to be i and $\frac{di}{dt}$ at $t = 0$. However, often in electrical problems, the initial current is known, but $\frac{di}{dt}$ at $t = 0$ must be calculated. Usually, the initial charge across the capacitor is given. Since $i = \frac{dq}{dt}$, the initial value of $\frac{di}{dt}$ can be determined by evaluating (1) at $t = 0$ to get

$$L\frac{di}{dt}(0) + Ri(0) + \frac{1}{C}q(0) = e(0).$$

Here $e(0)$ is the voltage of the source (battery) at $t = 0$.

Electrical Resonance

We consider an alternating voltage source

$$e(t) = E\sin\omega t.$$

Here, E is the maximum of the voltage source and ω is the (circular) frequency of the voltage source. The differential equation for the current in the RLC circuit follows from (2):

$$L\frac{d^2i}{dt^2} + R\frac{di}{dt} + \frac{1}{C}i = E\omega\cos\omega t. \tag{3}$$

If the electrical resistance can be neglected, $R = 0$, then the current in the circuit satisfies

$$L\frac{d^2i}{dt^2} + \frac{1}{C}i = E\omega\cos\omega t. \tag{4}$$

Solutions of the associated homogeneous equation oscillate with natural (circular) frequency $\omega_0 = 1/\sqrt{LC}$. Resonance occurs (without a resistor) if the forcing frequency equals the natural frequency

$$\omega = \omega_0 = \frac{1}{\sqrt{LC}}.$$

Electrical Response

With electrical resistance ($R > 0$), the general solution of (3) is in the form $i = i_p + i_h$, where i_p is a particular solution and i_h corresponds to the solution of the associated homogeneous equation. The solution of the associated homogeneous equation is derived from an exponential e^{rt}, which must correspond to an underdamped, critically damped, or overdamped oscillation (by analogy with the spring-mass system, $m = L$, $\delta = R$, $k = \frac{1}{C}$). The characteristic equation is

$$Lr^2 + Rr + \frac{1}{C} = 0.$$

The roots of the characteristic equation are $r = (-R \pm \sqrt{R^2 - 4L/C})/2L$. Since L, R, C are positive numbers, in all three cases (underdamped, critically damped, or overdamped), we have $i_h \to 0$ as $t \to \infty$. Thus, after some time, the general solution may be approximated by a particular solution. Using the method of undetermined coefficients, a particular solution may be found in the form

$$i_p = A\cos\omega t + B\sin\omega t.$$

We can obtain formulas for A and B and from them derive formulas for the amplitude of the response. Since we have just calculated this in Section 2.7 on forced mechanical vibrations, we will use the previously derived formula (28). In addition to $m \Rightarrow L$, $\delta \Rightarrow R$, $k \Rightarrow \frac{1}{C}$, we note that $F \Rightarrow E\omega$. The extra factor of ω will change some of the properties of the response diagram. Thus, from (26) and (27) of Section 2.7, we obtain

$$i_p = \frac{E\omega}{\sqrt{(\frac{1}{C} - L\omega^2)^2 + R^2\omega^2}}\cos(\omega t - \phi). \tag{5}$$

We will not be concerned with the phase angle ϕ here.

The amplitude I of the response current satisfies

$$I = \frac{E\omega}{\sqrt{(\frac{1}{C} - L\omega^2)^2 + R^2\omega^2}}. \tag{6}$$

As with a spring-mass system, the response depends on R, L, C, and the forcing circular frequency ω. The response is zero if $\omega = 0$. If $\omega \neq 0$, the response simplifies to

$$I = \frac{E}{\sqrt{(\frac{1}{\omega C} - L\omega)^2 + R^2}},$$

which is graphed in figure 2.8.2 for several values of $\gamma = R/(2\sqrt{L/C})$. The maximum of $|I|$ occurs at the minimum of the denominator, which occurs when $\frac{1}{\omega C} - L\omega = 0$

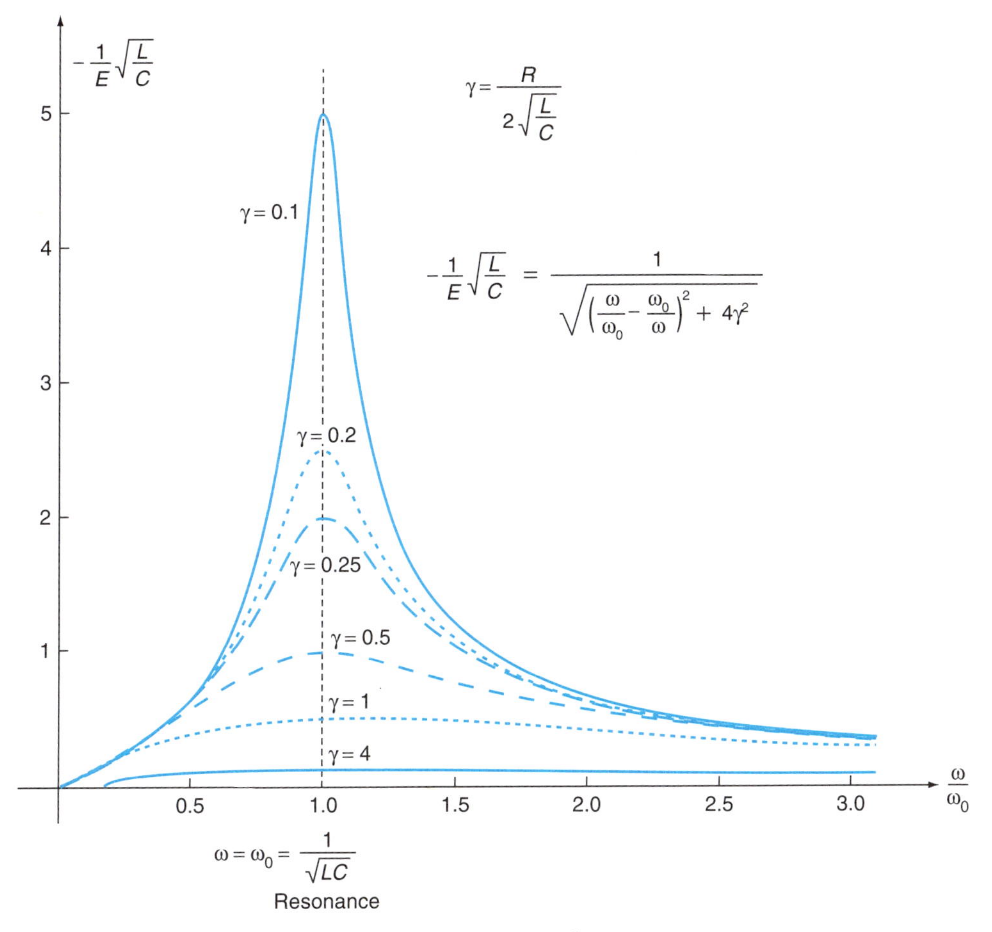

Figure 2.8.2 Response diagram.

or equivalently $\omega = \frac{1}{\sqrt{LC}} = \omega_0$. The amplitude response has a maximum when the system is forced at the natural frequency. Smaller values of γ correspond to smaller values of R. The case $R = \gamma = 0$, $\omega = \omega_0$, would be pure resonance.

Radios and Signals

Different radio stations transmit their signals at different frequencies. These signals act as inputs to electrical devices that receive the signals. Primitive radios operated like an RLC circuit. The input voltage source would be the sum of the input voltages from the various radio stations. Since the RLC circuit is linear, the principle of superposition applies, and the response will be the sum of the responses. Each station's signal gets amplified by the amount associated with the electrical response curve. The output will be the sum of signals of different frequencies. The largest amplitude response will be from the frequency nearest the resonant frequency for the RLC circuit. In primitive radios, the capacitance C was adjustable (via a knob). If you wanted to hear your favorite radio station with circular frequency ω^*, you adjusted the dial (changed the capacitance C) so that a resonant circular frequency $\frac{1}{\sqrt{LC}}$ equaled the known radio circular frequency ω^*:

$$\omega^{*2} = \frac{1}{LC}.$$

Actual radio receivers are more sophisticated, but perform essentially in this manner.

Exercises

1. An RLC circuit, given by figure 2.8.1, has a voltage source of $e(t) = 3\cos t$ volts. Values for the components are $R = 3\ \Omega$, $L = 0.5$ H, and $C = 0.4$ F. Initially, the charge on the capacitor is zero and the current in the resistor is 1 A. Find the charge on the capacitor and the current as functions of time.
2. An RLC circuit, given by figure 2.8.1, has a voltage source of $e(t) = 5\cos 2t$ volts. Values for the components are $R = 2\ \Omega$, $L = 1$ H, and $C = \frac{1}{17}$ F. Initially, the charge on the capacitor and the current in the resistor are zero. Find the charge on the capacitor and the current as functions of time.
3. An RLC circuit, given by figure 2.8.1, has a 1.5-V battery as a voltage source of $e(t) = 1.5$. The values for the components are $R = 1.5\ \Omega$, $L = 1$ H, and $C = 2$ F. Initially, the charge on the capacitor is 2C and the current in the resistor is 4 A. Find the charge on the capacitor and the current as functions of time.
4. An RLC circuit, given by figure 2.8.1, has a 9-V battery as a voltage source of $e(t) = 9$. The values for the components are $R = 5\ \Omega$, $L = 6$ H, and $C = 1$ F. Initially, the charge on the capacitor is 1C and the current in the resistor is zero. Find the charge on the capacitor and the current as functions of time.
5. An LC circuit ($R = 0$), given by figure 2.8.1, has $C = 0.1$ F and $e(t) = \sin \omega t$. Suppose that ω is a constant such that $e(t)$ has a frequency between 20 and 30 Hz. What values of L will not lead to resonance for any such ω?
6. The RLC circuit shown in figure 2.8.1 has $R = 20\ \Omega$, $L = 1$ H, and $C = 0.005$ F. The voltage source is shorted out ($e(t) = 0$). At time $t = 0$, there is a charge of 10 C on the capacitor and no current. Solve the differential equation for the charge, and put in phase amplitude form. How many seconds will it take the variable amplitude to be reduced 99%?
7. The RLC circuit in figure 2.8.1 has $R = 2\ \Omega$, $L = 1$ H, and $C = 0.5$ F. The voltage source $e(t)$ is a 1-V battery that is shorted out ($C = 0$) after π seconds. The initial charge on the capacitor is zero, and the initial current is zero. Find the charge on the capacitor and graph for $0 \le t \le 2\pi$ (seconds). (Assume that charge and current are continuous at $t = \pi$.)
8. The LC circuit in figure 2.8.1 ($R = 0$) has $L = 4$ H and $C = 0.25$ F. The voltage source is a 2-V battery that is shorted out ($e = 0$) after 4π seconds. The initial charge is zero and the initial current is 1 A. Find the charge on the capacitor and current in the inductor for $0 \le t \le 8\pi$. (Assume that the charge and current are continuous at $t = 4\pi$.)
9. Let $R = 0$ and $e(t) = 0$ in figure 2.8.1, so that we have an LC circuit with no voltage source. Let $E(i, q) = \left(\frac{L}{2}\right)i^2 + \left(\frac{1}{2C}\right)q^2$. Show that E is constant by verifying that $\frac{dE}{dt} = 0$. (This is the electrical equivalent of conservation of mechanical energy.)

10. An LC circuit ($R=0$) given by figure 2.8.1 has $L=8$ H and $C=2$ F. For what value of ω will the voltage source $e(t)=2\cos\omega t$ volts create resonance?
11. An LC circuit ($R=0$) given by figure 2.8.1 has $L=9$ H and $C=1$ F. The voltage source is $e(t)=4\cos 2t$. Since $R=0$, the free response will not be transient. However, there is one choice for $q(0)$, $i(0)$ for which the free response will be absent. What are the values of $q(0)$, $i(0)$?

For the next five exercises, use the correspondence

$$\text{inductance} \leftrightarrow \text{mass},$$
$$\text{resistance} \leftrightarrow \text{friction},$$

and

$$\frac{1}{\text{capacitance}} \leftrightarrow \text{spring constant}$$

between the circuit given by figure 2.8.1 and the spring-mass system in Section 2.5. (A mass of m in slugs or grams becomes an inductance of L henries, etc.)

12. Rewrite Exercise 2.5.18 in terms of this circuit and answer parts (a)–(c).
13. Rewrite Exercise 2.5.19 in terms of this circuit and answer parts (a)–(c).
14. Rewrite Exercise 2.5.20 in terms of this circuit and answer parts (a)–(c).
15. Rewrite Exercise 2.5.21 in terms of this circuit and answer parts (a)–(c).
16. Rewrite Exercise 2.5.22 in terms of this circuit and answer parts (a)–(c).

(**Input-Output Voltages**) Frequently it is helpful to consider circuits as input-output devices. Suppose for the circuit in figure 2.8.1 that $R>0$ so that the free response is transient. We assume that these transient terms are not important and consider the solution $q(t)$ to be given by the terms that are not transient, that is, the forced response. The **input voltage** will be taken as $e(t)$, and the **output voltage** as the voltage across the capacitor, which is $(1/C)q$.

17. Suppose that $E=1$, $R=6$, $L=1$, $C=\frac{1}{13}$, and $\omega=3$. Find I in (6).
18. Suppose that $E=1$, $R=7$, $L=12$, $C=1$, and $\omega=2$. Find I in (6).

2.9 Euler Equation

Linear differential equations with variable coefficients rarely have explicit solutions. Frequently, solutions are obtained numerically or by the method of power series: which is not covered in this book. However, in this section we study linear differential equations with variable coefficients of a specific kind where explicit solutions are not difficult to obtain. The second-order **Euler equation** is

$$at^2\frac{d^2x}{dt^2}+bt\frac{dx}{dt}+cx=0, \tag{1}$$

where a, b, c are constants and $a\neq 0$. The third-order Euler equation is

$$a_3t^3\frac{d^3x}{dt^3}+a_2t^2\frac{d^2x}{dt^2}+a_1t\frac{dx}{dt}+a_0x=0, \tag{2}$$

with a_3, a_2, a_1, a_0 constants and $a_3 \neq 0$. There is even a first-order Euler equation,

$$at\frac{dx}{dt} + bx = 0.$$

The Euler equation is also known as the Cauchy-Euler or equidimensional equation (see Exercise 15 at the end of this section). It arises, for example, in some drag problems in uniformly viscous flows. Euler equations are important not only because they arise in applications, but also because they are used in studying more advanced topics in differential equations such as what are called singular points.

We consider here the second-order Euler equation,

$$at^2x'' + btx' + cx = 0. \tag{3}$$

The variable coefficients of (3) are powers that increase by t for each derivative in (3). This suggests that solutions of (3) are likely to be powers of t,

$$x = t^r, \tag{4}$$

since powers have the property that their power decreases by one each time they are differentiated:

$$x' = rt^{r-1}, \tag{5}$$

$$x'' = r(r-1)t^{r-2}. \tag{6}$$

This is exactly what is needed to counteract the increased power in the variable coefficient of (3).

Substituting $x = t^r$ into $at^2x'' + btx' + cx = 0$ gives

$$at^2r(r-1)t^{r-2} + btrt^{r-1} + ct^r = 0,$$

$$[ar(r-1) + br + c]t^r = 0,$$

so that

$$\begin{aligned} ar(r-1) + br + c &= 0, \\ ar^2 + (b-a)r + c &= 0, \end{aligned} \tag{7}$$

which we call the **indicial equation**.

Case 1: Distinct Real Roots

If the polynomial $ar^2 + (b-a)r + c = 0$ has two distinct real roots, r_1, r_2, then t^{r_1}, t^{r_2} provide a fundamental set of solutions. In this case the general solution of (3) is

$$x = c_1t^{r_1} + c_2t^{r_2}. \tag{8}$$

Example 2.9.1 *Distinct Real Roots*

Find the general solution of

$$2t^2\frac{d^2x}{dt^2} + 7t\frac{dx}{dt} - 3x = 0, \quad t > 0. \tag{9}$$

• SOLUTION. The student is encouraged to rederive the indicial equation (7) the first few times. Letting $x = t^r$ in (9) and then dividing by t^r yields the indicial equation

$$2r(r-1) + 7r - 3 = 0$$

or

$$2r^2 + 5r - 3 = (2r-1)(r+3) = 0.$$

The roots are $r = \frac{1}{2}, -3$, and the general solution is

$$x = c_1 t^{1/2} + c_2 t^{-3}. \qquad \blacklozenge$$

Case 2: Complex Conjugate Roots

If the indicial equation has complex roots $r_1 = \alpha + i\beta$, $r_2 = \alpha - i\beta$, then the general solution is

$$x = \overline{c}_1 t^{\alpha+i\beta} + \overline{c}_2 t^{\alpha-i\beta} = t^{\alpha}(\overline{c}_1 t^{i\beta} + \overline{c}_2 t^{-i\beta}). \tag{10}$$

Because of the property of exponentials (for $t > 0$),

$$t^{i\beta} = e^{i\beta \ln t}.$$

Thus, (10) becomes

$$x = t^{\alpha}(\overline{c}_1 e^{i\beta \ln t} + \overline{c}_2 e^{-i\beta \ln t}).$$

We have shown using Euler's formula in Section 2.4 that a linear combination of $e^{\pm i\theta}$ is equivalent to a linear combination of $\cos\theta$ and $\sin\theta$. In this case, $\theta = \beta \ln t$. Thus, when the roots of the indicial equation for Euler's equation are complex numbers $\alpha \pm i\beta$, the general solution is

$$x = t^{\alpha}[c_1 \cos(\beta \ln t) + c_2 \sin(\beta \ln t)] \tag{11a}$$

or equivalently,

$$x = c_1 t^{\alpha} \cos(\beta \ln t) + c_2 t^{\alpha} \sin(\beta \ln t). \tag{11b}$$

Example 2.9.2 *Complex Conjugate Roots*

Find the general solution of

$$9t^2 x'' + 15tx' + 5x = 0, \qquad t > 0.$$

• SOLUTION. Letting $x = t^r$ yields the indicial equation $9r(r-1) + 15r + 5 = 9r^2 + 6r + 5 = (3r+1)^2 + 4 = 0$, so the roots are

$$r = -\frac{1}{3} \pm i\frac{2}{3}.$$

Since $t^r = t^{-\frac{1}{3} \pm i\frac{2}{3}} = t^{-\frac{1}{3}} t^{\pm i\frac{2}{3}}$, the general solution is

$$x = c_1 t^{-1/3} \cos\left(\frac{2}{3}\ln t\right) + c_2 t^{-1/3} \sin\left(\frac{2}{3}\ln t\right). \qquad \blacklozenge$$

Case 3: Repeated Real Roots

If the indicial equation $ar(r-1)+br+c=0$ has a repeated root $r=r_1$, then only one homogeneous solution is in the assumed form, $x=t^{r_1}$. The second independent homogeneous solution can always be obtained by reduction of order (Section 2.3). In the case of repeated roots for the Euler equation, the second solution is always

$$x=t^{r_1}\ln t. \tag{12}$$

We derive this fact with a specific example.

Example 2.9.3 *Using Reduction of Order*

Find the general solution of

$$t^2x''-7tx'+16x=0, \qquad t>0. \tag{13}$$

• SOLUTION. Letting $x=t^r$ yields the indicial equation $r(r-1)-7r+16=r^2-8r+16=(r-4)^2=0$, which has repeated roots $r=4, 4$. Since we have one solution $x=t^4$, the second solution in the fundamental set of solutions can be found by reduction of order. Let

$$x=vt^4.$$

Then (13) becomes, using the product rule,

$$t^2(v''t^4+2v'4t^3+v12t^2)-7t(v't^4+v4t^3)+16vt^4=0, \qquad t>0. \tag{14}$$

The coefficients of v cancel since $x=t^4$ is a solution, and (14) reduces to

$$v''t^6+v't^5=0.$$

Dividing by t^5 yields

$$tv''+v'=0. \tag{15}$$

In the method of reduction of order, we let

$$w=v',$$

so that (15) becomes a first-order linear differential equation,

$$tw'+w=0. \tag{16}$$

Separation yields

$$\frac{dw}{w}=-\frac{dt}{t}.$$

Thus (since we assume $t>0$),

$$\ln|w|=-\ln t+c.$$

After exponentiation,

$$w=\frac{c_1}{t}. \tag{17}$$

From (17) and $w = v'$, we have

$$v' = \frac{c_1}{t}.$$

Integrating gives

$$v = c_1 \ln t + c_2,$$

so that

$$x = t^4 v = t^4 (c_1 \ln t + c_2) = c_1 t^4 \ln t + c_2 t^4.$$

Thus, the new second independent solution is $t^4 \ln t$. ♦

From now on, to obtain the general solution of an Euler equation when there is a repeated root, just note that the second solution is $t^{r_1} \ln t$ and do not use reduction of order unless instructed to do so.

Example 2.9.4 *Repeated Roots*

Find the general solution of

$$t^2 x'' - t x' + x = 0, \quad t > 0.$$

• SOLUTION. Letting $x = t^r$ yields the indicial equation $r(r-1) - r + 1 = r^2 - 2r + 1 = (r-1)^2 = 0$, which has repeated roots 1, 1. Thus

$$x = c_1 t \ln t + c_2 t, \quad t > 0.$$ ♦

We summarize these three cases in the following theorem.

THEOREM 2.9.1 **Euler Equation** *The general solution of the second-order Euler equation $at^2 x'' + btx' + cx = 0$, $t > 0$, is given as follows. By substituting $x = t^r$, obtain the indicial equation $ar(r-1) + br + c = 0$. Find the roots r_1, r_2.*

1. *If $r_1 \neq r_2$ and r_1, r_2 are real, then*

$$x = c_1 t^{r_1} + c_2 t^{r_2}. \tag{18}$$

2. *If $r_1 = \alpha + \beta i$, $r_2 = \alpha - \beta i$, $\beta \neq 0$, then*

$$x = c_1 t^{\alpha} \cos(\beta \ln t) + c_2 t^{\alpha} \sin(\beta \ln t). \tag{19}$$

3. *If $r_1 = r_2$, then*

$$x = c_1 t^{r_1} + c_2 t^{r_1} \ln t. \tag{20}$$

Exercises

In Exercises 1–12, solve the second-order Euler equation for $t > 0$. If no initial conditions are given, find the general solution.

1. $t^2 x'' + t x' - x = 0.$
2. $t^2 x'' - 4t x' + 6x = 0.$

3. $t^2x'' + 3tx' + x = 0$.
4. $t^2x'' - tx' + 2x = 0$, $x(1) = 0$, $x'(1) = 2$.
5. $4t^2x'' + 8tx' + x = 0$.
6. $t^2x'' + tx' + x = 0$.
7. $t^2x'' + 4tx' + 2x = 0$, $x(1) = 1$, $x'(1) = 0$.
8. $9t^2x'' + 15tx' + 2x = 0$.
9. $t^2x'' + tx' + 4x = 0$.
10. $t^2x'' + 3tx' + 10x = 0$.
11. $t^2x'' + 3tx' + 8x = 0$.
12. $t^2x'' - 3tx' + 5x = 0$.
13. By looking at the answers to Exercises 1 through 12, find an Euler equation whose indicial equation has repeated roots. For this example, derive the second solution using reduction of order.
14. Find the general solution of $5t\left(\frac{dx}{dt}\right) + x = 0$ using separation. Compare your answer to the solution obtained by using the methods of this section.
15. (**Equidimensionality**) Let k be a constant, and perform the change of variables $t = ks$.

a) Show that the Euler equation $at^2\left(\frac{d^2x}{dt^2}\right) + bt\left(\frac{dx}{dt}\right) + cx = 0$ becomes

$$as^2\frac{d^2x}{ds^2} + bs\frac{dx}{ds} + cx = 0.$$

That is, the Euler equation is unaltered by a change of scale in the independent variable.

b) Contrast this with what happens when the same change of scale is performed on the constant coefficient equation $ax'' + bx' + cx = 0$.

c) Verify that, if $k > 0$ is a real constant and r_1, r_2 are distinct real constants, then $t = ks$ changes $c_1t^{r_1} + c_2t^{r_2}$ into $\tilde{c}_1s^{r_1} + \tilde{c}_2s^{r_2}$.

d) Verify that, if $k > 0$ and r_1 is a real constant, then $t = ks$ changes $c_1t^{r_1} + c_2t^{r_2}\ln t$ into $\tilde{c}_1s^{r_1} + \tilde{c}_2s^{r_1}\ln s$.

e) Verify that, if $k > 0$ and α, β are real constants, $\beta \neq 0$, then $t = ks$ changes $c_1t^{\alpha}\cos(\beta\ln t) + c_2t^{\alpha}\sin(\beta\ln t)$ into $\tilde{c}_1s^{\alpha}\cos(\beta\ln s) + \tilde{c}_2s^{\alpha}\sin(\beta\ln s)$.

16. (Alternative Method) Let $y = \ln t$ $(t = e^y)$.

a) Verify that

$$t\frac{dx}{dt} = \frac{dx}{dy}, \tag{21}$$

$$t\frac{d}{dt}\left(t\frac{dx}{dt}\right) = t^2\frac{d^2x}{dt^2} + t\frac{dx}{dt} = t^2\frac{d^2x}{dt^2} + \frac{dx}{dy} = \frac{d^2x}{dy^2}. \tag{22}$$

b) Show that the change of variables $y = \ln t$ changes

$$at^2\frac{d^2x}{dt^2} + bt\frac{dx}{dt} + cx = 0$$

into the constant coefficient equation

$$a\frac{d^2x}{dy^2} + (b - a)\frac{dx}{dy} + cx = 0. \tag{23}$$

Note that the indicial equation $ar^2 + (b-a)r + c = 0$ in Theorem 2.9.1 is just the characteristic equation of (23).

Euler equations of any order may be solved by either using $x = t^r$ to find the indicial polynomial or performing the change of variables $y = \ln t$ to make the differential equation linear constant coefficient. In Exercises 17–20, solve the Euler equations.

17. $t^4x'''' + 6t^3x''' + 7t^2x'' + tx' - x = 0$.
18. $t^4x'''' + 6t^3x''' + 9t^2x'' + 3tx' + x = 0$.
19. $t^3x''' + tx' - x = 0$.
20. $t^3x''' + 4t^2x'' = 0$.

2.10 Variation of Parameters (Second-Order)

In Section 2.6 the method of undetermined coefficients was used to find a particular solution of

$$x''(t) + p(t)x'(t) + q(t)x(t) = f(t). \tag{1}$$

There were two major restrictions on the method of undetermined coefficients. First, p and q had to be constants. Second, f had to be in a special form. This section will present a method for finding a particular solution of $x'' + p(t)x' + q(t)x = f$, provided we have first solved the associated homogeneous equation

$$x'' + p(t)x' + q(t)x = 0. \tag{2}$$

This new method will not require p, q to be constants nor f to be in a special form.

There are no general methods for obtaining a fundamental set of solutions $\{x_1, x_2\}$ of (2). In Section 2.3 we used reduction of order to obtain the second solution x_2 if one solution x_1 was known. Unfortunately, there are no general methods for obtaining one solution of (2) if p, q are not constants.

In this section, we will assume we have a fundamental set of solutions $\{x_1, x_2\}$ of (2). We shall first derive the method and then work several examples. We begin by looking for functions v_1 and v_2 such that

$$x = v_1x_1 + v_2x_2 \tag{3}$$

is a solution of (1),

$$x'' + px' + qx = f. \tag{4}$$

This is called the **method of variation of parameters** because the usual constants c_1, c_2 in the solution of the associated homogeneous equation are now varied. When (3) is substituted into (4), one differential equation results, involving two unknowns v_1 and v_2. In the method of variation of parameters, an ingenious (and unmotivated) observation is made. We choose v_1 and v_2 such that the first derivatives of x would be the same as if v_1 and v_2 were constants, even though v_1 and v_2 are not constants. That is, we want

$$x' = v_1 x_1' + v_2 x_2'. \tag{5}$$

However, x' should be calculated using the product rule from (3). Thus from (3) we have

$$x' = v_1' x_1 + v_1 x_1' + v_2' x_2 + v_2 x_2'.$$

Equation (5) has only two of these terms. Thus (5) can be valid only if the sum of the other two terms vanishes:

$$v_1' x_1 + v_2' x_2 = 0. \tag{6}$$

Thus (5) holds if (6) holds. The second derivative of x is calculated by differentiating (5) using the product rule twice. Using this x'' and x' from (5), we have that (4) is satisfied only if

$$v_1 x_1'' + v_1' x_2'' + v_2 x_2'' + v_2' x_2' + p(v_1 x_1' + v_2 x_2') + q(v_1 x_1 + v_2 x_2) = f.$$

Collecting terms yields

$$v_1(x_1'' + p x_1' + q x_1) + v_2(x_2'' + p x_2' + q x_2) + v_1' x_1' + v_2' x_2' = f. \tag{7}$$

Since x_1 and x_2 are homogeneous solutions satisfying $x'' + px' + qx = 0$, the v_1 and v_2 terms vanish and (7) simplifies to

$$v_1' x_1' + v_2' x_2' = f. \tag{8}$$

In summary, we have that, if v_1' and v_2' satisfy the linear algebraic equations (6) and (8), then

$$v_1' x_1 + v_2' x_2 = 0, \tag{9}$$

$$v_1' x_1' + v_2' x_2' = f, \tag{10}$$

and

$x = v_1 x_1 + v_2 x_2$ is a particular solution of the original differential equation (4).

The linear system (9) can be solved by elimination or Cramer's rule (see the summary at the end of this section) for v_1' and v_2'. To eliminate v_2', we multiply (9) by x_2' and multply (10) by x_2 and subtract:

$$v_1'(x_1 x_2' - x_2 x_1') = -f x_2.$$

We note that $x_1 x_2' - x_2 x_1'$ is the Wronskian of Section 2.3. Thus we can divide by it to get

$$v_1' = -\frac{f x_2}{x_1 x_2' - x_2 x_1'}.$$

We obtain v_2' from (9):

$$v_2' = \frac{f x_1}{x_1 x_2' - x_2 x_1'}.$$

By integration (indefinite or definite), we obtain v_1 and v_2. The particular solution is then determined from (3).

We begin by discussing two examples. An additional example appears after the summary. The first example illustrates just the solution technique.

Example 2.10.1 *Variation of Parameters*

From Section 2.9 $\{t, t^3\}$ is a fundamental set of solutions of $t^2x'' - 3tx' + 3x = 0$. Find the general solution of

$$t^2x'' - 3tx' + 3x = 4t^7.$$

• SOLUTION. We have $x_1 = t$, $x_2 = t^3$. Dividing by t^2, the differential equation is

$$x'' - \frac{3}{t}x' + \frac{3}{t^2}x = 4t^5.$$

Thus $f(t) = 4t^5$. We must solve (9), which is

$$v_1't + v_2't^3 = 0,$$
$$v_1'[t]' + v_2'[t^3]' = 4t^5,$$

or

$$v_1't + v_2't^3 = 0,$$
$$v_1' + v_2'3t^2 = 4t^5.$$

Solve the first equation for v_1'; $v_1' = -v_2't^2$, substitute into the second equation

$$-v_2't^2 + v_2'3t^2 = 4t^5,$$

and solve for v_2',

$$v_2' = 2t^3.$$

Then

$$v_1' = -v_2't^2 = -2t^5.$$

Thus, antidifferentiating v_1', v_2' gives

$$v_2 = \frac{t^4}{2}, \quad v_1 = -\frac{t^6}{3}.$$

A particular solution is

$$x_p = v_1x_1 + v_2x_2 = \left(-\frac{t^6}{3}\right)t + \left(\frac{t^4}{2}\right)t^3 = \frac{t^7}{6},$$

and the general solution is

$$x = \frac{t^7}{6} + c_1 t + c_2 t^3. \quad \blacklozenge$$

The next example illustrates both the derivation of (9) and the use of a definite integral.

Example 2.10.2 *Derive Variation of Parameters*

Find the general solution of

$$x'' + x = \frac{1}{t+1}. \tag{11}$$

Note that (11) is a differential equation with constant coefficients, but $1/(t+1)$ is not the kind of forcing function to which the method of undetermined coefficients can be applied. We shall solve (11) by the method of variation of parameters since we know how to obtain solutions of the associated homogeneous equations.

• SOLUTION. First we obtain solutions of the associated homogeneous equation by solving $x'' + x = 0$. By substituting $x = e^{rt}$, we obtain the characteristic equation $r^2 + 1 = 0$, which has roots $r = \pm i$. Thus, $\cos t$ and $\sin t$ form a fundamental set. According to the method of variation of parameters, we seek a solution of (11) in the form

$$x = v_1 \cos t + v_2 \sin t. \tag{12}$$

The derivative is the same as would occur if v_1 and v_2 were constants,

$$x' = -v_1 \sin t + v_2 \cos t. \tag{13}$$

However, by the product rule (13) is valid only if

$$v_1' \cos t + v_2' \sin t = 0. \tag{14}$$

Substituting (12) into (11) and taking the derivative of (13) using the product rule yields

$$-v_1' \sin t - v_1 \cos t + v_2' \cos t - v_2 \sin t + v_1 \cos t + v_2 \sin t = \frac{1}{t+1}. \tag{15}$$

After canceling the v_1 and v_2 terms, which should always occur, (15) becomes

$$-v_1' \sin t + v_2' \cos t = \frac{1}{t+1}. \tag{16}$$

The equations for (v_1', v_2') are (14) and (16). For students familiar with solving linear systems of equations, there are several options available, such as using augmented matrices or Cramer's rule (see formula (28) at the end of this section). We shall just solve for (v_1', v_2') using elimination. To eliminate v_2' we multiply (14) by $\cos t$, multiply (16) by $-\sin t$, and add the results:

$$v_1'(\cos^2 t + \sin^2 t) = -\frac{\sin t}{t+1}.$$

But $\cos^2 t + \sin^2 t = 1$, so that

$$v_1' = -\frac{\sin t}{t+1}. \tag{17}$$

We can now determine v_2' from (14):

$$v_2' = \frac{\cos t}{t+1}. \tag{18}$$

Since (17) and (18) cannot be antidifferentiated explicitly, we use the definite integral:

$$\begin{aligned} v_1 &= -\int_0^t \frac{\sin \bar{t}}{\bar{t}+1} d\bar{t} + c_1, \\ v_2 &= \int_0^t \frac{\cos \bar{t}}{\bar{t}+1} d\bar{t} + c_2. \end{aligned} \tag{19}$$

For convenience we have chosen the lower limit to be zero.

A solution is formed from (12):

$$\begin{aligned} x = {} & \sin t \int_0^t \frac{\cos \bar{t}}{\bar{t}+1} d\bar{t} - \cos t \int_0^t \frac{\sin \bar{t}}{\bar{t}+1} d\bar{t} \\ & + c_1 \cos t + c_2 \sin t. \end{aligned} \tag{20}$$

If the arbitrary constants c_1 and c_2 are kept, the general solution of (11) is obtained. If the constants are zero or any other specific value, then (20) is a particular solution. ♦

Influence Function

Although (20) is a correct and satisfactory general solution of the differential equation (11), some further algebraic manipulations yield an important and interesting result. Since $\bar{t}$ is a "dummy" variable of integration, the functions of t may be taken inside the integral and the two integrals combined:

$$x = \int_0^t \frac{\sin t \cos \bar{t} - \cos t \sin \bar{t}}{\bar{t}+1} d\bar{t} + c_1 \cos t + c_2 \sin t. \tag{21}$$

Furthermore, if we use the trigonometric addition formula, $\sin(a-b) = \sin a \cos b - \cos a \sin b$, then (21) becomes

$$x = \int_0^t \frac{\sin(t-\bar{t})}{\bar{t}+1} d\bar{t} + c_1 \cos t + c_2 \sin t. \tag{22}$$

The solution to the differential equation represents the response to the source $1/(t+1)$. It is seen that the solution at time t (called the response) is the sum (actually an integral part) of all sources from $\bar{t}=0$ (the initial time) to $\bar{t}=t$ (the present time). The function $\sin(t-\bar{t})$ is a weighting function. It is an **influence function** called the **Green's function**, which represents the weight of the contribution to the response at t due to the source at $\bar{t}$:

$$G(t,\bar{t})=\sin(t-\bar{t}). \tag{23}$$

Using this notation, the solution to the differential equation (11) is

$$x=\int_0^t G(t,\bar{t})\frac{1}{\bar{t}+1}d\bar{t}+c_1\cos t+c_2\sin t. \tag{24}$$

One advantage of solving the differential equation in this way is that if the source changes, then it is easy to change the solution accordingly. In fact, it can be seen that if the source was not described specifically, for example,

$$x''+x=f(t), \tag{25}$$

then the general solution could be written

$$x=\int_0^t G(t,\bar{t})f(\bar{t})d\bar{t}+c_1\cos t+c_2\sin t, \tag{26}$$

where the Green's function $G(t,\bar{t})$ is still given by (23).

It is also interesting to note the manner in which initial conditions are satisfied. Evaluating (22) at $t=0$ yields

$$x(0)=c_1.$$

In this example, it can be shown (but not as easily) that

$$x'(0)=c_2.$$

Summary of Variation of Parameters

Variation of parameters (second-order) is a method of calculating a particular solution of $x''+px'+qx=f$ given a fundamental set of solutions $\{x_1,x_2\}$ of the associated homogeneous equation $x''+px'+qx=0$. The method is as follows:

1. Find a fundamental set of solutions $\{x_1,x_2\}$ of $x''+px'+qx=0$.
2. After substituting $x=v_1x_1+v_2x_2$, solve the algebraic system of equations

$$\begin{aligned} v_1'x_1+v_2'x_2&=0,\\ v_1'x_1+v_2'x_2&=f \end{aligned} \tag{27}$$

   for the functions v_1', v_2'.
3. Antidifferentiate to find v_1, v_2.
4. Then $x=v_1x_1+v_2x_2$ is a particular solution of $x''+px'+qx=f$.
5. $x=v_1x_1+v_2x_2+c_1x_1+c_2x_2$ is the general solution of $x''+px'+qx=f$.

Note. If arbitrary constants are introduced in Step 3 when finding v_1, v_2, then $x = v_1x_1 + v_2x_2$ will be the general solution.
Note. The coefficients of v_1', v_2' in (27) are the entries of the matrix

$$\begin{bmatrix} x_1 & x_2 \\ x_1' & x_2' \end{bmatrix}$$

whose determinant is the Wronskian (Section 2.2). Since the Wronskian of a fundamental set of solutions is never zero, it follows from matrix theory (Cramer's rule; see below) that (27) can always be solved uniquely for v_1', v_2'. Thus the only difficulty is in Step 1.

Two comments are in order. First, a fact from algebra called Cramer's rule can be applied to the system of equations (27) to give formulas for v_1', v_2'. They are

$$v_1' = \frac{\det\begin{bmatrix} 0 & x_2 \\ f & x_2' \end{bmatrix}}{\det\begin{bmatrix} x_1 & x_2 \\ x_1' & x_2' \end{bmatrix}} \quad \text{and} \quad v_2' = \frac{\det\begin{bmatrix} x_1 & 0 \\ x_1' & f \end{bmatrix}}{\det\begin{bmatrix} x_1 & x_2 \\ x_1' & x_2' \end{bmatrix}}.$$

Evaluating the determinants on top gives

$$v_1' = -\frac{fx_2}{W[x_1, x_2]}, \quad v_2' = \frac{fx_1}{W[x_1, x_2]}, \tag{28}$$

where $W[x_1, x_2]$ is the Wronskian of x_1, x_2. Then

$$v_1 = -\int \frac{fx_2}{W[x_1, x_2]}dt, \quad v_2 \int = \frac{fx_1}{W[x_1, x_2]}dt, \tag{29}$$

and $x = v_1x_1 + v_2x_2$. The general solution results if two constants are added when integrating (29). If no integration constants are introduced, then a particular solution results.

In practice, it is probably quicker to use (28) to find v_1', v_2' than to solve (27) directly.

Example 2.10.3 *Variation of Parameters*

Find the general solution of

$$2x'' - 4x' + 2x = t^{-1}e^t, \quad t > 0. \tag{30}$$

• SOLUTION. Note that $t^{-1}e^t$ is not the kind of forcing term to which the method of undetermined coefficients can be applied. However, the differential equation has constant coefficients, so we know how to solve the associated homogeneous equation. We shall solve (30) by the method of variation of parameters. We rewrite (30) as

$$x'' - 2x' + x = \frac{1}{2}t^{-1}e^t,$$

so that $f(t) = \frac{1}{2}t^{-1}e^t$.

Step 1 We must solve $x'' - 2x' + x = 0$. By substituting $x = e^{rt}$, the characteristic equation is $r^2 - 2r + 1 = (r-1)^2 = 0$, which has roots $r = 1, 1$. Thus $\{e^t, te^t\}$ is a fundamental set of solutions. Let $x_1 = e^t$, $x_2 = te^t$.

Step 2 $x_1 = e^t$, $x_2 = te^t$,

$$W[e^t, te^t] = \det\begin{bmatrix} e^t & te^t \\ e^t & e^t + te^t \end{bmatrix} = e^{2t},$$

and $f = \frac{1}{2}t^{-1}e^t$. Thus (28) gives

$$v_1' = -\frac{t^{-1}e^t \cdot te^t}{2 \cdot e^{2t}} = -\frac{1}{2}$$

and

$$v_2' = \frac{t^{-1}e^t \cdot e^t}{2e^{2t}} = \frac{t^{-1}}{2}.$$

Step 3

$$v_1 = -\frac{t}{2}$$
$$v_2 = \frac{\ln t}{2}$$

Steps 4 and 5

$$\begin{aligned} x &= v_1x_1 + v_2x_2 + c_1x_1 + c_2x_2 \\ &= \frac{-t}{2}e^t + \frac{\ln t}{2}te^t + c_1e^t + c_2te^t \\ &= \frac{te^t \ln t}{2} + c_1e^t + \tilde{c}_2te^t \qquad \left(\tilde{c}_2 = c_2 - \frac{1}{2}\right) \end{aligned}$$

is the general solution of $2x'' - 4x' + 2x = t^{-1}e^t$. ♦

Exercises

In Exercises 1–14, find the general solution by the method of variation of parameters. Decide whether the method of undetermined coefficients could also have been used.

1. $x'' - x = e^{2t}$.
2. $x'' - 4x' + 4x = e^{2t}$.
3. $x'' + x = \dfrac{1}{\sin t}$.
4. $x'' + 4x = \dfrac{4}{\cos 2t}$.
5. $x'' + x = \tan t$.
6. $x'' + 2x' + x = \dfrac{e^{-t}}{1+t^2}$.
7. $x'' + 3x' + 2x = \dfrac{1}{1+e^{2t}}$.

8. $4x'' - x = t$.
9. $x'' - 6x' + 9x = e^{3t}t^{3/2}$, $x > 0$.
10. $x'' - 4x' + 3x = e^{-t}$.
11. $4x'' + 4x' + x = t^{-2}e^{-t/2}$.
12. $x'' + 5x' + 4x = e^{t}$.
13. $x'' - x' - 6x = e^{-2t}$.
14. $x'' - 3x' + 2x = e^{3t}\cos(e^{t})$.

In Exercises 15–18, a fundamental set of solutions is given for the associated homogeneous equation for $t > 0$. Solve the differential equation, using variation of parameters, and give the general solution. In each case the fundamental set of solutions could be found by the method in Section 2.9 for the Euler equation.

15. $t^2x'' - 2tx' + 2x = t^3$, $\{t, t^2\}$.
16. $t^2x'' + tx' - x = t^{1/2}$, $\{t, t^{-1}\}$.
17. $t^2x'' + 2tx' = t^{-1}$, $\{1, t^{-1}\}$.
18. $t^2x'' - 3tx' + 3x = t$, $\{t, t^3\}$.
19. Show that a particular solution of $x'' + px' + qx = f$ given by variation of parameters may be written as

$$x_p(t) = \int_0^t \frac{x_2(t)x_1(\bar{t}) - x_1(t)x_2(\bar{t})}{W[x_1, x_2](\bar{t})} f(\bar{t})d\bar{t}. \tag{31}$$

20. Show that (31) gives the unique solution of the initial value problem $x'' + px' + qx = f$, $x(0) = 0$, $x'(0) = 0$.

In Exercises 21–25, find the general solution by the method of variation of parameters. In these exercises a definite integral will be necessary as in Example 2.10.2.

21. $x'' - x = e^{-t^2}$.
22. $x'' - 4x = \sin(t^{1/3})$.
23. $x'' + 5x' + 6x = \dfrac{1}{t+1}$.
24. $x'' - 3x' + 2x = t^{1/5}$.
25. $t^2x'' + tx' - x = e^{t}$.

2.11 Variation of Parameters (nth-Order)

The method of variation of parameters given in Section 2.10 may also be used for nth-order linear differential equations.

THEOREM 2.11.1 Variation of Parameters for nth-Order Equations
Suppose that $\{x_1, \ldots, x_n\}$ is a fundamental set of solutions of the associated homogeneous equation for

$$a_n(t)x^{(n)}(t) + a_{n-1}(t)x^{(n-1)}(t) + \cdots + a_1(t)x'(t) + a_0(t)x(t) = f(t). \tag{1}$$

If $v_1'(t), \ldots, v_n'(t)$ satisfy the system of equations

$$\begin{array}{ccccccccc} v_1'(t)x_1(t) & + & v_2'(t)x_2(t) & + & \cdots & +v_n'(t)x_n(t) & = & 0 \\ v_1'(t)x_1'(t) & + & v_2'(t)x_2'(t) & + & \cdots & +v_n'(t)x_n'(t) & = & 0 \\ \vdots & & \vdots & & \vdots & & \vdots & \vdots \\ v_1'(t)x_1^{(n-1)}(t) & + & v_2'(t)x_2^{(n-1)}(t) & + & \cdots & +v_n'(t)x_n^{(n-1)}(t) & = & \frac{f(t)}{a_n(t)}, \end{array} \tag{2}$$

then $x_p = v_1x_1 + v_2x_2 + \cdots + v_nx_n$ is a particular solution of (1).

In general, solving (2) for $n > 2$ can become quite complicated. For small n, say 3 or 4, the system (2) can be solved by Cramer's rule. For larger n, computer programs doing symbolic manipulation can be used.

Cramer's rule applied to (2) takes the form

$$v_1' = \frac{\det\begin{bmatrix} 0 & x_2 & \cdots & x_n \\ \vdots & \vdots & & \vdots \\ 0 & x_2^{(n-2)} & & x_n^{(n-2)} \\ f/a_n & x_2^{(n-2)} & & x_n^{(n-2)} \end{bmatrix}}{W[x_1, \ldots, x_n]} = (-1)^{n+1} f \frac{\det\begin{bmatrix} x_2 & \cdots & x_n \\ \vdots & & \vdots \\ x_2^{(n-2)} & & x_n^{(n-2)} \end{bmatrix}}{a_n W[x_1, \ldots, x_n]}. \tag{3}$$

In general,

$$v_i' = (-1)^{n+i} \frac{f}{a_n} \frac{W_i}{W[x_1, \ldots, x_n]}. \tag{4}$$

Here W is the Wronskian of $\{x_1, \ldots, x_n\}$,

$$W[x_1, \ldots, x_n] = \det\begin{bmatrix} x_1 & x_2 & \cdots & x_n \\ x_1' & x_2' & \cdots & x_n' \\ \vdots & & & \\ x_1^{(n-1)} & x_2^{(n-1)} & \cdots & x_n^{(n-1)} \end{bmatrix},$$

and W_i is the Wronskian of the $n-1$ functions obtained by deleting x_i from the set $\{x_1, \ldots, x_n\}$ that appears in the numerator of (3). W and W_i generalize the Wronksian discussed in Section 2.2. The next example could be solved by undetermined coefficients, but we will use it to illustrate variation of parameters.

Example 2.11.1 *Variation of Parameters*

Solve

$$x''' - x'' = e^t, \tag{5}$$

using the method of variation of parameters.

• SOLUTION. The differential equation (5) has constant coefficients, and with $x = e^{rt}$ has characteristic equation $r^3 - r^2 = r^2(r-1) = 0$ with roots of $r = 0, 0, 1$.

Thus $\{1, t, e^t\}$ is a fundamental set of solutions of the associated homogeneous equation $x''' - x'' = 0$. Let $x_1 = 1$, $x_2 = t$, and $x_3 = e^t$. Then $a_3 = 1$, $n = 3$, and $f = e^t$. The equations (2) are

$$\begin{aligned} v_1' \cdot 1 + v_2' t + v_3' e^t &= 0, \\ v_1' \cdot 0 + v_2' \cdot 1 + v_3' e^t &= 0, \\ v_1' \cdot 0 + v_2' \cdot 0 + v_3' e^t &= e^t. \end{aligned} \tag{6}$$

This particular example may be easily solved to yield

$$v_3' = 1, \quad v_2' = -e^t, \quad v_1' = te^t - e^t, \tag{7}$$

or, upon antidifferentiation,

$$v_3 = t, \quad v_2 = -e^t, \quad v_1 = te^t - 2e^t.$$

Thus

$$x_p = v_1 x_1 + v_2 x_2 + v_3 x_3 = (te^t - 2e^t)1 + (-e^t)t + t(e^t) = te^t - 2e^t.$$

The general solution is

$$x = x_p + x_h = te^t - 2e^t + c_1 + c_2 t + c_3 e^t = te^t + c_1 + c_2 t + \tilde{c}_3 e^t.$$

Suppose, however, that instead of solving (6) directly, we had used Cramer's rule (4). Then

$$W[1, t, e^t] = \det \begin{bmatrix} 1 & t & e^t \\ 0 & 1 & e^t \\ 0 & 0 & e^t \end{bmatrix} = e^t,$$

and

$$\begin{aligned} v_1' &= (-1)^{3+1} \frac{e^t W[t, e^t]}{W[1, t, e^t]} = e^t \frac{\det \begin{bmatrix} t & e^t \\ 1 & e^t \end{bmatrix}}{e^t} = te^t - e^t, \\ v_2' &= (-1)^{3+2} \frac{e^t W[1, e^t]}{W[1, t, e^t]} = -e^t \frac{\det \begin{bmatrix} 1 & e^t \\ 0 & e^t \end{bmatrix}}{e^t} = -e^t, \\ v_3' &= (-1)^{3+3} \frac{e^t W[1, t]}{W[1, t, e^t]} = e^t \frac{\det \begin{bmatrix} 1 & t \\ 0 & 1 \end{bmatrix}}{e^t} = 1, \end{aligned}$$

which agrees with (7). ♦

Exercises

In Exercises 1–4, solve the differential equation by variation of parameters and give the general solution.

1. $x''' - x' = e^{2t}$.
2. $x''' + x' = \frac{1}{\sin t}$.

3. $x'''' - x''' = t$.
4. $x''' - 6x'' + 11x' - 6x = e^t$.
5. Verify that $W[e^{at}, e^{bt}, e^{ct}] = e^{(a+b+c)t}(b-a)(c-a)(c-b)$.

In Exercises 6–12, use Exercise 5 and variation of parameters to find the general solution.

6. $x''' - x'' - 4x' + 4x = e^{-t}$.
7. $x''' - 2x'' - x' + 2x = e^t$.
8. $x''' + x'' - 4x' - 4x = t$.
9. $x''' + 2x'' - x' - 2x = 1$.
10. $x''' - 3x'' - x' + 3x = \sin t$.
11. $x''' + 3x'' - x' - 3x = e^t$.
12. $x''' - 3x'' + 2x' = e^{-t}$.
13. Verify that $W[e^{\alpha t}, te^{\alpha t}, t^2e^{\alpha t}] = 2e^{3\alpha t}$.

In Exercises 14–17, use Exercise 13 and variation of parameters to find the general solution.

14. $x''' - 3x'' + 3x' - x = t^{1/2}e^t$.
15. $x''' + 3x'' + 3x' + x = t^{-3}e^{-t}$.
16. $x''' + 6x'' + 12x' + 8x = t^{-1}e^{-2t}$.
17. $x''' - 6x'' + 12x' - 8x = t^{7/2}e^{2t}$.

In Exercises 18–21, you are given a fundamental set of solutions for the associated homogeneous equation and f and a_n. Solve the differential equation using variation of parameters.

18. $\{1, t, t^2, t^3\}$, $f = t$, $a_4 = t$.
19. $\{e^t, te^t, t^2e^t\}$, $f = e^t$, $a_3 = 2$.
20. $\{1, t^2, t^3\}$, $f = t^{1/2}$, $a_3 = 1$.
21. $\{1, t, t^{1/2}\}$, $f = t^3$, $a_3 = t^2$.

CHAPTER 3

The Laplace Transform

3.1 Definition and Basic Properties

We have developed several methods for solving differential equations. In this chapter we shall introduce a different type of approach that is very important in many areas of applied mathematics. The idea is to use a *transformation* that changes one set of objects and operations into a different set of objects and operations. Our transformation will be the *Laplace transform*, and it will change a linear differential equation with constant coefficients into a problem in algebra. Many design procedures in such areas as circuit and control theory are based on the algebraic form of the problem provided by the Laplace transform. The Laplace transform is also especially well suited for handling discontinuous forcing functions and impulses. But first we need to develop the basic properties of the Laplace transform. In this chapter we will use t as the independent variable, since the Laplace transform is most often useful for initial value problems in time t, defined for $t \geq 0$.

Definition of Laplace Transform

Let $f(t)$ be a function defined on the interval $t \geq 0$. The **Laplace transform** of $f(t)$ is obtained by multiplying $f(t)$ by e^{-st} and integrating from $t=0$ to $t=\infty$. The Laplace transform of $f(t)$ is then a new function of the Laplace transform variable s, and it is given by

$$F(s) = \mathcal{L}[f(t)] = \int_0^\infty e^{-st} f(t)dt, \tag{1}$$

provided the improper integral exists. $\mathcal{L}[f(t)]$ is read "the Laplace transform of $f(t)$." The notation $F(s)$ emphasizes that the Laplace transform of $f(t)$ is a function of the variable s.

Throughout this chapter, lowercase letters will denote the function of t and capital letters will denote the Laplace transform. Thus,

$$\mathcal{L}[g(t)] = G(s) \text{ and } \mathcal{L}[y(t)] = Y(s).$$

The one exception is $H(t)$ for the Heaviside (unit step) function in Section 3.4.

Example 3.1.1 *Calculating Laplace Transform from Definition*

Calculate the Laplace transform of e^t.

• SOLUTION. Recall that, by definition, for any function $h(t)$ defined on $[0, \infty)$,

$$\int_0^\infty h(t)dt = \lim_{b\to\infty} \int_0^b h(t)dt,$$

and the integral is said to *converge* if this limit exists. If the limit does not exist, the integral is said to *diverge*. Thus, if $f(t) = e^t$, then

$$\begin{aligned}\mathcal{L}[e^t] &= \int_0^\infty e^{-st}e^t\,dt = \lim_{b\to\infty}\int_0^b e^{t(1-s)}dt \\ &= \lim_{b\to\infty}\begin{cases} b & \text{if } s=1, \\ \dfrac{e^{b(1-s)}}{1-s} - \dfrac{1}{1-s} & \text{if } s\neq 1.\end{cases}\end{aligned}$$

If $s < 1$, then $1 - s > 0$ and $\lim_{b\to\infty} e^{b(1-s)} = \infty$, so that the integral diverges. If $s = 1$, then $\lim_{b\to\infty} b = \infty$ and the integral diverges. On the other hand, if $s > 1$, then $1 - s < 0$, and the integral converges,

$$\int_0^\infty e^{-st}e^t\,dt = \lim_{b\to\infty}\frac{e^{b(1-s)}}{1-s} - \frac{1}{1-s} = \frac{0}{1-s} - \frac{1}{1-s} = \frac{1}{s-1}.$$

Thus $F(s) = \mathcal{L}[e^t] = 1/(s-1)$, and the domain of the Laplace transform of e^t is $1 < s < \infty$. ♦

Example 3.1.2 *Calculating Laplace Transform from Definition*

Let

$$f(t) = \begin{cases} 0 & \text{if } 0 \le t \le 2, \\ 3 & \text{if } 2 < t \le 4, \\ 0 & \text{if } 4 < t.\end{cases}$$

Calculate $\mathcal{L}[f(t)]$.

• SOLUTION.

$$\begin{aligned}\mathcal{L}[f(t)] &= \int_0^\infty e^{-st}f(t)dt \\ &= \int_0^2 e^{-st}f(t)\,dt + \int_2^4 e^{-st}f(t)\,dt + \int_4^\infty e^{-st}f(t)dt \\ &= \int_0^2 e^{-st}0\,dt + \int_2^4 e^{-st}3\,dt + \int_4^\infty e^{-st}0\,dt \\ &= 3\int_2^4 e^{-st}\,dt = \left.\frac{3e^{-st}}{-s}\right|_2^4 = \frac{3e^{-4s}}{-s} + \frac{3e^{-2s}}{s}.\end{aligned}$$

♦

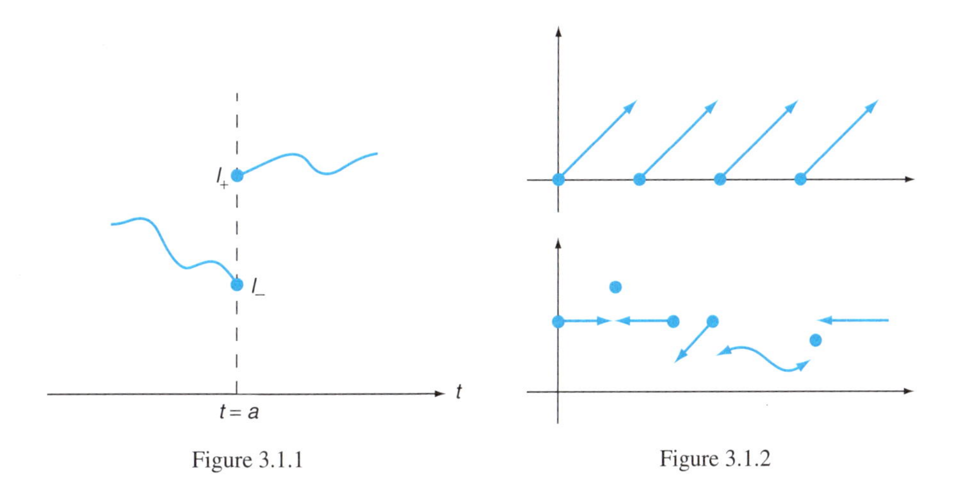

Figure 3.1.1

Figure 3.1.2

Obviously, we do not want to have to compute every Laplace transform from the definition. Tables of Laplace transforms have been developed. There is a short table (Table 3.1.1) at the end of this section, and a longer table (Table 3.2.1) appears in Section 3.2. But first we need to establish when the Laplace transform exists. One important feature of the Laplace transform is its ability to work with some functions, like $f(t)$ in Example 3.1.2, that are not everywhere continuous.

A function $f(t)$ defined on $[0, \infty)$ has a **jump discontinuity** at $a \in [0, \infty)$ if the one-sided limits

$$\lim_{t \to a^+} f(t) = l_+ \quad \text{and} \quad \lim_{t \to a^-} f(t) = l_-$$

exist but the function $f(t)$ is not continuous at $t = a$. The most important case is when $l_+ \neq l_-$, as illustrated in figure 3.1.1.

A function $f(t)$ is **piecewise continuous** on $[0, \infty)$ if, for every number $B > 0$, $f(t)$ is continuous on $[0, B]$ except possibly for a finite number of jump discontinuities. Note that a piecewise continuous function can have an infinite number of discontinuities, but there can be only a finite number of discontinuities on a finite interval. Two piecewise continuous functions are graphed in Figure 3.1.2.

A function $f(t)$ on $[0, \infty)$ is said to be of **exponential order** if there exist constants α, M, such that $|f(t)| \leq Me^{\alpha t}$ for $t \geq 0$. That is, as $t \to \infty$, $f(t)$ grows more slowly than a multiple of some exponential. Both examples in figure 3.1.2 are bounded functions and hence of exponential order (take $\alpha = 0$). Note that e^{t^2} is not of exponential order, since t^2 eventually grows faster than αt for any constant α (see Exercise 76 at the end of this section).

THEOREM 3.1.1 Existence of the Laplace Transform *If $f(t)$ is piecewise continuous and of exponential order on $[0, \infty)$, so that $|f(t)| \leq Me^{\alpha t}$, then $\mathcal{L}[f] = F(s)$ exists and is defined at least for $s > \alpha$.*

Example 3.1.3 *Laplace Transform of an Exponential*

Compute $\mathcal{L}[e^{at}]$, where a is a constant.

• SOLUTION

$$\begin{aligned}
\mathcal{L}[e^{at}] &= \int_0^{\infty} e^{-st} e^{at}\, dt \\
&= \lim_{b\to\infty} \int_0^{b} e^{t(a-s)}\, dt \\
&= \lim_{b\to\infty} \left. \frac{e^{t(a-s)}}{a-s} \right|_{t=0}^{t=b} \qquad (a \neq s) \\
&= \lim_{b\to\infty} \frac{e^{b(a-s)}}{a-s} - \frac{1}{a-s} \quad \left(= \left. \frac{e^{-t(s-a)}}{-(s-a)} \right|_{t=0}^{t=\infty} \right) \\
&= \begin{cases} -\dfrac{1}{a-s} & \text{if } s > a, \\ \text{diverges} & \text{if } s < a. \end{cases}
\end{aligned}$$

Thus

$$\mathcal{L}[e^{at}] = \frac{1}{s-a} \quad (s > a). \quad \text{(T2)} \tag{2}$$

Formulas which appear in the Laplace transform tables will be referred to by T plus the formula number in the table. ♦

An important special case of (2) is when $a = 0$. In this case,

$$\mathcal{L}[1] = \frac{1}{s}, \quad (s > 0), \quad \text{(T1)} \tag{3}$$

Inverse Laplace Transform

In the applications to be developed, sometimes we have to find a Laplace transform $F(s)$ given $f(t)$. At other times we will need to find a function $f(t)$ such that $\mathcal{L}[f(t)] = F(s)$ given $F(s)$. That is, we will have to invert the Laplace transform. In this situation, it is important to know that $f(t)$ is unique. Since the Laplace transform is given in terms of an integral, changing a few values of $f(t)$ will not change the Laplace transform (see Exercise 75). There exists a formula for the inverse Laplace transform, but it requires complex analysis.

THEOREM 3.1.2 Uniqueness of Inverse Laplace Transform *If $F(s)$ is given and there is a continuous function $f(t)$ such that $\mathcal{L}[f(t)] = F(s)$, then $f(t)$ is the only continuous function for which $\mathcal{L}[f(t)] = F(s)$.*

Since, for all practical purposes (at least ours), $F(s)$ uniquely determines $f(t)$, we may denote this by $f(t) = \mathcal{L}^{-1}[F(s)]$ and call $\mathcal{L}^{-1}$ the **inverse Laplace transform.**

Example 3.1.4 *Exponential Example*

In Example 3.1, we showed that $\mathcal{L}[e^{at}] = 1/(s-a)$. Thus,

$$\mathcal{L}^{-1}\left[\frac{1}{s-a}\right] = e^{at}.$$ ♦

It is helpful to memorize this result (and only a few others).

Linearity Property

The Laplace transform has several important properties. One is that it satisfies the linearity property. That is, the Laplace transform of a linear combination of functions equals the same linear combination of their Laplace transforms:

$$\mathcal{L}[c_1 f(t) + c_2 g(t)] = c_1 \mathcal{L}[f(t)] + c_2 \mathcal{L}[g(t)], \tag{4}$$

where c_1, c_2 are constants.

As a corollary, the inverse Laplace transform also satisfies the linearity property

$$\mathcal{L}^{-1}[c_1 F(s) + c_2 G(s)] = c_1 \mathcal{L}^{-1}[F(s)] + c_2 \mathcal{L}^{-1}[G(s)]. \tag{5}$$

Verification of (4)

$$\begin{aligned}\mathcal{L}[c_1 f(t) + c_2 g(t)] &= \int_0^\infty e^{-st}[c_1 f(t) + c_2 g(t)]dt \\ &= c_1 \int_0^\infty e^{-st} f(t)\, dt + c_2 \int_0^\infty e^{-st} g(t)\, dt \\ &= c_1 \mathcal{L}[f(t)] + c_2 \mathcal{L}[g(t)].\end{aligned}$$

Example 3.1.5 *Using Linearity When Taking Laplace Transforms*

Compute $\mathcal{L}[3e^t + 5e^{-2t} + 6]$ using (2), (3), and the linearity property (4).

• SOLUTION

$$\begin{aligned}\mathcal{L}[3e^t + 5e^{-2t} + 6] &= 3\mathcal{L}[e^t] + 5\mathcal{L}[e^{-2t}] + 6\mathcal{L}[1] \\ &= \frac{3}{s-1} + \frac{5}{s+2} + \frac{6}{s}.\end{aligned}$$ ♦

Example 3.1.6 *Using Linearity When Taking Inverse Laplace Transforms*

Given $F(s) = \frac{3}{s} + \frac{6}{s-3}$, find $f(t) = \mathcal{L}^{-1}[F(s)]$.

• SOLUTION. From (2) and (3), we have

$$e^{at} = \mathcal{L}^{-1}\left[\frac{1}{s-a}\right] \quad \text{and} \quad 1 = \mathcal{L}^{-1}\left[\frac{1}{s}\right].$$

Thus

$$\mathcal{L}^{-1}\left[\frac{3}{s}+\frac{6}{s-3}\right]=3\mathcal{L}^{-1}\left[\frac{1}{s}\right]+6\mathcal{L}^{-1}\left[\frac{1}{s-3}\right]$$
$$=3\cdot 1+6e^{st}=3+6e^{3t}.$$ ♦

Sinusoidal Functions

From Euler's formula, $e^{ibt}=\cos bt+i\sin bt$, we know that cosine and sine are linear combinations of imaginary exponentials:

$$\cos bt=\frac{1}{2}e^{ibt}+\frac{1}{2}e^{-ibt},$$
$$\sin bt=\frac{1}{2i}e^{ibt}-\frac{1}{2i}e^{-ibt}. \tag{6}$$

For example, we can derive the Laplace transform of $\cos bt$. Using (6) and the linearity property (4), we get

$$\mathcal{L}[\cos bt]=\frac{1}{2}\mathcal{L}[e^{ibt}]+\frac{1}{2}\mathcal{L}[e^{-ibt}]. \tag{7}$$

We have shown that $\mathcal{L}[e^{at}]=\frac{1}{s-a}$ for $s>a$, which is valid for real exponentials. If a is complex, we can repeat the derivation of its Laplace transform (see Example 3.1.3), and we find that there are no changes except that the restriction $s>a$ should be $s>\mathrm{Re}(a)$. That is, s must be greater than the real part of a. Thus, using a common denominator, we obtain

$$\mathcal{L}[\cos bt]=\frac{1}{2}\frac{1}{s-ib}+\frac{1}{2}\frac{1}{s+ib}=\frac{1}{2}\frac{s+ib+s-ib}{(s-ib)(s+ib)},$$

so that

$$\mathcal{L}[\cos bt]=\frac{s}{s^2+b^2} \quad \text{(T4)}. \tag{8}$$

In a similar manner, we can show

$$\mathcal{L}[\sin bt]=\frac{b}{s^2+b^2} \quad \text{(T3)}. \tag{9}$$

These are very fundamental formulas, and so they are presented in the tables. In Exercises 9 and 10 at the end of this section, these results are derived by explicit integration of the defining Laplace transform (1).

When using (T4), (T5), and several formulas from the next section to compute inverse Laplace transforms, we will often have to adjust the constants so that the expression corresponds to one in the table.

Example 3.1.7 *Inverse Laplace Transform: Denominator is s^2+b^2*

Given $F(s)=\frac{11}{s^2+17}$, find $f(t)=\mathcal{L}^{-1}[F(s)]$.

• SOLUTION From (9) we have $\mathcal{L}^{-1}[\frac{b}{s^2+b^2}]=\sin bt$. Then,

$$\frac{11}{s^2+17}=\frac{11}{s^2+(\sqrt{17})^2}=\frac{11}{\sqrt{17}}\frac{\sqrt{17}}{s^2+(\sqrt{17})^2},$$

which involves (9) with $b=\sqrt{17}$. Hence,

$$\mathcal{L}^{-1}\left[\frac{11}{s^2+17}\right]=\frac{11}{\sqrt{17}}\sin\sqrt{17}t.$$ ♦

Frequently, $F(s)$ will have to be expressed as a sum of several terms that can each be evaluated from known formulas.

Example 3.1.8 *Inverse Laplace Transform: Denominator is s^2+b^2*

Given $F(s)=\frac{2s+5}{s^2+9}$, find $f(t)=\mathcal{L}^{-1}[F(s)]$.

• SOLUTION

$$\begin{aligned}\mathcal{L}^{-1}\left[\frac{2s+5}{s^2+9}\right]&=\mathcal{L}^{-1}\left[2\frac{s}{s^2+9}+\frac{5}{3}\frac{3}{s^2+9}\right]\\&=2\mathcal{L}^{-1}\left[\frac{s}{s^2+9}\right]+\frac{5}{3}\,L^{-1}\left[\frac{3}{s^2+9}\right]\\&=2\cos 3t+\frac{5}{3}\sin 3t \quad \text{(by (8) and (9))}.\end{aligned}$$ ♦

Polynomials

Powers and polynomials also have elementary Laplace transforms. There are many ways to derive

$$\mathcal{L}[t^n]=\frac{n!}{s^{n+1}} \quad \text{(T9)}. \tag{10}$$

The derivation we present requires little knowledge of Laplace transforms and illustrates how many of the formulas can be directly derived using calculus. (Further properties of Laplace transforms discussed later in this section enable one to derive (10) more easily.)

Derivation of (10)

From the definition of the Laplace transform,

$$\mathcal{L}[t^n]=\int_0^\infty t^n e^{-st}\,dt.$$

We integrate by parts ($\int_a^b u\,dv=uv|_a^b-\int_a^b v\,du$) one time,

$$u=t^n, \qquad dv=e^{-st}dt,$$

$$du=nt^{n-1}dt, \qquad v=-\frac{1}{s}e^{-st},$$

so that

$$\mathcal{L}[t^n] = t^n \left(-\frac{1}{s}e^{-st}\right)\Bigg|_{t=0}^{\infty} + \frac{n}{s}\int_0^{\infty} t^{n-1}e^{-st}\,dt. \tag{11}$$

If $n \geq 1$, the first term on the right vanishes, (assuming $s > 0$), since

$$t^n \left(-\frac{1}{s}e^{-st}\right)\Bigg|_{t=0}^{\infty} = -\frac{1}{s}\lim_{b\to\infty}[b^n e^{-bs}] - 0 = 0.$$

Thus, (11) becomes a recursion formula for the Laplace transform

$$\mathcal{L}[t^n] = \frac{n}{s}\mathcal{L}[t^{n-1}]. \tag{12}$$

For example, evaluating (12) for $n = 1$ yields

$$\mathcal{L}[t] = \frac{1}{s}\mathcal{L}[1] = \frac{1}{s^2}, \tag{13}$$

using the fact (3) that $\mathcal{L}[1] = 1/s$. Evaluating (12) at subsequent values of n yields

$$\begin{aligned} n = 2: &\quad \mathcal{L}[t^2] = \frac{2}{s}\mathcal{L}[t] = \frac{2}{s^3}, \\ n = 3: &\quad \mathcal{L}[t^3] = \frac{3}{s}\mathcal{L}[t^2] = \frac{3\cdot 2}{s^4}. \end{aligned}$$

The pattern has now emerged, and we are able to conclude that (10) is valid.

Example 3.1.9 *Laplace Transform of a Polynomial*

Find the Laplace transform of $f(t) = t^4 + 3t^2 + 5$.

• SOLUTION. Using the linearity property and (10) yields

$$\mathcal{L}[t^4 + 3t^2 + 5] = \mathcal{L}[t^4] + 3\mathcal{L}[t^2] + 5\mathcal{L}[1] = \frac{4!}{s^5} + 3\frac{2!}{s^3} + 5\frac{1}{s}.$$ ♦

Example 3.1.10 *Inverse Laplace Transform: Denominator is s^n*

If $F(s) = 3/s^5$, find $f(t)$.

• SOLUTION. Using (10) or (T9) with $n = 4$,

$$f(t) = \mathcal{L}^{-1}[F(s)] = \mathcal{L}^{-1}\left[\frac{3}{s^5}\right] = \frac{3}{4!}\mathcal{L}^{-1}\left[\frac{4!}{s^5}\right] = \frac{3}{4!}t^4.$$ ♦

A short table (table 3.1.1) of elementary Laplace transforms and their corresponding inverse Laplace transform formulas appears later in this section. Table 3.2.1 in the next section is more extensive.

The first group of formulas in the tables give the Laplace transform of particular functions. The second group are generally operational formulas. Some of these will be developed shortly. The third group are additional formulas developed in the exercises and other sections. To simplify referencing, formulas have the same numbers in both tables; $H(t)$ is a step function, to be discussed in Section 3.4; and $\delta(t)$ is an impulse function, to be discussed in Section 3.6.

3.1.1 The shifting theorem (multiplying by an exponential)

Many of the more complex Laplace transform formulas are derived from simpler formulas using the shifting theorem.

THEOREM 3.1.3 Shifting Theorem

If $\mathcal{L}[f(t)] = F(s)$, *then*

$$\mathcal{L}[e^{ct} f(t)] = F(s-c). \quad \text{(T16)} \tag{14}$$

This shows that for Laplace transforms, multiplying $f(t)$ by e^{ct} is equivalent to shifting the transform variable from s to $s-c$.

Verification of (14)

$$\begin{aligned}\mathcal{L}[e^{ct} f(t)] &= \int_0^\infty e^{-st} e^{ct} f(t)dt \\ &= \int_0^\infty e^{-(s-c)t} f(t)dt \\ &= F(s-c),\end{aligned}$$

using (1).

Example 3.1.11 *Shifting Theorem*

Find the Laplace transform of $t^4 e^{2t}$.

• SOLUTION. We will apply the shifting theorem, (14), with $c=2$ and $f(t)=t^4$, so that $F(s)=4!/s^5$:

$$\mathcal{L}[e^{ct} f(t)] = F(s-c)$$

$$\text{so that} \quad \mathcal{L}[e^{2t} t^4] = F(s-2) = \frac{4!}{(s-2)^5}. \qquad \blacklozenge$$

Example 3.1.12 *Shifting Theorem*

Find the Laplace transform of $e^{-3t}\cos 6t$ using the Laplace transform of $\cos 6t$.

• SOLUTION. We will apply the shifting theorem (14), with $c=-3$ and $f(t)=\cos 6t$, so that $F(s)=s/(s^2+36)$. According to the shifting theorem,

$$\mathcal{L}[e^{ct} f(t)] = F(s-c)$$

$$\text{so that} \quad \mathcal{L}[e^{-3t}\cos 6t] = F(s+3) = \frac{s+3}{(s+3)^2+36}. \qquad \blacklozenge$$

Exponential times Sinusoidal Function

Important formulas for the Laplace transform of simple exponentials time sinusoidal functions may be obtained by generalizing the previous example:

$$\mathcal{L}[e^{at}\cos bt] = \frac{s-a}{(s-a)^2+b^2} \quad \text{(T6)} \tag{15}$$

and

$$\mathcal{L}[e^{at}\sin bt] = \frac{b}{(s-a)^2+b^2}. \quad \text{(T5)} \tag{16}$$

Exponential times Powers

The shifting theorem (14) may also be applied to the Laplace transform formula for powers (10):

$$\mathcal{L}[e^{at}t^n] = \frac{n!}{(s-a)^{n+1}}. \quad \text{(T11)} \tag{17}$$

This generalizes Example 3.1.1.

Exercises

1. Sketch the following function and explain why it is not piecewise continuous:

$$f(t) = \begin{cases} t, & 0 \le t < 2, \\ 3, & 2 = t, \\ \dfrac{1}{t-2}, & 2 < t. \end{cases}$$

In Exercises 2–4, sketch the function and explain why it is piecewise continuous.

2.

$$f(t) = \begin{cases} 1, & 0 \le t < 1, \\ -1, & 1 < t \le 2, \\ 1, & 2 < t. \end{cases}$$

3.

$$f(t) = \begin{cases} t, & 0 \le t < 1, \\ \frac{1}{2}, & 1 = t, \\ t-1, & 1 < t. \end{cases}$$

4.

$$f(t) = \begin{cases} e^{-t}, & 0 \le t < 1, \\ 3, & 1 = t, \\ e^{-t}, & 1 < t. \end{cases}$$

In Exercises 5–7, compute $\mathcal{L}[f(t)]$, using the formula $\mathcal{L}[f] = \int_0^\infty e^{-st} f(t)dt$.

5.

$$f(t) = \begin{cases} 1, & 0 \le t \le 1, \\ 0, & 1 < t. \end{cases}$$

6.

$$f(t) = \begin{cases} 1, & 0 \le t < 1, \\ -1, & 1 \le t < 2, \\ 0, & 2 \le t. \end{cases}$$

7.

$$f(t) = \begin{cases} t, & 0 \le t < 1, \\ 2-t, & 1 \le t \le 2, \\ 0, & 2 < t. \end{cases}$$

8. Verify that $\mathcal{L}[t] = 1/s^2$, using the formula $\mathcal{L}[f] = \int_0^\infty e^{-st} f(t)dt$.
9. Verify that $\mathcal{L}[\sin bt] = b/(s^2 + b^2)$, using the formula $\mathcal{L}[f] = \int_0^\infty e^{-st} f(t)dt$.
10. Verify that $\mathcal{L}[\cos bt] = s/(s^2 + b^2)$, using the formula $\mathcal{L}[f] = \int_0^\infty e^{-st} f(t)dt$.
11. Using((2), show that $\mathcal{L}[\sinh at] = a/(s^2 - a^2)$.
12. Using (2), show that $\mathcal{L}[\cosh at] = s/(s^2 - a^2)$.
13. Using (6), show that $\mathcal{L}[\sin bt] = b/(s^2 + b^2)$.

In Exercises 14–46, use (T1)–(T4) and (T9) (formulas (2), (3), (4), (8), (9), and (10)) to compute $\mathcal{L}[f]$.

14. $f(t) = 3t + 2$.
15. $f(t) = 3\cosh 2t$.
16. $f(t) = 4e^{3t} + 6e^{-t}$.
17. $f(t) = 5\sin 6t$.
18. $f(t) = 4\sin 3t + 5\cos 7t$.
19. $f(t) = -t + 3$.
20. $f(t) = e^t - e^{-t} + e^{2t}$.
21. $f(t) = 2 + \cos 5t$.
22. $f(t) = 2\sin 3t + 4\sin 5t$.
23. $f(t) = \sinh 3t$.
24. $f(t) = t + 3 - e^t$.
25. $f(t) = 2e^{-t} + 6e^{3t}$.
26. $f(t) = -4\sin 2t$.
27. $f(t) = 3t - 1 + \cosh 2t$.
28. $f(t) = 7e^{-5t} - 9e^{-3t} - 6$.
29. $f(t) = 7t^3 + 11t + 8$.
30. $f(t) = 2t^5 + 5t^2$.
31. $f(t) = 3t^4 + 4t^3$.
32. $f(t) = 2t^6 + 3t^3 + 6t^2$.

In Exercises 33–46, use the shifting theorem (14) or (T16) to compute $\mathcal{L}[f]$

33. $f(t) = e^{3t}\sin 5t$.

34. $f(t) = e^{5t} \sin 3t$.
35. $f(t) = e^{4t} \cos 7t$.
36. $f(t) = e^{7t} \cos 4t$.
37. $f(t) = e^{-3t} \sin 5t$.
38. $f(t) = e^{-5t} \sin 3t$.
39. $f(t) = e^{-4t} \cos 7t$.
40. $f(t) = e^{-7t} \cos 4t$.
41. $f(t) = e^{2t} t^6$.
42. $f(t) = e^{6t} t^4$.
43. $f(t) = e^{-2t} t^6$.
44. $f(t) = e^{-6t} t^4$.
45. $f(t) = e^{3t}(t^5 + t^3 + 1)$.
46. $f(t) = e^{-4t}(t^6 + 6t^2 + 5t + 4)$.

In Exercises 47–66, $F(s)$ is given. Use formulas (T1)–(T4) and (T9) to compute $f(t) = \mathcal{L}^{-1}[F(s)]$.

47. $\dfrac{1}{s^2} - \dfrac{1}{s}$.
48. $\dfrac{3}{s-3} + \dfrac{4}{s+3}$.
49. $\dfrac{1}{s^2+9}$.
50. $\dfrac{7s+1}{s^2+4}$.
51. $\dfrac{1+s}{s^2}$.
52. $\dfrac{1+s}{s^3}$.
53. $\dfrac{3}{s} - \dfrac{7}{s^2} + \dfrac{19}{s^2+1}$.
54. $\dfrac{2}{s} + \dfrac{3}{s+1} - \dfrac{7}{s-8}$.
55. $\dfrac{3s+7}{s^2+16}$.
56. $\dfrac{11}{s^2} - \dfrac{2}{s}$.
57. $\dfrac{5}{s+3} + \dfrac{7}{s-5}$.
58. $\dfrac{5s+4}{s^2+9}$.
59. s^{-6}.
60. $4s^{-9}$.
61. $\dfrac{2}{s^2+9}$.

62. $\dfrac{5s}{s^2+13}$.
63. $\dfrac{3}{s^{10}}$.
64. $\dfrac{2}{s^5}$.
65. $\dfrac{3}{2s^2+7}$.
66. $\dfrac{-s+1}{3s^2+11}$.

In Exercises 67–74, $F(s)$ is given. Use formulas (T1)–(T4) and (T9) to compute $f(t)=\mathcal{L}^{-1}[F(s)]$. The shifting theorem (T16) may be required.

67. $\dfrac{1}{(s-4)^3}$.
68. $\dfrac{5}{(s-3)^4}$.
69. $\dfrac{(s-5)}{(s-5)^2+9}$.
70. $\dfrac{(s+6)}{(s+6)^2+4}$.
71. $\dfrac{7}{(s-7)^2+16}$.
72. $\dfrac{8}{(s+4)^2+25}$.
73. $\dfrac{s+3}{(s+3)^2+5}$.
74. $\dfrac{6}{(s-2)^2+3}$.
75. Let

$$f(t)=\begin{cases}2 & \text{if } t=1,\\ t & \text{if } t\neq 1.\end{cases}$$

Show that $\mathcal{L}[f(t)]=\mathcal{L}[t]$.

76. Show that $\lim_{t\to\infty} e^{-\alpha t}e^{t^2}=\infty$ for any α. Use this to show that e^{t^2} is not of exponential order.
77. Determine if $e^{\sqrt{t}}$ is of exponential order.
78. Show that e^{t^β} is of exponential order if $0\le\beta\le 1$ and is not of exponential order if $\beta>1$.
79. Prove Theorem 3.1.1, using the following two facts:
a) If $\int_0^\infty |h(t)|dt$ converges, then $\int_0^\infty h(t)\,dt$ converges.
b) If $0<g(t)<r(t)$ and $\int_0^\infty r(t)\,dt$ converges, then $\int_0^\infty g(t)\,dt$ converges.
80. Verify by direct integration that $\mathcal{L}[e^{at}]=1/(s-a)$ if a is a complex number. Note that s must be greater than the real part of a.

3.1.2 Derivative Theorem (Multiplying by t)

Some additional results concerning Laplace transforms follow from the definition

$$F(s) = \int_0^\infty e^{-st} f(t)dt. \tag{18}$$

If we differentiate (18) with respect to s, we obtain

$$F'(s) = \frac{d}{ds}\int_0^\infty e^{-st} f(t)dt = \int_0^\infty \frac{\partial}{\partial s} e^{-st} f(t)dt = \int_0^\infty -te^{-st} f(t)dt$$

from the laws of differentiating integrals. Thus,

$$\mathcal{L}[tf(t)] = -F'(s). \quad \text{(T21)} \tag{19}$$

For Laplace transforms, multiplying $f(t)$ by t corresponds to taking minus the derivative of the transform $f(s)$ of $f(t)$.

Example 3.1.13 *Using the Derivative Theorem*

Compute the Laplace transform of different powers of t, $\mathcal{L}[t^n]$, using (19).

- SOLUTION. We already know that if $f(t) = 1$, then $F(s) = \frac{1}{s}$. Thus,

$$\mathcal{L}[1] = \frac{1}{s},$$

$$\mathcal{L}[t] = \mathcal{L}[t \cdot 1] = -\frac{d}{ds}\mathcal{L}[1] = -\frac{d}{ds}\left(\frac{1}{s}\right) = \frac{1}{s^2},$$

$$\mathcal{L}[t^2] = \mathcal{L}[t \cdot t] = -\frac{d}{ds}\mathcal{L}[t] = -\frac{d}{ds}\left(\frac{1}{s^2}\right) = \frac{2}{s^3},$$

$$\mathcal{L}[t^3] = \mathcal{L}[t \cdot t^2] = -\frac{d}{ds}\mathcal{L}[t^2] = -\frac{d}{ds}\left(\frac{2}{s^3}\right) = \frac{3!}{s^4}.$$

From this we can generalize that

$$\mathcal{L}[t^n] = \frac{n!}{s^{n+1}}. \quad \text{(T9)} \tag{20}$$

In this derivation successive usage of (19) has been made. Equation (20) was derived earlier (see (12)) using successive integration by parts. ♦

Example 3.1.14 *Using the Derivative Theorem*

Find $\mathcal{L}[te^{at}]$ using (19).

- SOLUTION. We use (19) with $f(t) = e^{at}$, so that $F(s) = \frac{1}{s-a}$. Thus,

$$\mathcal{L}[te^{at}] = -\frac{d}{ds}\frac{1}{s-a} = \frac{1}{(s-a)^2}.$$ ♦

TABLE 3.1.1
Short Table of Laplace Transforms

	$f(t)$	$\mathcal{L}[f(t)] = F(s) = \int_0^\infty e^{-st} f(t)dt$
(T1)	1	$\frac{1}{s}$
(T2)	e^{at}	$\frac{1}{s-a}$
(T3)	$\sin bt$	$\frac{b}{s^2+b^2}$
(T4)	$\cos bt$	$\frac{s}{s^2+b^2}$
(T9)	t^n	$\frac{n!}{s^{n+1}}$
(T15)	$f(t-c)H(t-c)$	$e^{-cs}F(s) \quad c>0$
(T16)	$e^{ct}f(t)$	$F(s-c)$
(T18)	$\frac{df}{dt} = f'(t)$	$sF(s) - f(0)$
(T19)	$\frac{d^2f}{dt^2} = f''(t)$	$s^2F(s) - sf(0) - f'(0)$
(T21)	$tf(t)$	$-F'(s)$
(T23)	$\int_0^t f(\tau)d\tau$	$\frac{F(s)}{s}$
(T24)	$\int_0^t f(\tau)g(t-\tau)d\tau$	$F(s)G(s)$

Higher powers of t can be obtained in the same way to give

$$\mathcal{L}[t^n e^{at}] = \frac{n!}{(s-a)^{n+1}}. \quad \text{(T11)} \tag{21}$$

This can also be derived from (20) using the shifting theorem (see (17)).

Example 3.1.15 *Using the Derivative Theorem*

Find $\mathcal{L}[t \sin bt]$.

• SOLUTION. We use (19) with $f(t) = \sin bt$, so that $F(s) = \frac{b}{s^2+b^2}$. Thus,

$$\mathcal{L}[t \sin bt] = -\frac{d}{ds}\left(\frac{b}{s^2+b^2}\right) = \frac{2bs}{(s^2+b^2)^2}, \quad \text{(T25)} \tag{22}$$

which we derive later in a different way. The corresponding result for cosines,

$$\mathcal{L}[t \cos bt] = \frac{s^2-b^2}{(s^2+b^2)^2}, \quad \text{(T26)} \tag{23}$$

can also be derived using (19). ♦

Exercises

In Exercises 81–90, compute the Laplace transform using the derivative theorem (19).

81. te^{5t}.
82. te^{-5t}.
83. $t\cos 5t$.
84. $t\sin 5t$.
85. $t^2\cos 3t$.
86. $t^2\sin 3t$.
87. $te^{5t}\sin 3t$.
88. $te^{5t}\cos 2t$.
89. $te^{-4t}\cos 5t$.
90. $te^{-4t}\sin 6t$.

In Exercises 91–100, compute the inverse Laplace transform using the derivative theorem (19).

91. $(s+3)^{-6}$.
92. $(s-2)^{-8}$.
93. $\dfrac{5s}{(s^2+9)^2}$.
94. $\dfrac{2s}{(s^2+25)^2}$.
95. $\dfrac{s^2-9}{(s^2+9)^2}$.
96. $\dfrac{s^2-25}{(s^2+25)^2}$.
97. $\dfrac{5}{(s^2+9)^2}$.
98. $\dfrac{3}{(s^2+25)^2}$.
99. $\dfrac{2s+1}{(s^2+1)^2}$.
100. $\dfrac{3s-5}{(s^2+4)^2}$.
101. Show that, if $\mathcal{L}[f(t)]=F(s)$, then $\mathcal{L}[f(ct)]=\frac{1}{c}F\left(\frac{s}{c}\right)$, where c is a positive constant.
102. Using (T21), verify by induction that $\mathcal{L}[t^n f(t)]=(-1)^n F^{(n)}(s)$. This is (T22).
103. Using Exercise 102 and (T2), show that

$$\mathcal{L}[t^n e^{at}]=\frac{n!}{(s-a)^{n+1}}.$$

This is (T11).

104. Partial fractions result in terms like $(As+B)(s^2+b^2)^{-2}$, which are not readily related to (22), (23). Thus (T27) is helpful. Using (22), (23), and (T3), show that

$$\mathcal{L}^{-1}\left[\frac{2b^2}{(s^2+b^2)^2}\right]=\frac{1}{b}\sin bt - t\cos bt. \quad \text{(T27)}$$

3.2 Inverse Laplace Transforms (Roots, Quadratics, and Partial Fractions)

Many inverse Laplace transforms can be obtained directly from the tables based on the standard Laplace transforms and properties of the previous section. However, in solving differential equations, often we will need the inverse Laplace transform of $F(s)$, where $F(s)$ is a rational function of s. That is, $F(s)$ is a ratio of polynomials in s. In this case partial fractions may be used to express $F(s)$ as a sum of simpler terms.

When the denominator of $F(s)$ has a quadratic factor, as^2+bs+c, we use the discriminant to distinguish between real roots ($b^2-4ac>0$), complex roots ($b^2-4ac<0$), and repeated real roots ($b^2-4ac=0$). We will discuss examples of these types.

Simple Real Roots

When the denominator of $F(s)$ only has simple real roots, then the inverse Laplace transform of the partial fraction decomposition is a straightforward application of the elementary exponential formula $\mathcal{L}[e^{at}]=\frac{1}{s-a}$. Also, the coefficients in the partial fraction expansion are easily determined.

Example 3.2.1 *Simple Real Roots in Denominator*

Given $F(s)=\frac{3}{s^2-4}$, find $f(t)=\mathcal{L}^{-1}[F(s)]$.

• SOLUTION. Using partial fractions, we have

$$\frac{3}{s^2-4}=\frac{3}{(s-2)(s+2)}=\frac{A}{s-2}+\frac{B}{s+2}.$$

Multiplying by the denominator gives $3=A(s+2)+B(s-2)$. Note that $s=2,-2$ are the roots of s^2-4. Letting $s=2,-2$ in $3=A(s+2)+B(s-2)$ gives $A=\frac{3}{4}$, $B=-\frac{3}{4}$. Thus,

$$\mathcal{L}^{-1}\left[\frac{3}{s^2-4}\right]=\frac{3}{4}\mathcal{L}^{-1}\left[\frac{1}{s-2}\right]-\frac{3}{4}\mathcal{L}^{-1}\left[\frac{1}{s+2}\right]$$

$$=\frac{3}{4}e^{2t}-\frac{3}{4}e^{-2t}.$$

♦

Example 3.2.2 *Simple Real Roots in Denominator*

Given $F(s) = \frac{2s+1}{s^3-3s^2+2s}$, find $f(t) = \mathcal{L}^{-1}[F(s)]$.

• SOLUTION. Using partial fractions, we have

$$\frac{2s+1}{s^3-3s^2+2s} = \frac{2s+1}{s(s-1)(s-2)} = \frac{A}{s} + \frac{B}{s-1} + \frac{C}{s-2}.$$

Multiplying by the denominator gives $2s+1 = A(s-1)(s-2) + Bs(s-2) + Cs(s-1)$. The roots are $s = 0, 1, 2$. Letting $s = 0, 1, 2$, gives $A = \frac{1}{2}$, $B = -3$, $C = \frac{5}{2}$. Thus,

$$\mathcal{L}^{-1}\left[\frac{2s+1}{s^3-3s^2+2s}\right] = \frac{1}{2}\mathcal{L}^{-1}\left[\frac{1}{s}\right] - 3\mathcal{L}^{-1}\left[\frac{1}{s-1}\right] + \frac{5}{2}\mathcal{L}^{-1}\left[\frac{1}{s-2}\right]$$
$$= \frac{1}{2} - 3e^t + \frac{5}{2}e^{2t}.$$

♦

Complex Roots for Quadratics (Completing the Square)

When working with Laplace transforms, we frequently get expressions where the denominator is a quadratic,

$$\frac{As+B}{as^2+bs+c},$$

when it has complex roots. If the denominator has purely imaginary roots (for example, the denominator is s^2+5), then the formula involving sines and cosines may be used directly as in Example 3.1.8. If the denominator has roots with nonzero real and imaginary parts ($b^2 - 4ac < 0$), then completing the square enables the inverse Laplace transform to be obtained.

Example 3.2.3 *Complex Roots: Completing the Square*

Given $F(s) = \frac{3s-17}{s^2-6s+13}$, find $f(t) = \mathcal{L}^{-1}[F(s)]$.

• SOLUTION. Since $b^2 - 4ac = 36 - 52 = -16 < 0$, the denominator does not have real factors. If we were to try to apply partial fractions, we would get the same expression back. $F(s)$ cannot be simplified further by partial fractions. Thus we must complete the square; $s^2 - 6s + 13 = (s-3)^2 + 4 = (s-3)^2 + 2^2$. There are two ways to proceed. One is to express the numerator $3s - 17$ in powers of $s - 3$ in order to use table 3.1.1 based on the shifting theorem (T16). Alternatively, we may go directly to the answer using the long table (table 3.2.1). We illustrate the second approach first. There are two formulas (T5), (T6) in table 3.2.1 with a denominator of $(s-3)^2 + 2^2$,

$$\frac{s-3}{(s-3)^2+2^2} \quad \text{and} \quad \frac{2}{(s-3)^2+2^2}.$$

Thus we wish to find constants A, B so that

$$\begin{aligned}\frac{3s-17}{(s-3)^2+2^2} &= A\frac{(s-3)}{(s-3)^2+2^2} + B\frac{2}{(s-3)^2+2^2} \\ &= \frac{A(s-3)+B2}{(s-3)^2+2^2}.\end{aligned} \tag{1}$$

The inverse Laplace transform of (1) is

$$f(t) = Ae^{3t}\cos 2t + Be^{3t}\sin 2t.$$

From (1), A, B must satisfy

$$3s - 17 = A(s-3) + 2B.$$

Equating powers of s gives $A = 3$. Letting $s = 3$ gives that $2B = -8$ or $B = -4$. Thus

$$f(t) = \mathcal{L}^{-1}\left[\frac{3s-17}{s^2-6s+13}\right] = 3e^{3t}\cos 2t - 4e^{3t}\sin 2t.$$

Alternatively, we can compute $f(t)$ using Table 3.1.1 and the shifting theorem (T16). We express $\frac{3s-17}{s^2-6s+13}$ as a function of $s-3$:

$$\frac{3s-17}{s^2-6s+13} = \frac{3(s-3)-8}{(s-3)^2+2^2}.$$

But, replacing $s-3$ by s, we compute that

$$\mathcal{L}^{-1}\left[\frac{3s-8}{s^2+2^2}\right] = \mathcal{L}^{-1}\left[3\frac{s}{s^2+2^2} - \frac{8}{2}\frac{2}{s^2+2^2}\right] = 3\cos 2t - 4\sin 2t.$$

Thus, by the shifting theorem (T16), we have

$$f(t) = \mathcal{L}^{-1}\left[\frac{3s-17}{s^2-6s+13}\right] = 3e^{3t}\cos 2t - 4e^{3t}\sin 2t.$$ ♦

The next two examples have simple real and simple complex roots.

Example 3.2.4 *Real and Complex Roots*

Given $F(s) = \frac{s-8}{(s-5)(s^2+4)}$, find $f(t) = \mathcal{L}^{-1}[F(s)]$.

• SOLUTION. Using partial fractions, we have

$$\frac{s-8}{(s-5)(s^2+4)} = \frac{A}{s-5} + \frac{Bs+C}{s^2+4}. \tag{2}$$

Finding the constants A, B, C often requires the most work of all the steps. Sometimes it is best to save that work until the end. If A, B, C were known, the inverse transform would be straightforward:

$$\mathcal{L}^{-1}\left[\frac{s-8}{(s-5)(s^2+4)}\right] = Ae^{5t} + B\cos 2t + C\frac{1}{2}\sin 2t.$$

From (2), we have $s-8 = A(s^2+4) + (s-5)(Bs+C)$. A is easy to obtain by evaluating at the real root $s=5$, so that $-3 = 29A$ or $A = -\frac{3}{29}$. The remaining coefficients (or all coefficients) may be obtained by equating like powers of s,

$$\begin{aligned} 1: &\quad -8 = 4A - 5C, \\ s: &\quad 1 = -5B + C, \\ s^2: &\quad 0 = A + B. \end{aligned}$$

Since we already know that $A = -\frac{3}{29}$, it follows from the s^2 equation that $B = \frac{3}{29}$ and from the s equation that $C = 1 + \frac{15}{29} = \frac{44}{29}$. A check on our arithmetic can be done by verifying that the remaining equation, the constant equation, holds: $-8 = -\frac{12}{29} - 5 - \frac{75}{29}$, which is valid. ♦

Example 3.2.5 *Real and Complex Roots*

Given $F(s) = \frac{s-8}{(s-5)(s^2+2s+5)}$, find $f(t) = \mathcal{L}^{-1}[F(s)]$.

• SOLUTION. The quadratic factor has complex roots since $b^2 - 4ac = -16 < 0$, so the denominator cannot be factored further. Using partial fractions,

$$\frac{s-8}{(s-5)(s^2+2s+5)} = \frac{A}{s-5} + \frac{Bs+C}{(s+1)^2+4}. \tag{3}$$

The first term in (3) is easily found in the table. The second term with the quadratic denominator can be determined by either of the techniques used in the last example. Using the first technique, we get

$$\frac{s-8}{(s-5)(s^2+4)} = \frac{A}{s-5} + B\frac{s+1}{(s+1)^2+4} + \widetilde{C}\frac{2}{(s+1)^2+4}. \tag{4}$$

Finding the constants A, B, $\widetilde{C}$ often requires the most work of all the steps. Sometimes it is best to save that work until the end. If A, B, $\widetilde{C}$ were known, the inverse transform would be straightforward:

$$\mathcal{L}^{-1}\left[\frac{s-8}{(s-5)(s^2+2s+5)}\right] = Ae^{5t} + Be^{-t}\cos 2t + \widetilde{C}e^{-t}\sin 2t.$$

We can either find A, B, C in (3), and then work a problem like the last example, or find A, B, $\widetilde{C}$ in (4). We shall do the second, since it is quicker. From (4), $s-8 = A(s^2+2s+5) + (s-5)[B(s+1) + 2\widetilde{C}]$. A is easy to obtain by evaluating at the real root $s = 5$, so that $-3 = 40A$ or $A = -\frac{3}{40}$. The remaining coefficients (or all coefficients) may be obtained by equating like powers of s,

$$\begin{aligned} 1: \quad & -8 = 5A - 5B - 10\widetilde{C}, \\ s: \quad & 1 = 2A - 5B + B + 2\widetilde{C}, \\ s^2: \quad & 0 = A + B. \end{aligned}$$

Since we already know that $A = -\frac{3}{40}$, it follows from the s^2 equation that $B = \frac{3}{40}$ and from the s equation that $2\widetilde{C} = 1 - 2A + 4B = 1 + \frac{6}{40} + \frac{12}{40} = 1\frac{18}{40}$. A check on our arithmetic can be done by verifying that the remaining equation, the constant equation, holds. Since $-8 = -\frac{15}{40} - \frac{15}{40} - 5 - \frac{90}{40}$, the check is valid. ♦

Repeated Real Roots

When $s = 0$ is a repeated root, we already know how to obtain the inverse Laplace transform:

$$\mathcal{L}[t^n] = \frac{n!}{s^{n+1}}. \quad \text{(T9)}$$

A direct application of the shifting theorem (T16),

$$\mathcal{L}[t^n e^{at}] = \frac{n!}{(s-a)^{n+1}}, \quad \text{(T11)} \tag{5}$$

enables us to determine the inverse Laplace transform when the repeated root is $s = a$. The inverse Laplace transform of a simple root is the corresponding elementary exponential. Equation (5) shows that the inverse Laplace transform of a multiple root is the corresponding exponential multiplied by one power of t for each time the root is repeated.

Example 3.2.6 *Repeated Real Roots*

If $F(s) = \frac{s^2}{(s+1)^3(s+5)}$, find $f(t)$.

• SOLUTION. Using partial fractions, we have

$$\begin{aligned} \frac{s^2}{(s+1)^3(s+5)} &= \frac{A}{s+5} + \frac{B}{(s+1)^3} + \frac{C}{(s+1)^2} + \frac{D}{s+1} \\ &= A\frac{1}{s+5} + \frac{B}{2!}\frac{2!}{(s+1)^3} + C\frac{1}{(s+1)^2} + D\frac{1}{s+1}. \end{aligned} \tag{6}$$

Determining the four coefficients can be the hardest part. Once the coefficients A, B, C, D are determined, the inverse Laplace transform can be obtained using the formulas for a simple root and the new formula for repeated roots:

$$\mathcal{L}^{-1}\left[\frac{s^2}{(s+1)^3(s+5)}\right] = Ae^{-5t} + B\frac{1}{2!}t^2e^{-t} + Cte^{-t} + De^{-t}.$$

Determining A, B is not difficult since $s = -5$ and $s = -1$ are roots. Multiplying (6) by the denominator $(s+5)(s+1)^3$ yields

$$s^2 = A(s+1)^3 + B(s+5) + C(s+1)(s+5) + D(s+5)(s+1)^2.$$

Evaluating at $s=-5$ yields $25=A(-4)^3$ or $A=\frac{5^2}{(-4)^3}=-\frac{25}{64}$. Evaluating at $s=-1$ yields $B=\frac{1}{4}$. Determining C, D is more difficult since $s=-1$ is a root of multiplicity 3. We equate coefficients of powers of s to get equations to solve for C, D. For example,

$$\begin{aligned} s^3: &\quad 0=A+D,\\ s^2: &\quad 1=3A+C+7D. \end{aligned}$$

Given $A=-\frac{25}{64}$ and $B=\frac{1}{4}$, we find that $D=\frac{25}{64}$ and $C=-\frac{9}{16}$. ♦

Repeated Imaginary Roots

Repeated imaginary roots are more difficult than repeated real roots. We will only derive results corresponding to repeated roots of multiplicity 2. Using Euler's formula, we get

$$\mathcal{L}[t\cos bt]=\frac{1}{2}\mathcal{L}[te^{ibt}]+\frac{1}{2}\mathcal{L}[te^{-ibt}].$$

Using (5) with $n=1$, we obtain

$$\mathcal{L}[t\cos bt]=\frac{1}{2}\frac{1}{(s-ib)^2}+\frac{1}{2}\frac{1}{(s+ib)^2}=\frac{1}{2}\frac{(s+ib)^2+(s-ib)^2}{(s^2+b^2)^2},$$

after using a common denominator. Squaring, adding, and canceling terms yields

$$\mathcal{L}[t\cos bt]=\frac{s^2-b^2}{(s^2+b^2)^2}. \quad \text{(T26)} \tag{7}$$

In a similar way we obtain

$$\mathcal{L}[t\sin bt]=\frac{2bs}{(s^2+b^2)^2}. \quad \text{(T27)} \tag{8}$$

These results can also be derived using (T21), as outlined in Example 3.1.15 in Section 3.1.2.

A useful inverse transform is obtained directly from (8):

$$\mathcal{L}^{-1}\left[\frac{s}{(s^2+b^2)^2}\right]=\frac{1}{2b}t\sin bt. \tag{9}$$

However, to obtain $\mathcal{L}^{-1}[\frac{1}{(s^2+b^2)^2}]$, we must do some further elementary calculations based on (7). Noting that $s^2-b^2=s^2+b^2-2b^2$ in (7) yields

$$\mathcal{L}[t\cos bt]=\frac{-2b^2}{(s^2+b^2)^2}+\frac{1}{s^2+b^2}.$$

Thus,

$$\mathcal{L}^{-1}\left[\frac{-2b^2}{(x^2+b^2)^2}\right]=t\cos bt-\frac{1}{b}\sin bt$$

or

$$\mathcal{L}^{-1}\left[\frac{1}{(s^2+b^2)^2}\right]=\frac{1}{2b^3}\sin bt-\frac{1}{2b^2}t\cos bt. \quad \text{(T27)} \tag{10}$$

Example 3.2.7 *Repeated Pure Imaginary Roots*

If $F(s)=\frac{s^3}{(s^2+9)^2}$, find $f(t)$.

• SOLUTION. Using partial fractions, we have

$$\begin{aligned}\frac{s^3}{(s^2+9)^2}&=\frac{As+B}{(s^2+9)^2}+\frac{Cs+D}{s^2+9}\\&=A\frac{s}{(s^2+9)^2}+B\frac{1}{(s^2+9)^2}+C\frac{s}{s^2+9}+\frac{D}{3}\left(\frac{3}{s^2+9}\right).\end{aligned} \tag{11}$$

Once the coefficients A, B, C, D are determined, the inverse Laplace transform requires the new formulas (9) and (10):

$$\begin{aligned}\mathcal{L}^{-1}\left[\frac{s^3}{(s^2+9)^2}\right]=A\frac{1}{2\cdot 3}t\sin 3t+B\left(\frac{1}{2\cdot 3^3}\sin 3t-\frac{1}{2\cdot 3^2}t\cos 3t\right)\\+C\cos 3t+\frac{D}{3}\sin 3t.\end{aligned}$$

The coefficients can be determined by multiplying both sides of (11) by $(s^2+9)^2$:

$$s^3=As+B+(s^2+9)(Cs+D).$$

Equating like powers of s yields

$$\begin{aligned}s^3:&\quad 1=C,\\ s^2:&\quad 0=D,\\ s:&\quad 0=A+9C,\\ 1=s^0:&\quad 0=B+9D.\end{aligned}$$

Thus, $A=-9$, $B=0$, $C=1$, $D=0$. ♦

Repeated Complex Roots

Formulas for repeated complex roots can be obtained using the shifting theorem (T16) directly from the formulas for repeated imaginary roots

$$\begin{aligned}\mathcal{L}^{-1}\left[\frac{s-a}{[(s-a)^2+b^2]^2}\right]&=\frac{1}{2b}te^{at}\sin bt, \quad \text{(T28)}\\ \mathcal{L}^{-1}\left[\frac{1}{[(s-a)^2+b^2]^2}\right]&=\frac{1}{2b^3}e^{at}\sin bt-\frac{1}{2b^2}te^{at}\cos bt. \quad \text{(T29)}\end{aligned} \tag{12}$$

Example 3.2.8 *Repeated Complex Roots*

If $F(s)=\frac{6s+7}{(s^2+6s+25)^2}$, find $f(t)$.

TABLE 3.2.1
Long Table of Laplace Transforms

	$f(t)$	$\mathcal{L}[f(t)] = F(s) = \int_0^\infty e^{-st} f(t)dt$
(T1)	1	$\frac{1}{s}$
(T2)	e^{at}	$\frac{1}{s-a}$
(T3)	$\sin bt$	$\frac{b}{s^2+b^2}$
(T4)	$\cos bt$	$\frac{s}{s^2+b^2}$
(T5)	$e^{at} \sin bt$	$\frac{b}{(s-a)^2+b^2}$
(T6)	$e^{at} \cos bt$	$\frac{s-a}{(s-a)^2+b^2}$
(T7)	$\sinh bt$	$\frac{b}{s^2-b^2}$
(T8)	$\cosh bt$	$\frac{s}{s^2-b^2}$
(T9)	t^n, n a positive integer	$\frac{n!}{s^{n+1}}$
(T10)	$t^p,\ p>0,$	$\frac{\Gamma(p+1)}{s^{p+1}}$
(T11)	$t^n e^{at}$	$\frac{n!}{(s-a)^{n+1}}$
(T12)	$\delta(t)$	1
(T13)	$\delta(t-a)$	$e^{-as}, \quad a>0$
(T14)	$H(t-c)$	$\frac{e^{-cs}}{s}, \quad c>0$
(T15)	$f(t-c)H(t-c),\ c>0$	$e^{-cs}F(s) \quad c>0$
(T16)	$e^{ct} f(t)$	$F(s-c)$
(T17)	$f(ct),\ c>0$	$\frac{1}{c}F\left(\frac{s}{c}\right), c>0$
(T18)	$f'(t)$	$sF(s) - f(0)$
(T19)	$f''(t)$	$s^2F(s) - sf(0) - f'(0)$
(T20)	$f^{(n)}(t)$	$s^n F(s) - s^{n-1} f(0) \cdots - sf^{(n-2)}(0) - f^{(n-1)}(0)$
(T21)	$tf(t)$	$-F'(s)$
(T22)	$t^2 f(t)$	$F''(s)$
(T22)	$t^n f(t),\ n>0$	$(-1)^n F^{(n)}(s)$
(T23)	$\int_0^t f(\tau)\, d\tau$	$\frac{F(s)}{s}$
(T24)	$\int_0^t f(\tau)g(t-\tau)d\tau = f * g$	$F(s)G(s)$

TABLE 3.2.2
Other Formulas Involving the Laplace Transform

(T25)	$\mathcal{L}[t\sin bt] = \dfrac{2bs}{(s^2+b^2)^2}$	
(T26)	$\mathcal{L}[t\cos bt] = \dfrac{s^2-b^2}{(s^2+b^2)^2}$	
(T27)	$\mathcal{L}^{-1}\left[\dfrac{2b^2}{(s^2+b^2)^2}\right] = \dfrac{1}{b}\sin bt - t\cos bt$	
(T28)	$\mathcal{L}[te^{at}\sin bt] = \dfrac{2b(s-a)}{[(s-a)^2+b^2]^2}$	
(T29)	$\mathcal{L}[te^{at}\cos bt] = \dfrac{(s-a)^2-b^2}{[(s-a)^2+b^2]^2}$	
(T30)	$\mathcal{L}^{-1}\left[\dfrac{2b^2}{[(s-a)^2+b^2]^2}\right] = \dfrac{1}{b}e^{at}(\sin bt - bt\cos bt)$	
(T31)	$\mathcal{L}[g(t)] = \dfrac{\int_0^T e^{-st}g(t)dt}{1-e^{-sT}} = \dfrac{\mathcal{L}[g(t)(1-H(t-T))]}{1-e^{-sT}}$	g periodic with period T
(T32)	$\mathcal{L}^{-1}\left[\dfrac{F(s)}{1-e^{-sT}}\right] = \mathcal{L}^{-1}\left[\sum_{n=0}^{\infty} e^{-nTs}F(s)\right] = \sum_{n=0}^{\infty} f(t-nT)H(t-nT)$	
(T33)	$\mathcal{L}^{-1}\left[\dfrac{F(s)}{1+e^{-sT}}\right] = \mathcal{L}^{-1}\left[\sum_{n=0}^{\infty}(-1)^n e^{-nTs}F(s)\right] = \sum_{n=0}^{\infty}(-1)^n f(t-nT)H(t-nT)$	

• SOLUTION. The roots are repeated. By the quadratic formula, the repeated complex roots are $s = -3 \pm 4i$. Completing the square, $s^2+6s+25 = (s+3)^2+16$, shows that transforms should be expressed as functions of $s+3$. Since $6s+7 = A(s+3)+B$ yields $6s+7 = 6(s+3)-11$, we have

$$\frac{6s+7}{(s^2+6s+25)^2} = \frac{6(s+3)-11}{[(s+3)^2+4^2]^2} = 6\frac{s+3}{[(s+3)^2+4^2]^2} - 11\frac{1}{[(s+3)^2+4^2]^2}.$$

Using (12), we obtain

$$\mathcal{L}^{-1}\left[\frac{6s+7}{(s^2+6s+25)^2}\right] = 6e^{-3t}\frac{1}{2\cdot 2}t\sin 2t - 11e^{-3t}\left(\frac{1}{2\cdot 2^3}\sin 2t - \frac{1}{2\cdot 2^2}t\cos 2t\right). \quad \blacklozenge$$

Many of the basic formulas are used in computing both Laplace and inverse Laplace transforms. However, some formulas tend to be used primarily for taking the Laplace transform and others for finding the inverse Laplace transform. This includes finding the inverse Laplace transform when there are repeated complex roots.

Exercises

In Exercises 1–68, $F(s)$ is given. Compute $f(t) = \mathcal{L}^{-1}[F(s)]$.

1. $\dfrac{1}{s^2-9}$.
2. $\dfrac{2}{s^2-16}$.
3. $\dfrac{5s-1}{s^2+7}$.
4. $\dfrac{7s-3}{s^2+5}$.
5. $\dfrac{s+2}{s^2-1}$.
6. $\dfrac{3}{s^2-s}$.
7. $\dfrac{2s-1}{s^2-s}$.
8. $\dfrac{5}{s^2+5s+6}$.
9. $\dfrac{-s-1}{s^2+s-2}$.
10. $\dfrac{2s-13}{s^2+8s+15}$.
11. $\dfrac{4}{s^2+s-6}$.
12. $\dfrac{s-1}{s^2+6s+5}$.
13. $\dfrac{s}{s^2-2s+26}$.
14. $\dfrac{2s}{s^2+4s+13}$.
15. $\dfrac{2s+5}{s^2-6s+18}$.
16. $\dfrac{17-3s}{s^2+2s+26}$.
17. $\dfrac{3s-2}{s^2+10s+26}$.
18. $\dfrac{3s-1}{s^2+2s+3}$.
19. $\dfrac{s}{s^2-5s+6}$.
20. $\dfrac{2}{s^2-13s+42}$.
21. $\dfrac{1}{s^3-3s^2+2s}$.

22. $\dfrac{2s+1}{s^3+3s^2+2s}$.
23. $\dfrac{s^2-2}{s^3+8s^2+7s}$.
24. $\dfrac{s^2+s}{s^3+4s^2-5s}$.
25. $\dfrac{s+3}{s^3-s}$.
26. $\dfrac{3}{s^3-4s}$.
27. $\dfrac{2s+1}{s^3+4s^2+13s}$.
28. $\dfrac{2s-5}{s^3-6s^2+18s}$.
29. $\dfrac{s^2-3}{s^3+2s^2+26s}$.
30. $\dfrac{3s-2}{s^3+10s^2+26s}$.
31. $\dfrac{s-8}{(s-5)(s^2+4)}$.
32. $\dfrac{s-5}{(s-8)(s^2+9)}$.
33. $\dfrac{s^2}{(s-3)(s^2-2s+26)}$.
34. $\dfrac{s+5}{(s+3)(s^2+2s+26)}$.
35. $\dfrac{s^3-1}{s^4+10s^2+9}$.
36. $\dfrac{1}{s^4+5s^2+4}$.
37. $\dfrac{s^3}{s^4-5s^2+4}$.
38. $\dfrac{s^2}{s^4-10s^2+9}$.
39. $\dfrac{s}{s^4-16}$.
40. $\dfrac{s^2-3}{s^4-1}$.

Repeated roots occur in Exercises 41–68.

41. $\dfrac{4}{(s+3)^6}$.
42. $\dfrac{2}{(s-3)^5}$.
43. $\dfrac{4}{(3s+1)^5}$.

44. $\dfrac{s-2}{(5-7s)^9}$.

45. $\dfrac{s^2}{(s+3)^2(s-3)^2}$.

46. $\dfrac{s-3}{(s+1)^2(s-1)^2}$.

47. $\dfrac{s}{(s+1)^2(s^2+1)}$.

48. $\dfrac{s^3}{(s-1)^2(s^2+4)}$.

49. $\dfrac{(s+5)}{(s+1)(s-1)^3}$.

50. $\dfrac{s+1}{(s-6)^3(s-4)}$.

51. $\dfrac{s^2-s}{(s^2+4)^2}$.

52. $\dfrac{2s^3-1}{(s^2+1)^2}$.

53. $\dfrac{s^3-1}{(s^2+9)^2}$.

54. $\dfrac{s^2+3s}{(s^2+16)^2}$.

55. $\dfrac{s^3-s^2}{(s^2+36)^2}$.

56. $\dfrac{s^3+s^2}{(s^2+25)^2}$.

57. $\dfrac{3s-1}{(s^2+2s+26)^2}$.

58. $\dfrac{s}{(s^2+4s+5)^2}$.

59. $\dfrac{s^2}{(s^2+6s+25)^2}$.

60. $\dfrac{s^3}{(s^2+6s+25)^2}$.

61. $\dfrac{s+5}{(s+1)(s-1)^3}$.

62. $\dfrac{s+8}{s(s+1)^3(s-1)}$.

63. $\dfrac{12}{s(s-8)^6}$.

64. $\dfrac{6s}{(s+1)(s-3)^5}$.

65. $\dfrac{s+5}{(s+1)^2(s-1)^2}$.

66. $\dfrac{s-1}{s^2(s-3)^2}$.

67. $\dfrac{s}{(s^2+1)(s-7)^4}$.

68. $\dfrac{5}{(s^2+4)(s-1)^5}$.

3.3 Initial Value Problems for Differential Equations

The formulas developed in the previous sections told how to take the Laplace transform (and inverse Laplace transform) of particular functions. This section will begin to show how Laplace transforms may be used in solving differential equations.

The next result is the key to solving linear constant coefficient differential equations using the Laplace transform. For convenience we will give two special versions and the general version.

THEOREM 3.3.1 Laplace Transforms of Derivatives

First Derivative Case

Suppose that $f(t)$ is continuous and of exponential order on $[0, \infty)$ and $f'(t)$ is piecewise continuous on $[0, \infty)$. Then $\mathcal{L}[f'(t)]$ exists and

$$\mathcal{L}[f'(t)] = s\mathcal{L}[f(t)] - f(0) = sF(s) - f(0). \quad \text{(T18)}$$

Second Derivative Case

Suppose that $f(t)$ and $f'(t)$ are continuous and of exponential order on $[0, \infty)$, and $f''(t)$ is piecewise continuous on $[0, \infty)$. Then $\mathcal{L}[f''(t)]$ exists and

$$\mathcal{L}[f''(t)] = s^2\mathcal{L}[f(t)] - sf(0) - f'(0) = s^2F(s) - sf(0) - f'(0). \quad \text{(T19)}$$

General Derivative Case

If $f(t), f'(t), \ldots, f^{(n-1)}(t)$ are continuous and of exponential order on $[0, \infty)$ and $f^{n)}(t)$ is piecewise continuous on $[0, \infty)$, then $\mathcal{L}[f^{(n)}(t)]$ exists and

$$\begin{aligned}\mathcal{L}[f^{(n)}(t)] &= s^n\mathcal{L}[f(t)] - s^{n-1}f(0) - s^{n-2}f'(0)\cdots - f^{(n-1)}(0)\\ &= s^nF(s) - s^{n-1}f(0) - s^{n-2}f'(0)\cdots - f^{(n-1)}(0). \quad \text{(T20)}\end{aligned}$$

The formulas for the first and second derivatives are included in our short table. Each derivative corresponds to multiplying by s with some affect of the initial condition given by the formulas. The formulas for the Laplace transform of time derivatives simplify if all initial conditions are zero.

VERIFICATION OF THE FIRST DERIVATIVE CASE. We calculate the Laplace transform of the first derivative using the definition of the Laplace transform

$$\mathcal{L}[f'(t)] = \int_0^\infty e^{-st} f'(t)dt.$$

Since $f'(t)$ appears in the integrand, we expect that integration by parts $\left(\int u dv = uv| - \int v du\right)$ may simplify the integral:

$$\begin{aligned} dv &= f'(t)dt, \quad & u &= e^{-st}, \\ v &= f(t), \quad & du &= -se^{-st}dt. \end{aligned}$$

Using integration by parts in this way yields

$$\mathcal{L}[f'(t)] = e^{-st} f(t)\Big|_{t=0}^{\infty} + s\int_0^\infty e^{-st} f(t)dt = s\mathcal{L}[f(t)] - f(0), \qquad (1)$$

since $e^{-st} f(t)\Big|_{t=0}^{\infty} = \lim_{b\to\infty} e^{-sb} f(b) - f(0) = -f(0)$ if $f(t)$ is of exponential order and s is large enough. The initial condition $f(0)$ is needed for the Laplace transform of the first derivative. The validity of integration by parts here requires that $f(t)$ be continuous. (Note Exercises 29 and 30.)

VERIFICATION FOR HIGHER DERIVATIVES. The Laplace transform of higher derivatives may be done by repeated use of (1). For example, to show the second-derivative case, compute as follows:

$$\begin{aligned} \mathcal{L}[f''(t)] &= \mathcal{L}[(f'(t))'] = s\mathcal{L}[f'(t)] - f'(0), && \text{(using (1))} \\ &= s[s\mathcal{L}[f(t)] - f(0)] - f'(0), && \text{(using (1) again)} \\ &= s^2\mathcal{L}[f(t)] - sf(0) - f'(0). && \text{as desired} \end{aligned}$$

Solving Differential Equations with Laplace Transforms

Linear differential equations with constant coefficients can be solved using Laplace transforms based on the formulas (T18) and (T19), (T20) for the Laplace transform of derivatives. Variables used to denote the solutions of differential equations vary between areas and applications. In this chapter we use y rather than x for the solution. We let $Y(s) = \mathcal{L}[y(t)]$.

Example 3.3.1 *Solving a Differential Equation*

Solve the initial value problem

$$y' - 3y = 4, \quad y(0) = 7.$$

• SOLUTION. Take the Laplace transform of both sides of the differential equation, and obtain (using the linearity of the Laplace transform),

$$\mathcal{L}[y'] - 3\mathcal{L}[y] = 4\mathcal{L}[1].$$

From the theorem (T18) on the Laplace transform of the first time derivative, we have

$$\mathcal{L}[y'] = sY(s) - y(0).$$

Thus, using the initial condition $y(0) = 7$, we have

$$sY(s) - 7 - 3Y(s) = \frac{4}{s}.$$

We solve for $Y(s)$:

$$(s-3)Y(s) = 7 + \frac{4}{s} = \frac{7s+4}{s},$$

and hence

$$Y(s) = \frac{7s+4}{s(s-3)}.$$

To find $y(t)$, we expand $Y(s)$ using partial fractions:

$$Y(s) = \frac{7s+4}{s(s-3)} = \frac{A}{s} + \frac{B}{s-3}. \tag{2}$$

In terms of the coefficients of the partial fraction expansion, the inverse Laplace transform is

$$y(t) = A + Be^{3t}.$$

To find A, B, multiply both sides of (2) by $s(s-3)$ and get

$$7s + 4 = A(s-3) + Bs.$$

Evaluating this at the roots $s = 0, 3$ of the denominator yields $A = -\frac{4}{3}$ and $B = \frac{25}{3}$, so that the solution of the initial value problem is

$$y(t) = -\frac{4}{3} + \frac{25}{3}e^{3t}.$$ ♦

Since this is a linear first-order equation with constant coefficients, it could also be easily solved by several techniques from Chapters 1 and 2.

This example motivates the general procedure. We express the procedure for all second-order linear differential equations with constant coefficients, before we do a number of examples.

Procedure for Solving Initial Value Problems Using Laplace Transforms

Consider the differential equation

$$ay'' + by' + cy = f(t), \tag{3}$$

with a, b, c constants, and $f(t)$ a function that has a Laplace transform. Assume that the initial conditions $y(0)$ and $y'(0)$ are given.

1. Take the Laplace transform of both sides (using linearity of the Laplace transform):
$$a\mathcal{L}[y''] + b\mathcal{L}[y'] + c\mathcal{L}[y] = \mathcal{L}[f(t)].$$

2. Apply the theorem on the Laplace transforms of derivatives ((T18) and (T19)) and use the initial conditions.

$$a[s^2Y(s) - sy(0) - y'(0)] + b[sY(s) - y(0)] + cY(s) = F(s).$$

3. Solve for $Y(s) = \mathcal{L}[y(t)]$:

$$Y(s)[as^2 + bs + c] = F(s) + a[sy(0) + y'(0)] + by(0),$$

or equivalently,

$$Y(s) = \frac{F(s) + a[sy(0) + y'(0)] + by(0)}{as^2 + bs + c}. \tag{4}$$

The factor $as^2 + bs + c$ arises because each derivative corresponds to multiplying by s.

4. In the specific examples, express $Y(s)$ in terms of functions of s appearing in the Laplace transform column of the Laplace transform table (Tables (3.1.1, 3.2.1, 3.2.2).
5. Compute $y(t)$ from $Y(s)$, using these formulas and the inverse Laplace transform techniques of Section 3.2.

We conclude from (4) that when we use Laplace transforms:

An ordinary differential equation with constant coefficients is always reduced to an algebraic equation for the inverse Laplace transform.

The denominator for the inverse Laplace transform $as^2 + bs + c$ is the same polynomial as the characteristic polynomial $ar^2 + br + c$ for the differential equation (3). Thus, roots of the denominator are the roots of the characteristic equation.

Example 3.3.2 *Second-Order Differential Equation*

Solve the initial problem

$$y'' + 3y' + 2y = 1, \quad y(0) = 0, \quad y'(0) = 2.$$

• SOLUTION. Take the Laplace transform of both sides of the differential equation, and obtain (using the linearity of the Laplace transform),

$$\mathcal{L}[y''] + 3\mathcal{L}[y'] + 2\mathcal{L}[y] = \mathcal{L}[1].$$

From Theorem 3.3 (or T18 and T19),

$$\mathcal{L}[y''] = s^2Y(s) - sy(0) - y'(0), \quad \mathcal{L}[y'] = sY(s) - y(0).$$

Thus, using the initial condition $y(0) = 0$, $y'(0) = 2$, we have

$$s^2Y(s) - 2 + 3sY(s) + 2Y(s) = \frac{1}{s}.$$

Now solve for $Y(s)$. Note that $Y(s)$ is multiplied by $s^2 + 3s + 2$, since each derivative corresponds to multiplying by s. You should always check that factor from the differential equation:

$$(s^2+3s+2)Y(s)=2+\frac{1}{s}=\frac{2s+1}{s}, \quad \text{or} \quad Y(s)=\frac{2s+1}{s(s^2+3s+2)}.$$

To find $y(t)$, expand $Y(s)$ using partial fractions:

$$Y(s)=\frac{2s+1}{s(s+1)(s+2)}=\frac{A}{s}+\frac{B}{s+1}+\frac{C}{s+2}.$$

To find A, B, C, multiply both sides by $s(s+1)(s+2)$ and get

$$2s+1=A(s+1)(s+2)+Bs(s+2)+Cs(s+1).$$

Evaluating this expression at the roots $s=0, -1, -2$ of the denominator yields

$$\begin{aligned} s=0: &\quad 1=A2, &\quad \text{so} \quad A=\frac{1}{2}, \\ s=-1: &\quad -1=B(-1)(1), &\quad \text{so} \quad B=1, \\ s=-2: &\quad -3=C(-2)(-1), &\quad \text{so} \quad C=-\frac{3}{2}, \end{aligned}$$

and hence

$$Y(s)=\frac{1}{2}\cdot\frac{1}{s}+\frac{1}{s+1}+\left(-\frac{3}{2}\right)\frac{1}{s+2}. \tag{5}$$

From (T2) and (T1),

$$1=\mathcal{L}^{-1}\left[\frac{1}{s}\right], \quad e^{-t}=\mathcal{L}^{-1}\left[\frac{1}{s+1}\right], \quad e^{-2t}=\mathcal{L}^{-1}\left[\frac{1}{s+2}\right],$$

so that from (5),

$$y(t)=\frac{1}{2}+e^{-t}-\frac{3}{2}e^{-2t}.$$

This problem, of course, could also be done by the method of undetermined coefficients discussed in Chapter 2. In the next section we shall consider problems for which the Laplace transform works more easily than these techniques. ♦

In concluding this section, three additional examples of solving differential equations using the Laplace transform will be given.

Example 3.3.3 *Repeated Roots*

Solve the initial value problem

$$y''-2y'+y=3e^t, \quad y(0)=1, \quad y'(0)=1.$$

• SOLUTION. Taking the Laplace transform of both sides gives

$$\mathcal{L}[y'']-2\mathcal{L}[y']+\mathcal{L}[y]=3\mathcal{L}[e^t].$$

Using (T18), (T19), and the initial conditions on the left, and (T2) on the right, we obtain

$$s^2Y(s) - s - 1 - 2[sY(s) - 1] + Y(s) = \frac{3}{s-1}.$$

Solving for $Y(s)$,

$$(s^2 - 2s + 1)Y(s) = \frac{3}{s-1} + s - 1 = \frac{s^2 - 2s + 4}{s-1},$$

$$Y(s) = \frac{s^2 - 2s + 4}{(s^2 - 2s + 1)(s-1)} = \frac{s^2 - 2s + 4}{(s-1)^3}.$$

Expanding $Y(s)$ by partial fractions,

$$Y(s) = \frac{s^2 - 2s + 4}{(s-1)^3} = \frac{A}{s-1} + \frac{B}{(s-1)^2} + \frac{C}{(s-1)^3}.$$

Multiply by $(s-1)^3$, obtaining

$$\begin{aligned} s^2 - 2s + 4 &= A(s-1)^2 + B(s-1) + C \\ &= A(s^2 - 2s + 1) + B(s-1) + C. \end{aligned}$$

Equating the coefficients of like powers of s,

$$\begin{aligned} 1: &\quad 4 = A - B + C, \\ s: &\quad -2 = -2A + B, \\ s^2: &\quad 1 = A, \end{aligned}$$

and, solving from the last equation up, we get $A = 1$, $B = 0$, $C = 3$. Thus,

$$Y(s) = \frac{1}{s-1} + \frac{3}{(s-1)^3}.$$

Now, from (T2) and (T11),

$$e^t = \mathcal{L}^{-1}\left[\frac{1}{s-1}\right] \quad \text{and} \quad t^2e^t = \mathcal{L}^{-1}\left[\frac{2}{(s-1)^3}\right].$$

Since $\frac{2}{(s-1)^3}$ appears in the table, but not $\frac{3}{(s-1)^3}$, we may need to rewrite $Y(s)$ as

$$Y(s) = \frac{1}{s-1} + \frac{3}{2} \cdot \frac{2}{(s-1)^3}$$

in order to get

$$y(t) = e^t + \frac{3}{2}t^2e^t.$$

♦

Example 3.3.4 *Pure Imaginary Roots*

Solve $y'' + 3y = \sin 5t, \quad y(0) = 0,\ y'(0) = 0.$

- SOLUTION. Taking the Laplace transform of both sides yields

$$\mathcal{L}[y''] + 3\mathcal{L}[y] = \mathcal{L}[\sin 5t]$$

or, by (T3), (T18), and (T19),

$$s^2 Y(s) + 3Y(s) = \frac{5}{s^2+25}.$$

Thus

$$Y(s) = \frac{5}{(s^2+3)(s^2+25)}.$$

By partial fractions,

$$Y(s) = \frac{5}{(s^2+3)(s^2+25)} = \frac{As+B}{s^2+3} + \frac{Cs+D}{s^2+25}.$$

Multiplying both sides by $(s^2+25)(s^2+3)$, we find that

$$5 = (As+B)(s^2+25) + (Cs+D)(s^2+3).$$

Equating like powers of s, we get

$$\begin{aligned} 1:&\quad 5 = 25B + 3D,\\ s:&\quad 0 = 25A + 3C,\\ s^2:&\quad 0 = B + D,\\ s^3:&\quad 0 = A + C. \end{aligned}$$

The second and fourth equations yield $A = C = 0$. The first and third equations give, after a little algebra,

$$B = \frac{5}{22}, \qquad D = -\frac{5}{22},$$

so that

$$Y(s) = \frac{5/22}{s^2+3} - \frac{5/22}{s^2+25}.$$

Formula (T3) gives

$$\mathcal{L}^{-1}\left[\frac{\sqrt{3}}{s^2+3}\right] = \sin\sqrt{3}t, \qquad \mathcal{L}^{-1}\left[\frac{5}{s^2+25}\right] = \sin 5t.$$

Thus, we may need to write $Y(s)$ as

$$Y(s) = \frac{5}{22\sqrt{3}} \cdot \frac{\sqrt{3}}{s^2+3} - \left(\frac{1}{22}\right)\frac{5}{s^2+25}$$

to conclude that

$$y(t) = \frac{5}{22\sqrt{3}} \sin\sqrt{3}t - \frac{1}{22}\sin 5t.$$ ♦

Example 3.3.5 *Resonance*

Solve $y'' + 4y = 3\sin 2t$ with $y(0) = 0$, $y'(0) = 0$.

TABLE 3.3.1

$f(t)$	$F(s)$	*Justification*
$\cos 2t$	$\dfrac{s}{s^2+4}$	(T4)
$t\cos 2t$	$-\dfrac{d}{ds}\left(\dfrac{s}{s^2+4}\right)=\dfrac{-(s^2+4)+2s^2}{(s^2+4)^2}$	(T21)
	$=\dfrac{s^2-4}{(s^2+4)^2}=\dfrac{(s^2+4)-8}{(s^2+4)^2}$	algebra
	$=\dfrac{-8}{(s^2+4)^2}+\dfrac{1}{s^2+4}$	algebra
$\frac{1}{2}\sin 2t$	$\dfrac{1}{s^2+4}$	(T3)
$t\cos 2t-\frac{1}{2}\sin 2t$	$\dfrac{-8}{(s^2+4)^2}$	subtraction

● SOLUTION. This is a resonance problem since the forcing function $\sin 2t$ is also a homogeneous solution. (The forcing frequency equals the natural frequency.) Taking the Laplace transform of both sides of the differential equation yields

$$\mathcal{L}[y'']+4\mathcal{L}[y]=3\frac{2}{s^2+4}.$$

Using the formulas for the Laplace transform of second derivative (T19) and applying the zero initial conditions gives

$$s^2Y(s)+4Y(s)=\frac{6}{s^2+4}.$$

Thus,

$$Y(s)=\frac{6}{(s^2+4)^2}=\frac{6}{(s^2+2^2)^2}=\frac{3}{4}\left[\frac{2\cdot 2^2}{(s^2+2^2)^2}\right], \tag{6}$$

which does not require further partial fractions decomposition. Equation (6) may be inverted using the long table or (T27). The solution is

$$y(t)=\frac{3}{4}\left(\frac{1}{2}\sin 2t-t\cos 2t\right).$$

To illustrate the use of some of the other formulas we shall also find the inverse Laplace transform of (6) using only the short table along with property (T21) for Laplace transforms. We summarize the calculation in table 3.3.1. Thus,

$$y(t)=\frac{3}{4}\left(\frac{1}{2}\sin 2t-t\cos 2t\right).$$ ♦

Computer programs have been written that do symbolic manipulations (they appear to manipulate formulas rather than numbers). This means that Laplace transform techniques and their accompanying algebra involving functions of s can be handled as simple as ordinary arithmetic on a pocket calculator.

Exercises

In Exercises 1–28 use Laplace transforms to solve the initial value problem.

1. $y'' - 4y = 1$, $y(0) = 0$, $y'(0) = 1$.
2. $y'' - 6y' + 5y = 2$, $y(0) = 0$, $y'(0) = -1$.
3. $y'' + 3y' - 4y = 0$, $y(0) = 1$, $y'(0) = 1$.
4. $y'' - 3y' - 4y = t$, $y(0) = 0$, $y'(0) = 0$.
5. $y'' + 5y' + 6y = 1$, $y(0) = 1$, $y'(0) = 1$.
6. $y'' + y = 1$, $y(0) = y'(0) = 2$.
7. $y'' - 3y' + 2y = 1$, $y(0) = 0$, $y'(0) = 1$.
8. $y'' - y = e^t$, $y(0) = 1$, $y'(0) = 0$.
9. $y'' + 9y = e^{-t}$, $y(0) = 1$, $y'(0) = 2$.
10. $y'' + 4y = 3\cos 5t$, $y(0) = 0$, $y'(0) = 3$.
11. $y'' + 4y' + 13y = 2$, $y(0) = 1$, $y'(0) = 0$.
12. $y'' - 2y' + 5y = 0$, $y(0) = 2$, $y'(0) = 1$.
13. $y^{(4)} - y = 1$, $y(0) = 3$, $y'(0) = 5$, $y''(0) = 0$, $y'''(0) = 0$.
14. $y' + y = e^t \sin t$, $y(0) = 1$.
15. $y' - y = 1 - t$, $y(0) = -1$.
16. $y'' + 2y' + 2y = e^{-t} \sin 2t$, $y(0) = 0$, $y'(0) = 0$.
17. $y'' + 2y' + y = e^{-t}$, $y(0) = 0$, $y'(0) = 1$.
18. $y''' + y'' - y' - y = 0$, $y(0) = 0$, $y'(0) = 0$, $y''(0) = 1$.
19. $y' + 2y = e^{-t} \cos t$, $y(0) = 0$.
20. $y' - 3y = t^2 e^t$, $y(0) = 1$.

Exercises 21–28 involve resonance.

21. $y'' + y = \sin t$, $y(0) = 0$, $y'(0) = 0$.
22. $y'' + y = \cos t$, $y(0) = 0$, $y'(0) = 1$.
23. $y'' + 9y = \cos 3t$, $y(0) = 0$, $y'(0) = 1$.
24. $y'' + 9y = \sin 3t$, $y(0) = 1$, $y'(0) = 0$.
25. $y'' + 4y = \cos 2t$, $y(0) = 0$, $y'(0) = 0$.
26. $y'' + 16y = \sin 4t$, $y(0) = 0$, $y'(0) = 0$.
27. $y'' + y = \sin t + \cos t$, $y(0) = 0$, $y'(0) = 0$.
28. $y'' + 4y = \sin 2t + \cos 2t$, $y(0) = 0$, $y'(0) = 0$.
29. Suppose that $f(t)$ and $f'(t)$ are continuous and of exponential order for $t > 0$, and that $\lim_{t\to 0^+} f(t)$ exists and is denoted $f(0^+)$. Show that $\mathcal{L}[f'(t)] = sF(s) - f(0^+)$.
30. Suppose that $f(t)$, $f'(t)$, $f''(t)$ are continuous and of exponential order for $t > 0$. Suppose that $\lim_{t\to 0^+} f(t)$ exists and is denoted $f(0^+)$. Suppose also that $\lim_{t\to 0^+} f'(t)$ exists and is denoted $f'(0^+)$. Show that $\mathcal{L}[f''(t)] = s^2 F(s) - f'(0^+) - sf(0^+)$.

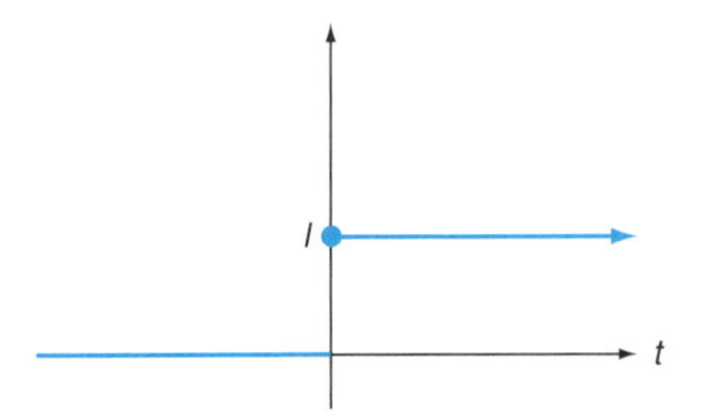

Figure 3.4.1 Graph of Heaviside function.

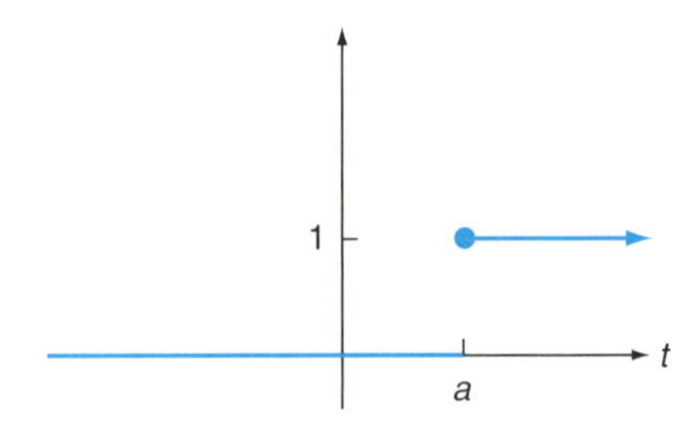

Figure 3.4.2 Graph of $H(t-a)$.

3.4 Discontinuous Forcing Functions

In the previous section we saw how the Laplace transform could be used to solve linear differential equations with constant coefficients. All the examples given, however, could easily have been solved by the method of undetermined coefficients. In this section, we shall examine problems with discontinuous forcing where the Laplace transform works better.

The key is a notational one. We need a way to write a piecewise continuous function as a simple formula so that it may be handled the way the simpler functions $\cos t$, $\sin 5t$, and $e^{3t}\sin 2t$ were treated. The needed notation involves the unit-step or Heaviside function. Since manipulating the unit step function often gives students difficulty, we shall discuss it very carefully.

DEFINITION OF THE HEAVISIDE FUNCTION. The unit step, or Heaviside function $H(t)$, is defined as

$$H(t) = \begin{cases} 1 & \text{if } t \geq 0, \\ 0 & \text{if } t < 0. \end{cases}$$

Its graph is given in figure 3.4.1. It corresponds to 1 being turned on at $t = 0$. ♦

The graph of $H(t-a)$ (figure 3.4.2) is just a translate of that of $H(t)$. Thus, corresponding to turning 1 on at $t = a$,

$$H(t-a) = \begin{cases} 0 & \text{if } t < a, \\ 1 & \text{if } t \geq a. \end{cases}$$

The notations $u(t)$, $u_0(t)$, or $1(t)$ are sometimes also used to denote the Heaviside function. Other notations are also used.

Example 3.4.1 *Graphing Expressions with Heaviside Functions*

Graph $f(t) = H(t-1) - H(t-3)$.

• SOLUTION. Since the formula for $H(t-1)$ changes at $t = 1$, and the formula for $H(t-3)$ changes at $t = 3$, the formula for $f(t)$ will undergo changes at $t = 1$ and $t = 3$. Accordingly, we divide the t interval into the intervals $(-\infty, 1)$, $[1, 3)$, and $[3, \infty)$.

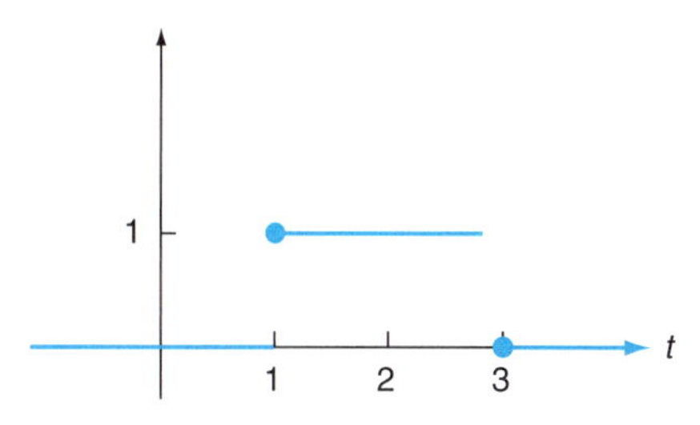

Figure 3.4.3 Graph of $H(t-1)-H(t-3)$.

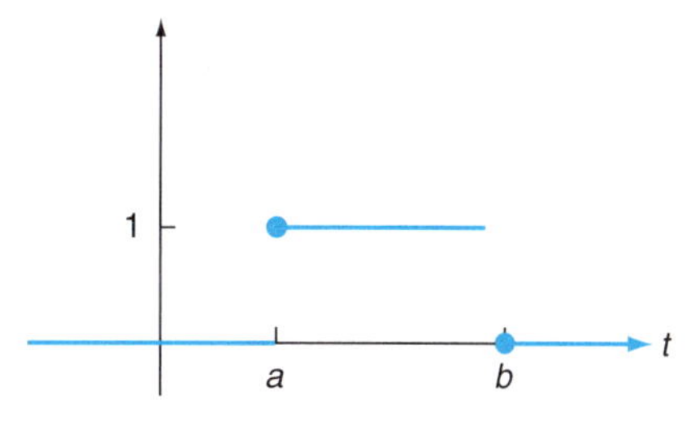

Figure 3.4.4 Graph of $H(t-a)-H(t-b)$.

If $t<1$, then $f(t) = H(t-1)-H(t-3) = 0-0$, since $H(t-1) = 0$ if $t<1$ and $H(t-3)=0$ if $t<3$.
If $1 \le t < 3$, then $f(t) = H(t-1)-H(t-3) = 1-0 = 1$, since $H(t-1)=1$ if $t \ge 1$ and $H(t-3)=0$ if $t<3$.
If $3 \le t$, then $f(t) = H(t-1)-H(t-3) = 1-1 = 0$, since $H(t-1)=1$ if $t \ge 1$ and $H(t-3)=1$ if $t \ge 3$.

Figure 3.4.3 gives the graph of $f(t) = H(t-1)-H(t-3)$. ♦

From this example we make the following deduction.

USEFUL OBSERVATION. If $0<a<b$, then

$$H(t-a)-H(t-b) = \begin{cases} 0 & \text{if } t<a, \\ 1 & \text{if } a \le t < b, \\ 0 & \text{if } b \le t, \end{cases} \tag{1}$$

and $H(t-a)-H(t-b)$ has the graph in figure 3.4.4.

Now let $g(t)$ be some function of t. Then if $a<b$,

$$g(t)[H(t-a)-H(t-b)] = \begin{cases} 0 & \text{if } t<a, \\ g(t) & \text{if } a \le t < b, \\ 0 & \text{if } b \le t. \end{cases}$$

The function $g(t)$ is turned on at $t=a$ and then turned off at $t=b$.

Example 3.4.2 *Graphing Expressions with Heaviside Functions*

Let $f(t) = t^2[H(t-1)-H(t-2)]$. Then

$$t^2[H(t-1)-H(t-2)] = \begin{cases} 0 & \text{if } t<1, \\ t^2 & \text{if } 1 \le t < 2, \\ 0 & \text{if } 2 \le t, \end{cases}$$

and the graph of $t^2[H(t-1)-H(t-2)]$ is given in Figure 3.4.5. The formula $t^2[H(t-1)-H(t-2)]$ picks out that part of the graph of t^2 between 1 and 2, and sets the rest equal to zero. ♦

We are now ready to write a piecewise continuous function in terms of unit step functions.

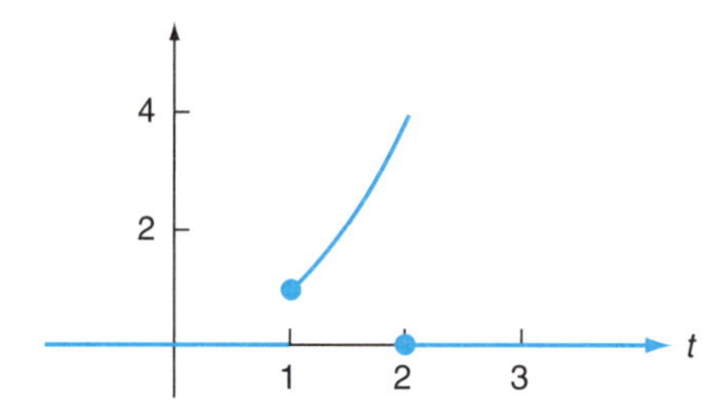

Figure 3.4.5 Graph of $t^2[H(t-1)-H(t-2)]$.

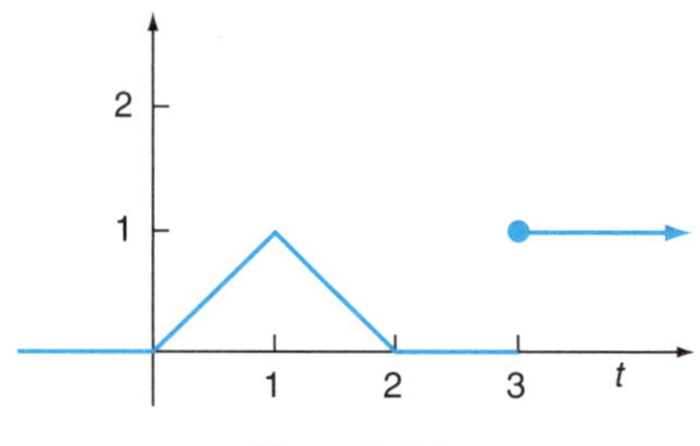

Figure 3.4.6

Example 3.4.3 *Writing in Terms of Heaviside Functions*

Write in terms of unit step functions the function $f(t)$ given by the graph in figure 3.4.6.

• SOLUTION. The function $f(t)$ is

$$f(t)=\begin{cases} t & \text{if } 0\le t<1, \\ 2-t & \text{if } 1\le t<2, \\ 0 & \text{if } 2\le t<3, \\ 1 & \text{if } 3\le t. \end{cases}$$

Thus

$$f(t)=\underbrace{t[H(t)-H(t-1)]}+\underbrace{(2-t)[H(t-1)-H(t-2)]}+\underbrace{1H(t-3)}. \qquad (2)$$

Note that the first term in (2) is zero except when $0\le t<1$, and then it takes on the value t. The second term in (2) is zero except when $1\le t<2$, and then it takes on the value $2-t$. The third term in (2) is zero except when $3\le t$, and then it takes on the value 1. Note: since we consider only $t\ge 0$, the first term in (2) could have been written $t[1-H(t-1)]$. ♦

Example 3.4.4 *Graphing Expressions with Heaviside Functions*

Graph $f(t)=2H(t)+tH(t-1)+(3-t)H(t-2)-3H(t-4)$ for $t\ge 0$.

• SOLUTION. Looking at the Heaviside functions, we see that the formula for $f(t)$ undergoes changes at $t=1$, $t=2$, and $t=4$. Thus we break up the interval $[0,\infty)$

TABLE 3.4.1
Computing values for Example 3.4.4.

t interval	$f(t)=2H(t)+tH(t-1)+(3-t)H(t-2)-3H(t-4)$
$0\le t<1$	$f(t)=2\cdot 1+t\cdot 0+(3-t)\cdot 0-3\cdot 0=2$
$1\le t<2$	$f(t)=2\cdot 1+t\cdot 1+(3-t)\cdot 0-3\cdot 0=2+t$
$2\le t<4$	$f(t)=2\cdot 1+t\cdot 1+(3-t)\cdot 1-3\cdot 0=5$
$4\le t$	$f(t)=2\cdot 1+t\cdot 1+(3-t)\cdot 1-3\cdot 1=2$

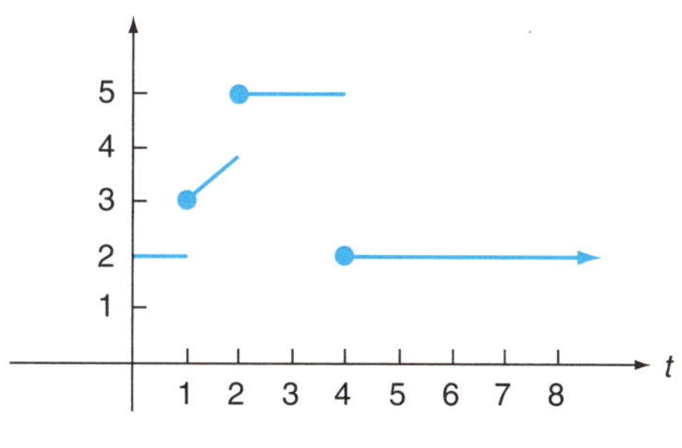

Figure 3.4.7

into four subintervals $[0, 1)$, $[1, 2)$, $[2, 4)$, $[4, \infty)$, and see what $f(t)$ looks like on each subinterval. Using the definition of the Heaviside function, we form Table 3.4.1. The graph of this function is given in figure 3.4.7. ♦

Laplace Transforms of Discontinuous Functions

We have seen that functions that have jump discontinuities can be represented in terms of the unit step function $H(t-c)$. Frequently, the Laplace transform can be calculated using (T15):

$$\mathcal{L}[f(t-c)H(t-c)]=e^{-cs}F(s), \quad \text{for } c>0. \quad \text{(T15)} \tag{3}$$

The formula for the Laplace transform of a step function is a special case of (3). We let $f(t-c)=1$, in which case $f(t)=1$. Then $F(s)=1/s$, so that (3) becomes

$$\mathcal{L}[H(t-c)]=\frac{e^{-cs}}{s} \quad \text{for } c>0. \quad \text{(T14)} \tag{4}$$

Formula (4) (T14) can be independently derived. Formula (3) (T15) is trickier to use in other cases, so several examples will be given after its verification. For taking Laplace transforms, it is the only formula in the short table that involves the Heaviside step function $H(t-c)$. For taking inverse Laplace transforms, it is the only formula we have obtained so far that involves e^{-cs}, exponentials in the transform variable.

VERIFICATION OF (3) FOR $c>0$. We use the definition of the Laplace transform and properties of the Heaviside step function, $H(t-c)=0$ for $t<c$ and $H(t-c)=1$ for $t>c$,

$$\begin{aligned}\mathcal{L}[f(t-c)H(t-c)] &= \int_0^\infty e^{-st} f(t-c)H(t-c)dt \\ &= \int_0^c e^{-st} f(t-c)H(t-c)dt + \int_c^\infty e^{-st} f(t-c)H(t-c)dt \\ &= \int_0^c e^{-st} f(t-c)\, 0\, dt + \int_c^\infty e^{-st} f(t-c)\, 1\, dt \\ &= \int_c^\infty e^{-st} f(t-c)dt.\end{aligned}$$

We now make the change of variables $\tau = t - c$ so the $d\tau = dt$:

$$\begin{aligned}\mathcal{L}[f(t-c)H(t-c)] &= \int_0^\infty e^{-s(\tau+c)} f(\tau)d\tau \\ &= e^{-cs}\int_0^\infty e^{-s\tau} f(\tau)d\tau = e^{-cs}F(s).\end{aligned}$$

Example 3.4.5 *Laplace Transform of Expressions with Heaviside Functions*

Compute $\mathcal{L}[t^2 H(t-1)]$.

• SOLUTION. In order to use (3) (T15), we must take

$$t^2 H(t-1) = f(t-c)H(t-c).$$

Thus $c = 1$ and $t^2 = f(t-1)$. To compute $e^{-cs}F(s)$ in (3), we need $F(s) = \mathcal{L}[f(t)]$. But we have $f(t-1)$, which is not $f(t)$. To compute $f(t)$, introduce a new variable, say τ, and let $\tau = t - 1$, or $t = \tau + 1$. Then

$$\begin{aligned}f(t-1) &= t^2, \\ f(\tau) &= (\tau+1)^2 = \tau^2 + 2\tau + 1,\end{aligned}$$

so that

$$f(t) = t^2 + 2t + 1. \tag{5}$$

Formula (5) may be clear directly from $f(t-1) = t^2$. Thus from (T9),

$$\begin{aligned}F(s) = \mathcal{L}[f(t)] &= \mathcal{L}[t^2 + 2t + 1] \\ &= \frac{2}{s^3} + \frac{2}{s^2} + \frac{1}{s},\end{aligned}$$

and (3) gives that

$$\mathcal{L}[t^2 H(t-1)] = e^{-cs}F(s) = e^{-s}\left[\frac{2}{s^3} + \frac{2}{s^2} + \frac{1}{s}\right]. \qquad \blacklozenge$$

Example 3.4.6 *Laplace Transform of Expressions with Heaviside Functions*

Compute $\mathcal{L}[(e^t + 1)H(t-2)]$ using (3).

• SOLUTION. Taking

$$(e^t + 1)H(t-2) = f(t-2)H(t-2)$$

gives $c=2$ and $f(t-2)=e^t+1$. Let $\tau=t-2$ or $\tau+2=t$. Then $f(\tau)=e^{\tau+2}+1=e^2e^\tau+1$. Thus, $f(t)=e^2e^t+1$, so that

$$F(s)=\mathcal{L}[e^2e^t+1]=e^2\frac{1}{s-1}+\frac{1}{s},$$

and from (3)

$$\mathcal{L}[(e^t+1)H(t-2)]=e^{-cs}F(s)=e^{-2s}\left[\frac{e^2}{s-1}+\frac{1}{s}\right]. \qquad \blacklozenge$$

Example 3.4.7 *Inverse Laplace Transform Involving e^{-cs}*

If $Y(s)=\frac{e^{-3s}}{s^2}$, find $y(t)=\mathcal{L}^{-1}[Y(s)]$.

• SOLUTION. The presence of e^{-3s} in $Y(s)$ is an indication that we should use (3) (T15). Taking

$$\frac{e^{-3s}}{s^2}=e^{-cs}F(s),$$

we see that $c=3$ and $F(s)=1/s^2$. Thus $f(t)=\mathcal{L}^{-1}[F(s)]=t$. According to (T15), the result for $y(t)$ involves a shift and involves a Heaviside step function:

$$y(t)=\mathcal{L}^{-1}[e^{-cs}F(s)]=f(t-c)H(t-c)=(t-3)H(t-3). \qquad \blacklozenge$$

Example 3.4.8 *Inverse Laplace Transform Involving e^{-cs}*

If $Y(s)=e^{-2s}\left[\frac{3}{s}+\frac{s}{s^2+4}\right]$, find $y(t)=\mathcal{L}^{-1}[Y(s)]$.

• SOLUTION. Again the e^{-2s} suggests that we should use formula (T15). Taking

$$e^{-2s}\left[\frac{3}{s}+\frac{s}{s^2+4}\right]=e^{-cs}F(s),$$

we see that $c=2$ and $F(s)=\frac{3}{s}+\frac{s}{(s^2+4)}$. Thus $f(t)=3+\cos 2t$. Hence shifting and introducing a Heaviside step function gives

$$y(t)=\mathcal{L}^{-1}[e^{-cs}F(s)]=f(t-2)H(t-2)=[3+\cos[2(t-2)]]H(t-2). \qquad \blacklozenge$$

3.4.1 Solution of Differential Equations

We can now use the Laplace transform to solve linear constant coefficient differential equations with piecewise continuous forcing functions.

Example 3.4.9 *Piecewise Continuous Forcing*

Solve thc differential equation

$$y''+3y'+2y=g(t), \quad y(0)=0, \quad y'(0)=1, \tag{6}$$

using the Laplace transform where $g(t)$ has the graph in figure 3.4.8.

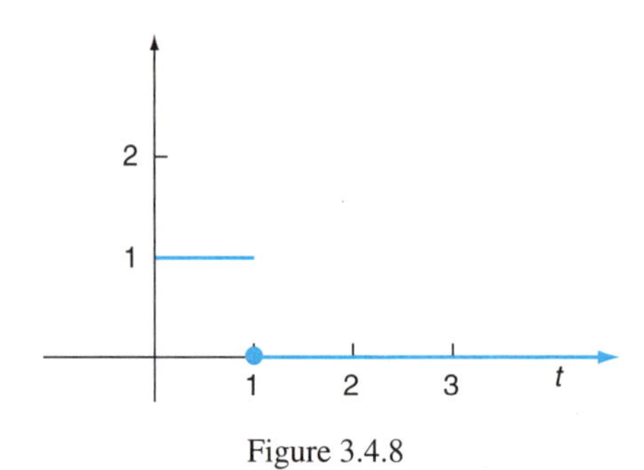

Figure 3.4.8

• SOLUTION. The function $g(t)$ is given by

$$g(t) = \begin{cases} 1 & \text{if } 0 \leq t < 1, \\ 0 & \text{if } 1 \leq t. \end{cases} \tag{7}$$

Thus $g(t) = [1 - H(t-1)]$ (or equivalently, $H(t) - H(t-1)$), and the differential equation is

$$y'' + 3y' + 2y = 1 - H(t-1), \quad y(0) = 0, \quad y'(0) = 1.$$

Taking the Laplace transform of both sides of this differential equation gives

$$s^2 Y(s) - sy(0) - y'(0) + 3[sY(s) - y(0)] + 2Y(s) = \frac{1}{s} - \frac{e^{-s}}{s},$$

so that

$$Y(s) = \frac{s+1}{s(s^2+3s+2)} - e^{-s}\frac{1}{s(s^2+3s+2)}. \tag{8}$$

First, we shall find the inverse Laplace transform of the first term in $Y(s)$. Note that

$$\frac{s+1}{s(s^2+3s+2)} = \frac{s+1}{s(s+1)(s+2)} = \frac{1}{s(s+2)}.$$

By partial fractions,

$$\frac{1}{s(s+2)} = \frac{A}{s} + \frac{B}{s+2},$$

so that

$$1 = A(s+2) + Bs.$$

Evaluating at $s = 0$ and $s = -2$ yields $A = \frac{1}{2}$, $B = -\frac{1}{2}$. Thus

$$\mathcal{L}^{-1}\left[\frac{1}{s(s+2)}\right] = \mathcal{L}^{-1}\left[\frac{1/2}{s} + \frac{-1/2}{s+2}\right] = \frac{1}{2} - \frac{1}{2}e^{-2t}. \tag{9}$$

Now the second term of (8) is in the form $e^{-cs}F(s)$, where $c=1$, so that (3), or (T15), should be used. Here

$$F(s)=\frac{1}{s(s^2+3s+2)}=\frac{1}{s(s+2)(s+1)}.$$

Again using partial fractions, we have

$$F(s)=\frac{1}{s(s+2)(s+1)}=\frac{A}{s}+\frac{B}{s+2}+\frac{C}{s+1},$$

so that

$$1=A(s+2)(s+1)+Bs(s+1)+Cs(s+2).$$

Since the denominator has distinct linear factors, evaluating at their roots $s=0$, $s=-1$, $s=-2$ yields $A=\frac{1}{2}$, $C=-1$, and $B=\frac{1}{2}$. Hence

$$F(s)=\frac{1/2}{s}+\frac{1/2}{s+2}+\frac{-1}{s+1},$$

and

$$f(t)=\frac{1}{2}+\frac{1}{2}e^{-2t}-e^{-t}.$$

Thus

$$\begin{aligned}\mathcal{L}^{-1}[e^{-s}F(s)]&=f(t-1)H(t-1)\\&=\left[\frac{1}{2}+\frac{1}{2}e^{-2(t-1)}-e^{-(t-1)}\right]H(t-1).\end{aligned}\tag{10}$$

Finally, combining the two terms (9), (10) gives

$$y(t)=\left[\frac{1}{2}-\frac{1}{2}e^{-2t}\right]-\left[\frac{1}{2}+\frac{1}{2}e^{-2(t-1)}-e^{-(t-1)}\right]H(t-1).\tag{11}$$

The approach of this section should be contrasted with the examples in Section 1.7.4, where we had to match initial conditions with terminal values at every discontinuity of the "forcing function" $g(t)$. ♦

A PHYSICAL INTERPRETATION. As explained in Section 2.8, under certain operating conditions, equation (6) of the preceding example is a mathematical model for the RLC circuit in figure 3.4.9.

In figure 3.4.9 g is a voltage source that varies with time, $y(0)=0$ is the initial charge in the capacitor, and $y'(0)=1$ is the initial current in the capacitor. There are several ways to interpret $g(t)$ as given by (7). One is that it is the voltage across nodes A and B in figure 3.4.9, and these nodes are connected to other circuit elements not shown. Alternatively, g can be thought of as a 1-V battery that is shorted out for $t>1$. Intuitively, one expects that, because of the relatively large resistance (the circuit is overdamped, in the terminology of Section 2.8), the capacitor will charge

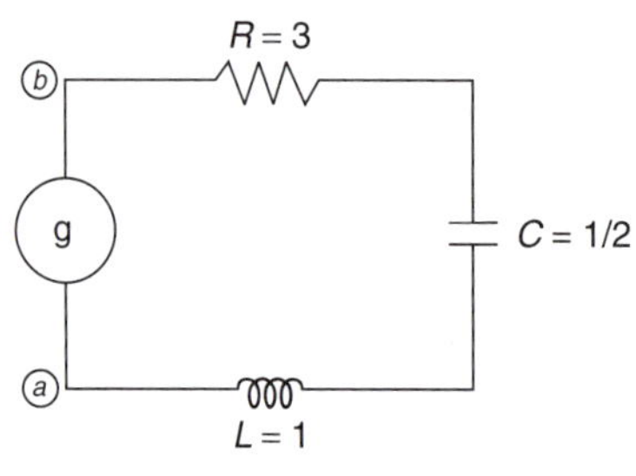

Figure 3.4.9

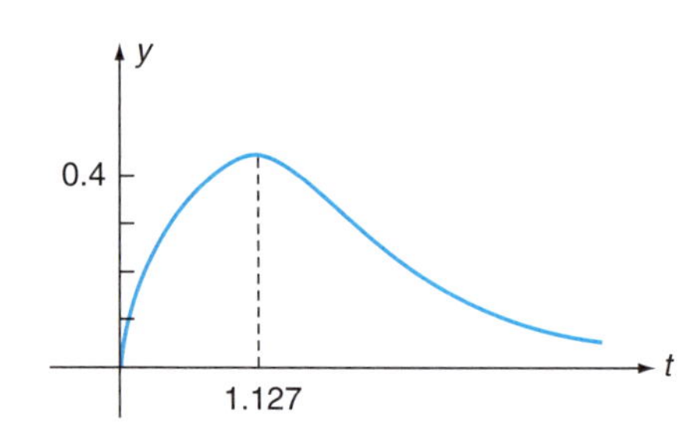

Figure 3.4.10 Graph of the solution (11) of (b).

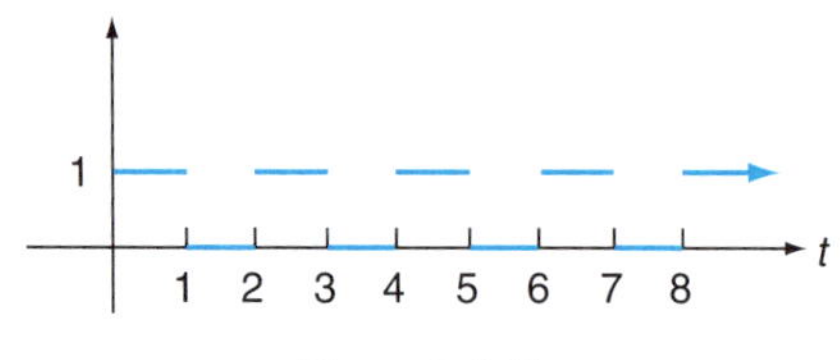

Figure 3.4.11

for $0 \leq t < 1$ and discharge for $t \geq 1$. Figure 3.4.10 is the graph of $y(t)$. Note that, in fact, the capacitor keeps charging for 1.127 time units before beginning to discharge. That is because the inductor gives the current charging the capacitor the electrical equivalent of momentum. It takes a while for the current charging the capacitor to stop and reverse direction.

Example 3.4.10 *Periodic Forcing*

Solve the differential equation

$$y'' + 3y' + 2y = g(t), \quad y(0) = 0, \quad y'(0) = 1, \tag{12}$$

using the Laplace transform, where $g(t)$ has the graph shown in figure 3.4.11.

This is the same as the previous example, except that $g(t)$ is now a periodic function. In the next section, we shall discuss periodic functions in more detail, since they are very important.

• SOLUTION. First note that

$$\begin{aligned} g(t) &= [1 - H(t-1)] + [H(t-2) - H(t-3)] + [H(t-4) - H(t-5)] + \cdots \\ &= 1 - H(t-1) + H(t-2) - H(t-3) + H(t-4) - H(t-5) + \cdots \\ &= \sum_{n=0}^{\infty} (-1)^n H(t-n). \end{aligned}$$

While $g(t)$ is given by an infinite series, for any specific value of t only a finite number of terms in the series are nonzero. For example, if $t = 13.7$, then $H(13.7 - n)$ is zero

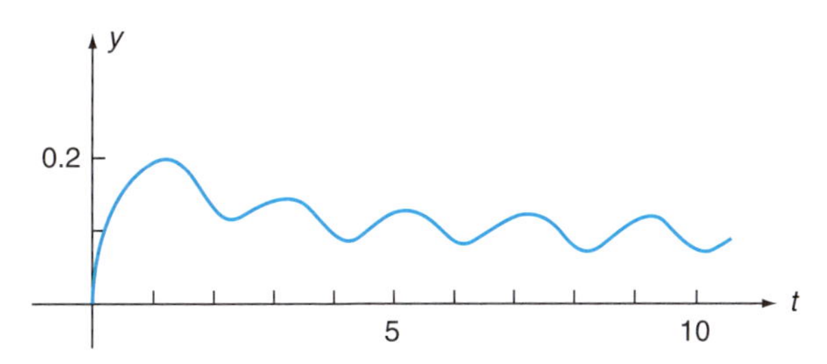

Figure 3.4.12 Graph of (16).

if $n > 13.7$. The differential equation is now written as

$$y'' + 3y' + 2y = \sum_{n=0}^{\infty}(-1)^n H(t-n). \tag{13}$$

It can be shown that it makes sense to take the Laplace transform of the series in (13), which can be done term by term. Taking the Laplace transform of both sides of (13) gives

$$s^2 Y(s) - 1 + 3sY(s) + 2Y(s) = \sum_{n=0}^{\infty}(-1)^n \frac{e^{-ns}}{s}.$$

Solving for $Y(s)$ gives

$$Y(s) = \frac{1}{s^2+3s+2} + \sum_{n=0}^{\infty}(-1)^n e^{-ns}\left[\frac{1}{s(s^2+3s+2)}\right] \tag{14}$$

$$= \frac{1}{s+1} + \frac{-1}{s+2} + \sum_{n=0}^{\infty}(-1)^n e^{-ns}\left[\frac{1}{s(s^2+3s+2)}\right]. \tag{15}$$

From the preceding example, we know that

$$\mathcal{L}^{-1}\left[e^{-ns}\frac{1}{s(s^2+3s+2)}\right] = f(t-n)H(t-n)$$

and

$$f(t) = \mathcal{L}^{-1}\left[\frac{1}{s(s^2+3s+2)}\right] = \frac{1}{2} + \frac{1}{2}e^{-2t} - e^{-t}.$$

Thus (15) gives that

$$y(t) = e^{-t} - e^{-2t} + \sum_{n=0}^{\infty}(-1)^n\left[\frac{1}{2} + \frac{1}{2}e^{-2(t-n)} - e^{-(t-n)}\right]H(t-n). \tag{16}$$

Again, for any specific value of t, only a finite number of terms in the series in (16) are nonzero. Figure 3.4.12 is the graph of (16). ♦

Exercises

In Exercises 1–3, write the function defined on $[0, \infty)$ in terms of unit-step functions and graph.

1.
$$f(t) = \begin{cases} 0 & 0 \le t < 2, \\ 3 & 2 \le t < 5, \\ t & 5 \le t. \end{cases}$$

2.
$$f(t) = \begin{cases} 1 & 0 \le t < 1, \\ -1 & 1 \le t < 2, \\ 1 & 2 \le t < 3, \\ -1 & 3 \le t. \end{cases}$$

3.
$$f(t) = \begin{cases} \sin t & 0 \le t < \pi, \\ 0 & \pi \le t < 2\pi, \\ \sin t & 2\pi \le t < 3\pi, \\ 0 & 3\pi \le t. \end{cases}$$

In Exercises 4–9, write the function defined on $[0, \infty)$ in terms of unit step functions.

4.
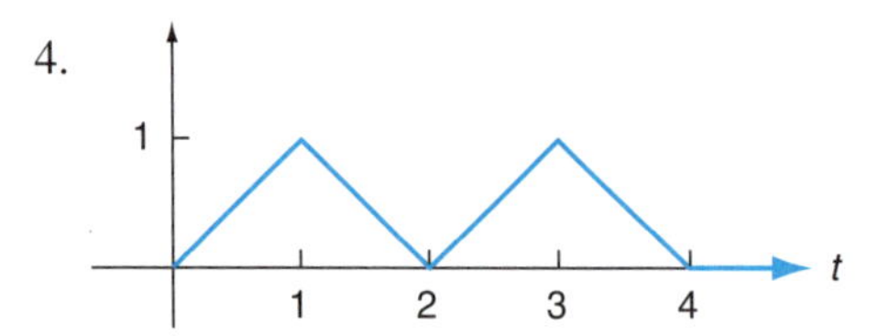

5.
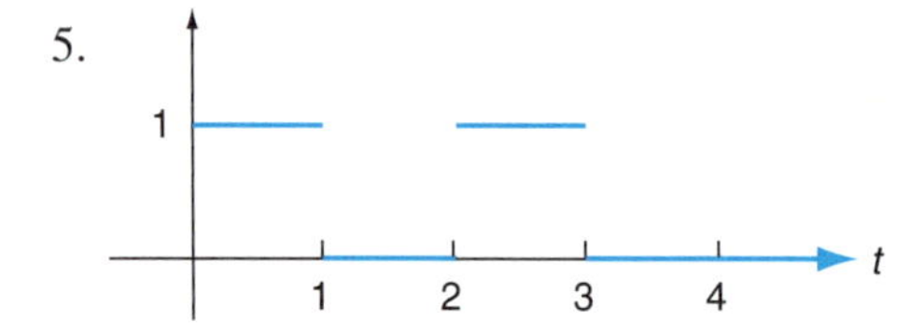

6.
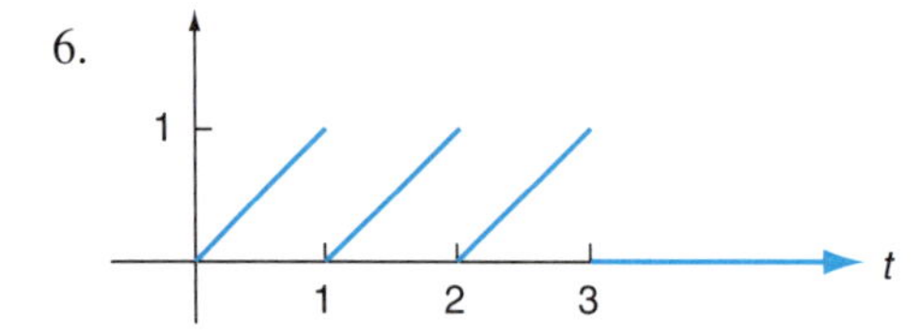

7.
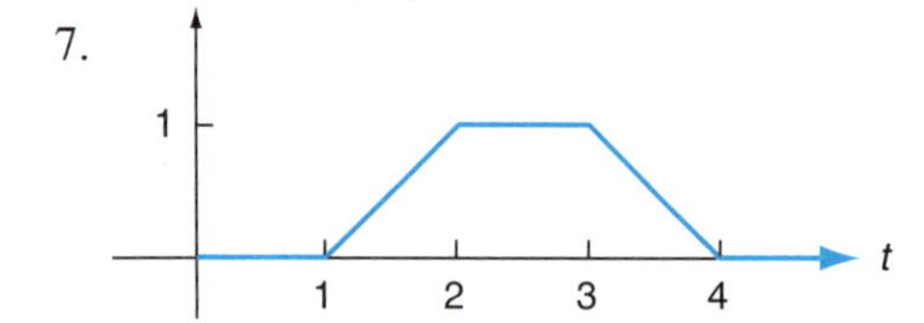

8.

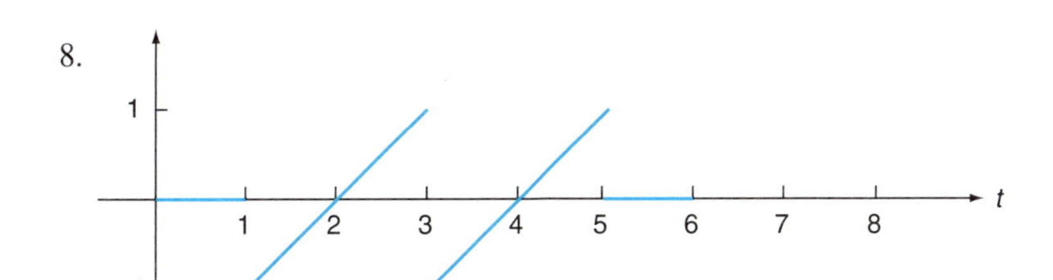

9.

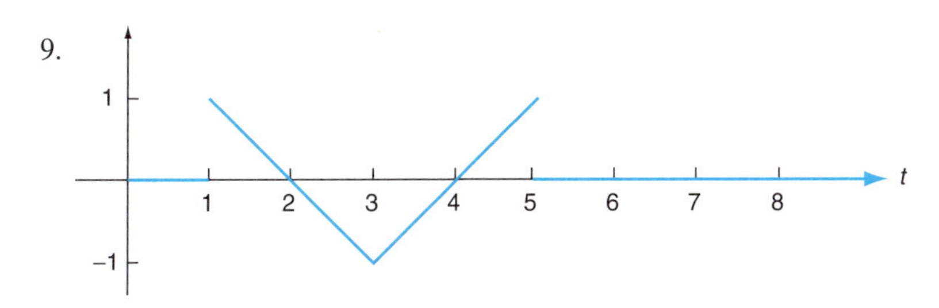

In Exercises 10–21, compute the Laplace transform $Y(s)$ of $y(t)$.

10. $y(t) = 2 - 5H(t-1) + 6H(t-3)$.
11. $y(t) = tH(t-2)$.
12. $y(t) = (t-2)H(t-2)$.
13. $y(t) = (t^3+1)H(t-1)$.
14. $y(t) = e^{3t}H(t-2) + 6H(t-3)$.
15. $y(t) = \sin t\, H(t-\pi)$.
16. $y(t) = \sin t\, H(t-\pi/2)$.
17. $y(t) = \cos t\, H(t-2\pi)$.
18. $y(t) = (t^3+t)H(t-2)$.
19. $y(t) = e^{2t}H(t-3)$.
20. $y(t) = t^2e^{-3t}H(t-1)$.
21. $y(t) = te^{5t}H(t-2)$.

In Exercises 22–29, compute the inverse Laplace transform $y(t)$ of $Y(s)$.

22. $Y(s) = \dfrac{e^{-3s}}{s^3}$.
23. $Y(s) = e^{-2s}\dfrac{1}{s^2+s} + e^{-3s}\dfrac{1}{s^2+s}$.
24. $Y(s) = \dfrac{1}{s^2} + \dfrac{e^{-s}}{s^2+4} + e^{-2s}\dfrac{s}{s^2+9}$.
25. $Y(s) = \dfrac{e^{-3s}}{s^2+2s+2}$.
26. $Y(s) = \dfrac{e^{-2s}}{(s+4)^5}$.

27. $Y(s) = \dfrac{e^{-s}}{s^2+1} - \dfrac{e^{-2s}}{s^2+4}$.

28. $Y(s) = e^{-3s}\left[\dfrac{1}{s} + \dfrac{1}{s-2}\right]$.

29. $Y(s) = e^{-s}\left[\dfrac{4}{s} + \dfrac{6s}{s^2+9}\right]$.

30. Using only the definitions of the Laplace transform and the Heaviside function, verify that

$$\mathcal{L}[f(t-c)H(t-c)] = e^{-cs}F(s)$$

if $c > 0$, $f(t)$ is piecewise-continuous of exponential order, and $\mathcal{L}[f(t)] = F(s)$.

In Exercises 31–48, solve the differential equation using the Laplace transform.

31. $y'' + y = g(t)$, $y(0) = y'(0) = 0$, where

$$g(t) = \begin{cases} 0, & 0 \le t < 2, \\ 1, & 2 \le t < 5, \\ 0, & 5 \le t. \end{cases}$$

32. $y'' - y = g(t)$, $y(0) = y'(0) = 0$, where

$$g(t) = \begin{cases} 0, & 0 \le t < 1, \\ 1, & 1 \le t < 3, \\ 2, & 3 \le t. \end{cases}$$

33. $y' - 3y = g(t)$, $y(0) = 1$, $g(t)$ given by

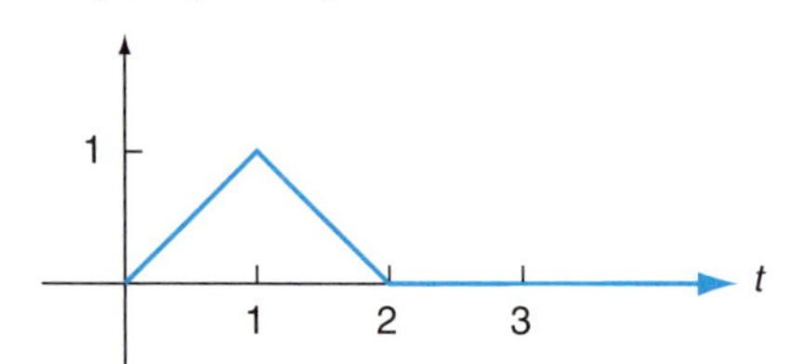

34. $y' + y = g(t)$, $y(0) = 0$, $g(t)$ given by

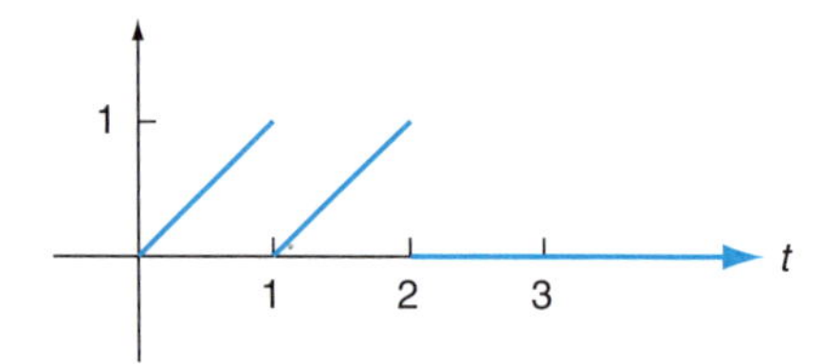

35. $y'' + 2y' + 10y = g(t)$, $y(0) = 1$, $y'(0) = 0$, where

$$g(t) = \begin{cases} 1, & 0 \le t < 3, \\ 0, & 3 \le t. \end{cases}$$

36. $y'' + 5y' + 6y = g(t)$, $y(0) = y'(0) = 0$, where

$$g(t) = \begin{cases} 0, & 0 \le t < \pi/2, \\ \cos t, & \pi/2 \le t < 3\pi/2, \\ 0, & 3\pi/2 \le t. \end{cases}$$

37. $y'' + 4y = g(t)$, $y(0) = 0$, $y'(0) = 0$, g given by the graph in Exercise 5.
38. $y'' - 2y' + y = g(t)$, $y(0) = 1$, $y'(0) = 1$, where

$$g(t) = \begin{cases} t^2, & 0 \le t < 1, \\ 0, & 1 \le t. \end{cases}$$

39. $y'' + 9y = g(t)$, $y(0) = 0$, $y'(0) = 0$, where

$$g(t) = \begin{cases} 1, & 0 \le t < 2, \\ e^{-t}, & t \ge 2. \end{cases}$$

40. $y'' - 9y = g(t)$, $y(0) = 0$, $y'(0) = 0$, where

$$g(t) = \begin{cases} e^{-t}, & 0 \le t < 4, \\ 1, & t \ge 4. \end{cases}$$

41. $y' - 5y = g(t)$, $y(0) = 0$, where

$$g(t) = \begin{cases} e^{3t}, & 0 \le t < 4, \\ 0, & t \ge 4. \end{cases}$$

42. $y' + 5y = g(t)$, $y(0) = 0$, where

$$g(t) = \begin{cases} 0, & 0 \le t < 4, \\ e^{3t}, & t \ge 4. \end{cases}$$

43. $y' + 3y = g(t)$, $y(0) = 0$, where

$$g(t) = \begin{cases} 0, & 0 \le t < \frac{\pi}{2}, \\ \cos t, & t \ge \frac{\pi}{2}. \end{cases}$$

44. $y' - 4y = g(t)$, $y(0) = 0$, where

$$g(t) = \begin{cases} \cos t, & 0 \le t < \frac{\pi}{2}, \\ 0, & t \ge \frac{\pi}{2}. \end{cases}$$

45. $y' + 2y = g(t)$, $y(0) = 0$, where

$$g(t) = \begin{cases} 0, & 0 \le t < 3, \\ \sin t, & t \ge 3. \end{cases}$$

46. $y' + 4y = g(t)$, $y(0) = 0$, where

$$g(t) = \begin{cases} \sin t, & 0 \le t < 5, \\ 0, & t \ge 5. \end{cases}$$

47. $y'' + 9y = g(t)$, $y(0) = 0$, $y'(0) = 1$, where

$$g(t) = \begin{cases} \sin t, & 0 \le t < 4, \\ 0, & t \ge 4. \end{cases}$$

48. $y'' + 4y = g(t)$, $y(0) = 1$, $y'(0) = 0$, where

$$g(t) = \begin{cases} 0, & 0 \le t < 2, \\ \cos t, & t \ge 2. \end{cases}$$

In Exercises 49 and 50, use the method given in Example 3.4.10 to solve the differential equation using Laplace transforms.

49. $y'' + 4y = g(t)$, $y(0) = 0$, $y'(0) = 0$, where $g(t)$ is periodic and given by

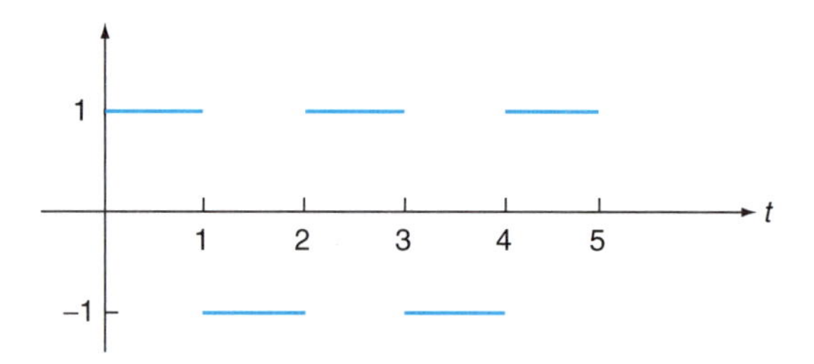

50. $y' + 2y = g(t)$, $y(0) = 1$, where $g(t)$ is periodic and given by

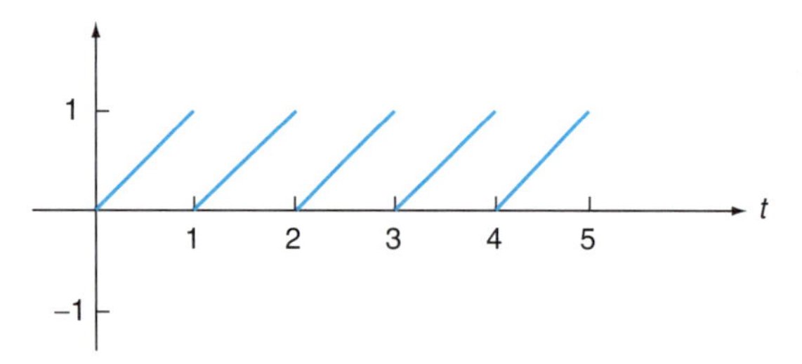

3.5 Periodic Functions

A function $g(t)$ defined on $[0, \infty)$ is **periodic with period** T if $g(t+T) = g(t)$ for all $t > 0$. Periodic functions appear in many applications. We have already seen examples of periodic functions and the Laplace transform. This section will consider them in more detail.

The basic formula is

If $g(t)$ is periodic on $[0, \infty)$ with period T and has a Laplace transform, then

$$\mathcal{L}[g(t)] = \frac{\int_0^T e^{-st} g(t)dt}{1 - e^{-sT}}. \tag{1}$$

VERIFICATION. Suppose $g(t)$ is periodic with period T and has a Laplace transform. Then

$$\mathcal{L}[g(t)] = \int_0^\infty e^{-st} g(t)dt = \int_0^T e^{-st} g(t)dt + \int_T^\infty e^{-st} g(t)dt$$

(let $t = T + \tau$ in the second integral)

$$= \int_0^T e^{-st} g(t)dt + \int_0^\infty e^{-s(T+\tau)} g(T+\tau)d\tau$$

$$= \int_0^T e^{-st} g(t)dt + e^{-sT} \int_0^\infty e^{-st} g(\tau)d\tau,$$

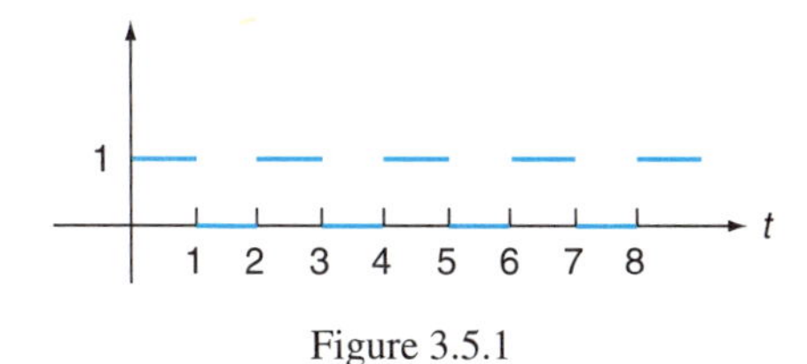

Figure 3.5.1

(since $g(t)$ has period T). Thus

$$\mathcal{L}[g(t)] = \int_0^T e^{-st} g(t)dt + e^{-sT}\mathcal{L}[g(t)].$$

Solving this equation for $\mathcal{L}[g(t)]$ gives (1). Since $g(t)$ is periodic, we only need $0 < t < T$.

Example 3.5.1 *Laplace Transform of Periodic Function*

Find $\mathcal{L}[g(t)]$, where

$$g(t) = \begin{cases} 1 & \text{if } 0 \le t < 1, \\ 0 & \text{if } 1 \le t < 2; \end{cases} \quad \text{period 2.}$$

Figure 3.5.1 gives the graph of $g(t)$.

• SOLUTION. Since the period is 2, formula (1) gives

$$\begin{aligned}\mathcal{L}[g(t)] &= \frac{1}{1-e^{-2s}} \int_0^2 e^{-st} g(t)dt \\ &= \frac{1}{1-e^{-2s}} \int_0^1 e^{-st} dt \\ &= \frac{1}{1-e^{-2s}} \left(\frac{e^{-st}}{-s} \Bigg|_{t=0}^{t=1} \right) \\ &= \frac{1}{1-e^{-2s}} \left(\frac{e^{-s}}{-s} + \frac{1}{s} \right),\end{aligned}$$

which can be simplified to

$$\frac{1}{s(1+e^{-s})}. \qquad \blacklozenge$$

The most difficult part of (1) is computing the integral $\int_0^T e^{-st} g(t)dt$. This can be done using the Laplace transform tables, as follows. If

$$\hat{g}(t) = \begin{cases} g(t) & \text{for } 0 \le t \le T, \\ 0 & \text{for } t > T, \end{cases}$$

then

$$\int_0^T e^{-st} g(t)dt = \mathcal{L}[\hat{g}(t)], \tag{2}$$

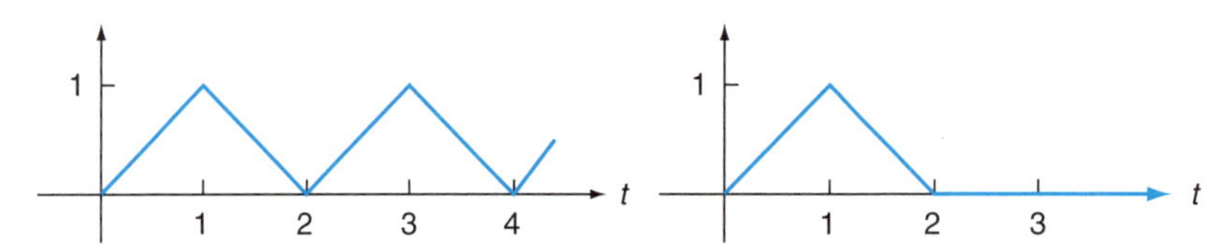

Figure 3.5.2 g from Example 3.5.2 and $\tilde{g}$ from Example 3.5.2.

since

$$\int_0^T e^{-st} g(t)dt = \int_0^T e^{-st}\hat{g}(t)dt$$
$$= \int_0^\infty e^{-st}\hat{g}(t)dt = \mathcal{L}[\hat{g}(t)].$$

Note that $\hat{g}(t) = g(t)[1 - H(t - T)]$. This approach is (T31).

Example 3.5.2 *Laplace Transform of Periodic Function*

Let $g(t)$ have period 2, where $g(t)$ is graphed in Figure 3.5.2 and is given by

$$g(t) = \begin{cases} t & \text{for } 0 \le t \le 1, \\ 2 - t & \text{for } 1 \le t \le 2. \end{cases}$$

Find $G(s)$.

• SOLUTION. Let

$$\hat{g}(t) = t[1 - H(t-1)] + (2-t)[H(t-1) - H(t-2)]$$
$$= t + 2(1-t)H(t-1) + (t-2)H(t-2).$$

Then $\hat{g}(t) = g(t)$ for $0 \le t \le 2$ and $\hat{g}(t) = 0$ for $t > 2$. Graphs of g and $\hat{g}$ are given in figure 3.5.2. Thus by (1), (2),

$$\mathcal{L}[g(t)] = \frac{1}{1-e^{-2s}} \int_0^2 e^{-st} g(t)dt$$
$$= \frac{1}{1-e^{-2s}} \mathcal{L}[\hat{g}(t)]$$
$$= \frac{1}{1-e^{-2s}} \mathcal{L}[t + 2(1-t)H(t-1) + (t-2)H(t-2)]$$
$$= \frac{1}{1-e^{-2s}} \left[\frac{1}{s^2} - 2\frac{e^{-s}}{s^2} + \frac{e^{-2s}}{s^2}\right].$$ ♦

Inverse Laplace transforms are not usually computed using (1). One method to compute them is by the use of series. Recall that, for $|x| < 1$,

$$\frac{1}{1-x} = 1 + x + x^2 + \cdots = \sum_{n=0}^{\infty} x^n. \tag{3}$$

If $s > 0$, $T > 0$, then $e^{-sT} < 1$. Letting $x = e^{-sT}$, (3) becomes

$$\frac{1}{1-e^{-st}} = 1 + e^{-sT} + e^{-2sT} + \cdots = \sum_{n=0}^{\infty} e^{-nsT}. \tag{4}$$

Thus, if $\mathcal{L}[f(t)] = F(s)$,

$$\frac{1}{1-e^{-sT}} F(s) = \sum_{n=0}^{\infty} e^{-nsT} F(s).$$

Taking the inverse Laplace transform term by using (T15) with $c = nT$ gives

$$\mathcal{L}^{-1}\left[\frac{1}{1-e^{-sT}} F(s)\right] = \sum_{n=0}^{\infty} f(t-nT)H(t-nT). \quad (T32) \tag{5}$$

Similarly, letting $x = -e^{-sT}$ in (3) gives

$$\frac{1}{1+e^{-sT}} = \sum_{n=0}^{\infty} (-1)^n e^{-nsT},$$

and

$$\mathcal{L}^{-1}\left[\frac{1}{1+e^{-sT}} F(s)\right] = \sum_{n=0}^{\infty} (-1)^n f(t-nT)H(t-nT). \quad (T33) \tag{6}$$

Example 3.5.3 *Inverse Laplace Transform with $1 - e^{-sT}$ in the Denominator*

Let $G(s) = \frac{e^{-2s}-1}{s^2(1-e^{-3s})}$. Find $g(t) = \mathcal{L}^{-1}[G(s)]$.

• SOLUTION. Using (4), we get

$$\begin{aligned} G(s) &= \frac{1}{s^2}(e^{-2s} - 1) \sum_{n=0}^{\infty} e^{-3ns} \\ &= \sum_{n=0}^{\infty} \frac{1}{s^2} e^{-(2+3n)s} - \sum_{n=0}^{\infty} \frac{1}{s^2} e^{-3ns}. \end{aligned}$$

Now, by (T15), $\mathcal{L}^{-1}[\frac{1}{s^2} e^{-cs}] = (t-c)H(t-c)$. Thus,

$$g(t) = \sum_{n=0}^{\infty} (t-2-3n)H(t-2-3n) - \sum_{n=0}^{\infty} (t-3n)H(t-3n).$$

This example also illustrates that, if $G(s)$ has a $1/[1-e^{-sT}]$-factor, it is not necessarily the case that $g(t)$ is periodic. ♦

ALTERNATIVE SOLUTION. We could also do Example 3.5.3 using formula (5) with $T=3$ and $F(s)=-\frac{1}{s^2}+\frac{1}{s^2}e^{-2s}$. Then $f(t)=-t+(t-2)H(t-2)$, and (5) is

$$\begin{aligned}
&\sum_{n=0}^{\infty} f(t-3n)H(t-3n)\\
&\quad=\sum_{n=0}^{\infty}[-(t-3n)+(t-3n-2)H(t-3n-2)]H(t-3n)\\
&\quad=\sum_{n=0}^{\infty}-(t-3n)H(t-3n)+(t-3n-2)H(t-3n-2)H(t-3n)\\
&\quad=\sum_{n=0}^{\infty}-(t-3n)H(t-3n)+(t-3n-2)H(t-3n-2).
\end{aligned}$$

The last equality follows, since

$$H(t-a)H(t-b)=H(t-c),$$

where c is the larger of a, b.

Exercises

In Exercises 1–8, sketch $g(t)$ and find $G(s)$ using formula (1), or (1) and (2).

1. $g(t)=|\sin t|$ (rectified sine).
2. $g(t)=|\cos t|$ (rectified cosine).
3. $$g(t)=\begin{cases} t^2 & \text{for } 0\le t<1,\\ 2-t^2 & \text{for } 1\le t<2, \end{cases}\quad \text{period 2.}$$
4. $g(t)=t$ for $0\le t<1$, period 1.
5. $g(t)=e^t$ for $0\le t<1$, period 1.
6. $g(t)=t^3$ for $0\le t<1$, period 1.
7. $$g(t)=\begin{cases} 1 & \text{for } 0\le t<1,\\ -1 & \text{for } 1\le t<2, \end{cases}\quad \text{period 2.}$$
8. $$g(t)=\begin{cases} t & \text{for } 0\le t<1,\\ 1 & 1\le t<2,\\ -1 & \text{for } 2\le t<3, \end{cases}\quad \text{period 3.}$$

In Exercises 9–20, find $g(t)$, given $G(s)$.

9. $G(s)=\dfrac{1}{1-e^{-s}}\left[\dfrac{1}{s^2}+\dfrac{1}{s^3}\right].$

10. $G(s) = \dfrac{1}{1+e^{-3s}}\left[\dfrac{2s}{s^2+6s+13}\right].$

11. $G(s) = \dfrac{1}{1-e^{-s}}\left[\dfrac{s}{s^2+4}\right].$

12. $G(s) = \dfrac{1}{s}\tanh\dfrac{s}{2}.$

13. $G(s) = \dfrac{1}{1-e^{-2s}}\left[\dfrac{1}{s^2}+\dfrac{e^{-s}}{s^3}\right].$

14. $G(s) = \dfrac{1}{s}\operatorname{sech} s.$

15. $G(s) = \dfrac{1}{1-e^{-\pi s}}\left[\dfrac{1}{s}+\dfrac{e^{-\pi s/2}}{s^2+1}\right].$

16. $G(s) = \dfrac{1}{1-e^{-3s}}\left[\dfrac{1}{s}+\dfrac{e^{-s}}{s^2}+\dfrac{e^{-2s}}{s^3}\right].$

17. $G(s) = \dfrac{1}{1+e^{-5s}}\left[\dfrac{1}{s^3}+\dfrac{e^{-2s}}{s^4}\right].$

18. $G(s) = \dfrac{1}{1+e^{-s}}\left[\dfrac{1}{s^2-1}+\dfrac{e^{-2s}}{s^2-4}\right].$

19. $G(s) = \dfrac{1}{s^2}\operatorname{csch} s.$

20. $G(s) = \dfrac{1}{1+e^{-3s}}\left[\dfrac{1}{s+2}+\dfrac{e^{-2s}}{s-2}\right].$

21. Verify that $H(t-a)H(t-b) = H(t-c)$, where c is the larger of a, b

In Exercises 22–28, solve the initial value problem, using the Laplace transform.

22. $y' + y = g(t)$, $y(0) = 0$, $g(t)$ from Exercise 1.
23. $y'' + y = |\sin 2t|$, $y(0) = 0$, $y'(0) = 0$.
24. $y'' - y = \sin t - |\sin t|$, $y(0) = 0$, $y'(0) = 1$.
25. $y' + y = g(t)$, $y(0) = 1$, $g(t)$ as in Exercise 5.
26. $y' - 2y = g(t)$, $y(0) = 0$, $g(t)$ as in Exercise 4.
27. $y'' - 4y = g(t)$, $y(0) = y'(0) = 0$, $g(t)$ as in Exercise 7.
28. $y'' + 2y' + y = g(t)$, $y(0) = y'(0) = -1$, $g(t)$ as in Exercise 8.
29. Suppose that the series $\sum_{n=0}^{\infty} t^n/n!$ for e^t can be Laplace-transformed term by term. Show that the resulting series gives $\mathcal{L}[e^t]$, which is $1/(s-1)$.

3.6 Integrals and the Convolution Theorem

We have seen that the Laplace transform of the derivative of a function is obtained by multiplying the transform of that function by s, since

$$\mathcal{L}[g'(t)] = sG(s) - g(0). \quad \text{(T18)} \tag{1}$$

Thus, it should not be a surprise that the Laplace transform of an integral is obtained by dividing by s. To be precise,

$$\mathcal{L}\left[\int_0^t f(\tau)d\tau\right] = \frac{F(s)}{s}. \quad \text{(T23)} \tag{2}$$

DERIVATION OF (2). We use (1) with $g(t) = \int_0^t f(\tau)d\tau$, so that $g'(t) = f(t)$ and $g(0) = 0$:

$$\mathcal{L}[g'(t)] = F(s) = sG(s) - g(0) = s\mathcal{L}\left[\int_0^t f(\tau)d\tau\right], \tag{3}$$

since $g(0) = 0$. Equation (2) follows from (3) by algebraically solving for $\mathcal{L}[\int_0^t f(\tau)d\tau]$.

Example 3.6.1 *Inverse Laplace Transform Using (T23)*

Although it is better to use partial fractions, calculate $\mathcal{L}^{-1}[\frac{3}{s(s+5)}]$ using (2).

• SOLUTION. We let $F(s) = \frac{3}{s+5}$, so that $f(t) = 3e^{-5t}$. According to (2),

$$\mathcal{L}^{-1}\left[\frac{F(s)}{s}\right] = \mathcal{L}^{-1}\left[\frac{3}{s(s+5)}\right] = \int_0^t f(\tau)d\tau$$

$$= \int_0^t 3e^{-5\tau}d\tau = 3\frac{e^{-5t}}{-5}\bigg|_0^t = \frac{3}{5}(1 - e^{-5\tau}). \qquad \blacklozenge$$

Finding the inverse Laplace transform of a product of transforms is a question which arises frequently. The following theorem is quite useful. The integral theorem (2) for Laplace transforms is a special case of this theorem.

THEOREM 3.6.1 Convolution *Suppose that $f(t)$ and $g(t)$ are functions with Laplace transforms $\mathcal{L}[f(t)] = F(s)$ and $\mathcal{L}[g(t)] = G(s)$. Then*

$$\mathcal{L}\left[\int_0^t f(\tau)g(t-\tau)d\tau\right] = F(s)G(s), \quad \text{(T24)} \tag{4}$$

and thus

$$\mathcal{L}^{-1}[F(s)G(s)] = \int_0^t f(\tau)g(t-\tau)d\tau. \tag{5}$$

A derivation of (4) is given later in this section. The inverse of a product of transforms equals $\int_0^t f(\tau)g(t-\tau)d\tau$, which is a new function called the **convolution of**

f and g. A convenient notation for the convolution of f and g is $f * g$,

$$(f * g)(t) = \int_0^t f(\tau)g(t-\tau)d\tau. \tag{6}$$

Fortunately it does not matter whether we have the convolution of f and g or the convolution of g and f, since they are the same,

$$f * g = g * f. \tag{7}$$

VERIFICATION OF (7). By the definition (6) of the convolution of f and g,

$$f * g = \int_0^t f(\tau)g(t-\tau)d\tau. \tag{8}$$

We make a change of variables (t fixed) in the integral in (8), $w = t - \tau$. Then $dw = -d\tau$ and (8) becomes

$$f * g = \int_t^0 f(t-w)g(w)(-dw) = \int_0^t f(t-w)g(w)dw = g * f.$$

Thus in the key convolution theorems (4) and (5) it does not matter whether $\int_0^t f(\tau)g(t-\tau)d\tau$ or $\int_0^t g(\tau)f(t-\tau)d\tau$ is used, since they are equal.

Example 3.6.2 *Convolution*

Suppose $f(t) = e^t$ and $g(t) = e^{4t}$. Compute $f * g$.

- SOLUTION. Using (6), the convolution $f * g$ is just another function of t:

$$\begin{aligned} e^t * e^{4t} &= \int_0^t e^{\tau} e^{4(t-\tau)} d\tau \\ &= e^{4t} \int_0^t e^{-3\tau} d\tau \\ &= e^{4t} \left. \frac{e^{-3\tau}}{-3} \right|_{\tau=0}^{\tau=t} = \frac{1}{3} e^{4t}(1 - e^{-3t}) = \frac{1}{3}(e^{4t} - e^t). \end{aligned}$$

♦

Example 3.6.3 *Inverse Laplace Transform Using the Convolution*

Use the convolution theorem (5) to evaluate $\mathcal{L}^{-1}\left[\frac{1}{(s+1)(s+2)}\right]$.

- SOLUTION. Let $F(s) = \frac{1}{s+1}$, $G(s) = \frac{1}{s+2}$. Then

$$f(t) = \mathcal{L}^{-1}[F(s)] = \mathcal{L}^{-1}\left[\frac{1}{s+1}\right] = e^{-t},$$

$$g(t) = \mathcal{L}^{-1}\left[\frac{1}{s+2}\right] = e^{-2t}.$$

Formula (5) now gives

$$\mathcal{L}^{-1}\left[\frac{1}{(s+1)(s+2)}\right]=e^{-t}*e^{-2t}=\int_0^t e^{-\tau}e^{-2(t-\tau)}d\tau$$
$$=e^{-2t}\int_0^t e^{\tau}d\tau=e^{-2t}(e^t-1)=e^{-t}-e^{-2t}.$$

Of course, this example could also have been done using a partial-fractions expansion ♦

Application of the Convolution Theorem to Differential Equations

Consider the initial value problem

$$y''+b^2y=f(t), \tag{9}$$

with $y(0)=0$ and $y'(0)=0$.

Taking the Laplace transform of both sides of the differential equation yields

$$s^2Y(s)+b^2Y(s)=F(s),$$

since $y(0)=0$ and $y'(0)=0$. Here $Y(s)$ is the Laplace transform of $y(t)$ and $F(s)$ is the Laplace transform of $f(t)$. Solving for the Laplace transform of the solution gives

$$Y(s)=\frac{F(s)}{s^2+b^2}.$$

We note that we can represent $Y(s)$ as a product of transforms of known functions

$$Y(s)=F(s)G(s),$$

where

$$G(s)=\frac{1}{s^2+b^2}.$$

We have introduced $G(s)$ whose inverse transform can be obtained easily (if necessary using (T3))

$$g(t)=\frac{1}{b}\sin bt.$$

Since $Y(s)$ equals a product of transforms $F(s)G(s)$, $y(t)$ may be obtained using the convolution theorem (5)

$$y(t)=\int_0^t f(\tau)g(t-\tau)d\tau=\int_0^t f(\tau)\frac{1}{b}\sin b(t-\tau)d\tau. \tag{10}$$

The function $(1/b)\sin b(t-\tau)$ is an **influence function** expressing how much the input at τ influences the output at t. The relatively simple solution (10) of (9) was obtained using the convolution theorem for Laplace transforms.

Without using Laplace transforms, the usual method to solve (9) is the method of variation of parameters. To show how (10) could have been obtained using variation of parameters, we use the trigonometric addition formula $\sin(a-b) = \sin a \cos b - \cos a \sin b$, so that (10) becomes

$$y = v_1 \cos bt + v_2 \sin bt, \tag{11}$$

where $v_1 = -\int_0^t f(\tau)\frac{1}{b}\sin b\tau d\tau$ and $v_2 = \int_0^t f(\tau)\frac{1}{b}\cos b\tau d\tau$. Equation (11) is in the form assumed in the method of variation of parameters.

3.6.1 Derivation of the Convolution Theorem (optional)

The convolution theorem (T24)

$$\mathcal{L}^{-1}[F(s)G(s)] = \int_0^t f(\tau)g(t-\tau)d\tau \tag{12}$$

is very important, but not easy to verify. We will be able to recognize the inverse transform of a product of transforms $F(s)G(s)$ if we consider different dummy variables of integration for the defining Laplace transforms, $F(s) = \int_0^\infty e^{-s\tau} f(\tau)d\tau$ and $G(s) = \int_0^\infty e^{-su} g(u)du$, so that

$$\begin{aligned} F(s)G(s) &= \int_0^\infty e^{-s\tau} f(\tau)d\tau \int_0^\infty e^{-su} g(u)du \\ &= \int_0^\infty \int_0^\infty e^{-s(u+\tau)} f(\tau)g(u)\, du d\tau. \end{aligned} \tag{13}$$

We make a change of variables in the u-integral, and let $t = u + \tau$, as motivated by the exponent in (13). We integrate over t (rather than u) for fixed τ, so that $dt = du$:

$$F(s)G(s) = \int_{\tau=0}^\infty \int_{t=\tau}^\infty e^{-st} f(\tau)g(t-\tau)\, dt d\tau. \tag{14}$$

The Laplace transforms of $f(t)$ and $g(t)$ are only defined for $t \geq 0$. If we define

$$g(Q) = 0 \text{ for } Q < 0, \tag{15}$$

then we can extend the t integral in (14) to begin from $t = 0$ instead of $t = \tau$, since this only adds 0 to the integral. Thus, switching orders

$$\begin{aligned} F(s)G(s) &= \int_{\tau=0}^\infty \int_{t=0}^\infty e^{-st} f(\tau)g(t-\tau)dt d\tau \\ &= \int_{t=0}^\infty \int_{\tau=0}^\infty e^{-st} f(\tau)g(t-\tau)d\tau dt \\ &= \int_{t=0}^\infty e^{-st} \left[\int_{\tau=0}^\infty f(\tau)g(t-\tau)d\tau\right] dt. \end{aligned}$$

We now recognize that the function in brackets above must be the function whose Laplace transform is $F(s)G(s)$. Thus,

$$\mathcal{L}\left[\int_{\tau=0}^\infty f(\tau)g(t-\tau)d\tau\right] = F(s)G(s). \tag{16}$$

However, $g(Q)$ is zero (see (15)) for negative arguments which occur for $\tau > t$. Thus, there are no contributions to the integral in (16) for $\tau > t$ so that

$$\mathcal{L}\left[\int_{\tau=0}^{t} f(\tau)g(t-\tau)d\tau\right] = F(s)G(s). \tag{17}$$

Equation (17) is the convolution theorem for Laplace transforms, and it is equivalent to (12).

Exercises

1. If $f(t)=t$, $g(t)=e^t$, find $f*g$ from the definition (1).
2. If $f(t)=e^t$, $g(t)=\cos t$, find $f*g$ from the definition (1).
3. Compute $1*1$.
4. Compute $1*t$.
5. Using formula (5), compute
$$\mathcal{L}^{-1}\left[\frac{1}{s^2}\cdot\frac{1}{(s^2+1)}\right].$$
6. Using formula (5), compute
$$\mathcal{L}^{-1}\left[\frac{s}{(s^2+1)^2}\right].$$

In Exercises 7–16, solve the initial value problem using the convolution theorem of Laplace transforms. Here $f(t)$ is an unspecified function which has a Laplace transform.

7. $y''-5y'+4y=f(t)$, $y(0)=0$, $y'(0)=0$.
8. $y''+7y'+10y=f(t)$, $y(0)=0$, $y'(0)=0$.
9. $y''-9y=f(t)$, $y(0)=0$, $y'(0)=0$.
10. $y''-16y=f(t)$, $y(0)=0$, $y'(0)=0$.
11. $y''+4y=f(t)$, $y(0)=0$, $y'(0)=7$.
12. $y''+25y=f(t)$, $y(0)=3$, $y'(0)=0$.
13. $y'+7y=f(t)$, $y(0)=2$.
14. $y'-3y=f(t)$, $y(0)=-6$.
15. $y''-2y'+10y=f(t)$, $y(0)=0$, $y'(0)=0$.
16. $y''+6y'+25y=f(t)$, $y(0)=0$, $y'(0)=0$.
17. $y''+4y'+13y=f(t)$, $y(0)=0$, $y'(0)=0$.

An example of an **integral equation** is

$$\begin{aligned} x(t) &= \int_0^t h(t-\tau)x(\tau)d\tau + g(t) \\ &= (h*x)(t) + g(t), \end{aligned}$$

where h, g are known functions and $x(t)$ is an unknown function. This integral equation can be solved by taking the Laplace transform, solving for $X(s)$, and then computing $\mathcal{L}^{-1}[X(s)]$.

In Exercises 18–25, solve the given integral equation using the Laplace transform.

18. $x(t) = \dfrac{1}{2}\displaystyle\int_0^t x(s)ds + 1.$

19. $x(t) = \displaystyle\int_0^t \cos(t-\tau)x(\tau)d\tau + \sin t.$

20. $x(t) = \displaystyle\int_0^t (t-\tau)x(\tau)d\tau + 1.$

21. $x(t) = \displaystyle\int_0^t e^{-t+\tau}x(\tau)d\tau + 2.$

22. $6x(t) = \displaystyle\int_0^t (t-\tau)^3 x(\tau)d\tau + t.$

23. $x(t) = \displaystyle\int_0^t 2\sin(2t-2\tau)x(\tau)d\tau + \sin t.$

24. $x(t) = \displaystyle\int_0^t \cos(2t-2\tau)x(\tau)d\tau + e^{3t}.$

25. $x(t) = -\displaystyle\int_0^t \sinh(t-\tau)x(\tau)d\tau + 3.$

26. Suppose that f, g, h arc piecewise continuous functions on $[0, \infty)$, and α, β are constants. Using the definition (1) of convolution, verify that
    a) $(\alpha f + \beta g) * h = \alpha(f * h) + \beta(g * h)$.
    b) $f * (\alpha g + \beta h) = \alpha(f * g) + \beta(f * h)$.
    c) Using (5), verify that $f * (g * h) = (f * g) * h$.
27. Suppose that f is piecewise continuous, of exponential order ($|f(t)| \leq Me^{\alpha t}$). Show that, for $s > \alpha$,

$$|F(s)| \leq \frac{M}{s-\alpha}.$$

28. Using Exercise 27 and the Laplace transform, show that there do not exist f, g both piecewise continuous, of exponential order, such that $f * g = 1$.
29. The following example illustrates one way in which integral equations are "nicer" than differential equations. Let

$$g(t) = \begin{cases} e^t & \text{for } 0 \leq t < 1, \\ -1 + e^t + e^{t-1} & \text{for } 1 \leq t. \end{cases}$$

    a) Show that $g(t)$ is continuous for $t \geq 0$, but $g'(1)$ is not defined. Also show that $g(t)$ is a solution of $y' - y = H(t-1)$, $y(0) = 1$ for $t \geq 0, t \neq 1$.
    b) Show that $g(t)$ satisfies the integral equation $y(t) = \int_0^t [y(\tau) + H(\tau - 1)]d\tau + 1$ for all $t \geq 0$, including $t = 1$.
30. Suppose that $x(t)$ is a piecewise continuous function for $-\infty < t < \infty$. Think of t as being the location across the real axis. That is, t measures a position

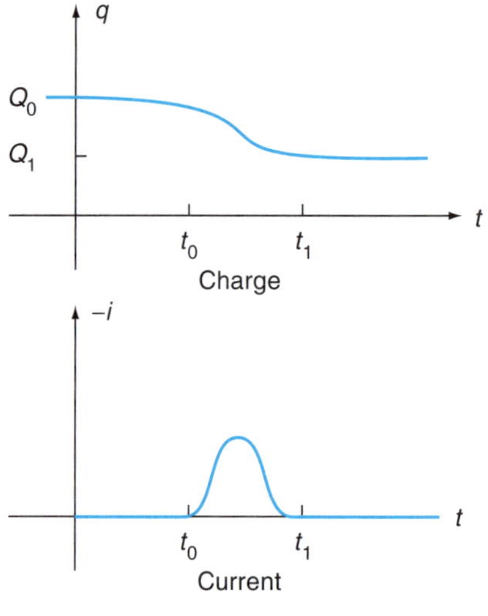

Figure 3.7.1

rather than time. Let $x(t)$ be a quantity that varies along the t-axis, such as light intensity or frequency. A measurement of $x(t)$ would actually be a (possibly weighted) average of x over an interval, say $[t-a, t+a]$, which contains t. The *unweighted average* over $[t-a, t+a]$ would be

$$\frac{1}{2a}\int_{-a}^{a} x(t+\tau)d\tau. \tag{18}$$

Let $(1*x)(z)$ be the value of $(1*x)$ evaluated at $t=z$. Show that (18) can be written as

$$\frac{1}{2a}[(1*x)(t+a)-(1*x)(t-a)].$$

3.7 Impulses and Distributions

Suppose that a capacitor having charge Q_0 for time $t \leq t_0$ is suddenly partially discharged (think of the spark in a spark plug) to a lower charge Q_1 at time t_1. The capacitor then maintains the charge Q_1 for $t \geq t_1$. The amount of charge lost is $Q_1 - Q_0$. If this discharge happens quickly, then for a very brief period of time, the current $i(t)$, which is the rate of change of the charge $q(t)$, must be very large. The faster the discharge, the higher the current gets. The relationship between change in charge and current is

$$\Delta Q = Q_1 - Q_0 = \int_{t_0}^{t_1} \frac{dq}{dt}dt = \int_{t_0}^{t_1} i(t)dt. \tag{1}$$

Typical graphs of $q(t)$ and $i(t)$ are shown in figure 3.7.1. If the time interval is shortened, we have the graphs shown in figure 3.7.2.

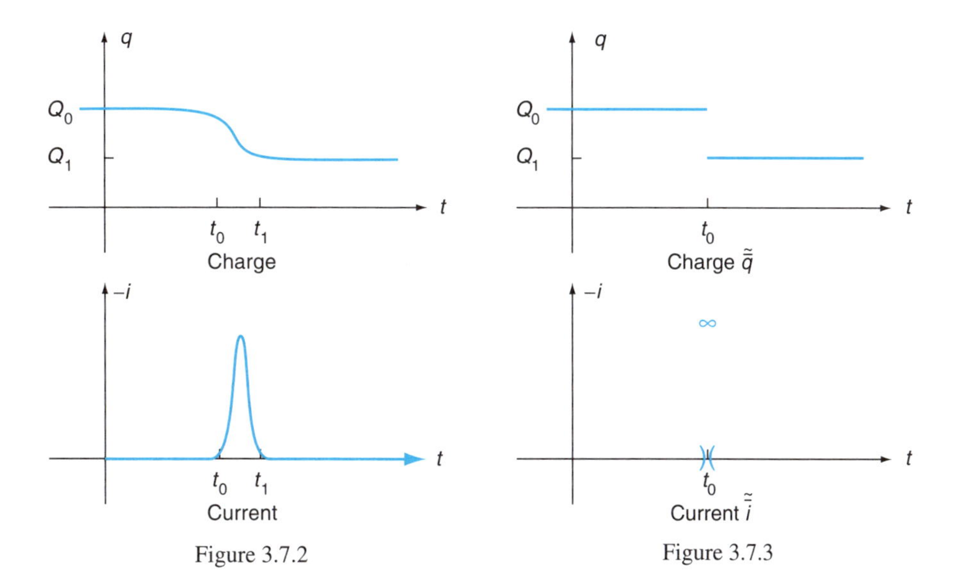

Figure 3.7.2

Figure 3.7.3

Since $i(t)$ is zero outside $[t_0, t_1]$, (1) may be replaced by

$$\Delta Q = \int_{-\infty}^{\infty} i(t)dt.$$

Similarly, one may think of a mass m being acted on by a large force for a brief period of time, leading to a change in momentum. In this case the momentum mv plays the role of the charge, and the instantaneous force $(mv)'$ plays the role of the current.

Suppose we take the limit as the length of the time interval $[t_0, t_1]$ goes to zero by letting $t_1 \to t_0^+$. The charge $q(t)$ approaches the function (graphed in figure 3.7.3)

$$\tilde{q}(t) = Q_0 - (Q_0 - Q_1)H(t - t_0),$$

where H is the Heaviside function. Let $\tilde{i}(t)$ be the limit of $i(t)$ as $t_1 \to t_0^+$. But $i(t)$ goes to zero for $t \neq t_0$ and $i(t_0) \to \infty$. On the other hand, we still want (1) to hold:

$$\Delta Q = Q_1 - Q_0 = \int_{-\infty}^{\infty} \tilde{i}(t)dt,$$

which means $\tilde{i}(t)$ should have a finite area. Clearly, this limiting current $\tilde{i}(t)$ is not a function in the sense with which we are familiar. It is an example of a **distribution** or **impulse function**. These generalized functions are a very convenient mathematical notation.

Definition of Delta Function

$\delta(t)$ is a mathematical object known as a **delta function**. It is an example of a **distribution** or **generalized function**. It has the following properties:

i) $\delta(t) = 0$ if $t \neq 0$.

ii) $\delta(0)$ is not defined.
iii) $\int_{-\infty}^{\infty} \delta(t)dt = 1$.
iv) If $g(t)$ is a continuous function on $(-\infty, \infty)$, then

$$\int_{-\infty}^{\infty} g(t)\delta(t)dt = g(0).$$ ♦

It is possible to build a logical, rigorous definition of $\delta(t)$, but we shall not do so. Intuitively, we may think of $\delta(t)$ as an approximation of a physical impulse of magnitude 1 at time $t = 0$. For example, it could be the rapid transfer of one unit of charge at time zero. It can be shown that if a is a constant, then

v) $\delta(t-a) = 0$ if $t \neq a$.
vi) $\int_{-\infty}^{\infty} \delta(t-a)dt = 1$.
vii) If $g(t)$ is a continuous function on $(-\infty, \infty)$, then

$$\int_{-\infty}^{\infty} g(t)\delta(t-a)dt = g(a).$$

From these formulas, we get

$$\int_{-\infty}^{t} \delta(t)dt = \begin{cases} 1 & \text{if } t > 0, \\ 0 & \text{if } t < 0. \end{cases}$$

Thus, formally,

$$\int_{-\infty}^{t} \delta(t)dt = H(t),$$

and $\delta(t)$ may be considered, in some sense, to be the derivative of the Heaviside function.

Laplace Transform of a Delta Function

One nice property of the Laplace transform is that it works almost as easily for distributions and impulses as it does for ordinary functions. Proceeding formally,

$$\mathcal{L}[\delta(t-a)] = \int_0^{\infty} e^{-st}\delta(t-a)dt = e^{-sa}. \quad \text{(by (vii) above)}$$

Thus

$$\mathcal{L}[\delta(t-a)] = e^{-as}. \quad \text{(T13)}$$

In particular,

$$\mathcal{L}[\delta(t)] = 1. \quad \text{(T12)}$$

There are many other distributions, for example, "derivatives" of $\delta(t)$, but they will not be considered here (note Exercise 11). ♦

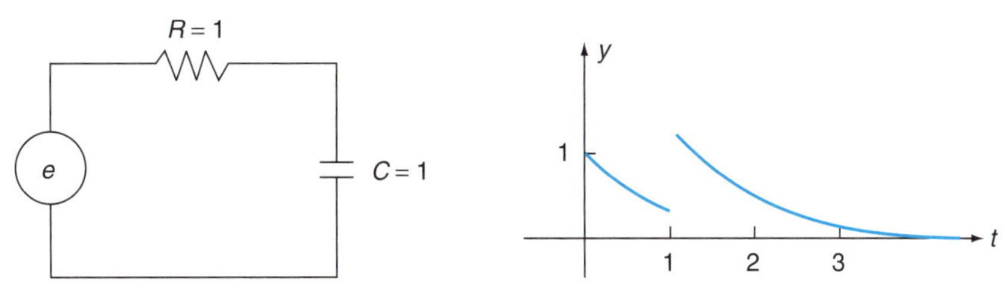

Figure 3.7.4

Figure 3.7.5 Graph of (2).

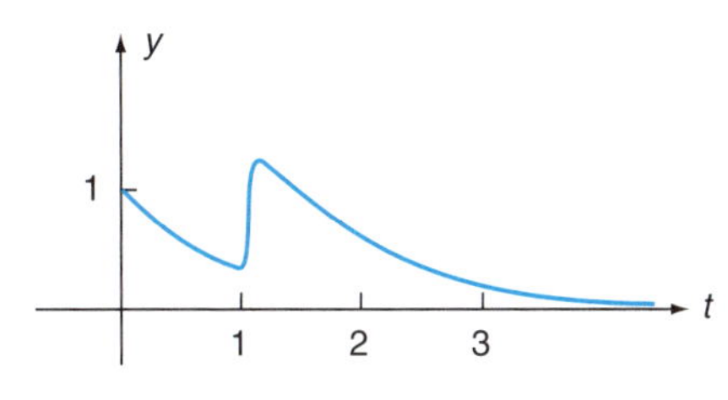

Figure 3.7.6

Example 3.7.1 *Impulsive Forcing Function*

Solve $y' + y = \delta(t-1)$, $y(0) = 1$.

• SOLUTION. Taking the Laplace transform of both sides using (T13) gives

$$sY(s) - y(0) + Y(s) = e^{-s}.$$

Solving for $Y(s)$, we have

$$Y(s) = \frac{1}{s+1} + \frac{e^{-s}}{s+1},$$

so that, by (T2) and (T15),

$$y(t) = \mathcal{L}^{-1}\left[\frac{1}{s+1}\right] + \mathcal{L}^{-1}\left[e^{-s}\frac{1}{s+1}\right] = e^{-t} + e^{-(t-1)}H(t-1). \qquad (2)$$

♦

Physically, this example could be viewed as the simple linear RC circuit in figure 3.7.4, where y is the charge on the capacitor at time t, and there is an initial charge of one on the capacitor. For $0 \le t < 1$, the voltage e is zero and the capacitor is discharging. At time $t = 1$, there is a voltage impulse, that is, a very large voltage applied for a brief period, which recharges the capacitor. Then the voltage is again zero, and the capacitor resumes discharging.

The graph of (2) is given in figure 3.7.5. This graph should be interpreted as meaning that, in a real problem, $y(t)$ would be given by a function like that in figure 3.7.6.

Needless to say, in real problems involving impulses, some care should be taken to ensure that the equations being used still provide accurate models in the presence of the large, but brief, values given by the impulse.

Exercises

In Exercises 1–8, solve the differential equation.

1. $y' + 8y = \delta(t-1) + \delta(t-2), \ y(0) = 0.$
2. $y'' + 2y' + 2y = \delta(t-5), y(0) = 1, \ y'(0) = 0.$
3. $y'' + 6y' + 109y = \delta(t-1) - \delta(t-7), y(0) = 0, \ y'(0) = 0.$
4. $y'' + 3y' + 2y = -2\delta(t-1), y(0) = 1, \ y'(0) = 0.$
5. $y'' + 4y' + 3y = 1 + \delta(t-3), y(0) = 0, \ y'(0) = 1.$
6. $y' + 4y = 1 - \delta(t-4), y(0) = 1.$
7. $y'' + y = 1 + \delta(t-2\pi), y(0) = 1, \ y'(0) = 0.$
8. $y'' + 4y = t + 4\delta(t-4\pi), y(0) = y'(0) = 1$

Exercises 9 and 10 require a personal computer or access to computer facilities.

9. For Exercises 1–4 above,
   a) Sketch the solution of the differential equation.
   b) Obtain the graph of the solution, and compare it to your sketch.
10. For Exercises 5–8 above,
   a) Sketch the solution of the differential equation.
   b) Obtain the graph of the solution, and compare it to your sketch.
11. Let $\delta^{(n)}(t)$, $n \geq 0$, n an integer, have the properties that
   i) $\delta^{(n)}(t) = 0$ if $t \neq 0$.
   ii) $\int_{-\infty}^{\infty} \delta^{(n)}(t-a)g(t)dt = (-1)^n g^{(n)}(a)$ for continuous g.
   Show formally that $\mathcal{L}[\delta^{(n)}(t-a)] = s^n e^{-as}$.

CHAPTER 4

An Introduction to Linear Systems of Differential Equations and Their Phase Plane

4.1 Introduction

Up to this point we have considered only single differential equations with one dependent variable. In many applications, however, one is led to simultaneously consider several ordinary differential equations with several dependent variables and one independent variable. Such **systems of differential equations** may be linear or nonlinear. Nonlinear systems are studied in Chapter 6. If all the differential equations are linear in the dependent variables, the resulting linear systems of differential equations are most naturally studied using vector notation and matrix theory. In this chapter, in order to introduce the basic ideas, we only consider two linear differential equations which will yield 2×2 matrices. In Section 4.2, we show how to solve such linear differential equations using eigenvalues and their corresponding eigenvectors. In Section 4.3, we show how to understand these solutions of linear differential equations using the phase plane.

For simplicity, in this chapter we will consider linear homogeneous systems of two differential equations with constant coefficients, such as

$$\frac{dx}{dt} = 2x + 3y, \tag{1a}$$

$$\frac{dy}{dt} = 4x - 5y. \tag{1b}$$

Our motivation is that this type of equation arises in physical applications. We briefly consider two motivating examples, one from mixtures and the other from electrical engineering. Another motivation for linear systems is that most numerical software for solving differential equations is written for systems. Furthermore, in Chapter 6 we show that interesting nonlinear systems of differential equations can be approximated by these types of linear systems in the neighborhood of equilibria.

MULTILOOP CIRCUITS. One application involves multiloop circuits. Consider, for example, the circuit in figure 4.1.1. The voltage law applied to the top and bottom

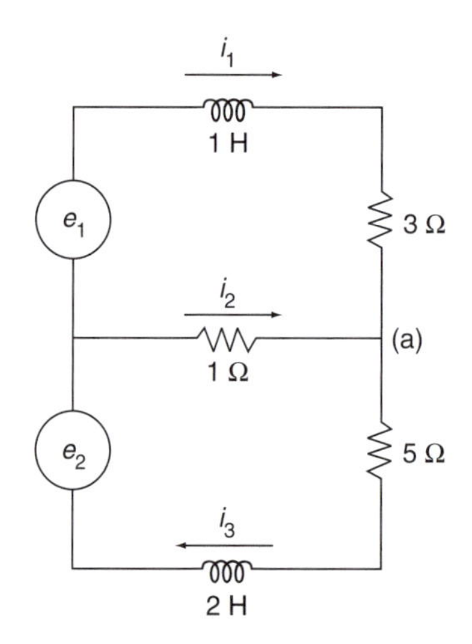

Figure 4.1.1

loops of this circuit yields

$$\frac{di_1}{dt} + 3i_1 - i_2 = e_1(t), \tag{2}$$

$$2\frac{di_3}{dt} + 5i_3 + i_2 = e_2(t), \tag{3}$$

respectively. The current law at node (a) is

$$i_1 + i_2 - i_3 = 0. \tag{4}$$

Using this equation to eliminate i_2 from (2) and (3) gives the system

$$\frac{di_1}{dt} + 4i_1 - i_3 = e_1(t), \tag{5a}$$

$$2\frac{di_3}{dt} - i_1 + 6i_3 = e_2(t). \tag{5b}$$

The equations for the currents are nonhomogeneous if at least one of the voltage sources is nonzero.

Example 4.1.1 *Constant Volume Mixture or Compartmental Model*

Consider the mixture problem for two tanks diagrammed in figure 4.1.2. Here, we wish to determine the system of differential equations that describes the amount of salt in each well-mixed tank. Two 4-liter tanks are initially filled with saltwater. It is given that tank A initially has 0.1 kilograms of salt, and tank B initially has 0.5 kilograms of salt. Pure water flows into tank A from an outside source at a rate of 15 liters/hour. Saltwater flows from tank A to tank B at 16 liters/hour. Saltwater is piped back into tank A at the rate of 1 liter/hour, and in addition 15 liters/hour of the saltwater mixture escapes entirely from tank B.

• SOLUTION. Note that we have made a mixture problem in which the amount of water in each tank is constant, 4 liters. Tank A has 4 liters of the mixture, since water

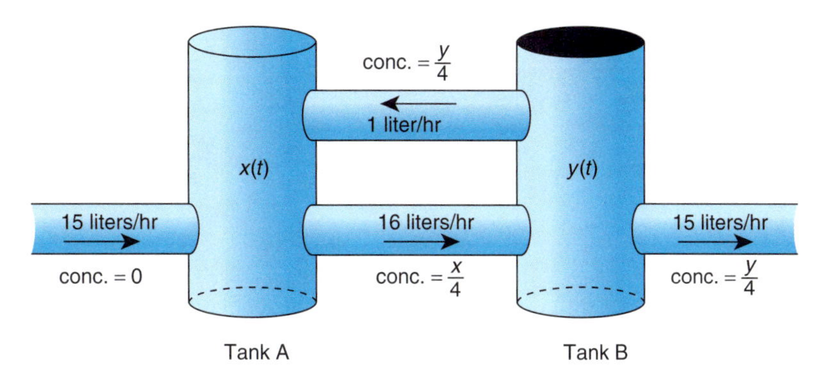

Figure 4.1.2

in entering and leaving at the same rate of $15+1=16$ liters per hour. Similarly, Tank B remains at 4 liters. Let x be the amount of salt in tank A at time t, and y the amount of salt in tank A at time t. Then

$$\frac{dx}{dt} = \begin{bmatrix} \text{inflow rate} \\ \text{salt from} \\ \text{outside} \end{bmatrix} + \begin{bmatrix} \text{inflow rate} \\ \text{salt from} \\ \text{tank B} \end{bmatrix} - \begin{bmatrix} \text{outflow rate} \\ \text{salt from} \\ \text{tank A} \end{bmatrix},$$

$$\frac{dy}{dt} = \begin{bmatrix} \text{inflow rate} \\ \text{salt from} \\ \text{tank A} \end{bmatrix} - \begin{bmatrix} \text{outflow rate} \\ \text{salt from} \\ \text{tank B to tank A} \end{bmatrix} - \begin{bmatrix} \text{outflow rate} \\ \text{salt to be} \\ \text{discarded} \end{bmatrix},$$

or

$$\frac{dx}{dt} = 0 \cdot 15 + 1 \cdot \frac{y}{4} - 16\frac{x}{4}, \tag{6a}$$

$$\frac{dy}{dt} = 16\frac{x}{4} - 1 \cdot \frac{y}{4} - 15\frac{y}{4}. \tag{6b}$$

We have, then, the linear system of differential equations

$$\frac{dx}{dt} = -4x + \frac{y}{4}, \tag{7a}$$

$$\frac{dy}{dt} = 4x - 4y. \tag{7b}$$

These differential equations should be solved subject to the initial conditions $x(0) = 0.1$ and $y(0) = 0.5$. ♦

ELIMINATION. Much of this book is devoted to solving systems using matrix methods. However, we should say that for small-sized systems elimination often is straightforward. There are many ways to solve (7a) and (7b) by elimination. For example, we can eliminate y from (7a),

$$y = 4\frac{dx}{dt} + 16x, \tag{8}$$

and substitute into (7b) to obtain

$$4\frac{d^2x}{dt^2} + 16\frac{dx}{dt} = 4x - 16\frac{dx}{dt} - 64x. \tag{9}$$

Collecting terms and dividing by 4 yields

$$\frac{d^2x}{dt^2} + 8\frac{dx}{dt} + 15x = 0. \tag{10}$$

This second-order differential equation with constant coefficients can be solved by letting $x = e^{rt}$, so that the characteristic equation is $r^2 + 8r + 15 = (r+5)(r+3) = 0$. The general solution is

$$x(t) = c_1e^{-3t} + c_2e^{-5t}. \tag{11}$$

To find $y(t)$, we substitute this into (8) and obtain, after some algebra,

$$y(t) = 4c_1e^{-3t} - 4c_2e^{-5t}. \tag{12}$$

We will use matrix methods because they work better for more complicated problems.

SECOND-ORDER EQUATIONS HAVE EQUIVALENT SYSTEMS. One reason that numerical software is written for first-order systems is that all differential equations can be rewritten as equivalent first-order systems. Consider the following second-order differential equation:

$$\frac{d^2x}{dt^2} + 11\frac{dx}{dt} + 43x = 0. \tag{13}$$

There are many ways to write it as a system. Usually we keep x as one variable and introduce y as the derivative $y = \frac{dx}{dt}$. Then the system using the differential equation (13) is

$$\frac{dx}{dt} = y, \tag{14a}$$

$$\frac{dy}{dt} = \frac{d^2x}{dt^2} = -43x - 11\frac{dx}{dt} = -43x - 11y. \tag{14b}$$

4.2 Introduction to Linear Systems of Differential Equations

Physical problems are sometimes formulated in terms of linear systems of differential equations. In the introduction to this chapter, we have shown that one specific mixture problem with two containers satisfies the following system of linear differential equations:

$$\frac{dx}{dt} = -4x + \frac{1}{4}y, \tag{1a}$$

$$\frac{dy}{dt} = 4x - 4y. \tag{1b}$$

4.2.1 Solving Linear Systems Using Eigenvalues and Eigenvectors of the Matrix

LINEAR SYSTEMS OF DIFFERENTIAL EQUATIONS WITH VECTOR AND MATRIX NOTATION. We consider linear systems of two homogeneous differential equations with constant coefficients,

$$\frac{dx}{dt} = ax + by, \tag{2a}$$

$$\frac{dy}{dt} = cx + dy. \tag{2b}$$

Sometimes coupled systems can be shown to be equivalent to a second-order differential equation by elimination and solved in this way. Here we wish to instead use methods to solve the differential equations based on vectors and matrices. Here we just introduce these ideas using some simple examples. We introduce the (boldface) column vector $\mathbf{x}$:

$$\mathbf{x} = \begin{bmatrix} x \\ y \end{bmatrix}, \tag{3}$$

and the 2×2 matrix of constants $\mathbf{A}$:

$$\mathbf{A} = \begin{bmatrix} a & b \\ c & d \end{bmatrix}. \tag{4}$$

We note that

$$\frac{d\mathbf{x}}{dt} = \begin{bmatrix} \frac{dx}{dt} \\ \frac{dy}{dt} \end{bmatrix} \tag{5}$$

represents the left-hand side of (2). Matrix multiplication of a 2×2 matrix and a 2×1 column vector yields a 2×1 column vector defined by

$$\begin{bmatrix} a & b \\ c & d \end{bmatrix} \begin{bmatrix} x \\ y \end{bmatrix} = \begin{bmatrix} ax + by \\ cx + dy \end{bmatrix}. \tag{6}$$

Using this notation, the linear system (2a)–(2b) becomes

$$\begin{bmatrix} \frac{dx}{dt} \\ \frac{dy}{dt} \end{bmatrix} = \frac{d\mathbf{x}}{dt} = \begin{bmatrix} a & b \\ c & d \end{bmatrix} \mathbf{x} = \mathbf{A}\mathbf{x}. \tag{7}$$

We will solve the system mostly in the form (2a)–(2b) without using matrix multiplication. Matrix multiplication is used extensively in more advanced treatments involving more equations.

SOLUTION OF LINEAR SYSTEMS USING EIGENVALUES AND EIGENVECTORS. We explain how to solve linear systems of differential equations,

$$\frac{dx}{dt} = ax + by, \tag{8a}$$

$$\frac{dy}{dt} = cx + dy, \tag{8b}$$

or equivalently

$$\frac{d\mathbf{x}}{dt} = \begin{bmatrix} a & b \\ c & d \end{bmatrix} \mathbf{x} = \mathbf{A}\mathbf{x}, \tag{9}$$

using eigenvalues and eigenvectors. We know that elementary exponentials e^{rt} are solutions of constant coefficient homogeneous differential equations, where we determine r by substitution into the differential equation. Equation (8a)–(8b) or (9) is a linear system with two unknowns, x and y. We assume there are elementary solutions in which both x and y are proportional to the same exponential:

$$\mathbf{x}(t) = \begin{bmatrix} x(t) \\ y(t) \end{bmatrix} = e^{\lambda t} \begin{bmatrix} u \\ v \end{bmatrix} = e^{\lambda t}\mathbf{v}, \tag{10}$$

calling the unknown λ instead of r. Here $\mathbf{v} = [\begin{smallmatrix} u \\ v \end{smallmatrix}]$ is a vector of constants, which we must also determine.

Substituting (10) into the differential equation (9), we obtain

$$\begin{bmatrix} \frac{dx}{dt} \\ \frac{dy}{dt} \end{bmatrix} = \lambda e^{\lambda t} \begin{bmatrix} u \\ v \end{bmatrix} = e^{\lambda t} \begin{bmatrix} a & b \\ c & d \end{bmatrix} \begin{bmatrix} u \\ v \end{bmatrix}, \tag{11}$$

or, equivalently,

$$\frac{dx}{dt} = \lambda e^{\lambda t} u = e^{\lambda t}(au + bv) \tag{12a}$$

and

$$\frac{dy}{dt} = \lambda e^{\lambda t} v = e^{\lambda t}(cu + dv), \tag{13}$$

or, after combining these two equations,

$$\frac{d\mathbf{x}}{dt} = \lambda e^{\lambda t}\mathbf{v} = \mathbf{A}e^{\lambda t}\mathbf{v}. \tag{14}$$

We first cancel the common factor $e^{\lambda t}$ and get that the vector $\mathbf{v} = [\begin{smallmatrix} u \\ v \end{smallmatrix}]$ satisfies the linear homogeneous system of algebraic equations,

$$au + bv = \lambda u, \tag{15a}$$

$$cu + dv = \lambda v, \tag{15b}$$

or

$$\mathbf{A}\mathbf{v} = \lambda \mathbf{v}. \tag{16}$$

It is usual to rewrite the system

$$(a - \lambda)u + bv = 0, \tag{17a}$$

$$cu + (d - \lambda)v = 0, \tag{17b}$$

or

$$(\mathbf{A} - \lambda \mathbf{I})\mathbf{v} = 0, \tag{18}$$

where $\mathbf{I} = \left[\begin{smallmatrix} 1 & 0 \\ 0 & 1 \end{smallmatrix}\right]$ is the identity matrix. For 2×2 systems this becomes

$$\begin{bmatrix} a-\lambda & b \\ c & d-\lambda \end{bmatrix} \begin{bmatrix} u \\ v \end{bmatrix} = 0. \tag{19}$$

We note that for the linear system of algebraic equations, $u = 0$, $v = 0$ solves the system (19) for all λ. We call $(0, 0)$ the trivial solution. This corresponds to what we call the trivial solution $x(t) = 0$, $y(t) = 0$ of the linear system of homogeneous differential equations. We wish to determine nontrivial solutions. We need some knowledge of elementary linear algebra, explained in the next paragraph. We will show that for nontrivial solutions, the determinant must be zero:

$$\det \begin{bmatrix} a-\lambda & b \\ c & d-\lambda \end{bmatrix} = \det(\mathbf{A} - \lambda \mathbf{I}) = (a-\lambda)(d-\lambda) - bc = 0. \tag{20}$$

The values of λ which satisfy (20) are called **eigenvalues** of the matrix **A**. After determining the eigenvalues, the nontrivial $\mathbf{v} = \left[\begin{smallmatrix} u \\ v \end{smallmatrix}\right]$ must be determined from (17a)–(17b) and is called an **eigenvector** corresponding to that specific eigenvalue. We will do some examples shortly to illustrate these ideas.

ELEMENTARY LINEAR ALGEBRA. We review some elementary algebra. The equations we have are

$$(a - \lambda)u + bv = 0, \tag{21a}$$

$$cu + (d - \lambda)v = 0. \tag{21b}$$

We are given a, b, c, d and are trying to solve for λ and u and v. If λ were known, then (21a)–(21b) would be two linear equations, which define two straight lines going through the origin $u = 0$, $v = 0$. Since the lines both go through the origin, there are two possibilities. If the two equations (21a)–(21b) represent two different lines (if the equations are not constant multiples of each other), then there is only one intersection, $u = 0$, $v = 0$. If the two equations (21a)–(21b) represent the same straight line, then the two equations are constant multiples of each other, and either straight line solution may be used as their intersection. Either there is a unique solution $u = 0$, $v = 0$ or there are an infinite number of solutions.

However, we do not know λ yet, so we need to derive a way to find the values of λ such that there are an infinite number of nonzero solutions. We proceed as follows. v can be eliminated from (21a), (21b) by multiplying (21a) by $d - \lambda$, multiplying (21b) by b, and subtracting. In this case, we get

$$u[(a - \lambda)(d - \lambda) - bc] = 0.$$

If the bracketed term is nonzero, then u must equal zero, which results in $v = 0$, and we get only the trivial solution of the linear system of differential equations. Thus in order for there to be nonzero solutions of the system of linear differential equations,

$$(a - \lambda)(d - \lambda) - bc = 0. \tag{22}$$

This condition determines λ for which there are nonzero solutions of the differential equation. Recall that the determinant of a 2×2 matrix is given by

$$\det \begin{bmatrix} q & r \\ s & t \end{bmatrix} = qt - rs.$$

We rewrite the condition (22) in terms of the determinant equaling zero:

$$\det \begin{bmatrix} a-\lambda & b \\ c & d-\lambda \end{bmatrix} = 0. \tag{23}$$

We do not want only $u=0$, $v=0$. Thus, in order for there to be an infinite number of solutions, the determinant must equal zero, and we use that condition (22) or (23) to determine the values of λ: called the **eigenvalues** of the matrix. In this case, the two equations (21a)–(21b) represent the same straight line, and either equation may be used to find u, v, which are called the **eigenvector** $[\begin{smallmatrix} u \\ v \end{smallmatrix}]$ corresponding to the eigenvalue λ.

DETERMINING THE EIGENVALUES. Since (18) is a linear homogeneous system, the eigenvalues can be determined from the determinant condition:

$$\det(\mathbf{A} - \lambda\mathbf{I}) = 0, \tag{24}$$

or equivalently

$$\det \begin{bmatrix} a-\lambda & b \\ c & d-\lambda \end{bmatrix} = (a-\lambda)(d-\lambda) - bc = 0. \tag{25}$$

Evaluating this determinant gives a quadratic equation for the two eigenvalues:

$$\lambda^2 - (a+d)\lambda + ad - bc = 0. \tag{26}$$

Often in examples the roots of the quadratic (eigenvalues) are simple to compute and use. In general we determine the eigenvalues using the quadratic formula:

$$\lambda = \frac{a+d \pm \sqrt{(a+d)^2 - 4(ad-bc)}}{2}. \tag{27}$$

4.2.2 Solving Linear Systems if the Eigenvalues are Real and Unequal

Example 4.2.1 *Mixture Example*

We consider the system of two first-order differential equations from Section 4.2 corresponding to two interconnected fluid tanks:

$$\frac{dx}{dt} = -4x + \frac{1}{4}y, \tag{28a}$$

$$\frac{dy}{dt} = 4x - 4y, \tag{28b}$$

or

$$\begin{bmatrix} \frac{dx}{dt} \\ \frac{dy}{dt} \end{bmatrix} = \frac{d\mathbf{x}}{dt} = \begin{bmatrix} -4 & \frac{1}{4} \\ 4 & -4 \end{bmatrix} \mathbf{x} = \mathbf{A}\mathbf{x}. \tag{29}$$

• SOLUTION. For systems, we substitute:

$$\begin{bmatrix} x(t) \\ y(t) \end{bmatrix} = e^{\lambda t} \begin{bmatrix} u \\ v \end{bmatrix}, \tag{30}$$

and obtain (17a)–(17b) which is

$$\begin{bmatrix} -4-\lambda & \frac{1}{4} \\ 4 & -4-\lambda \end{bmatrix} \begin{bmatrix} u \\ v \end{bmatrix} = 0. \tag{31}$$

FIND EIGENVALUES. The eigenvalues satisfy the determinant condition

$$\begin{aligned} \det(\mathbf{A} - \lambda \mathbf{I}) &= (-4-\lambda)(-4-\lambda) - 4\frac{1}{4} \\ &= \lambda^2 + 8\lambda + 15 = (\lambda+3)(\lambda+5) = 0. \end{aligned} \tag{32}$$

This is the same as the characteristic equation obtained by elimination in the introduction. We see there are two eigenvalues $\lambda = -3, -5$. There is a different eigenvector corresponding to each eigenvalue, which we now determine.

FIND EIGENVECTORS CORRESPONDING TO $\lambda = -3$. To obtain the eigenvector corresponding to $\lambda = -3$, we substitute $\lambda = -3$ into the homogeneous system (31) and obtain

$$-u + \frac{1}{4}v = 0, \tag{33a}$$

$$4u - v = 0. \tag{33b}$$

As explained earlier, since there is a nonzero solution, these two equations must give the same result. Here the second equation is a constant multiple -4 of the first equation. (If the two equations give different results, then you've made an algebraic error or your value of λ is not really an eigenvalue.) Since the equations are equivalent, we only use one of them, it doesn't matter which:

$$-u + \frac{1}{4}v = 0. \tag{34}$$

There is a constant multiple in the general solution, so that we can take any solution of (34) other than $u = v = 0$. For simplicity, we take $u = 1$, so that $v = 4$, and

$$\text{an eigenvector corresponding to } \lambda = -3 \text{ is } \begin{bmatrix} 1 \\ 4 \end{bmatrix}. \tag{35}$$

This means that the system of differential equations has the following elementary solution corresponding to the eigenvalue $\lambda = -3$:

$$\mathbf{x}(t) = \begin{bmatrix} x \\ y \end{bmatrix} = \begin{bmatrix} 1 \\ 4 \end{bmatrix} e^{-3t}. \tag{36}$$

FIND EIGENVECTORS CORRESPONDING TO $\lambda = -5$. We follow the same procedure, so that from (31) when $\lambda = -5$ the eigenvector satisfies

$$u + \frac{1}{4}v = 0, \tag{37a}$$

$$4u + v = 0. \tag{37b}$$

These two equations give the same result, as they must, and we use either one. We choose

$$u+\frac{1}{4}v=0. \tag{38}$$

We may take $u=1$ and $v=-4$, so that

$$\text{an eigenvector corresponding to } \lambda=-5 \text{ is } \begin{bmatrix} 1 \\ -4 \end{bmatrix}. \tag{39}$$

This means that the system of differential equations has the following elementary solution corresponding to the eigenvalue $\lambda=-5$:

$$\mathbf{x}(t)=\begin{bmatrix} x \\ y \end{bmatrix}=\begin{bmatrix} 1 \\ -4 \end{bmatrix} e^{-5t}. \tag{40}$$

FIND GENERAL SOLUTION OF LINEAR SYSTEM OF DIFFERENTIAL EQUATIONS. As in our earlier study of linear differential equations, it can be shown that the general solution of a linear system of differential equations is a linear combination of the two elementary solutions, here corresponding to each eigenvalue:

$$\mathbf{x}(t)=\begin{bmatrix} x \\ y \end{bmatrix}=c_1\begin{bmatrix} 1 \\ 4 \end{bmatrix} e^{-3t}+c_2\begin{bmatrix} 1 \\ -4 \end{bmatrix} e^{-5t}. \tag{41}$$

This gives $x(t)=c_1e^{-3t}+c_2e^{-5t}$ and $y(t)=4c_1e^{-3t}-4c_2e^{-5t}$, which is the same as (11)–(12) which was obtained in the introduction of this chapter by elimination. ♦

Example 4.2.2 *Second-Order Equation as a System*

Find the general solution of

$$\frac{d^2x}{dt^2}+7\frac{dx}{dt}+12x=0. \tag{42}$$

• SOLUTION. We can solve this differential equation without converting it to a system. If we substitute $x=e^{rt}$, we obtain the characteristic equation $r^2+7r+12=0$, which is easily factored as $(r+3)(r+4)=0$. The roots of the characteristic equation are $r=-4$ and $r=-3$, so that the general solution is a linear combination of these two simple exponential solutions:

$$x(t)=c_1e^{-4t}+c_2e^{-3t}. \tag{43}$$

We wish to obtain the same answer using eigenvalues and eigenvectors for linear systems. First we must convert the second-order differential equation to a system. If we let $y=\frac{dx}{dt}$, then we obtain an equivalent system of two first-order differential equations:

$$\frac{dx}{dt}=y, \tag{44a}$$

$$\frac{dy}{dt}=-12x-7y. \tag{44b}$$

or

$$\frac{d\mathbf{x}}{dt}=\begin{bmatrix} 0 & 1 \\ -12 & -7 \end{bmatrix}\mathbf{x}=\mathbf{A}\mathbf{x}. \tag{45}$$

For systems we substitute

$$\begin{bmatrix} x(t) \\ y(t) \end{bmatrix} = e^{\lambda t} \begin{bmatrix} u \\ v \end{bmatrix}, \tag{46}$$

and obtain

$$\begin{bmatrix} -\lambda & 1 \\ -12 & -7-\lambda \end{bmatrix} \begin{bmatrix} u \\ v \end{bmatrix} = 0. \tag{47}$$

♦

FIND EIGENVALUES. The eigenvalues satisfy the determinant condition

$$\det(\mathbf{A} - \lambda \mathbf{I}) = \lambda^2 + 7\lambda + 12 = (\lambda + 4)(\lambda + 3) = 0. \tag{48}$$

This is the same as the characteristic equation, and we see there are two eigenvalues $\lambda = -4, -3$. There is a different eigenvector corresponding to each eigenvalue.

FIND EIGENVECTORS CORRESPONDING TO $\lambda = -4$. To obtain the eigenvector corresponding to $\lambda = -4$, we use the homogeneous system (47) and obtain

$$4u + v = 0, \tag{49a}$$

$$-12u - 3v = 0. \tag{49b}$$

These two equations must give the same result (one equation is a constant multiple of the other), so that we only use one of them (if the two equations give different results then you've made an algebraic error):

$$4u + v = 0. \tag{50}$$

There is a constant multiple in the general solution, so that we can take any solution of (50). For simplicity, we take $u = 1$, which gives $v = -4$, so that the eigenvector corresponding to $\lambda = -4$ is $\left[\begin{smallmatrix} 1 \\ -4 \end{smallmatrix}\right]$. This means that the system of differential equations has the following elementary solution corresponding to the eigenvalue $\lambda = -4$:

$$\mathbf{x}(t) = \begin{bmatrix} x \\ y \end{bmatrix} = \begin{bmatrix} 1 \\ -4 \end{bmatrix} e^{-4t}. \tag{51}$$

FIND EIGENVECTORS CORRESPONDING TO $\lambda = -3$. We follow the same procedure, so that from (47), when $\lambda = -3$ the eigenvector satisfies

$$3u + v = 0, \tag{52a}$$

$$-12u - 4v = 0. \tag{52b}$$

These two equations give the same result, as they must, and we use either one:

$$3u + v = 0. \tag{53}$$

We may take $u = 1$, which gives $v = -3$, so that the eigenvector corresponding to $\lambda = -3$ is $\left[\begin{smallmatrix} 1 \\ -3 \end{smallmatrix}\right]$. This means that the system of differential equations has the following elementary solution corresponding to the eigenvalue $\lambda = -3$:

$$\mathbf{x}(t) = \begin{bmatrix} x \\ y \end{bmatrix} = \begin{bmatrix} 1 \\ -3 \end{bmatrix} e^{-3t}. \tag{54}$$

FIND GENERAL SOLUTION. The general solution will be a linear combination of the two elementary solutions corresponding to each eigenvalue:

$$\mathbf{x}(t)=\begin{bmatrix} x \\ y \end{bmatrix}=c_1\begin{bmatrix} 1 \\ -4 \end{bmatrix}e^{-4t}+c_2\begin{bmatrix} 1 \\ -3 \end{bmatrix}e^{-3t}. \tag{55}$$

This gives $x(t)=c_1e^{-4t}+c_2e^{-3t}$, which is equivalent to (43). We also obtain $y(t)=-4c_1e^{-4t}-3c_2e^{-3t}$ which in this case is the same as $y=\frac{dx}{dt}$. ♦

GENERAL SOLUTION OF LINEAR SYSTEM IF TWO EIGENVALUES ARE REAL AND UNEQUAL: REAL DISTINCT EIGENVALUES. Suppose we have a 2×2 matrix with real distinct eigenvalues, which we call λ_1 and λ_2. Suppose the eigenvector $\mathbf{v}_1$ corresponds to the eigenvalue λ_1. That means that $e^{\lambda_1 t}\mathbf{v}_1$ satisfies the linear system. Similarly, the eigenvector $\mathbf{v}_2$ corresponding to the eigenvalue λ_2 gives the solution $e^{\lambda_2 t}\mathbf{v}_2$. Since the system of differential equations is linear, we can obtain the general solution using a linear combination of these two, or

$$\mathbf{x}(t)=\begin{bmatrix} x \\ y \end{bmatrix}=c_1e^{\lambda_1 t}\mathbf{v}_1+c_2e^{\lambda_2 t}\mathbf{v}_2. \tag{56}$$

4.2.3 Finding General Solutions of Linear Systems in the Case of Complex Eigenvalues

Consider the linear system

$$\frac{d\mathbf{x}}{dt}=\mathbf{A}\mathbf{x}.$$

The eigenvalues of $\mathbf{A}$ arise by substituting $\mathbf{x}=e^{\lambda t}\mathbf{v}$. If the eigenvalues $\lambda=\alpha\pm i\beta$ of a matrix $\mathbf{A}$ are complex, then $e^{\lambda t}=e^{(\alpha\pm i\beta)t}$. Due to Euler's formula, solutions involve $e^{\alpha t}\cos\beta t$ and $e^{\alpha t}\sin\beta t$, as in the general solution of linear constant coefficient differential equations with complex roots of the characteristic equation. The key fact we need is the following:

If $\mathbf{A}$ is a real matrix with a complex eigenvalue λ and eigenvector $\mathbf{v}$, then

$$\lambda=\alpha+i\beta,\ \mathbf{v}=\mathbf{a}+i\mathbf{b},$$

then $\bar{\lambda}$ is another eigenvalue (recall that $\bar{\lambda}$ is called the **complex conjugate** of λ) and $\bar{\mathbf{v}}$ its associated eigenvector, where

$$\bar{\lambda}=\alpha-i\beta,\ \bar{\mathbf{v}}=\mathbf{a}-i\mathbf{b}.$$

That $\bar{\lambda}$ is also an eigenvalue follows from the fact that $0=\det(\mathbf{A}-\lambda\mathbf{I})$ is a quadratic polynomial with real coefficients if $\mathbf{A}$ is real, and thus the complex roots of the characteristic polynomial will occur in conjugate pairs. Since $\mathbf{A}\lambda=\lambda\mathbf{v}$, then taking conjugates of both sides gives

$$\overline{\mathbf{A}\mathbf{v}}=\overline{\lambda\mathbf{v}}$$

or

$$\mathbf{A}\overline{\mathbf{v}} = \overline{\lambda}\overline{\mathbf{v}},$$

since $\mathbf{A}$ is real, so that $\overline{\mathbf{v}}$ is an eigenvector for the eigenvalue $\overline{\lambda}$.

This fact is very helpful since it means we actually have to compute only the eigenvectors for *one* of the eigenvalues in a conjugate pair.

Example 4.2.3 *Complex Eigenvalues and Eigenvectors*

Find the eigenvalues and eigenvectors for

$$\mathbf{A} = \begin{bmatrix} 1 & -1 \\ 1 & 1 \end{bmatrix}. \tag{57}$$

• SOLUTION. The eigenvalues and eigenvectors satisfy

$$\begin{bmatrix} 1-\lambda & -1 \\ 1 & 1-\lambda \end{bmatrix} \begin{bmatrix} u \\ v \end{bmatrix} = 0. \tag{58}$$

The characteristic equation is

$$\det(\mathbf{A} - \lambda\mathbf{I}) = \det \begin{bmatrix} 1-\lambda & -1 \\ 1 & 1-\lambda \end{bmatrix} = \lambda^2 - 2\lambda + 2 = 0. \tag{59}$$

The roots of the characteristic polynomial are the eigenvalues, and they can be found using the quadratic formula:

$$\lambda = \frac{2 \pm \sqrt{4-8)}}{2} = 1 \pm i. \tag{60}$$

The eigenvalues are $\lambda_1 = 1+i, \lambda_2 = 1-i$. First, we find the eigenvector for λ_1. Solving (58) gives

$$\begin{bmatrix} -i & -1 \\ 1 & -i \end{bmatrix} \begin{bmatrix} u \\ v \end{bmatrix} = 0. \tag{61}$$

The two linear equations must be equivalent (since that is the property of eigenvalues). For the complex case, it may not be very obvious that the two linear homogeneous equations are equivalent. Here we see that the second row is i times the first row. Thus, we only use the first equation (row). Thus, $-iu - v = 0$ or $v = -iu$. There are several ways to proceed. We note that u is arbitrary, and

$$\mathbf{v} = \begin{bmatrix} u \\ -iu \end{bmatrix} = u \begin{bmatrix} 1 \\ -i \end{bmatrix}. \tag{62}$$

We may take $u = 1$, in which case the eigenvector corresponding to the eigenvalue $\lambda_1 = 1+i$ is

$$\mathbf{v}_1 = \begin{bmatrix} 1 \\ -i \end{bmatrix}. \tag{63}$$

Then, the complex conjugate will be the eigenvector associated with $\lambda_2 = 1-i$:

$$\mathbf{v}_2 = \overline{\mathbf{v}}_1 = \begin{bmatrix} 1 \\ i \end{bmatrix}. \tag{64}$$

♦

GENERAL SOLUTION (COMPLEX EIGENVALUES): USING EIGENVALUES AND EIGENVECTORS. It is helpful to find real solutions. Since the eigenvalues are complex conjugates of each other, the eigenvectors are also complex conjugates of each other. Because of this, it can be shown that the real and imaginary parts of one complex solution form an independent set. We assume a complex eigenvalue is $\lambda = \alpha + i\beta$ and the corresponding complex eigenvector $\mathbf{v} = \mathbf{a} + i\mathbf{b}$. Using Euler's formula, a complex solution would satisfy

$$e^{\lambda t}\mathbf{v} = e^{(\alpha+i\beta)t}(\mathbf{a} + i\mathbf{b}) = e^{\alpha t}(\cos\beta t + i\sin\beta t)(\mathbf{a} + i\mathbf{b}). \tag{65}$$

Multiplying out and collecting real and imaginary parts, yields

$$e^{(\alpha+i\beta)t}(\mathbf{a} + i\mathbf{b}) = e^{\alpha t}(\mathbf{a}\cos\beta t - \mathbf{b}\sin\beta t) + ie^{\alpha t}(\mathbf{a}\sin\beta t + \mathbf{b}\cos\beta t). \tag{66}$$

The real part

$$e^{\alpha t}(\mathbf{a}\cos\beta t - \mathbf{b}\sin\beta t) \tag{67}$$

and the imaginary part

$$e^{\alpha t}(\mathbf{a}\sin\beta t + \mathbf{b}\cos\beta t) \tag{68}$$

are independent solutions of $\frac{d\mathbf{x}}{dt} = \mathbf{A}\mathbf{x}$. In summary, we have

THEOREM 4.2.1 *General Solution of the Differential Equation for Complex Eigenvalues*

If $\mathbf{A}$ is a real matrix with complex eigenvalue $\lambda = \alpha + i\beta$ and associated eigenvector $\mathbf{v} = \mathbf{a} + i\mathbf{b}$, then the general solution can be written as a linear combination of these two:

$$\mathbf{x}(t) = c_1 e^{\alpha t}(\mathbf{a}\sin\beta t + \mathbf{b}\cos\beta t) + c_2 e^{\alpha t}(\mathbf{a}\cos\beta t - \mathbf{b}\sin\beta t). \tag{69}$$

Example 4.2.4 *Solution of $\frac{dx}{dt} = Ax$ with Complex Eigenvalues and Eigenvectors*

Find the general solution of

$$\frac{dx}{dt} = x - y, \tag{70a}$$

$$\frac{dy}{dt} = x + y. \tag{70b}$$

• SOLUTION. Let

$$\mathbf{x} = \begin{bmatrix} x \\ y \end{bmatrix}, \quad \mathbf{A} = \begin{bmatrix} 1 & -1 \\ 1 & 1 \end{bmatrix}, \tag{71}$$

so that (70a)–(70b) is $\frac{d\mathbf{x}}{dt} = \mathbf{A}\mathbf{x}$. This $\mathbf{A}$ is the same as in the previous example (4.2.3), so that we have $\lambda = \alpha + i\beta = 1 + i$ is an eigenvalue and the associated eigenvector can be taken to be

$$\mathbf{v} = \mathbf{a} + i\mathbf{b} = \begin{bmatrix} 1 \\ -i \end{bmatrix} = \begin{bmatrix} 1 \\ 0 \end{bmatrix} + i \begin{bmatrix} 0 \\ -1 \end{bmatrix}.$$

Thus,

$$\alpha = 1, \beta = 1, \mathbf{a} = \begin{bmatrix} 1 \\ 0 \end{bmatrix}, \mathbf{b} = \begin{bmatrix} 0 \\ -1 \end{bmatrix}.$$

The formulas (67) and (68) become

$$\mathbf{x}_1(t) = e^t \left[\cos t \begin{bmatrix} 1 \\ 0 \end{bmatrix} - \sin t \begin{bmatrix} 0 \\ -1 \end{bmatrix} \right] = \begin{bmatrix} e^t \cos t \\ e^t \sin t \end{bmatrix}, \tag{72a}$$

$$\mathbf{x}_2(t) = e^t \left[\cos t \begin{bmatrix} 0 \\ -1 \end{bmatrix} + \sin t \begin{bmatrix} 1 \\ 0 \end{bmatrix} \right] = \begin{bmatrix} e^t \sin t \\ -e^t \cos t \end{bmatrix}, \tag{72b}$$

and $\mathbf{x}_1(t), \mathbf{x}_2(t)$ is a fundamental set of solutions of (70a) and (70b). The general solution of (70a) and (70b) is $c_1\mathbf{x}_1 + c_2\mathbf{x}_2$ or

$$x(t) = c_1 e^t \cos t + c_2 e^t \sin t,$$
$$y(t) = c_1 e^t \sin t - c_2 e^t \cos t.$$

♦

4.2.4 Special Systems with Complex Eigenvalues (optional)

Problems with complex eigenvalues can be complicated. In the following specific problem, we show that solving the system simplifies in polar coordinates. Unfortunately this is not a general result. However, it can be shown that problems with complex eigenvalues always simplify in this way if one first introduces a very special change of coordinates. Here we just consider one example in which a change of coordinates turns out not to be necessary (for which polar coordinates simplifies the analysis) The problem is

$$\frac{dx}{dt} = 2x - 5y, \tag{73a}$$

$$\frac{dy}{dt} = 5x + 2y. \tag{73b}$$

The eigenvalues satisfy

$$\det \begin{bmatrix} 2-\lambda & -5 \\ 5 & 2-\lambda \end{bmatrix} = (2-\lambda)^2 + 25 = 0, \tag{74}$$

so that the eigenvalues are complex, $\lambda = 2 \pm 5i$. For this example (but not necessarily other examples) the differential equation simplifies introducing the complex representation $z(t) = x(t) + iy(t)$, since

$$\frac{dz}{dt} = \frac{dx}{dt} + i\frac{dy}{dt} = 2x - 5y + i(5x + 2y) = 2(x + iy) + 5i(x + iy) = (2 + 5i)z. \tag{75}$$

Using polar coordinates, $z(t) = x + iy = r\cos\theta + ir\sin\theta = r(t)e^{i\theta(t)}$, we obtain

$$\frac{dz}{dt} = \frac{dr}{dt}e^{i\theta} + ire^{i\theta}\frac{d\theta}{dt}. \tag{76}$$

Setting these two expression for $\frac{dz}{dt}$ equal yields

$$\frac{dr}{dt}e^{i\theta} + ire^{i\theta}\frac{d\theta}{dt} = (2+5i)re^{i\theta}, \tag{77}$$

and simplifying (by dividing by $e^{i\theta}$), we have

$$\frac{dr}{dt} + ir\frac{d\theta}{dt} = 2r + 5ir. \tag{78}$$

By equating the real and imaginary parts, we obtain a simple system of differential equations

$$\frac{dr}{dt} = 2r, \; \frac{d\theta}{dt} = 5, \tag{79}$$

which has the general solution

$$r(t) = r(0)e^{2t}, \theta(t) = 5t + \theta(0). \tag{80}$$

Returning to cartesian coordinates, we have

$$x(t) = r(0)e^{2t}\cos(5t+\theta(0)), \;\; y(t) = r(0)e^{2t}\sin(5t+\theta(0)). \tag{81}$$

The solution involves a linear combination of the correct functions $e^{2t}\cos 5t$ and $e^{2t}\sin 5t$, since $\lambda = 2 \pm 5i$.

An alternate way to use polar coordinates is to note that $r^2 = x^2 + y^2$ and $\tan\theta = \frac{y}{x}$. In the next section (introduction to the phase plane) we will see that polar coordinates are helpful in understanding the solution when the eigenvalues are complex. Thus,

$$r\frac{dr}{dt} = x\frac{dx}{dt} + y\frac{dy}{dt} = x(2x-5y) + y(5x+2y) = 2(x^2+y^2) = 2r^2, \tag{82}$$

using the differential equation (and dividing by 2). Thus,

$$\frac{dr}{dt} = 2r, \text{ so that, } r(t) = r(0)e^{2t}. \tag{83}$$

The equation for the polar angle is a little harder to obtain

$$\sec^2\theta\frac{d\theta}{dt} = \frac{x\frac{dy}{dt} - y\frac{dx}{dt}}{x^2} = \frac{x(5x+2y) - y(2x-5y)}{x^2} = \frac{5(x^2+y^2)}{x^2}. \tag{84}$$

Since $\sec^2\theta = \frac{1}{\cos^2\theta} = \frac{r^2}{x^2}$, we have

$$\frac{d\theta}{dt} = 5, \text{ so that } \theta(t) = 5t + \theta(0). \tag{85}$$

The solution to the system of differential equations is

$$x(t) = r(0)e^{2t}\cos(5t+\theta(0)), \;\; y(t) = r(0)e^{2t}\sin(5t+\theta(0)). \tag{86}$$

4.2.5 General Solution of a Linear System if the Two Real Eigenvalues are Equal (Repeated) Roots

Consider the linear system $\frac{d\mathbf{x}}{dt} = \mathbf{A}\mathbf{x}$. If the two real eigenvalues are repeated, then sometimes there are two independent eigenvectors, and the solution can be still represented by (56). This is very rare with 2×2 systems, but happens more often with larger systems.

Example 4.2.5 *Elementary Example with Repeated Real Eigenvalues with Two Independent Eigenvectors*

Consider

$$\frac{dx}{dt} = 6x, \tag{87a}$$

$$\frac{dy}{dt} = 6y, \tag{87b}$$

or

$$\frac{d\mathbf{x}}{dt} = \begin{bmatrix} 6 & 0 \\ 0 & 6 \end{bmatrix} \mathbf{x} = \mathbf{A}\mathbf{x}. \tag{88}$$

- SOLUTION. For systems, we substitute:

$$\begin{bmatrix} x(t) \\ y(t) \end{bmatrix} = e^{\lambda t} \begin{bmatrix} u \\ v \end{bmatrix} \tag{89}$$

and obtain

$$\begin{bmatrix} 6-\lambda & 0 \\ 0 & 6-\lambda \end{bmatrix} \begin{bmatrix} u \\ v \end{bmatrix} = 0. \tag{90}$$

The eigenvalues satisfy the determinant condition, $(6-\lambda)^2 = 0$, so that the eigenvalues are repeated reals $\lambda = 6, 6$. The eigenvectors corresponding to the eigenvalue $\lambda = 6$ satisfies (90)

$$\begin{bmatrix} 0 & 0 \\ 0 & 0 \end{bmatrix} \begin{bmatrix} u \\ v \end{bmatrix} = 0. \tag{91}$$

Thus, $0u + 0v = 0$ and both u and v are arbitrary (constants), and the solution of the system of differential equation is

$$x(t) = ue^{6t},\ y(t) = ve^{6t}, \tag{92}$$

where u and v are arbitrary constants. This is the same result as obtained by directly solving the original uncoupled system (87a)–(87b). In the language of eigenvectors, the eigenvector corresponding to $\lambda = 6$ satisfies

$$\mathbf{v} = \begin{bmatrix} u \\ v \end{bmatrix} = u \begin{bmatrix} 1 \\ 0 \end{bmatrix} + v \begin{bmatrix} 0 \\ 1 \end{bmatrix},$$

where u and v are arbitrary constants. Here we say there are two independent eigenvectors corresponding to the eigenvalues 6, and for convenience we call the eigenvectors $\left[\begin{smallmatrix} 1 \\ 0 \end{smallmatrix}\right]$ and $\left[\begin{smallmatrix} 0 \\ 1 \end{smallmatrix}\right]$.

If the two real eigenvalues are equal, but as is usual there is only one independent eigenvector $\mathbf{v}$, then only one independent solution to the differential equation can be represented in the simple way $e^{\lambda t}\mathbf{v}$. The second solution is more complicated, but as in second-order equations, involves $te^{\lambda t}$ in some way.

Example 4.2.6 *Elementary Example with Repeated Real Eigenvalues with One Independent Eigenvector*

Consider the system of differential equations

$$\frac{dx}{dt} = 7x + y, \tag{93a}$$

$$\frac{dy}{dt} = 7y, \tag{93b}$$

or

$$\frac{d\mathbf{x}}{dt} = \begin{bmatrix} 7 & 1 \\ 0 & 7 \end{bmatrix} \mathbf{x} = \mathbf{A}\mathbf{x}. \tag{94}$$

• SOLUTION. For systems, we substitute

$$\begin{bmatrix} x(t) \\ y(t) \end{bmatrix} = e^{\lambda t} \begin{bmatrix} u \\ v \end{bmatrix} \tag{95}$$

and obtain

$$\begin{bmatrix} 7-\lambda & 1 \\ 0 & 7-\lambda \end{bmatrix} \begin{bmatrix} u \\ v \end{bmatrix} = 0. \tag{96}$$

The eigenvalues satisfy the determinant condition $(7-\lambda)^2 = 0$, so that the eigenvalues are repeated reals $\lambda = 7, 7$. The eigenvectors corresponding to the eigenvalue $\lambda = 7$ satisfies (96)

$$\begin{bmatrix} 0 & 1 \\ 0 & 0 \end{bmatrix} \begin{bmatrix} u \\ v \end{bmatrix} = 0. \tag{97}$$

Thus, $v = 0$ and $0u + 0v = 0$, so that only u is arbitrary (constant). The special solution of the system of differential equation corresponding to the eigenvalue $\lambda = 7$ is

$$x(t) = ue^{7t}, \; y(t) = 0, \tag{98}$$

where u is an arbitrary constant. In the language of eigenvectors, the eigenvector corresponding to $\lambda = 7$ satisfies $\mathbf{v} = \left[\begin{smallmatrix} u \\ v \end{smallmatrix}\right] = u\left[\begin{smallmatrix} 1 \\ 0 \end{smallmatrix}\right]$, where u is an arbitrary constant. Here we say there is only one independent eigenvector corresponding to the repeated eigenvalue 7, and for convenience we call the eigenvector $\left[\begin{smallmatrix} 1 \\ 0 \end{smallmatrix}\right]$. For this example, it is relatively easy to obtain all solutions of the linear system of differential equations (and not just the above solutions corresponding to the eigenvector) because $y(t)$ itself just solves an elementary first-order homogeneous differential equation with constant coefficients. Thus, we first find easily the general solution for $y(t)$, namely, $y(t) = y(0)e^{7t}$, and then $x(t)$ solves a (nonhomogeneous) linear first-order differential equation

$$\frac{dx}{dt} - 7x = y = y(0)e^{7t}. \tag{99}$$

Linear first-order equations can always be solved by an integrating factor $e^{\int p(t)dt}$. Here $p(t) = -7$ and the integrating factor is e^{-7t} (see Section 1.6.3). Multiplying both sides of (99) by e^{-7t} yields

$$e^{-7t}\left(\frac{dx}{dt} - 7x\right) = \frac{d}{dt}(e^{-7t}x(t)) = y(0). \tag{100}$$

Definite integrating from $t = 0$ to $t = t$ yields

$$e^{-7t}x(t) - x(0) = y(0)t. \tag{101}$$

Solving for $x(t)$ gives $x(t) = y(0)te^{7t} + x(0)e^{7t}$. Thus, the solution of the initial value problem for the system (93) is

$$x(t) = y(0)te^{7t} + x(0)e^{7t}, \quad y(t) = y(0)e^{7t}, \tag{102}$$

which involves e^{7t} and te^{7t}. This is the same idea that occurs for constant coefficient homogeneous differential equations with real repeated roots. ♦

4.2.6 Eigenvalues and Trace and Determinant (optional)

The eigenvalues satisfy the quadratic equation (26), $\lambda^2 - (a+d)\lambda + ad - bc = 0$, obtained from the determinant condition. It is sometimes helpful to organize the information concerning the eigenvalues in terms of the trace (tr) and determinant (det) of the matrix $\mathbf{A} = \left[\begin{smallmatrix} a & b \\ c & d \end{smallmatrix}\right]$:

$$\operatorname{tr}\mathbf{A} = a + d, \tag{103}$$

$$\det\mathbf{A} = ad - bc. \tag{104}$$

We first reexpress the quadratic in terms of the trace and determinant:

$$\lambda^2 - (\operatorname{tr}\mathbf{A})\lambda + \det\mathbf{A} = 0. \tag{105}$$

Using the quadratic formula gives

$$\lambda = \frac{\operatorname{tr}\mathbf{A} \pm \sqrt{(\operatorname{tr}\mathbf{A})^2 - 4\det\mathbf{A}}}{2}. \tag{106}$$

If $4\det\mathbf{A} \leq (\operatorname{tr}\mathbf{A})^2$, then the eigenvalues are real, and if $4\det\mathbf{A} > (\operatorname{tr}\mathbf{A})^2$, then the eigenvalues are complex.

MORE ON EIGENVALUES AND TRACE AND DETERMINANT. Usually, we can obtain the eigenvalues explicitly. However, the following result is sometimes useful (as a check, perhaps). The quadratic will factor with its two roots λ_1 and λ_2, so that in general

$$(\lambda-\lambda_1)(\lambda-\lambda_2)=\lambda^2-(\lambda_1+\lambda_2)\lambda+\lambda_1\lambda_2=0. \tag{107}$$

By comparing (105) and (107), we see that the sum of the eigenvalues equals the trace of the matrix and the product of the eigenvalues equals the determinant:

$$\lambda_1+\lambda_2=\operatorname{tr}\mathbf{A}, \tag{108}$$

$$\lambda_1\lambda_2=\det\mathbf{A}. \tag{109}$$

Example 4.2.7

Consider the matrix $\begin{bmatrix} -4 & \frac{1}{4} \\ 4 & -4 \end{bmatrix}$. (a) Find the eigenvalues; (b) show that they are consistent with the trace and determinant conditions.

• SOLUTION. (a) This matrix corresponds to our mixing problem, where we showed (by factoring the quadratic equation) that the eigenvalues were $\lambda=-3$ and $\lambda=-5$.

(b) The trace of the matrix is -8 should equal the sum of the eigenvalues $-3-5=-8$. The determinant of the matrix $16-1=15$ should equal the product of the eigenvalues $(-3)(-5)=15$. ♦

TRACE AND DETERMINANT FOR $n\times n$ MATRICES. The result (108)–(109) also can be shown to be valid for $n\times n$ matrices:

$$\sum_i \lambda_i=\operatorname{tr}\mathbf{A}, \tag{110}$$

$$\prod_i \lambda_i=\det\mathbf{A}. \tag{111}$$

Exercises

In Exercises 1–12, find eigenvalues of the matrix $\mathbf{A}$:

$$\mathbf{A}=\begin{bmatrix} a & b \\ c & d \end{bmatrix}. \tag{112}$$

If the eigenvalues are real, find the corresponding eigenvectors. When checking your answers with those in the back of the book, keep in mind that any nonzero multiple of the given eigenvector may be used.

1. $a=0, b=1, c=-4, d=0$.
2. $a=1, b=3, c=1, d=-1$.
3. $a=2, b=1, c=1, d=2$.
4. $a=-3, b=-2, c=1, d=-5$.
5. $a=3, b=-1, c=1, d=2$.
6. $a=-2, b=0, c=0, d=-3$.
7. $a=1, b=0, c=1, d=-3$.

8. $a=-1, b=3, c=1, d=1$.
9. $a=4, b=-3, c=1, d=0$.
10. $a=-1, b=2, c=-2, d=-1$.
11. $a=3, b=2, c=0, d=4$.
12. $a=1, b=0, c=0, d=-3$.

In Exercises 13–16, find eigenvalues only and show that the eigenvalues are consistent with the trace and determinant condition.

13. $a=0, b=1, c=-4, d=0$.
14. $a=1, b=3, c=1, d=-1$.
15. $a=2, b=1, c=1, d=2$.
16. $a=-3, b=-2, c=1, d=-5$.

In Exercises 17–28, determine eigenvalues and eigenvectors if the eigenvalues are real. Also, determine the general solution of the following linear systems of differential equations:

$$\frac{dx}{dt} = ax + by,$$
$$\frac{dy}{dt} = cx + dy. \tag{113}$$

As a hint, problems with * have complex eigenvalues. When checking your answers with those in the back of the book, keep in mind that any nonzero multiple of the given eigenvector may be used.

17*. $a=0, b=1, c=-4, d=0$.
18. $a=1, b=3, c=1, d=-1$.
19. $a=2, b=1, c=1, d=2$.
20*. $a=-3, b=-2, c=1, d=-5$.
21*. $a=3, b=-1, c=1, d=2$.
22. $a=-2, b=0, c=0, d=-3$.
23. $a=1, b=0, c=1, d=-3$.
24. $a=-1, b=3, c=1, d=1$.
25. $a=4, b=-3, c=1, d=0$.
26*. $a=-1, b=2, c=-2, d=-1$.
27. $a=3, b=2, c=0, d=4$.
28. $a=1, b=0, c=0, d=-3$.

In exercises 29–44, determine eigenvalues and eigenvectors if the eigenvalues are real. Also, determine the general solution of the following linear systems. As a hint, problems with * have complex eigenvalues. When checking your answers with those in the back of the book, keep in mind that any nonzero multiple of the given eigenvector may be used.

29. $\frac{dx}{dt} = x, \ \frac{dy}{dt} = x + 2y$.

30. $\frac{dx}{dt} = 2x - y, \ \frac{dy}{dt} = 3x - 2y$.

31*. $\dfrac{dx}{dt} = -x - 5y,\ \dfrac{dy}{dt} = x + y.$

32. $\dfrac{dx}{dt} = 2x - y,\ \dfrac{dy}{dt} = 2x + 5y.$

33*. $\dfrac{dx}{dt} = x - y,\ \dfrac{dy}{dt} = x + y.$

34*. $\dfrac{dx}{dt} = -2x + 2y,\ \dfrac{dy}{dt} = -x.$

35. $\dfrac{dx}{dt} = -5x - 4y,\ \dfrac{dy}{dt} = 2x + y.$

36*. $\dfrac{dx}{dt} = x + 5y,\ \dfrac{dy}{dt} = -2x - y.$

37. $\dfrac{dx}{dt} = y,\ \dfrac{dy}{dt} = 2x + y.$

38*. $\dfrac{dx}{dt} = -x - 2y,\ \dfrac{dy}{dt} = 2x - y.$

39. $\dfrac{dx}{dt} = -5x - y,\ \dfrac{dy}{dt} = 3x - y.$

40*. $\dfrac{dx}{dt} = x + 2y,\ \dfrac{dy}{dt} = -4x - 3y.$

41*. $\dfrac{dx}{dt} = -x + 4y,\ \dfrac{dy}{dt} = -4x - y.$

42*. $\dfrac{dx}{dt} = 3x + 2y,\ \dfrac{dy}{dt} = -2x + 3y.$

43. $\dfrac{dx}{dt} = 4x + 3y,\ \dfrac{dy}{dt} = 3x + 4y.$

44. $\dfrac{dx}{dt} = 2x + 3y,\ \dfrac{dy}{dt} = 3x + 2y.$

In Exercises 45–50, you are given the eigenvalues and eigenvectors for a 2×2 matrix $\mathbf{A}$. Write down the general solution of $\frac{d\mathbf{x}}{dt} = \mathbf{A}\mathbf{x}$.

45. $\left\{3i, \begin{bmatrix} 2 \\ i \end{bmatrix}\right\}, \left\{-3i, \begin{bmatrix} 2 \\ -i \end{bmatrix}\right\}.$

46. $\left\{2i, \begin{bmatrix} 2 \\ 3i \end{bmatrix}\right\}, \left\{-2i, \begin{bmatrix} 2 \\ -3i \end{bmatrix}\right\}.$

47. $\left\{1+2i, \begin{bmatrix} 1+i \\ 3i \end{bmatrix}\right\}, \left\{1-2i, \begin{bmatrix} 1-i \\ -3i \end{bmatrix}\right\}.$

48. $\left\{2+i, \begin{bmatrix} 1+2i \\ 3i \end{bmatrix}\right\}, \left\{-2i, \begin{bmatrix} 1-2i \\ -3i \end{bmatrix}\right\}.$

49. $\left\{5i, \begin{bmatrix} 1+i \\ i \end{bmatrix}\right\}, \left\{-5i, \begin{bmatrix} 1-i \\ -i \end{bmatrix}\right\}.$

50. $\left\{2i, \begin{bmatrix} 2 \\ 3i \end{bmatrix}\right\}, \left\{-2i, \begin{bmatrix} 2 \\ -3i \end{bmatrix}\right\}.$

4.3 The Phase Plane for Linear Systems of Differential Equations

4.3.1 Introduction to the Phase Plane for Linear Systems of Differential Equations

In this section, we continue to analyze only linear systems of differential equations,

$$\frac{dx}{dt} = ax + by, \tag{1a}$$

$$\frac{dy}{dt} = cx + dy. \tag{1b}$$

This section is devoted to explaining the important concept of the **plane plane**. But in this section we only consider the phase plane for linear systems (1a)–(1b). The phase plane for nonlinear systems is discussed in Chapter 6.

Solutions $x(t)$, $y(t)$ of differential equations were previously each graphed as functions of time. Here instead we introduce the x, y-plane. Each value of t corresponds to a point x, y in the plane. A solution of the differential equation $x(t)$, $y(t)$ satisfying a given initial condition now traces out a curve in the x, y-plane. This parameterized curve (along with an indication of the direction the solution moves along the curve as time t increases) is called a **trajectory** or **orbit**. (The term **solution curve** is also sometimes used.) The set of trajectories (corresponding to all initial conditions) in the x, y-plane, together with an indication of the solutions direction as time increases, is the **phase plane**. Usually only a few representative solutions are drawn. (This sketch is sometimes called a phase portrait.)

Example 4.3.1 *Trajectory in the Phase Plane*

We consider the linear system

$$\frac{dx}{dt} = y, \tag{2a}$$

$$\frac{dy}{dt} = -x. \tag{2b}$$

• SOLUTION. Although we can solve this system by matrix methods, in this example, elimination is perhaps simpler. Substituting $y = \frac{dx}{dt}$ into (2b) yields the second-order differential equation with constant coefficients

$$\frac{d^2x}{dt^2} = -x. \tag{3}$$

The characteristic equation obtained by substituting $x = e^{rt}$ is $r^2 = -1$, so that $r = \pm i$ and the general solution is

$$x = c_1 \cos t + c_2 \sin t, \tag{4a}$$

$$y = -c_1 \sin t + c_2 \cos t. \tag{4b}$$

We wish to consider the solution associated with one initial condition, and for simplicity we choose $x(0) = 1$ and $y(0) = 0$. In this case $c_1 = 1$ and $c_2 = 0$, so the solution

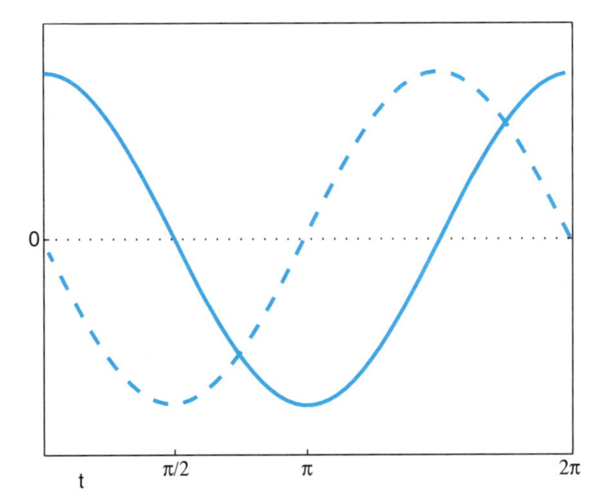

Figure 4.3.1 Solution of initial value problem $x(t) = \cos t$, $y(t) = -\sin t$.

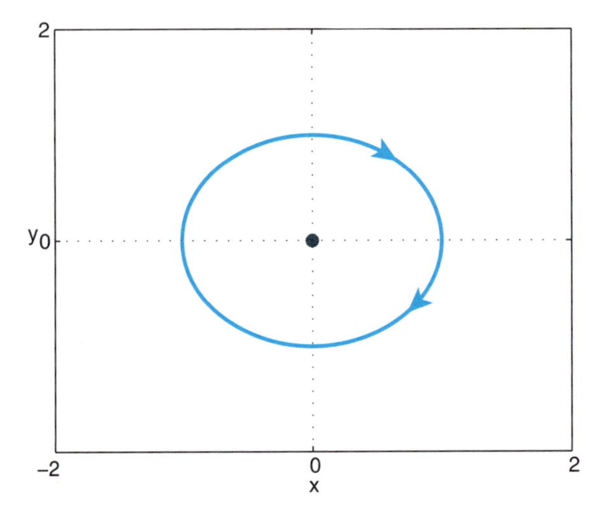

Figure 4.3.2 Trajectory in the phase plane.

of this initial value problem is

$$x = \cos t \tag{5a}$$

$$y = -\sin t. \tag{5b}$$

In figure 4.3.1we graph (5a)–(5b) in the traditional way $x = x(t)$ and $y = y(t)$, giving two elementary trigonometric functions. The phase plane (x, y) results by just plotting the set of points $(x(t), y(t))$ for different values of t. Thus the figure does not show t. The trajectory for this initial condition is the circle (graphed in figure 4.3.2) with radius 1, since

$$x^2 + y^2 = 1. \tag{6}$$

By plotting points for different t values, we see the trajectory moves clockwise. For example, $t = 0$ corresponds to $x = 1$, $y = 0$ and $t = \pi/2$ corresponds to $x = 0$, $y = -1$. The reason the solution moves clockwise in this example is that here the usual polar angle θ is $\theta = -t$, and thus the polar angle decreases in time. Arrows indicating how the solution moves clockwise, as time t increases is also included in figure 4.3.2. We will soon give further discussion on the direction of the orbits in the phase plane. ♦

PHASE PLANES ON COMPUTER SCREENS. It is quite common now for phase planes to be readily determined and vividly displayed on your computer screen using software such as Matlab, Scilab, Mathematica, or MAPLE. You will probably learn more about the subject if you have such a system available to you, or at least have access to a graphics program for phase planes for differential equations. These programs solve the system of differential equations numerically subject to given initial conditions. It is then easy for the program to take the numerical time-dependent solutions for $x(t)$ and $y(t)$ and use them directly to graph the phase plane (x, y). The time-dependent solutions of the differential equation are the parametric description of the curve shown on your screen which we call the phase plane.

NUMERICAL SOLUTIONS. Here we are not particularly interested in what numerical method the program uses. That is a separate interesting specialized mathematical subject. Those interested in numerical methods for differential equations should consult longer books on differential equations or specialized books on numerical methods for ordinary differential equations.

DIRECTION FIELD. The phase plane can be graphically constructed directly from the system of differential equations (1a)–(1b) without solving the system of differential equation. The vector $\mathbf{x} = \left[\begin{smallmatrix} x \\ y \end{smallmatrix}\right]$ has the representation as a row vector $\mathbf{x} = (x, y)$ or $\mathbf{x} = x\mathbf{i} + y\mathbf{j}$, so that in calculus it is shown that $\frac{d\mathbf{x}}{dt}$ is a tangent vector to the orbit or trajectories. As with using the slope to help sketch the solutions for first-order differential equations (see Section 1.4), a grid of points is chosen. At each (x, y) point in the grid, the right hand side of the system of differential equations (1a)–(1b) immediately determines the vector

$$\begin{bmatrix} \frac{dx}{dt} \\ \frac{dy}{dt} \end{bmatrix} = \begin{bmatrix} ax + by \\ cx + dy \end{bmatrix},$$

which is tangent to the solution curves in the phase plane in the direction that time increases.

If software plots these vectors, the plot is called a **vector field** plot. However, often some of the vectors are small, and others large, so that it can be hard to tell what is happening to solutions in some regions. Accordingly, many programs have the ability to plot the **direction field** for systems, including the linear systems we study in this chapter. This is easy to implement on computers or even graphing calculators. It is similar to a velocity vector, and everywhere (at each point in the grid) it has magnitude and direction. Usually the program draws a vector of equal small size so as not to interfere with neighboring vectors, and we call this the **direction field.** By having all vectors the same size it is easier to visualize the solutions. The phase plane may be

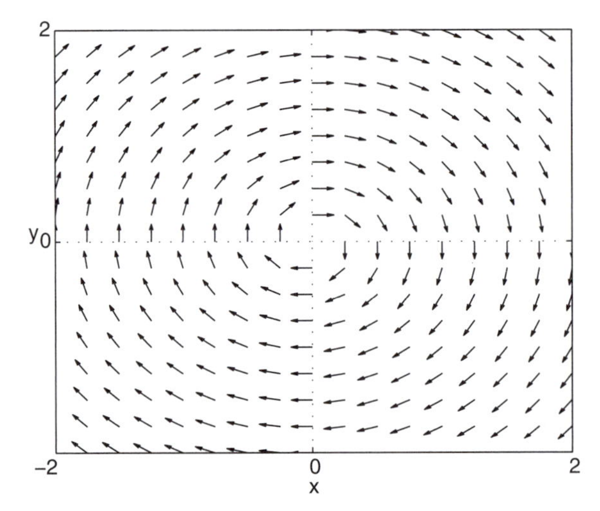

Figure 4.3.3 Direction field for (2a)–(2b).

approximated by forming curves tangent to the vectors of the direction field. We will do some examples.

EXAMPLE OF DIRECTION FIELD. Consider the previous example of a linear system (2a)–(2b). The direction field for this linear system is shown in figure 4.3.3 from some computer output. Since this is the first example for which we do a direction field, we want to explain very carefully. Let us take, for example, $x = 1$ and $y = 1$. From the differential equation, at that point the direction field should be a vector in the direction $\begin{bmatrix} 1 \\ -1 \end{bmatrix}$. Look at figure 4.3.3 at the point $x = 1$, $y = 1$, and note the vector there is in that direction. For example, for the point $x = -1$ and $y = -1$, from the differential equation the direction field should be a vector in the direction $\begin{bmatrix} -1 \\ 1 \end{bmatrix}$. We also compare the phase plane of the exact solution (circle) with the direction field and we see that the direction field suggests motion similar to the circle. The direction field is tangential to the circular orbits in the phase plane.

EQUILIBRIUM SOLUTION. The origin $x(t) = 0$, $y(t) = 0$ is a constant or **equilibrium** solution (does not depend on t) of any linear (homogeneous) system (1a)–(1b). Solutions of the linear system whose initial conditions are at the origin stay at the origin. The phase plane representation for the equilibrium solution is just a point that does not move, the equilibrium at the origin. For solutions of the system of differential equations that are not equilibria, the solutions will move in time. Then the trajectory or orbits in the phase plane will be curves.

STABLE OR UNSTABLE EQUILIBRIUM. If all solutions of the linear system stay near the equilibrium for all initial conditions near the equilibrium, then we say the equilibrium is **stable**. If there is at least one initial condition for which the solution goes away from the equilibrium, then we say the equilibrium is unstable. In this section, we will determine conditions for which the equilibrium (the origin) for the linear system is stable or unstable.

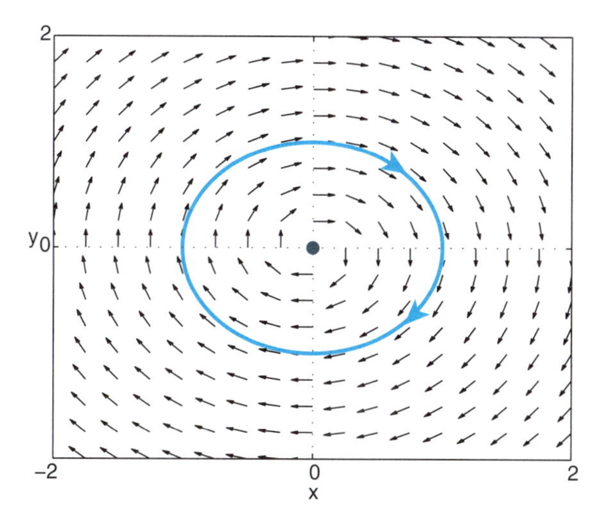

Figure 4.3.4 Trajectory in phase plane sketched from direction field.

OVERVIEW. In this section our interest is in explaining or directly constructing the phase planes using mathematical ideas. We will show how to obtain these trajectories in the phase plane using explicit solutions of the system of linear differential equations by using the matrix methods for solutions to linear systems of differential equations we have just studied in Section 4.2. Eigenvalues and eigenvectors will be very important. We will learn how to solve the phase plane for linear systems by doing a large number of examples. We begin with some very elementary examples where we do not need matrix methods. Then, we proceed to systematically investigate most cases of linear systems of differential equations including those involving real and complex eigenvalues.

In the problems of this section, $\frac{d\mathbf{x}}{dt}$ does not depend explicitly on t. Thus if two trajectories were to intersect, then at that point there would be two solutions which satisfy the same initial condition which would violate the uniqueness of solutions. Thus

Trajectories cannot intersect other trajectories or cross themselves.

This fact will be used repeatedly in what follows.

Example 4.3.2

For the following linear system, (a) identify its equilibrium; (b) sketch the direction field using software; (c) explain the direction field using the system of differential equations; and (d) sketch the phase plane using solutions of the system of differential equations;

$$\frac{dx}{dt} = -3x, \tag{7a}$$

$$\frac{dy}{dt} = -y. \tag{7b}$$

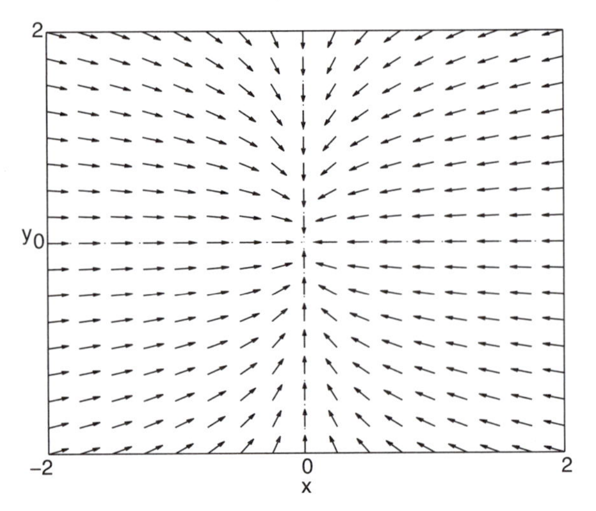

Figure 4.3.5 Direction field for (7a)–(7b).

• SOLUTION. (a) Here $x = y = 0$ solves the system of differential equations, so $(0, 0)$ is an equilibrium point. It is the only solution of $-3x = 0$, $-y = 0$, so it is the only equilibrium solution.

(b) Figure 4.3.5 shows the direction field (based on some readily available computer program) for the time-dependent system (7a)–(7b). (c) Since $\frac{dx}{dt} = -3x$, in the right half-plane (where $x > 0$) trajectories satisfy $\frac{dx}{dt} < 0$, so $x(t)$ decreases as time increases and the solution flows to the left. Similarly, in the left half-plane (where $x < 0$) trajectories flow to the right. Since $\frac{dy}{dt} = -y$, trajectories in the upper half-plane (where $y > 0$) have $\frac{dy}{dt} < 0$, so $y(t)$ decreases as time increases and the solution flows downward. In the lower half-plane ($y < 0$) the solution moves upward at time increases. From the figure we see that all solutions "flow into" the equilibrium $(0, 0)$. Such an equilibrium is called **asymptotically stable**.

(d) We also can explain the trajectories in the phase plane using the explicit solution of the system. The system (7a)–(7b) constitutes an uncoupled pair of linear differential equations whose solutions are

$$x(t) = c_1 e^{-3t}, \quad y(t) = c_2 e^{-t}. \tag{8}$$

If $c_2 = 0$, the solution in the phase plane is simple, namely, the line $y = 0$. Since $x(t) = c_1 e^{-t}$, the trajectories along $y = 0$ approach the origin. More precisely, the trajectories correspond to two rays approaching the origin (one with positive x and one with negative x) and the equilibrium. Similarly, the solution corresponding to $c_1 = 0$ corresponds to two rays approaching the origin (in opposite directions) along $x = 0$. These are the four straight rays sketched in figure 4.3.6. If both $c_1 \neq 0$ and $c_2 \neq 0$, then the trajectory is more complicated. Certainly, all solutions approach the origin as $t \to +\infty$. As $t \to +\infty$, $e^{-3t} \to 0$ much faster than e^{-t}, so that these other trajectories must approach and be tangent to the line $x = 0$ as $t \to +\infty$, as shown in figure 4.3.6 This kind of equilibrium is an example of a **stable node**. We will discuss other examples of this later in this section. ♦

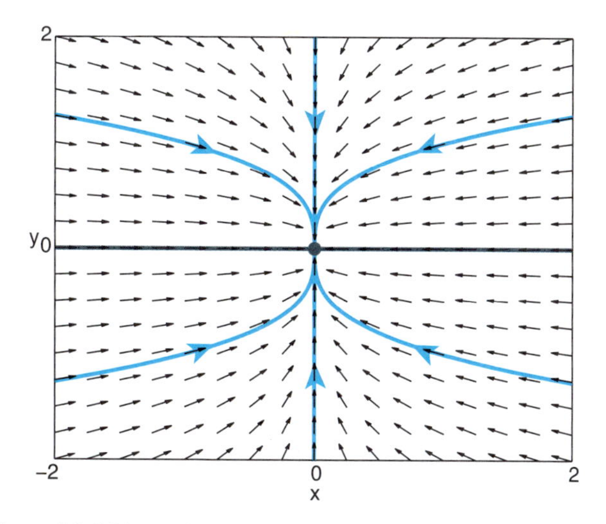

Figure 4.3.6 Phase plane for (7a)–(7b) sketched with direction field.

Example 4.3.3

Sketch the direction field and the phase plane for the system

$$\frac{dx}{dt} = 3x, \tag{9a}$$

$$\frac{dy}{dt} = y, \tag{9b}$$

and describe the behavior of the solutions near the equilibrium (0, 0).

• SOLUTION. It turns out that the direction field and phase plane for this example is almost identical to the previous one. The direction field and phase plane are the same but the directions of the arrows in time are reversed. To show that, we let $t = -\tau$. Since, for example, by the chain rule $\frac{dx}{dt} = \frac{dx}{d\tau}\frac{d\tau}{dt} = -\frac{dx}{dt}$. The system of differential equations becomes

$$\frac{dx}{d\tau} = -3x, \tag{10a}$$

$$\frac{dy}{d\tau} = -y. \tag{10b}$$

The trajectories are the same, but their directions are reversed in time. Now all solutions flow away from the equilibrium (0, 0); the equilibrium is **unstable**. This kind of equilibrium is called an **unstable node**. All solutions (except the equilibrium itself) go away from the equilibrium (origin) as time increases, and we note that the solutions go to infinity as $t \to +\infty$. It is most important to note that all solutions approach the origin backward in time (as $t \to -\infty$). A more subtle result is to note that backward in time the solution approaches the origin tangent to $x = 0$. ♦

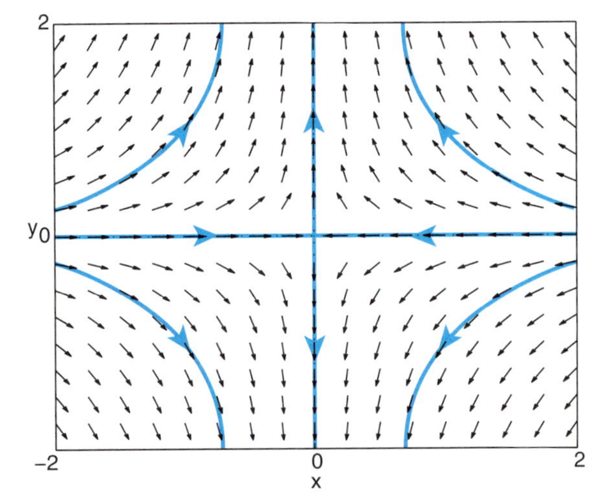

Figure 4.3.7 Phase plane for (11a)–(11b) sketched with direction field.

Example 4.3.4

Sketch the direction field using software and sketch the phase plane for the linear system of differential equations using solutions of the system of differential equations

$$\frac{dx}{dt} = -x, \tag{11a}$$

$$\frac{dy}{dt} = 2y. \tag{11b}$$

• SOLUTION. The equilibrium is again the origin $x = y = 0$ or $(0, 0)$ The direction field is sketched in figure 4.3.7. The solution to the system is $x(t) = c_1 e^{-t}$ and $y(t) = c_2 e^{2t}$. If $c_2 = 0$, the solution in the phase plane is again the line $y = 0$. Since $x(t) = c_1 e^{-t}$, the trajectories along $y = 0$ approach the origin. More precisely, the trajectories correspond to two rays approaching the origin, exactly the same as Example 4.3.2. The solution corresponding to $c_1 = 0$ corresponds to the vertical line $x = 0$, but arrows are introduced along $x = 0$ away (outgoing) from the origin, since y is exponentially increasing and going toward $\pm$ infinity. This solution $x = 0$ with $y(t) = c_2 e^{2t}$ approaches the origin backward in time (as $t \to -\infty$). There are four straight line rays sketched in figure 4.3.7, two going toward the origin but two away from the origin. If both $c_1 \neq 0$ and $c_2 \neq 0$, then the trajectory is more complicated. These solutions approach infinity as $t \to +\infty$ in the direction $x = 0$. These solutions also approach infinity backward in time as $t \to -\infty$, but along $y = 0$, as shown in figure 4.3.7. This kind of equilibrium is an example of a **saddle point**. We will discuss other examples of this later in this section. ♦

An equilibrium is defined to be stable if all initial conditions near the equilibrium stay near the equilibrium. (A more technical definition is needed for nonlinear systems.) In this example, initial conditions along $y = 0$ approach the equilibrium in a stable-like manner. However, the definition requires this to occur for all initial conditions. We can see from the figure that most initial conditions go away from the equilibrium, so we say that **all saddle points are unstable equilibrium**.

The last three examples were intended as motivation for a more general discussion of phase plane analysis for linear systems of differential equations

$$\frac{dx}{dt} = ax + by, \tag{12a}$$

$$\frac{dy}{dt} = cx + dy. \tag{12b}$$

4.3.2 Phase Plane for Linear Systems of Differential Equations

For the rest of this section, we continue to study linear (homogeneous) systems of differential equations

$$\frac{dx}{dt} = ax + by, \tag{13a}$$

$$\frac{dy}{dt} = cx + dy. \tag{13b}$$

We note that $x = y = 0$ corresponding to the origin (0, 0) is always an equilibrium point.

An understanding of the phase plane of linear systems comes from their explicit solution. In Section 4.2 we showed how to use eigenvalues and eigenvectors to solve linear system. We substitute

$$\begin{bmatrix} x(t) \\ y(t) \end{bmatrix} = e^{\lambda t} \begin{bmatrix} u \\ v \end{bmatrix}, \tag{14}$$

and obtain

$$(a - \lambda)u + bv = 0, \tag{15a}$$

$$cu + (d - \lambda)v = 0, \tag{15b}$$

or what we will use from now on,

$$\begin{bmatrix} a - \lambda & b \\ c & d - \lambda \end{bmatrix} \begin{bmatrix} u \\ v \end{bmatrix} = 0. \tag{16}$$

The pair of equations has a nonzero solution for u, v if and only if the determinant of the coefficient matrix is zero, that is,

$$\det \begin{bmatrix} a-\lambda & b \\ c & d-\lambda \end{bmatrix} = (a-\lambda)(d-\lambda) - bc = \lambda^2 - (a + d)\lambda + ad - bc = 0. \tag{17}$$

This is called the **characteristic equation** for the linear system of differential equations (13a)–(13b). This is also the condition for λ to be an **eigenvalue** of the coefficient matrix $\mathbf{A} = \begin{bmatrix} a & b \\ c & d \end{bmatrix}$. The eigenvector $\begin{bmatrix} u \\ v \end{bmatrix}$ corresponding to the eigenvalue λ satisfies (15a)–(15b) or (16).

There are a number of questions we wish to answer:

1. Is the equilibrium $x = y = 0$ stable or unstable?
2. What happens to the solution as time increases, and in particular what happens for long time ($t \to +\infty$)?
3. What do trajectories of the solutions look like in the phase plane?

It turns out that the behavior of the solutions of linear systems of differential equations is linked to the nature of the eigenvalues λ_1 and λ_2 of the coefficient matrix. Different behavior occurs if the eigenvalues are both positive, both negative, of opposite signs, or are complex (with positive, negative, or zero real part). We consider most cases in detail by first considering a special example and then indicating what happens in general. For your convenience, at the end of this section we summarize these results in Theorem 1.

4.3.3 Real Eigenvalues

Case 1: λ_1, λ_2 *Real, Distinct, and Positive (Unstable Node)*

Example 4.3.5

We first do an example similar to Example 4.3.1, which has both eigenvalues real, distinct, and positive:

$$\frac{dx}{dt} = x, \tag{18a}$$

$$\frac{dy}{dt} = 4y. \tag{18b}$$

• SOLUTION. The eigenvalues (roots of the characteristic equation) are 1 and 4, and the solution to the system is

$$x(t) = c_1 e^t,\ y(t) = c_2 e^{4t}. \tag{19}$$

This can be rewritten as a vector, which helps in understanding the trajectories in the phase plane:

$$\mathbf{x}(t) = \begin{bmatrix} x \\ y \end{bmatrix} = \begin{bmatrix} c_1 e^t \\ c_2 e^{4t} \end{bmatrix} = c_1 \begin{bmatrix} 1 \\ 0 \end{bmatrix} e^t + c_2 \begin{bmatrix} 0 \\ 1 \end{bmatrix} e^{4t}. \tag{20}$$

For example, (see figure 4.3.8), if $c_2 = 0$, the solution in the phase plane is $c_1 \begin{bmatrix} 1 \\ 0 \end{bmatrix} e^t$. This solution is in the direction $\begin{bmatrix} 1 \\ 0 \end{bmatrix}$ which corresponds to the x-axis ($y = 0$), corresponding to two rays (depending on the sign of c_1) which go to infinity as time increases (and approach the origin as $t \to -\infty$). The other straight line trajectory going to infinity as time increases corresponds to the solution $c_2 \begin{bmatrix} 0 \\ 1 \end{bmatrix} e^{4t}$ and is in the vertical direction. The other solutions shown in figure 4.3.8 go away from the origin and go to infinity as time increases. We say the equilibrium (0, 0) is an **unstable node**. These trajectories also go to the origin as $t \to -\infty$ in the direction tangent to $y = 0$. ♦

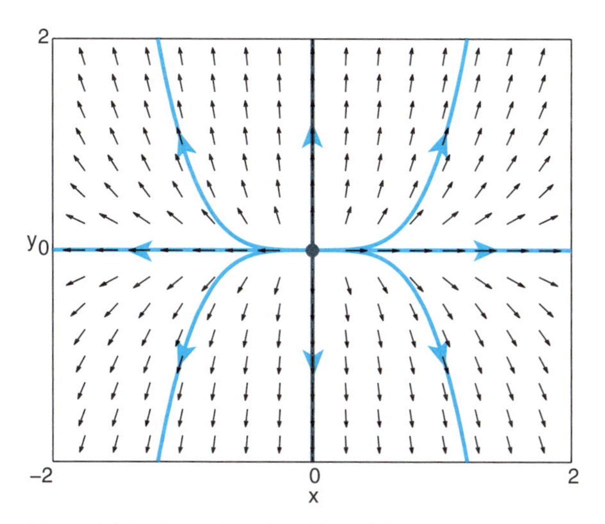

Figure 4.3.8 Phase plane for (18a)–(18b), an unstable node.

WE NOW DISCUSS CASE I IN GENERAL WITH REAL DISTINCT POSITIVE EIGENVALUES $\lambda_1 > 0$ AND $\lambda_2 > 0$ (UNSTABLE NODE). Suppose $\mathbf{v}_1$ is an eigenvector corresponding to the eigenvalue λ_1, and $\mathbf{v}_2$ is an eigenvector corresponding to the eigenvalue λ_2, as shown in Section 4.2. That means there are two elementary solutions of the system, $\mathbf{v}_1 e^{\lambda_1 t}$ and $\mathbf{v}_2 e^{\lambda_2 t}$. From this we obtain the **general solution**,

$$\mathbf{x}(t) = \begin{bmatrix} x \\ y \end{bmatrix} = c_1 \mathbf{v}_1 e^{\lambda_1 t} + c_2 \mathbf{v}_2 e^{\lambda_2 t}. \tag{21}$$

If $c_2 = 0$, the solution is $\mathbf{x}(t) = \left[\begin{smallmatrix} x \\ y \end{smallmatrix}\right] = c_1 \mathbf{v}_1 e^{\lambda_1 t}$, with $\lambda_1 > 0$. In the phase plane this solution is in the direction of the eigenvector $\mathbf{v}_1$, since the solution is time-dependent multiples of the eigenvector. In the phase plane, the trajectory of this solution is a straight line in the direction of the eigenvector $\mathbf{v}_1$, going away from the origin (since $\lambda_1 > 0$) toward infinity. These correspond to the two outward going rays in the direction of the eigenvector $\mathbf{v}_1$ shown in figure 4.3.9.

Similarly, if $c_1 = 0$, the solution is $\mathbf{x}(t) = \left[\begin{smallmatrix} x \\ y \end{smallmatrix}\right] = c_2 \mathbf{v}_2 e^{\lambda_2 t}$, with $\lambda_2 > 0$. In the phase plane, this solution also corresponds to two outward going straight lines (since $\lambda_2 > 0$) trajectories in the direction of the other eigenvector $\mathbf{v}_2$, as seen in figure 4.3.9. The other trajectories in the phase plane move away in time from the origin and go to infinity. Backward in time, solutions approach the origin. **In general, when the roots are distinct, real, and positive, the origin is an unstable node whose phase plane resembles figure 4.3.9.** It can be shown (subtle) that backward in time the solution approaches the eigenvector direction of the smallest positive eigenvalue (since as $t \to -\infty$ the exponential $e^{\lambda t}$ corresponding to the largest eigenvalue goes to zero fastest). Computer-generated phase planes for linear systems are often incomplete, without the straight line trajectories corresponding to the eigenvector directions.

Case 2: λ_1, λ_2 Real, Distinct, and Negative (Stable Node)

We again start with an example.

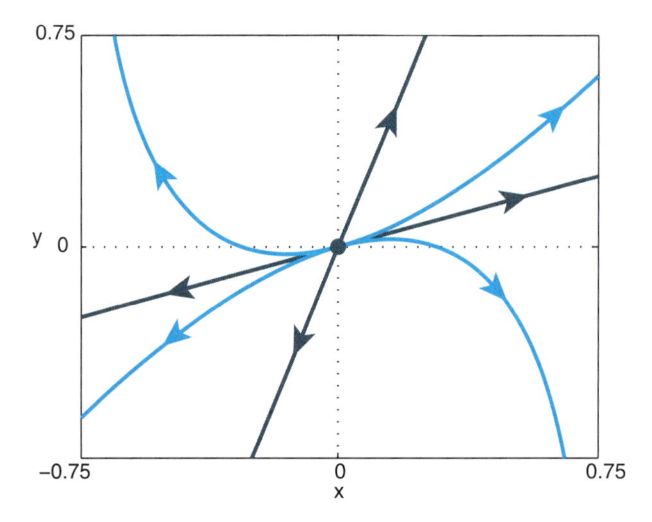

Figure 4.3.9 Phase plane for unstable node, assuming $\lambda_2 > \lambda_1 > 0$.

Example 4.3.6

We do an example similar to Example 4.2.1 with both eigenvalues real, distinct, and negative:

$$\frac{dx}{dt} = -x, \tag{22a}$$

$$\frac{dy}{dt} = -4y. \tag{22b}$$

• SOLUTION. The eigenvalues (roots of the characteristic equation) are -4 and -1, and the general solution to the system is

$$x(t) = c_1 e^{-t}, \quad y(t) = c_2 e^{-4t}. \tag{23}$$

We can rewrite this as a vector:

$$\mathbf{x}(t) = \begin{bmatrix} x \\ y \end{bmatrix} = \begin{bmatrix} c_1 e^{-t} \\ c_2 e^{-4t} \end{bmatrix} = c_1 \begin{bmatrix} 1 \\ 0 \end{bmatrix} e^{-t} + c_2 \begin{bmatrix} 0 \\ 1 \end{bmatrix} e^{-4t}. \tag{24}$$

A phase plane diagram for this system is given in figure 4.3.10. If $c_2 = 0$, the solution in the phase plane is $c_1 \left[\begin{smallmatrix} 1 \\ 0 \end{smallmatrix}\right] e^{-t}$. This solution is two straight line rays going toward the origin along the x-axis ($y = 0$). If $c_1 = 0$, the solution in the phase plane is two vertical rays moving toward the origin. These four rays are marked in blue in figure 4.3.10. In this example, the trajectories are moving toward the origin. For this reason, the origin is called an **asymptotically stable** equilibrium. Most trajectories approach the origin along curves that are tangent to the x-axis ($y = 0$). The equilibrium at the origin is a **stable node**. ♦

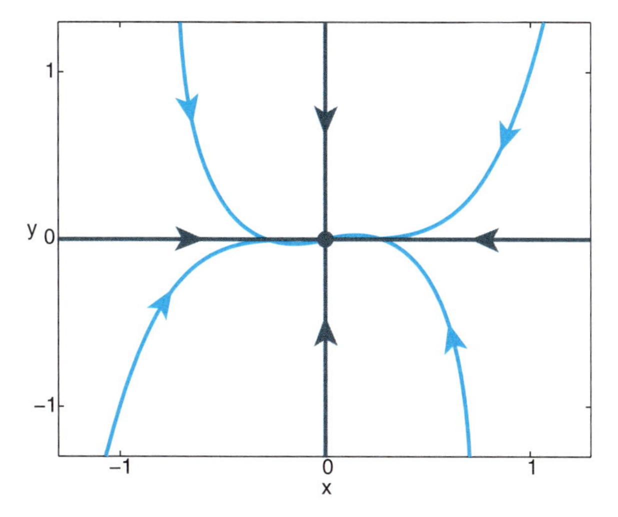

Figure 4.3.10 Phase plane for (22a)–(22b), a stable node.

Example 4.3.7 *Case 2. λ_1, λ_2 Real, Distinct, and Negative (Stable Node)*

Classify the equilibrium at the origin and sketch a phase plane for the linear system

$$\frac{dx}{dt} = -2x - 1y, \tag{25a}$$

$$\frac{dy}{dt} = -x - 2y. \tag{25b}$$

• SOLUTION. We substitute

$$\begin{bmatrix} x(t) \\ y(t) \end{bmatrix} = e^{\lambda t} \begin{bmatrix} u \\ v \end{bmatrix}, \tag{26}$$

and obtain

$$\begin{bmatrix} -2-\lambda & -1 \\ -1 & -2-\lambda \end{bmatrix} \begin{bmatrix} u \\ v \end{bmatrix} = 0. \tag{27}$$

The eigenvalues satisfy the determinant condition

$$\lambda^2 + 4\lambda + 3 = (\lambda + 3)(\lambda + 1) = 0. \tag{28}$$

The origin is a stable node since the eigenvalues (roots) are $-3, -1$. To sketch the trajectories in the phase plane, we determine the solution using eigenvalues and eigenvectors. The eigenvector corresponding to $\lambda = -3$ satisfies $u - v = 0$. We choose $u = 1$, so that $v = 1$, and the eigenvector corresponding to $\lambda = -3$ is $\left[\begin{smallmatrix} 1 \\ 1 \end{smallmatrix}\right]$. In this way, we obtain the elementary solution $\mathbf{x}(t) = \left[\begin{smallmatrix} x \\ y \end{smallmatrix}\right] = c_1 \left[\begin{smallmatrix} 1 \\ 1 \end{smallmatrix}\right] e^{-3t}$. The trajectories in the phase plane are two straight line rays ($y = x$) approaching the origin in the direction $\left[\begin{smallmatrix} 1 \\ 1 \end{smallmatrix}\right]$, as shown in figure 4.3.11. The eigenvector corresponding to $\lambda = -1$ satisfies $-u - v = 0$. We choose $u = 1$ and $v = -1$, so that the eigenvector

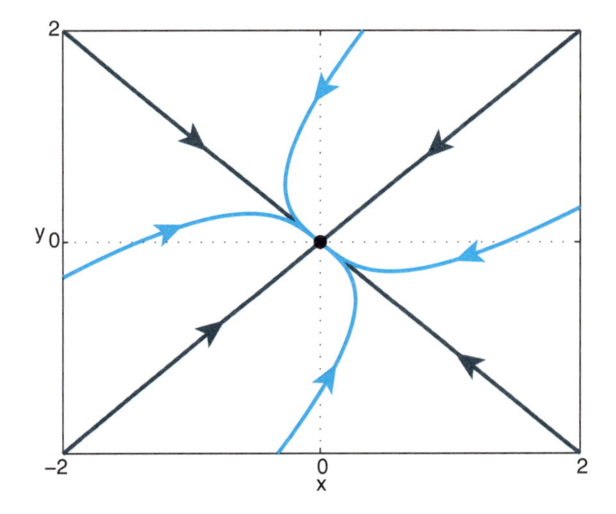

Figure 4.3.11 Phase plane for (25a)-(25b), a stable node.

corresponding to $\lambda = -1$ is $\begin{bmatrix} 1 \\ -1 \end{bmatrix}$. The elementary solution corresponding to the eigenvalue $\lambda = -1$ is $\begin{bmatrix} x \\ y \end{bmatrix} = c_2 \begin{bmatrix} 1 \\ -1 \end{bmatrix} e^{-t}$, which is the direction $y = -x$ of the two straight line rays approaching the origin in figure 4.3.11. From this we obtain the general solution:

$$\begin{bmatrix} x \\ y \end{bmatrix} = c_1 \begin{bmatrix} 1 \\ 1 \end{bmatrix} e^{-3t} + c_2 \begin{bmatrix} 1 \\ -1 \end{bmatrix} e^{-t}. \tag{29}$$

The solutions approach the origin as time increases. More specifically, the non-straight line solutions approach the origin tangent to $y = -x$ (same as the vector $\begin{bmatrix} 1 \\ -1 \end{bmatrix}$) since e^{-3t} is much smaller than e^{-t} as $t \to +\infty$. ♦

WHEN THE ROOTS (EIGENVALUES) ARE DISTINCT, REAL, AND NEGATIVE, CASE 2 THE ORIGIN IS A STABLE NODE. The phase plane near the origin resembles that in figure 4.3.11). Cases 1 and 2 are quite similar. They differ mainly in that when the roots (eigenvalues) are both positive, the trajectories move away from the origin (the origin is unstable), while when the roots are both negative, the trajectories approach the origin (the origin is stable). The general solution is again given by (21).

Case 3: λ_1, λ_2 *Real, Opposite Signs (Saddle Point)*

Example 4.3.8

A simple example of Case 3 with eigenvalues with real opposite signs is the system

$$\frac{dx}{dt} = 3x, \tag{30a}$$

$$\frac{dy}{dt} = -y. \tag{30b}$$

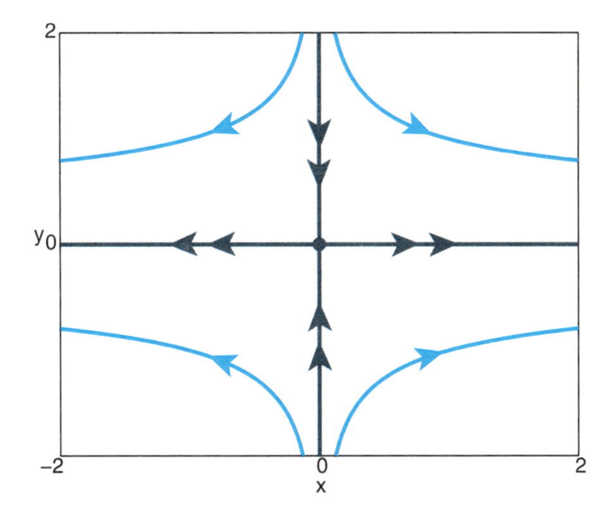

Figure 4.3.12 Phase plane for (30a)–(30b), a saddle point.

• SOLUTION. The eigenvalues (roots of characteristic equation) are 3 and -1, and the general solution to the system is

$$x(t) = c_1 e^{3t}, \quad y(t) = c_2 e^{-t}. \tag{31}$$

We can rewrite this as a vector:

$$\mathbf{x}(t) = \begin{bmatrix} x \\ y \end{bmatrix} = \begin{bmatrix} c_1 e^{3t} \\ c_2 e^{-t} \end{bmatrix} = c_1 \begin{bmatrix} 1 \\ 0 \end{bmatrix} e^{3t} + c_2 \begin{bmatrix} 0 \\ 1 \end{bmatrix} e^{-t}. \tag{32}$$

The trajectories in the phase plane for this system are given in figure 4.3.12. If $c_2 = 0$, the solution in the phase plane is $c_1 \left[\begin{smallmatrix} 1 \\ 0 \end{smallmatrix}\right] e^t$. This solution is two straight line rays going away from the origin along the x-axis $y = 0$. If $c_1 = 0$, the solution in the phase plane is two vertical rays moving toward the origin inward along the y-axis ($x = 0$). These four rays are marked in figure 4.3.12. If $c_1 \neq 0$ and $c_2 \neq 0$, then as time increases the solution must go away from the origin. Specifically, as $t \to +\infty$, $y(t) \to 0$ but $|x(t)| \to \infty$. As time goes backward, these solutions also go away from the origin and go to infinity. Specifically, as $t \to -\infty$, $x(t) \to 0$ but $|y(t)| \to \infty$. Trajectories come in along the positive and negative y-axis and go out along the positive and negative x-axis. The origin is unstable, since there are trajectories that start near the origin but eventually move away. (Note, there are two trajectories on the y-axis which do approach the origin, but they are the only ones.) The phase plane attained with the help of the direction field is shown in figure 4.3.12. The origin is said to be a **saddle point**. It is called a saddle point, because a property of a saddle of a horse is that in one direction the saddle goes away from the seat and in the other direction it goes toward the seat. The phase plane also looks like the topographic map for a pass through a mountain. ♦

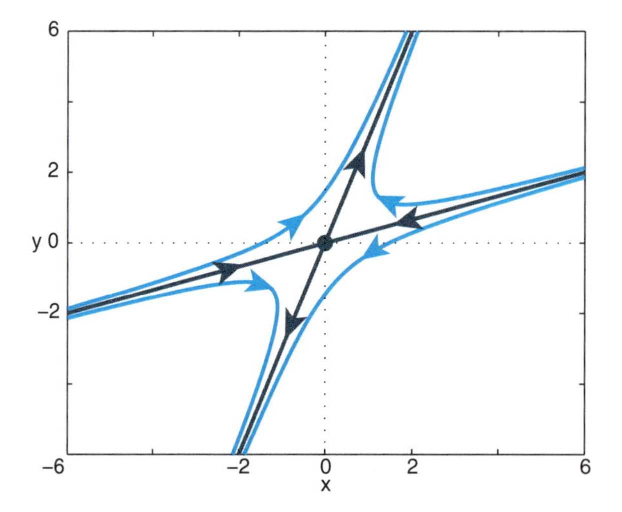

Figure 4.3.13 Phase plane for a saddle point.

CASE 3: WHEN EIGENVALUES (ROOTS) ARE REAL AND OF OPPOSITE SIGNS, THE ORIGIN IS AN UNSTABLE SADDLE POINT. The general solution is again given by (21). Saddle points are always unstable. In general, the trajectories come in alongside one eigenvector (corresponding to the negative eigenvalue) and go out alongside the other eigenvector (corresponding to the positive eigenvalue). This is illustrated in figures 4.3.13. The next example helps our understanding of the phase plane for a saddle point.

Example 4.3.9 *Case 3. λ_1, λ_2 Real, Opposite Signs (phase plane of a saddle point)*

Classify the equilibrium at the origin and sketch a phase diagram for the linear system

$$\frac{dx}{dt} = -7x + 6y, \tag{33a}$$

$$\frac{dy}{dt} = 6x + 2y. \tag{33b}$$

• SOLUTION. For systems we substitute:

$$\begin{bmatrix} x(t) \\ y(t) \end{bmatrix} = e^{\lambda t} \begin{bmatrix} u \\ v \end{bmatrix} \tag{34}$$

and obtain

$$\begin{bmatrix} -7-\lambda & 6 \\ 6 & 2-\lambda \end{bmatrix} \begin{bmatrix} u \\ v \end{bmatrix} = 0. \tag{35}$$

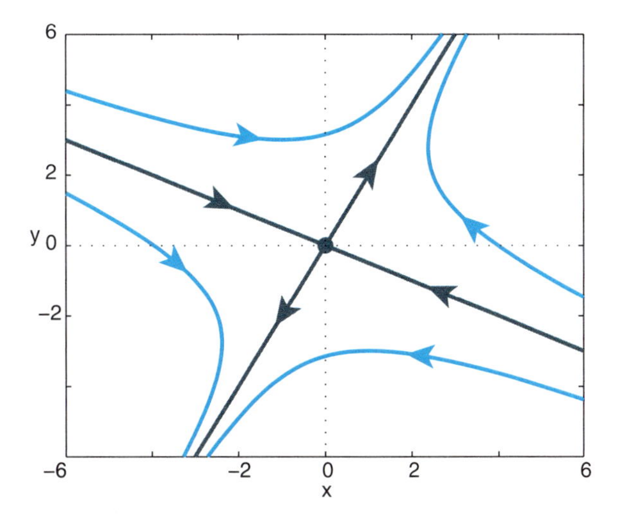

Figure 4.3.14 Phase plane for (33a)–(33b), which is a saddle point

The eigenvalues satisfy the determinant condition

$$\lambda^2 + 5\lambda - 50 = (\lambda + 10)(\lambda - 5) = 0. \tag{36}$$

The origin is an (unstable) saddle point, since the eigenvalues (roots) have opposite signs 5, -10. To sketch the trajectories in the phase plane, we determine the solution using eigenvalues and eigenvectors. The eigenvector corresponding to $\lambda = 5$ satisfies $-12u + 6v = 0$. We choose $u = 1$, so that $v = 2$, and the eigenvector corresponding to $\lambda = 5$ is $\left[\begin{smallmatrix}1\\2\end{smallmatrix}\right]$. In this way, we obtain the elementary solution $\mathbf{x}(t) = \left[\begin{smallmatrix}x\\y\end{smallmatrix}\right] = c_1\left[\begin{smallmatrix}1\\2\end{smallmatrix}\right]e^{5t}$. The trajectories in the phase plane are two straight line rays ($y = 2x$) going away from the origin toward infinity, in the direction $\left[\begin{smallmatrix}1\\2\end{smallmatrix}\right]$, as shown in figure 4.3.14. The eigenvector corresponding to $\lambda = -10$ satisfies satisfies $3u + 6v = 0$. We choose $v = 1$ and $u = -2$, so that the eigenvector corresponding to $\lambda = -10$ is $\left[\begin{smallmatrix}-2\\1\end{smallmatrix}\right]$. The elementary solution corresponding to the eigenvalue $\lambda = -10$ is $\left[\begin{smallmatrix}x\\y\end{smallmatrix}\right] = c_2\left[\begin{smallmatrix}-2\\1\end{smallmatrix}\right]e^{-10t}$, which is the direction $y = -x/2$ of the two straight line rays approaching the origin in Figure 4.3.14. Saddle points for linear systems are characterized by two opposite rays approaching the equilibrium but two opposite side rays going away from the equilibrium. From these solutions, we obtain the general solution:

$$\begin{bmatrix}x\\y\end{bmatrix} = c_1\begin{bmatrix}1\\2\end{bmatrix}e^{5t} + c_2\begin{bmatrix}-2\\1\end{bmatrix}e^{-10t}. \tag{37}$$

The non-straight line trajectories have the following properties. These trajectories all go away from the origin as time increases. Specifically, the non-straight line solutions approach infinity along the direction of the ray (eigenvector) associated with the positive eigenvalue, as shown in figure 4.3.14. Backward in time, these solutions approach infinity along the direction of the ray (eigenvector) associated with the negative eigenvalue. ♦

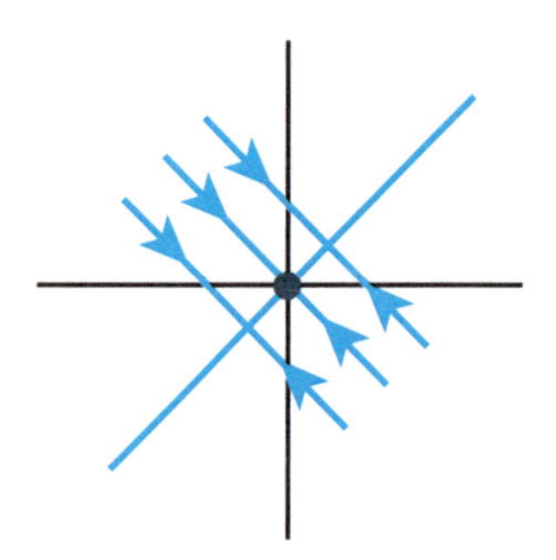

Figure 4.3.15 Case 5. Phase plane with one zero eigenvalue.

Case 4: Equal Eigenvalues (repeated roots) $\lambda_1 = \lambda_2$

We don't think this case is as important as the others, so we omit phase portraits of the two possibilities (one independent eigenvector and two independent eigenvectors).

Case 5: One Eigenvalue Zero ($\lambda_1 = 0$)

There are two cases, depending on whether $\lambda_2 < 0$ or $\lambda_2 > 0$. We just consider one example. Suppose $\lambda_1 = 0$ with eigenvector $\left[\begin{smallmatrix}1\\2\end{smallmatrix}\right]$ and $\lambda_2 = -10$ with eigenvector $\left[\begin{smallmatrix}1\\-2\end{smallmatrix}\right]$. Then the general solution would be

$$\begin{bmatrix} x \\ y \end{bmatrix} = c_1 \begin{bmatrix} 1 \\ 2 \end{bmatrix} + c_2 \begin{bmatrix} 1 \\ -2 \end{bmatrix} e^{-10t}. \tag{38}$$

If $c_1 = 0$, solutions along the eigenvector $\left[\begin{smallmatrix}1\\-2\end{smallmatrix}\right]$ move to the origin as time increases. If $c_2 = 0$, solutions along the eigenvector $\left[\begin{smallmatrix}1\\2\end{smallmatrix}\right]$ do not move in time. Other solutions approach the line $c_1\left[\begin{smallmatrix}1\\2\end{smallmatrix}\right]$ parallel to the direction $\left[\begin{smallmatrix}1\\-2\end{smallmatrix}\right]$ as shown in figure 4.3.15. If the negative eigenvalue were positive, trajectories would move in the opposite direction away from the line $c_1\left[\begin{smallmatrix}1\\2\end{smallmatrix}\right]$.

4.3.4 Complex Eigenvalues

Cases 6 and 7: Complex Eigenvalues (nonzero real part) $\lambda = \alpha \pm i\beta$*: Spirals*

We now consider the case of complex roots $\lambda = \alpha \pm i\beta$, where $\alpha \neq 0$ and $\beta \neq 0$. (Pure imaginary roots, $\lambda = \pm i\beta$, are discussed in Case 8.) For complex roots, because of Euler's formula, the solutions involve $e^{\alpha t}\cos(\beta t)$ and $e^{\alpha t}\sin(\beta t)$. It is clear that when $\alpha > 0$, the trajectories will travel away from the origin. Hence, the origin is unstable. But when $\alpha < 0$, the trajectories approach the origin and the origin is asymptotically stable. The result in the phase plane is a **spiral** centered at the origin, and the proof in some important special cases is given below. **Case 6:** If $\alpha > 0$, then the equilibrium is an **unstable spiral**. **Case 7:** If $\alpha < 0$, then the equilibrium is a **stable spiral**. The solutions may spiral either clockwise or counterclockwise but that is not usually an important consideration. Figure 4.3.16 shows all four possibilities. The figure shows only one spiral in each case corresponding to one initial condition. To account for all initial conditions, one should visualize an infinite number of spirals.

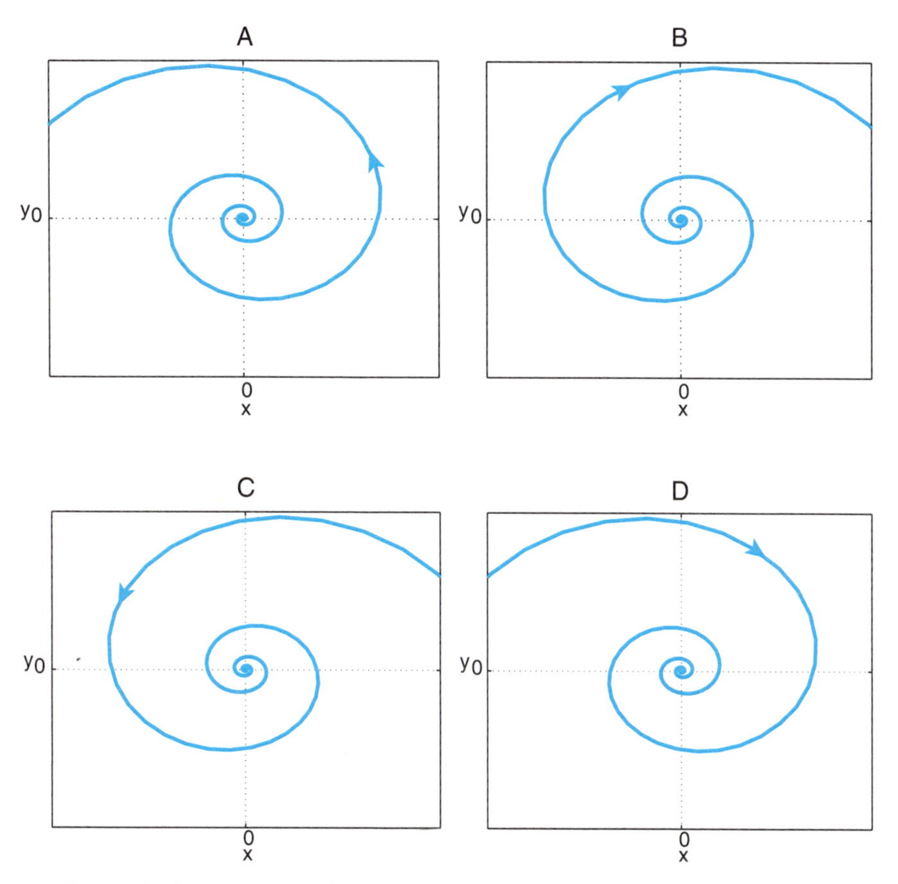

Figure 4.3.16 (a) Unstable spiral (counterclockwise), (b) unstable spiral (clockwise), (c) stable spiral (counterclockwise), (d) stable spiral (clockwise).

Example 4.3.10 *Phase Plane of a Spiral*

Determine the phase plane for

$$\frac{dx}{dt} = 2x + y, \tag{39a}$$

$$\frac{dy}{dt} = -x + 2y. \tag{39b}$$

• SOLUTION. The eigenvalues satisfy

$$\det\begin{bmatrix} 2-\lambda & 1 \\ -1 & 2-\lambda \end{bmatrix} = \lambda^2 - 4\lambda + 5 = 0, \tag{40}$$

so that the eigenvalues (roots) are complex, $\lambda = 2 \pm i$. Since $\alpha = 2 > 0$, we have an unstable spiral. To determine whether it is clockwise or counterclockwise, we just take one simple nonzero point on the x or y axis. For example, say $x = 1$, $y = 0$, in which case the tangent vector is $[\frac{dx}{dt}, \frac{dx}{dt}] = [2, -1]$, which points down and to the right from $(1, 0)$. Thus, the unstable spiral is clockwise as in figure 4.3.16b. ♦

SOME VERY IMPORTANT EXAMPLES OF COMPLEX EIGENVALUES: CASES 6 AND 7. These examples simplify using polar coordinates. But this does not always work. To better describe the behavior of trajectories, we analyze the specific system

$$\frac{dx}{dt} = \alpha x - \beta y, \tag{41a}$$

$$\frac{dy}{dt} = \beta x + \alpha y. \tag{41b}$$

The eigenvalues satisfy

$$\det \begin{bmatrix} \alpha - \lambda & -\beta \\ \beta & \alpha - \lambda \end{bmatrix} = (\alpha - \lambda)^2 + (\beta)^2 = 0, \tag{42}$$

so that the eigenvalues (roots) are complex, $\lambda = \alpha \pm i\beta$.

For this example (but not necessarily other examples), the differential equation simplifies using the polar coordinates, $r^2 = x^2 + y^2$ and $\tan\theta = \frac{y}{x}$. We obtain

$$r\frac{dr}{dt} = x\frac{dx}{dt} + y\frac{dy}{dt} = x(\alpha x - \beta y) + y(\beta x + \alpha y) = \alpha(x^2 + y^2) = \alpha r^2, \tag{43}$$

dividing by 2 and using the differential equation. We have

$$\frac{dr}{dt} = \alpha r, \text{ with the general solution } r(t) = r(0)e^{\alpha t}. \tag{44}$$

We next obtain the differential equation for the polar angle,

$$\sec^2\theta\frac{d\theta}{dt} = \frac{x\frac{dy}{dt} - y\frac{dx}{dt}}{x^2} = \frac{x(\beta x + \alpha y) - y(\alpha x - \beta y)}{x^2} = \frac{\beta(x^2 + y^2)}{x^2}. \tag{45}$$

Since $\sec^2\theta = \frac{1}{\cos^2\theta} = \frac{r^2}{x^2}$, we obtain

$$\frac{d\theta}{dt} = \beta, \text{ with the general solution obtained by integration, } \theta(t) = \beta t + \theta(0). \tag{46}$$

Again, we see that if $\alpha > 0$, the solution moves away from the origin, since $r \to +\infty$ as $t + \infty$. But, because θ is a linear function of t, the trajectories spiral around the origin. Similarly, when $\alpha < 0$, the solutions spiral in toward the origin. We therefore refer to the origin as a **spiral point**. It is a **stable spiral** if $\alpha < 0$, and an **unstable spiral** if $\alpha > 0$. We can determine the direction of rotation of the spiral from (46). If $\beta > 0$, then θ (polar angle) increases in time, and the spiral is counterclockwise (see figure 4.3.16a or c). If $\beta < 0$, then θ decreases in time, and the spiral is clockwise (see figure 4.3.16b or d). The direction of the spiral can also be determined in a simpler way directly from the original system. For example, setting $y = 0$ in (41b), we find that $\frac{dy}{dt} = \beta x$. Now, if $\beta > 0$, then as the trajectory crosses the positive x-axis, $y = 0$, we have $\frac{dy}{dt} > 0$, and consequently, the trajectory is spiraling counterclockwise. Similarly, $\beta < 0$ the trajectory spirals clockwise. These spirals are **logarithmic spirals** because $\ln r(t) = \ln r(0) + \alpha t = \ln r(0) + \frac{\alpha}{\beta}(\theta(t) - \theta(0))$.

In the general case, the spirals may be distorted, but the stability criteria persists; when the roots $\lambda = \alpha \pm i\beta$ are complex numbers, the origin is a spiral point that is unstable when $\alpha > 0$ (see figure 4.3.16 a or b) and asymptotically stable when $\alpha < 0$ (see figure 4.3.16 c or d).

Example 4.3.11 *Case 6. Complex Eigenvalues (roots) $\lambda = \alpha \pm i\beta$: Unstable Spirals.*

Classify the equilibrium at the origin and sketch the phase plane for the system

$$\frac{dx}{dt} = 2x + y, \tag{47a}$$

$$\frac{dy}{dt} = -x + 2y. \tag{47b}$$

• SOLUTION. The eigenvalues (characteristic equation) for this system is

$$\det \begin{bmatrix} 2-\lambda & 1 \\ -1 & 2-\lambda \end{bmatrix} = (2-\lambda)^2 + 1 = 0,$$

which has complex eigenvalues (roots), $\lambda = 2 \pm i$. Thus, the origin is an unstable spiral, since $\alpha = 2 > 0$. Setting $y = 0$, $x = 1$ in (47b), we get the tangent vector

$$\begin{bmatrix} \frac{dx}{dt} \\ \frac{dy}{dt} \end{bmatrix} = \begin{bmatrix} 2 \\ -1 \end{bmatrix},$$

which points down and to the right from $(1, 0)$. Thus the unstable spiral is clockwise, and we have precisely the case of figure 4.3.16 b. ♦

Case 8: Purely Imaginary Eigenvalues (roots) $\lambda = \pm i\beta$: Centers

Example 4.3.12 *Undamped Spring-Mass System*

Sketch the trajectories in the phase plane for the first-order system corresponding to the unforced undamped spring mass system in Section 2.5:

$$m\frac{d^2x}{dt^2} + kx = 0. \tag{48}$$

• SOLUTION. We will solve this oscillation of a spring mass system in a number of different ways.

SOLUTION USING SECOND-ORDER DIFFERENTIAL EQUATION METHODS. We can determine the trajectories in the phase plane by solving the second-order linear differential equation with constant coefficients associated with the spring-mass system (48). The amplitude and phase form of the general solution are particularly helpful here:

$$x(t) = A\sin(\omega t + \phi), \tag{49}$$

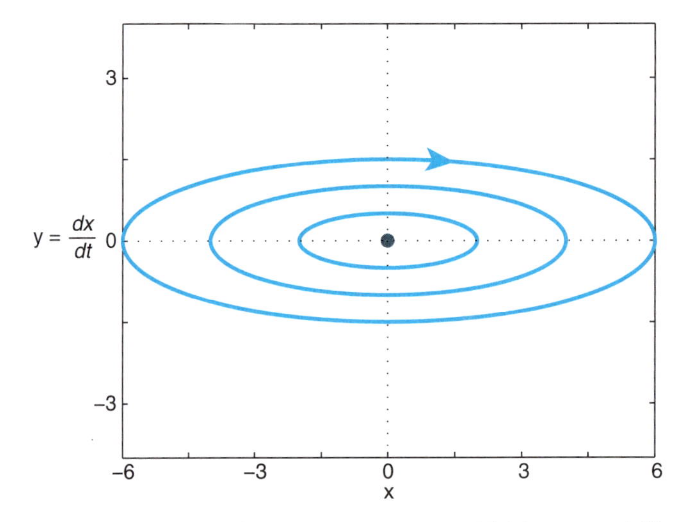

Figure 4.3.17 Phase plane for spring-mass system which is a center (ellipse).

where the natural frequency satisfies $\omega = \sqrt{\frac{k}{m}}$. In this case, by taking the derivative, the velocity $y = \frac{dx}{dt}$ satisfies

$$y(t) = \frac{dx}{dt} = A\omega \cos(\omega t + \phi). \tag{50}$$

Time can be eliminated from (49) and (50), giving directly the equation for the trajectories in the phase plane:

$$x^2 + \frac{y^2}{\omega^2} = A^2 \sin^2 + A^2 \cos^2 = A^2. \tag{51}$$

The phase plane consists of a family of **ellipses** shown in figure 4.3.17. Trajectories move clockwise since, for example, at $x = 0$, $y = 1$ from (50) we see that $\frac{dx}{dt} = 1$ so that x is increasing in time there. The solutions orbit periodically (cyclically) through the same points, with the same velocities. We call the equilibrium (0, 0) a **center**.

SOLUTION USING DIRECTION FIELD. By introducing the velocity $y = \frac{dx}{dt}$, this equation (48) can be converted to a first-order system:

$$\frac{dx}{dt} = y, \tag{52a}$$

$$\frac{dy}{dt} = -\frac{k}{m}x. \tag{52b}$$

The equilibrium is the origin $x = y = 0$. The direction field for $\frac{k}{m} = 1$ corresponds to our earlier problem, (2a), (2b). With $\omega = \sqrt{\frac{k}{m}} = 1$ the ellipses become circles as shown in figures 4.3.3 and 4.3.4. In that case the trajectories appear to be either circles or spirals which encircle the origin clockwise. They can't spiral because the exact time-dependent solutions are periodic. They are ellipses here because of (51).

SOLUTION USING SYSTEM OF FIRST-ORDER DIFFERENTIAL EQUATIONS. For the spring-mass system, (52a)–(52b). For systems, we substitute:

$$\begin{bmatrix} x(t) \\ y(t) \end{bmatrix} = e^{\lambda t} \begin{bmatrix} u \\ v \end{bmatrix}, \tag{53}$$

and obtain

$$\begin{bmatrix} -\lambda & 1 \\ -\frac{k}{m} & -\lambda \end{bmatrix} \begin{bmatrix} u \\ v \end{bmatrix} = 0. \tag{54}$$

The eigenvalues satisfy the determinant condition

$$\lambda^2 + \frac{k}{m} = 0. \tag{55}$$

The eigenvalues (roots) are purely imaginary $\lambda = \pm i\sqrt{\frac{k}{m}}$, and by Euler's formula the solution must involve $\cos\sqrt{\frac{k}{m}}t$ and $\sin\sqrt{\frac{k}{m}}t$, which we already knew. ♦

Example 4.3.13 *Case 8. Purely Imaginary Eigenvalues (Roots) $\lambda = \pm i\beta$: Centers*

Another example we study is the special system

$$\frac{dx}{dt} = -\beta y, \tag{56a}$$

$$\frac{dy}{dt} = -\beta x, \tag{56b}$$

which is just system (41a)–(41b) with $\alpha = 0$.

• SOLUTION. Let us first determine the eigenvalues before we rely on previously obtained results. To find the eigenvalues, we substitute

$$\begin{bmatrix} x(t) \\ y(t) \end{bmatrix} = e^{\lambda t} \begin{bmatrix} u \\ v \end{bmatrix}, \tag{57}$$

and obtain

$$\begin{bmatrix} -\lambda & -\beta \\ \beta & -\lambda \end{bmatrix} \begin{bmatrix} u \\ v \end{bmatrix} = 0. \tag{58}$$

The eigenvalues satisfy the determinant condition

$$\lambda^2 + \beta^2 = 0. \tag{59}$$

The eigenvalues (roots) are purely imaginary $\lambda = \pm i\beta$, and then, by Euler's formula, the solution involves $\cos\beta t$ and $\sin\beta t$. To analyze solutions in the phase plane, we use previous results for polar coordinates and substitute $\alpha = 0$ into case 6 and 7. In this case from (44) and (46), we obtain

$$\frac{dr}{dt} = 0, \ \frac{d\theta}{dt} = \beta. \tag{60}$$

Hence $r(t) = r(0)$ and $\theta(t) = \beta t + \theta(0)$. Since r is constant (but arbitrary), the trajectories are concentric circles about the origin. The circular orbits are counterclockwise if $\beta > 0$ (see figure 4.3.18a) and clockwise $\beta < 0$ (see figure 4.3.18b). Thus, the

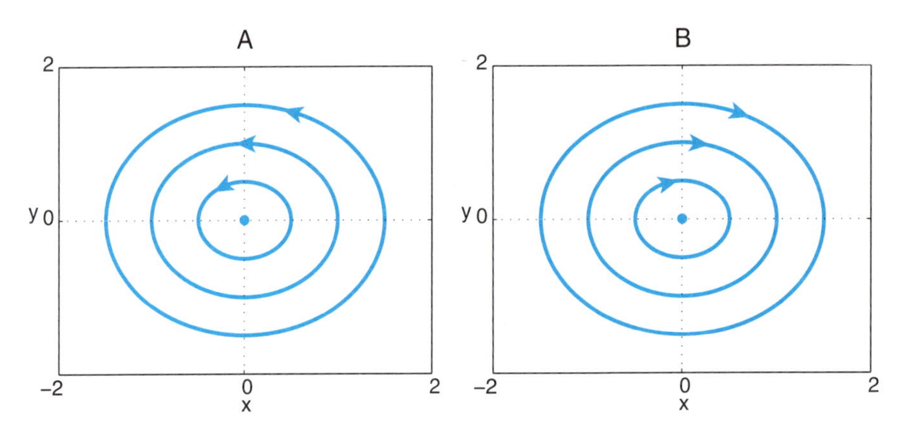

Figure 4.3.18 (a) Center: counterclockwise circles ($\beta > 0$),
(b) center: clockwise circles ($\beta < 0$)

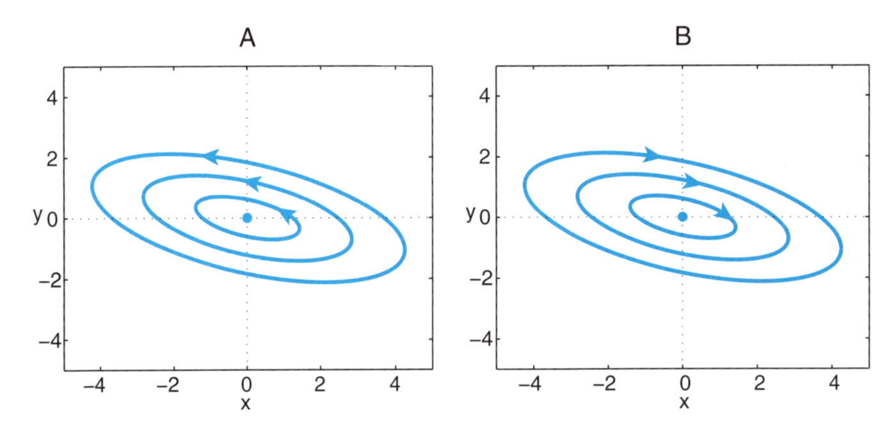

Figure 4.3.19 (a) Center: Counterclockwise skewed ellipses,
(b) center: clockwise skewed ellipses.

motion is a periodic rotation around a circle centered at the origin. Appropriately, the origin is called a **center** and is a stable equilibrium. ♦

IN GENERAL, WHEN THE EIGENVALUES ARE PURELY IMAGINARY, THE ORIGIN IS A STABLE CENTER. The trajectories are "skewed ellipses" centered at the origin with axes of the ellipse not necessarily $x = 0$ and $y = 0$ as in our examples. Motion is periodic. The equilibrium is **stable** since nearby solutions do not move very far away. However, since the solutions do not approach the equilibrium, the equilibrium is not asymptotically stable. Typical phase diagrams are shown in Figure 4.3.19.

4.3.5 General Theorems

THEOREM 4.3.1 *Theorem on Phase Portraits and Stability of Linear Systems. We now summarize the phase plane behavior of a linear system of differential equations and the stability of the equilibrium at the origin:*

$$\frac{dx}{dt} = ax + by, \tag{61a}$$

$$\frac{dy}{dt} = cx + dy, \tag{61b}$$

in terms of the coefficient matrix

$$\mathbf{A} = \begin{bmatrix} a & b \\ c & d \end{bmatrix}. \tag{62}$$

The eigenvalues of the coefficient matrix satisfy the determinant condition, which is

$$\det \begin{bmatrix} a-\lambda & b \\ c & d-\lambda \end{bmatrix} = \lambda^2 - (a+d)\lambda + (ad-bc) = 0. \tag{63}$$

Case 1: Two positive eigenvalues: unstable node (see figure 4.3.9)
Case 2: Two negative eigenvalues: stable node (see figure 4.3.11)
Case 3: One positive and one negative eigenvalue: unstable saddle point (see figure 4.3.13)
Case 4: Equal eigenvalues (repeated roots)
Case 5: One eigenvalue zero: stable if $\lambda_2 < 0$ (see figure 4.3.15) and unstable if $\lambda_2 > 0$
Case 6: Complex eigenvalues (positive real part): unstable spiral (see figure 4.3.16 a and b)
Case 7: Complex eigenvalues (negative real part): stable spiral (see figure 4.3.16 c and d)
Case 8: Complex eigenvalues (zero real part): stable center (see figure 4.3.19)

CLASSIFICATION OF STABILITY OF LINEAR SYSTEMS. Here we will classify the stability of the zero solution of linear systems in terms of the trace and determinant of the matrix. These results have nice extensions when we include the phase plane. The eigenvalues of the matrix solve the following quadratic equation (105) in term of the trace and determinant of the matrix:

$$\lambda^2 - \operatorname{tr}\mathbf{A}\lambda + \det\mathbf{A} = 0. \tag{64}$$

Using the quadratic formula, we have

$$\lambda = \frac{\operatorname{tr}\mathbf{A} \pm \sqrt{(\operatorname{tr}\mathbf{A})^2 - 4\det\mathbf{A}}}{2}. \tag{65}$$

If $4\det\mathbf{A} > (\operatorname{tr}\mathbf{A})^2$, then the eigenvalues are complex (the phase plane will be spirals), stable (spirals) if $\operatorname{tr}\mathbf{A} < 0$ and unstable (spirals) if $\operatorname{tr}\mathbf{A} > 0$. It is helpful to remember (108) and (109), so that $\lambda_1\lambda_2 = \det\mathbf{A}$ and $\lambda_1 + \lambda_2 = \operatorname{tr}\mathbf{A}$. If one eigenvalue is positive ($\lambda_1 > 0$) and the other negative ($\lambda_2 < 0$) (saddle points), the zero solution is automatically unstable, and this corresponds to $\det\mathbf{A} < 0$. The other regions will have real eigenvalues of the same sign (nodes) with $\det\mathbf{A} = \lambda_1\lambda_2 > 0$; the unstable case (nodes) satisfy $\lambda_1 + \lambda_2 = \operatorname{tr}\mathbf{A} > 0$ while the stable case (nodes) satisfy $\lambda_1 + \lambda_2 = \operatorname{tr}\mathbf{A} < 0$. This classification of the stability of the zero solution (including phase plane) for linear

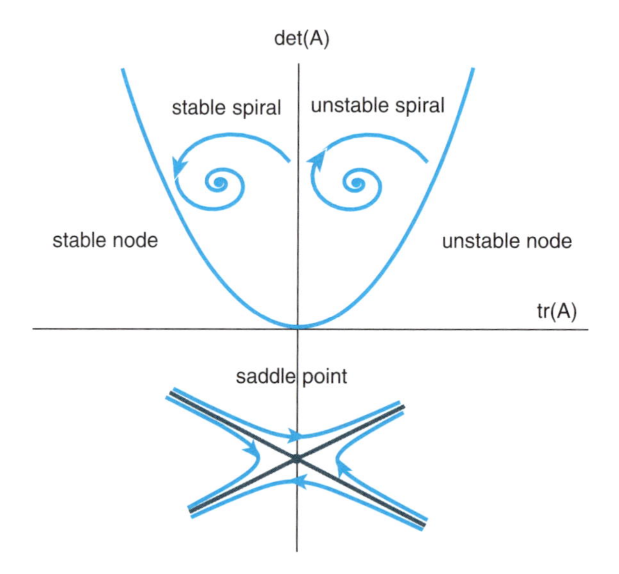

Figure 4.3.20 Phase plane and stability for linear systems classified by determinant and trace.

systems using trace and determinants is graphed in figure 4.3.20. From the figure, we see an interesting theorem: **the solution $\mathbf{x} = 0$ is stable for linear systems if and only if** $\det \mathbf{A} > 0$ **and** $\operatorname{tr} A < 0$. If $\det \mathbf{A} = 0$, at least one eigenvalue is zero.

Exercises

In Exercises 1–6, find the direction field using software,

$$\frac{d\mathbf{x}}{dt} = \begin{bmatrix} a & b \\ c & d \end{bmatrix} \mathbf{x} = \mathbf{A}\mathbf{x}, \tag{66}$$

where

1. $a = 0, b = 1, c = -4, d = 0$.
2. $a = 1, b = 3, c = 1, d = -1$.
3. $a = 2, b = 1, c = 1, d = 2$.
4. $a = -3, b = -2, c = 1, d = -5$.
5. $a = 3, b = -1, c = 1, d = 2$.
6. $a = -2, b = 0, c = 0, d = -3$.

In Exercises 7–22, determine the eigenvalues and eigenvectors if the eigenvalues are real (or use results from exercises from Section 4.2. if you have covered those exercises). Also classify the system (state whether stable or unstable node, stable or unstable spiral, center, saddle point) and in all cases sketch the phase plane of the linear system. (As a hint, problems with * have complex eigenvalues.) When checking your answers with those in the back of the book, keep in mind that any nonzero multiple of the given eigenvector may be used.

7. $\dfrac{dx}{dt} = x, \dfrac{dy}{dt} = x + 2y.$

8. $\dfrac{dx}{dt} = 2x - y, \dfrac{dy}{dt} = 3x - 2y.$

9*. $\dfrac{dx}{dt} = -x - 5y, \dfrac{dy}{dt} = x + y.$

10. $\dfrac{dx}{dt} = 2x - y, \dfrac{dy}{dt} = 2x + 5y.$

11*. $\dfrac{dx}{dt} = x - y, \dfrac{dy}{dt} = x + y.$

12*. $\dfrac{dx}{dt} = -2x + 2y, \dfrac{dy}{dt} = -x.$

13. $\dfrac{dx}{dt} = -5x - 4y, \dfrac{dy}{dt} = 2x + y.$

14*. $\dfrac{dx}{dt} = x + 5y, \dfrac{dy}{dt} = -2x - y.$

15. $\dfrac{dx}{dt} = y, \dfrac{dy}{dt} = 2x + y.$

16*. $\dfrac{dx}{dt} = -x - 2y, \dfrac{dy}{dt} = 2x - y.$

17. $\dfrac{dx}{dt} = -5x - y, \dfrac{dy}{dt} = 3x - y.$

18*. $\dfrac{dx}{dt} = x + 2y, \dfrac{dy}{dt} = -4x - 3y.$

19*. $\dfrac{dx}{dt} = -x + 4y, \dfrac{dy}{dt} = -4x - y.$

20*. $\dfrac{dx}{dt} = 3x + 2y, \dfrac{dy}{dt} = -2x + 3y.$

21. $\dfrac{dx}{dt} = 4x + 3y, \dfrac{dy}{dt} = 3x + 4y.$

22. $\dfrac{dx}{dt} = 2x + 3y, \dfrac{dy}{dt} = 3x + 2y.$

In Exercises 23–35, determine the eigenvalues and eigenvectors if the eigenvalues are real (or use results from exercises from Section 4.2 if you have covered those exercises), classify the system (state whether stable or unstable node, stable or unstable spiral, center, saddle point) and in all cases sketch the phase plane of the linear system. (As a hint, problems with * have complex eigenvalues.) When checking your answers with those in the back of the book, keep in mind that, any nonzero multiple of the given eigenvector may be used:

$$\frac{d\mathbf{x}}{dt} = \begin{bmatrix} a & b \\ c & d \end{bmatrix} \mathbf{x} = \mathbf{A}\mathbf{x}, \tag{67}$$

where

23*. $a = 0, b = 1, c = -4, d = 0.$

24. $a=1, b=3, c=1, d=-1$.
25. $a=2, b=1, c=1, d=2$.
26*. $a=-3, b=-2, c=1, d=-5$.
27*. $a=3, b=-1, c=1, d=2$.
28. $a=-2, b=0, c=0, d=-3$.
29. $a=1, b=0, c=1, d=-3$.
30. $a=-1, b=3, c=1, d=1$.
31. $a=4, b=-3, c=1, d=0$.
32*. $a=-1, b=2, c=-2, d=-1$.
33. $a=2, b=-1, c=1, d=0$.
34. $a=3, b=2, c=0, d=4$.
35. $a=1, b=0, c=0, d=-3$.
36. For Exercises 23–28 without finding the eigenvalues, classify the system (stable or unstable node, stable or unstable spiral, center, saddle point), determine using the trace and determinant condition.
37. For Exercises 29–35 without finding the eigenvalues, classify the system (stable or unstable node, stable or unstable spiral, center, saddle point), determine using the trace and determinant condition.

In Exercises 38–41, graph the phase portrait given the eigenvalues and the eigenvectors:

38. $\lambda_1=1, \begin{bmatrix}1\\1\end{bmatrix}, \quad \lambda_2=-1, \begin{bmatrix}1\\-1\end{bmatrix}$

39. $\lambda_1=1, \begin{bmatrix}2\\1\end{bmatrix}, \quad \lambda_2=-1, \begin{bmatrix}1\\-2\end{bmatrix}$

40. $\lambda_1=1, \begin{bmatrix}1\\1\end{bmatrix}, \quad \lambda_2=2, \begin{bmatrix}1\\-1\end{bmatrix}$

41. $\lambda_1=-1, \begin{bmatrix}1\\1\end{bmatrix}, \quad \lambda_2=-2, \begin{bmatrix}1\\-1\end{bmatrix}$

CHAPTER 5

Mostly Nonlinear First-Order Differential Equations

5.1 First-Order Differential Equations

In this chapter, we will solve and analyze first-order differential equations. Earlier in Sections 1.3 and 1.6 we studied first-order differential equations

$$\frac{dx}{dt} = f(x, t),$$

when the differential equation was separable,

$$\frac{dx}{dt} = h(t)g(x),$$

or the differential equation was linear,

$$\frac{dx}{dt} + p(t)x = q(t).$$

For separable differential equations, explicit solutions can be obtained by direct integration. For linear differential equations, the general solution, $x(t) = x_p(t) + cx_h(t)$, is always in the form of a particular solution $x_p(t)$ plus an arbitrary multiple of a homogeneous solution $x_h(t)$. Explicit solutions of first-order linear differential equations can always be obtained using an integrating factor, $\exp(\int p(t)dt)$. If $p(t)$ is a constant (in which case the linear differential equation has constant coefficients) and $q(t)$ is a polynomial, exponential, or sinusoidal function, or a product or sum of these, it is usually easier to obtain a particular solution by the method of undetermined coefficients.

In this chapter we will discuss other properties of first-order differential equations, mostly for differential equations which are nonlinear. In Sections 5.2 and 5.3, we will discuss qualitative properties of solutions of first-order differential equations which are called **autonomous**:

$$\frac{dx}{dt} = f(x). \tag{1}$$

First, we will study equilibrium solutions and their stability. Although (1) is separable, we will introduce elementary geometric methods to describe the qualitative behavior of the solution, obtaining different information than that obtained by integration. Applications of these ideas to nonlinear models of population dynamics will be given in Section 5.4.

5.2 Equilibria and Stability

The first-order differential equation

$$\frac{dx}{dt} = f(x) \tag{1}$$

is called **autonomous** (meaning self-governing). Autonomous equations have the property that the rate of change $\frac{dx}{dt}$ depends only on the value of the unknown solution x and not on time t. Many physical systems have this property. If the nine o'clock laboratory class is expected to get the same experimental results when measuring the speed of a reaction as the ten o'clock class, then the physical experiment can be quite complicated, but the results do not depend on the starting time. We have, of course, assumed that both classes have the same initial experimental configuration, and are measuring the reaction rate at the same concentration of chemicals x.

5.2.1 Equilibrium

A specific solution of an autonomous differential equation,

$$\frac{dx}{dt} = f(x), \tag{2}$$

is called an **equilibrium** or **steady-state** solution if the solution is constant in time (it does not depend on time t):

$$x(t) = a = \text{constant}.$$

In order for a constant to be an equilibrium solution, it must satisfy the differential equation. Since $x = a$ does not depend on time, $\frac{da}{dt} = 0$ and letting $x = a$ in (2) gives $0 = f(a)$. Thus

$$x = a \text{ is an equilibrium of } \frac{dx}{dt} = f(x) \text{ precisely when } 0 = f(a). \tag{3}$$

Thus, all zeros (or roots) of the function $f(x)$ are equilibrium solutions. The equilibrium $x = a$ are determined from $f(a) = 0$.

Think of (2) as modeling a physical process, and assume we are at an equilibrium $x = a$, that is, $x(t_0) = a$. Then $\frac{dx}{dt} = 0$ says that everything is in perfect balance (the forces, pressures, etc. balance) and the process $x(t)$ stays at $x = a$. We say $x = a$ is an equilibrium solution.

Example 5.2.1 *Equilibria*

Find the equilibria of the first-order nonlinear autonomous differential equation

$$\frac{dx}{dt} = x - x^3. \tag{4}$$

• SOLUTION. In order for a constant $x = a$ to be an equilibrium solution, it must satisfy the differential equation (4). Letting $x = a$ in (4) gives

$$0 = x - x^3 = x(1 - x^2).$$

In this case, there are three equilibria solutions:

$$x = 0 \quad \text{and} \quad x = \pm 1.$$

♦

5.2.2 Stability

A physical system in perfect balance may be very difficult to achieve (a pencil standing on its point is unstable) or easy to maintain (a marble at the bottom of a bowl is stable). This section introduces the fundamental idea of stability which plays an important role in such areas as control systems and fluid mechanics. Solutions in perfect balance, that is, equilibrium solutions, may be stable or unstable. What we mean by this is that if the initial conditions are nudged a little bit, changed a small amount from the equilibrium, then the solution of the differential equation may remain near the equilibrium or the differential equation may take the solution away from the equilibrium. There are more precise mathematical definitions of these ideas, but we will be satisfied with the following definition of stability of an equilibrium:

An equilibrium $x = a$ is **stable**:
∗ if all solutions starting near $x = a$ stay nearby and
∗ the closer the solution starts to $x = a$ the closer nearby it stays.

Otherwise the equilibrium is said to be **unstable**. To be stable, the solution should stay near the equilibrium for all nearby initial conditions. For example, if the solution does not stay near the equilibrium for even one nearby initial condition, we say the equilibrium is unstable. Although these ideas are somewhat vague, we will develop more mathematical details shortly.

Sometimes it is helpful to define a stronger sense of stability.

An equilibrium $x = a$ is **asymptotically stable**:
∗ if $x = a$ is stable and
∗ the solution approaches the equilibrium for all nearby initial conditions, that is, $x(t) \to a$ as $t \to \infty$.

There were simple examples in Chapter 4 in which the equilibrium is stable, but not asymptotically stable.

5.2.3 Review of Linearization

To analyze the stability of an equilibrium, we must consider initial conditions that are near to the equilibrium $x = a$. One way to do this analysis is to approximate the differential equation in the neighborhood of the equilibrium. In this way we hope to simplify the differential equation. Mathematically, if we wish to restrict our attention to be near some point, then the function $f(x)$ that appears in the differential equation may be approximated by a polynomial, the Taylor polynomial used in Taylor series analysis. Recall from calculus that the Maclaurin polynomial for approximating a function $f(x)$ around $x = 0$ is

$$f(x) \approx f(0) + f'(0)x + \frac{f''(0)}{2}x^2 + \cdots.$$

(The Maclaurin polynomial is the first few terms of the Maclaurin series for $f(x)$ at $x = 0$.) This can be generalized to approximating a function around $x = a$ and is called the **Taylor polynomial**,

$$f(x) \approx f(a) + f'(a)(x-a) + \frac{f''(a)}{2}(x-a)^2 + \cdots,$$

which is the first few terms of the Taylor series expansion of $f(x)$. The linear term, $f(a) + f'(a)(x-a)$, including the constant $f(a)$, is the **linearization** or **tangent line approximation** to the function $f(x)$ in the neighborhood of the equilibrium $x = a$. Most of the time, when we are near the equilibrium, we will be satisfied with approximating our function by its linearization:

$$f(x) \approx f(a) + f'(a)(x-a). \tag{5}$$

In using these approximations, we assume that f is twice continuously differentiable at and near $x = a$. It is important that $f(a)$, $f'(a)$, and $f''(a)$ are all constant since they are evaluated at $x = a$.

5.2.4 Linear Stability Analysis

An equilibrium $x = a$ is stable or unstable depending on the behavior of the solution of the differential equation

$$\frac{dx}{dt} = f(x) \tag{6}$$

near the equilibrium. Recall that the equilibrium $x = a$ satisfies $f(a) = 0$. We first assume x is near a and, as an approximation, replace $f(x)$ by its linearization:

$$\frac{dx}{dt} = f(a) + f'(a)(x-a). \tag{7}$$

The differential equation (7) simplifies because $x = a$ is an equilibrium, $f(a) = 0$, so that

$$\frac{dx}{dt} = f'(a)(x - a). \tag{8}$$

The differential equation (8) is actually quite simple since $f'(a)$ is a constant. It is linear with constant coefficients. However, that may not be particularly apparent. It is usual to introduce a new dependent variable

$$y \equiv x - a,$$

which is the **displacement from an equilibrium** $x = a$. Letting $x = y + a$, the differential equation (8) becomes

$$\frac{dy}{dt} = f'(a)y, \tag{9}$$

since $x = a$ is a constant and hence $\frac{dy}{dt} = \frac{dx}{dt}$.

Equation (9) is a linear differential equation with constant coefficient $f'(a)$. We easily obtain its general solution to be

$$y = x - a = ce^{f'(a)t}. \tag{10}$$

From (10), we see that the behavior of the solution near the equilibrium depends in a very simple way on the sign of $f'(a)$. If $f'(a) > 0$, the solutions exponentially grow in time away from the equilibrium, and we say that the equilibrium $x = a$ is unstable (sometimes we say the equilibrium $y = 0$ is unstable). If $f'(a) < 0$, the solutions exponentially decay in time toward the equilibrium, in which case we say that the equilibrium $x = a$ is stable (sometimes we say the equilibrium $y = 0$ is stable). In fact, if $f'(a) < 0$, the equilibrium $x = a$ is asymptotically stable, since from (10), $x(t) \to a$ as $t \to \infty$. We call this a **linearized stability analysis**:

Suppose $x = a$ is an equilibrium of $\frac{dx}{dt} = f(x)$.
If $f'(a) < 0$, the equilibrium $x = a$ is stable.
If $f'(a) > 0$, the equilibrium $x = a$ is unstable.
If $f'(a) = 0$, the equilibrium $x = a$ may be stable or unstable.

Example 5.2.2 *Linear Stability*

We reconsider the first-order nonlinear differential equation (4) from Example 5.2.1,

$$\frac{dx}{dt} = f(x) = x - x^3, \tag{11}$$

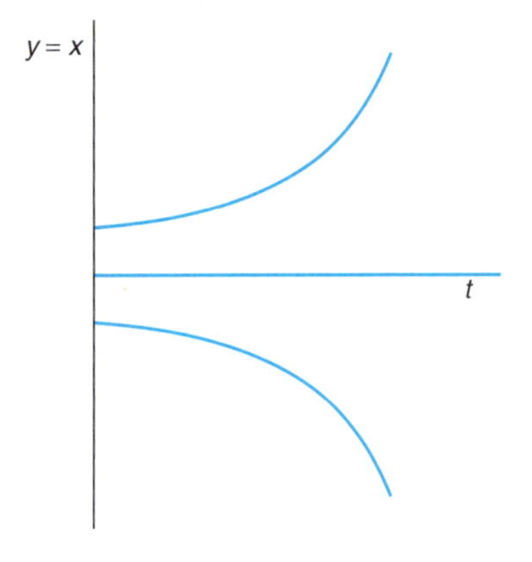

Figure 5.2.1 Near $x = 0$. $x = 0$ is unstable.

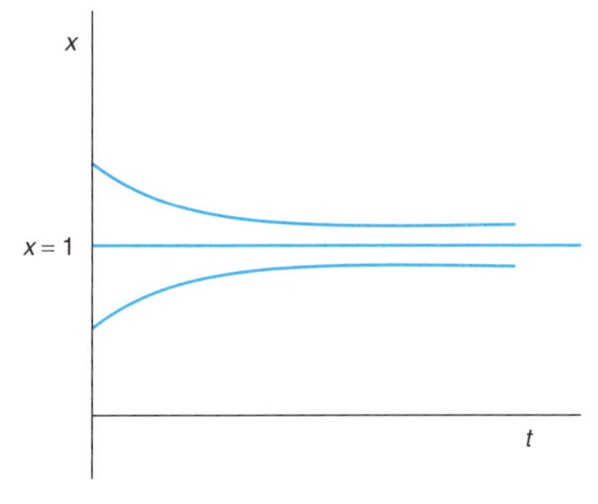

Figure 5.2.2 Near $x = 1$. $x = 1$ is stable.

which has three equilibria, $x = 0$ and $x = \pm 1$. From the linear stability analysis, the stability of the equilibria is determined from the sign of $f'(a)$. In general, $f'(x) = 1 - 3x^2$. Since $f'(0) = 1 > 0$, we conclude that the equilibrium $x = 0$ is unstable since the linearized equation $y' = y$ has time-dependent solutions $y = ce^t$ which grow with t as shown in figure 5.2.1. Figure 5.2.1 shows how x depends on t near $x = 0$ and shows that $x = 0$ is unstable.

The other two equilibria are stable since $f'(\pm 1) = -2 < 0$. They each have linearized equation $y' = -2y$ which has the solution $y = ce^{-2t}$ shown in figure 5.2.2 for the case $x = 1$. Figure 5.2.2 shows how x depends on t near $x = 1$ and shows that $x = 1$ is stable. Here $\lim_{t \to \infty} y = 0$. A similar behavior occurs near $x = -1$.

The solutions to (11) are graphed later in figure 5.3.4. Note in that figure that, although $x = 1$ is stable, if we start far enough from that equilibrium, say $x_0 = -1.5$, then $x(t)$ need not go to the equilibrium $x = 1$. We still refer to $x = 1$ as being stable since stability (as we describe it) only depends on nearby initial conditions. ♦

Validity of Linearization and the Borderline Cases (Neutral Stability)

If $f'(a) = 0$, the equilibrium $x = a$ is on the borderline between the clearly stable and unstable cases. This case is of less importance in a first exposure to the concepts of stability, and it is somewhat subtle and may lead to some confusion. Technically, for the *linear* differential equation (9), since $f'(a) = 0$, the linearized differential equation is $y' = 0$. The equilibrium $y = 0$ is stable (but not asymptotically stable), since from the solution of the linear differential equation (10), $y(t) = c$ or $x(t) = a + c$. In other words, if the solution starts near the equilibrium, it doesn't move at all according to the *linear* differential equation since $y' = 0$. (It thus stays near the equilibrium according to the *linear* equation since it is assumed to start near the equilibrium.)

It will be shown in the next section that our conclusions concerning the stability of an equilibrium $x = a$ based on the linearization are always valid for the real nonlinear problem in the clear cases $f'(a) > 0$ and $f'(a) < 0$. To be more explicit, the equilibrium $x = a$ is unstable if $f'(a) > 0$ and stable if $f'(a) > 0$. However, in the borderline case between stable and unstable in which $f'(a) = 0$, the stability of the real nonlinear problem may differ from the simplistic conclusion reached from its linearization. When $f'(a) = 0$, we will show that the equilibrium solution of the real problem may be stable or unstable. Its stability is not determined by the linearization, but is determined instead by keeping the first nonzero higher-order nonlinear terms in the Taylor polynomial approximation of $f(x)$ around its equilibrium. This case which is linearly stable (but not asymptotically stable) $f'(a) = 0$ is sometimes called **neutrally stable**.

Summary of Linear Stability Analysis

We summarize these results which we refer to as a **linearized stability analysis.**

Suppose $x = a$ is an equilibrium of $\frac{dx}{dt} = f(x)$.
If $f'(a) < 0$, the equilibrium $x = a$ is stable.
If $f'(a) > 0$, the equilibrium $x = a$ is unstable.
If $f'(a) = 0$, the equilibrium $x = a$ may be stable or unstable depending on the nonlinear terms

Exercises

In Exercises 1–7, determine all equilibria.

1. $\dfrac{dx}{dt} = x(x-1)$.
2. $\dfrac{dx}{dt} = x + x^3$.
3. $\dfrac{dx}{dt} = (x-1)(x-2)^2$.
4. $\dfrac{dx}{dt} = (x^2-1)(x^2-3)$.
5. $\dfrac{dx}{dt} = \sin x$.
6. $\dfrac{dx}{dt} = \cos x$.
7. $\dfrac{dx}{dt} = x(a - bx)$, where a and b are positive constants.

In Exercises 8–15, determine all equilibria and their linearized stability. If linear stability is valid, solve the linearized differential equation and graph this time-dependent solution. If the linearized analysis is inconclusive, say so.

8. $\dfrac{dx}{dt} = x(x-1)$.
9. $\dfrac{dx}{dt} = x(x-1)^2$.
10. $\dfrac{dx}{dt} = (x-1)(x-2)^2$.

11. $\dfrac{dx}{dt} = (x^2 - 1)(x^2 - 3)$.
12. $\dfrac{dx}{dt} = \sin x$.
13. $\dfrac{dx}{dt} = \tan x$.
14. $\dfrac{dx}{dt} = e^{3x} - 7$.
15. $\dfrac{dx}{dt} = e^{-3x} - 5$.

5.3 One-Dimensional Phase Lines

In the previous section, the nonlinear autonomous first-order differential equation

$$\frac{dx}{dt} = f(x) \tag{1}$$

was discussed by analyzing the linear stability of equilibrium solutions. It is usually easy to determine whether each equilibrium is stable or unstable. However, the linear analysis only describes the behavior of the solution near the equilibrium. In the present section, we will introduce a geometric method to understand the behavior of solutions of (1) which is valid not only near an equilibrium but also far away from all equilibria.

The differential equation (1) states that $\frac{dx}{dt}$ is a function of x. That relationship can often be graphed easily since $f(x)$ is given. A typical functional relationship is graphed in figure 5.3.1. At time t, the solution $x(t)$ is a point on the x-axis, and as time increases, $x(t)$ moves along the x-axis. We introduce arrows to the right ($\rightarrow$) if $x(t)$ is an increasing function of time t, and arrows to the left ($\leftarrow$) if $x(t)$ is decreasing. The differential equation easily determines how these arrows should be introduced. In the upper half-plane, where $f(x) > 0$, $\frac{dx}{dt} > 0$, and it follows that x is an increasing function of time. So the arrows in the graph point to the right ($\rightarrow$), indicating that $x(t)$ moves to the right as time increases. Similarly, where $f(x) < 0$, in the lower half-plane, $\frac{dx}{dt} < 0$, and the arrows show motion to the left ($\leftarrow$) as time increases since x is a decreasing function of time there. These results can be summarized in the **one-dimensional phase line** sketched in figure 5.3.1.

Places where the curve intersects the x-axis are the equilibrium points since $f(x) = 0$ and $\frac{dx}{dt} = 0$ there. Not only can we determine graphically where the equilibrium is located, but it is easy to determine in all cases if the equilibrium is stable or unstable. If the arrows indicate that all solutions near an equilibrium move toward the equilibrium as time increases, then the equilibrium is stable, as marked in figure 5.3.1. If some solutions near an equilibrium move away from that equilibrium as time increases, then that equilibrium is unstable.

The stability of an equilibrium can be easily determined from the one-dimensional phase line, whether or not the stability of the equilibrium can be determined by linearization. In the situation assumed in figure 5.3.1, two of the equilibria are clearly stable and one unstable, as marked in the figure. If the slope of the curve is positive when it crosses the x-axis at the equilibrium ($f'(a) > 0$), then the equilibrium is

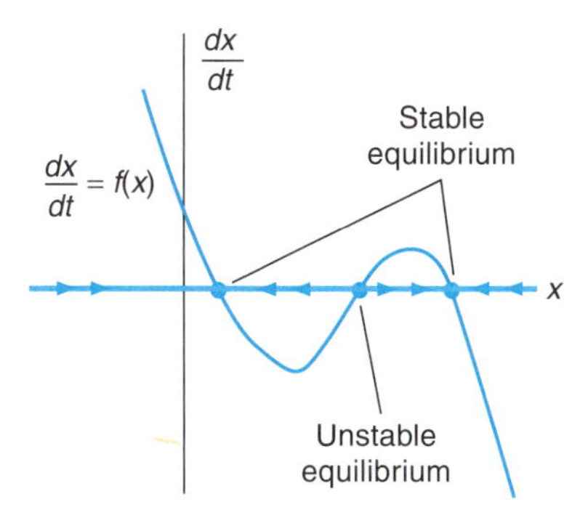

Figure 5.3.1 Phase line.

unstable. If the slope of the curve is negative when it crosses the x-axis ($f'(a) < 0$), then the equilibrium is stable. Thus, the geometric analysis has verified the conclusions of the linear stability analysis in the usual simple cases in which $f'(a) \neq 0$. In the example graphed in figure 5.3.1, we assumed that the three equilibria all satisfied the simple criteria. However, as is developed further in the Exercises, this geometric method can also be used to determine the stability of an equilibrium in the more difficult case in which the linearization does not determine the stability ($f'(a) = 0$).

Furthermore, we have easily determined the qualitative behavior of the solution to the differential equation corresponding to all initial conditions. We know qualitatively how the solution approaches an equilibrium if the solution starts near an equilibrium, and we even know how the solution behaves when it starts far away from an equilibrium. If there is more than one stable equilibrium, we can determine which initial conditions approach which equilibrium.

We discuss two illustrative examples.

Example 5.3.1 *Two Stable Equilibria*

As a specific example to illustrate the geometric method of a one-dimensional phase line, reconsider the first-order nonlinear differential equation

$$\frac{dx}{dt} = f(x) = x - x^3 = -x(x^2 - 1) = -x(x-1)(x+1),$$

which has three equilibria, $x = 0$ and $x = \pm 1$. The phase line requires the graph of the cubic $f(x) = x - x^3$, which is given in figure 5.3.2. The graph sketched in figure 5.3.2 has two critical points at $x = \pm\frac{1}{\sqrt{3}}$, one a local maximum and the other a local minimum. The one-dimensional phase line shows that the equilibrium $x = 0$ is clearly unstable, while the equilibria $x = \pm 1$ are clearly stable. There are two stable equilibria. Since $f'(0) = 1 > 0$, we conclude from a linear stability analysis that the equilibrium $x = 0$ is unstable. The other two equilibria are stable, since $f'(\pm 1) = -2 < 0$. This shows that the stability of the equilibrium determined from the phase line agrees with the linear stability analysis.

We see that we do not need to sketch $f(x)$ in order to determine the phase line. All we need is the sign of $f(x)$, perhaps from an elementary sign chart as shown in figure 5.3.3.

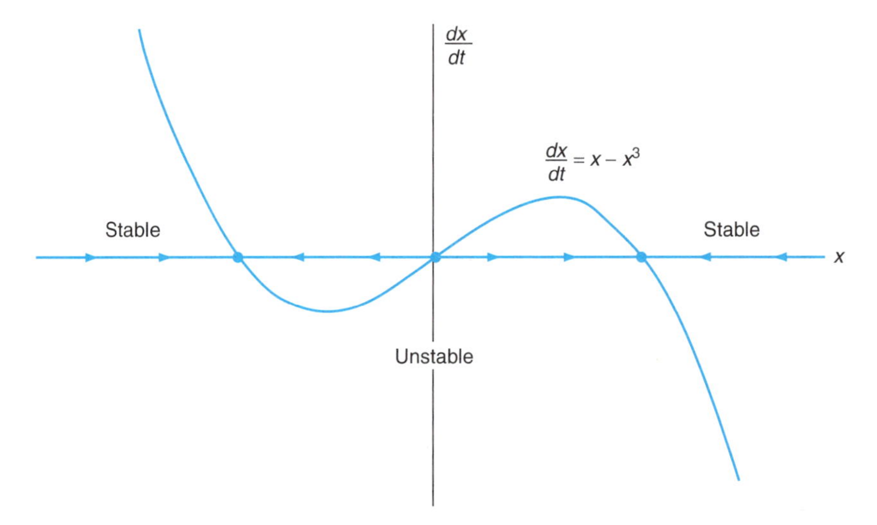

Figure 5.3.2 Phase line.

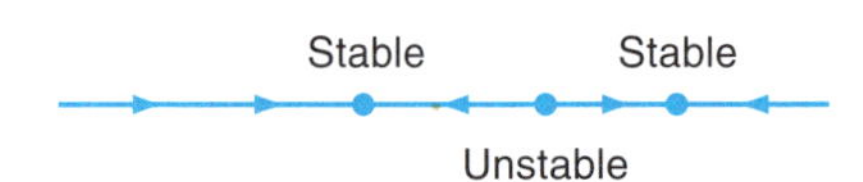

Figure 5.3.3 Phase line from sign of $f(x)$.

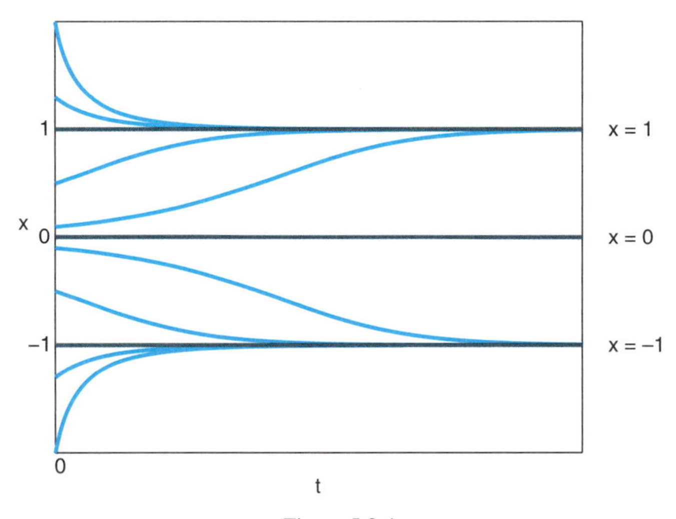

Figure 5.3.4

In figure 5.3.4, we have sketched the qualitative behavior of x as a function of time for various initial conditions. (This was called the direction field in Section 1.4.) We begin with three equilibrium, $x = 0, +1, -1$. For example, we see that the solution approaches 1 as $t \to +\infty$ for all initial conditions which satisfy $x(0) > 0$. For initial conditions such that $x(0) < 0$, $x(t) \to -1$ as $t \to +\infty$. For the initial condition $x(0) = 0$, the unstable equilibrium, the solution stays at the equilibrium for all time,

$x(t) = x(0) = 0$. $x = \pm 1$ are stable. From this method we can only determine the qualitative behavior of these curves. The precise dependence of x on t can be determined by integration of this separable equation (or accurately approximated using contemporary numerical methods). ♦

The advantage of the present geometric method is the simplicity by which we obtain an understanding of fundamental relationships of equilibrium, stability, and the manner in which the solution depends on the initial conditions. If the differential equation is more complicated, for example, by depending on some parameter, then the qualitative method based on the one-dimensional phase line is particularly effective. If more quantitative details are needed, the one-dimensional phase line can always be supplemented by further analytic work or numerical computations.

Two graphs of solutions cannot intersect. If two different solutions intersected in a finite time, it would violate the uniqueness theorem of the initial value problem because there would be two solutions with the same initial conditions. This proves, for example, that the solution takes an infinite amount of time to reach $x = 1$, since $x = 1$ is a solution of the differential equation.

In general, for first-order autonomous equations only very simple qualitative behavior is possible. If a solution approaches an equilibrium, it does so directly without any oscillations.

Example 5.3.2 *An Explosion*

Let us consider the simple nonlinear autonomous first-order differential equation

$$\frac{dx}{dt} = x^2. \tag{2}$$

• SOLUTION. Although this problem can be explicitly solved relatively easily by separation (see the Exercises), it is interesting to note that the one-dimensional phase line is a quick and easy method that probably gives a better **understanding** of some of the behavior of the solutions to this differential equation. In figure 5.3.5, $\frac{dx}{dt}$ is easily graphed as a function of x. From this figure, we see immediately that $x = 0$ is an unstable equilibrium. If the initial conditions are less than zero, $x(0) < 0$, then the solution approaches 0 as $t \to \infty$. (It cannot reach the equilibrium, $x = 0$, in a finite time because of the uniqueness theorem.) For these initial conditions, the equilibrium appears stable. However, to be stable, the solutions for all nearby initial conditions must approach the equilibrium, not just those initial conditions less than the equilibrium. From figure 5.3.5, we see that $x(t)$ goes away from zero for initial conditions that are positive, $x(0) > 0$. Thus, $x = 0$ is an unstable equilibrium. Furthermore, for these initial conditions, $x(0) > 0$, it is clear that the solution approaches infinity, $x(t) \to \infty$. However, from the phase line one cannot know that the solution explodes in a finite time (has a vertical asymptote). This can most easily be seen from the explicit solution obtained by separation in Exercise 20.

The stability of equilibria $x = a$ is usually determined by the linearization. However, in this example, $f(x) = x^2$, so that $f'(x) = 2x$, and $f'(0) = 0$. The linearization approximation in the neighborhood of the equilibrium $x = 0$ is

$$\frac{dx}{dt} = f'(0)x = 0x = 0.$$

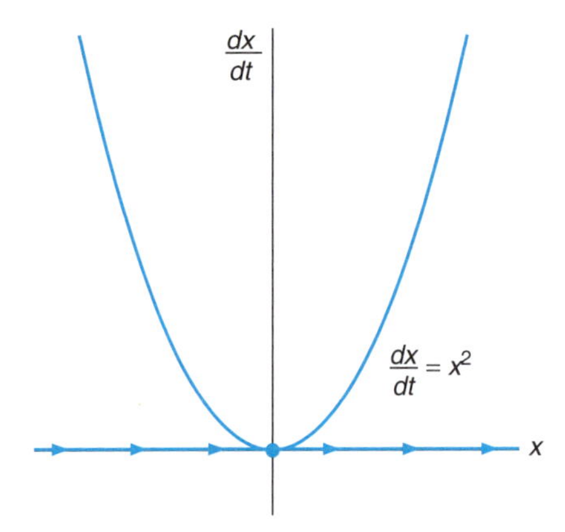

Figure 5.3.5 Phase line.

In this case, $f'(a)=0$, and the linearization is on the border between being stable and unstable. The linearization is neutrally stable, and the stability of $x=0$ cannot be determined from the linearization. We have shown above that the equilibrium $x=0$ is unstable based on the geometric one-dimensional phase line analysis of the nonlinear differential equation. ♦

Exercises

In Exercises 1–11, graph the one-dimensional phase lines. In doing so, determine all equilibria and their stability.

1. $\frac{dx}{dt}=x(x-1)$.
2. $\frac{dx}{dt}=x(x-1)^2$.
3. $\frac{dx}{dt}=(x-1)(x-2)^2$.
4. $\frac{dx}{dt}=(x^2-1)(x^2-3)$.
5. $\frac{dx}{dt}=\sin x$.
6. $\frac{dx}{dt}=\tan x$.
7. $\frac{dx}{dt}=x^2+1$.
8. $\frac{dx}{dt}=-x^2$.
9. $\frac{dx}{dt}=x^3$.
10. $\frac{dx}{dt}=-x^3$.

11. $\dfrac{dx}{dt} = (x-3)^4$.

In Exercises 12–13 determine stability using the one-dimensional phase line.

12. $\dfrac{dx}{dt} = (x-1)(x-2)(x-3)$.

13. $\dfrac{dx}{dt} = (x-1)(x-2)(x-3)(x-4)$.

In Exercises 14–17, sketch time-dependent solutions based on the one-dimensional phase line.

14. $\dfrac{dx}{dt} = x(x-1)$.

15. $\dfrac{dx}{dt} = x(x-1)^2$.

16. $\dfrac{dx}{dt} = (x-1)(x-2)^2$.

17. $\dfrac{dx}{dt} = (x^2-1)(x^2-3)$.

18. In this problem, we will consider equilibria $x=a$ for the differential equation $\frac{dx}{dt} = f(x)$ in the case $f'(a)=0$.
  a) Determine the stability of $x=a$, if $f'(a)=0$, but $f''(a)\neq 0$.
  b) Determine the stability of $x=a$, if $f'(a)=0$, $f''(a)=0$, but $f'''(a)\neq 0$.
  c) Generalize the results of part (b) to even and odd n.
  d) Generalize the results of part (b) based on sign changes of $f(x)$.
  e) Simplify the criteria in (d) by applying the first derivative test for a maximum or minimum to $-\int f(x)dx$.
19. Use separation to determine the stability of $x=0$ for $\frac{dx}{dt}=x^2$. Compare your results to those obtained in this section.
20. Use separation to show that solutions of $\frac{dx}{dt}=x^2$ explode in a finite time if the initial condition is positive.
21. Use separation to show that solutions of $\frac{dx}{dt}=x^2$ do not have a vertical asymptote but are well behaved for all time if the initial condition is negative. Show that the solution approaches $x=0$, but never reaches $x=0$ in a finite time.

5.4 Application to Population Dynamics: The Logistic Equation

The simplest model for the growth of a population x results from assuming the growth rate k is a positive constant

$$\frac{dx}{dt} = kx.$$

In this case, the population exponentially grows, $x(t)=x(0)e^{kt}$, and eventually becomes indefinitely large. A more realistic model of population growth is needed when the population becomes sufficiently large. Experiments having shown that there is a crowding effect. The growth rate $\frac{1}{x}\frac{dx}{dt}$ is not a constant, but depends on the population itself, diminishing as the population increases. For simplicity, we choose to model the growth rate itself as a linear function of the population:

$$\frac{1}{x}\frac{dx}{dt} = a - bx.$$

Equivalently we have

the logistic equation:

$$\frac{dx}{dt} = x(a - bx) = ax - bx^2. \tag{1}$$

Equation (1) is a first-order nonlinear differential equation which is autonomous. The constants a and b are assumed to be positive. The growth rate, $a - bx$, depends on the population x. When the population is zero, the growth rate is a. This represents the growth rate without the crowding effect of environmental influences (and corresponds to the growth rate we called k previously). The constant b is the decrease in the growth rate per individual. According to this model, if the population becomes sufficiently large, the growth rate becomes negative. This model was first investigated in the late 1830s by Verhulst and is known as the **logistic equation**.

Although the logistic equation (1) is separable, the qualitative behavior of the solutions can be more readily determined using the one-dimensional phase line and other ideas of this chapter. To analyze a nonlinear equation such as (1), we first look for equilibrium populations (equilibrium solutions of the differential equation). The rate of change of the population will be zero if $0 = x(a - bx)$. The population can be in equilibrium in two different ways:

$$x = 0 \quad \text{and} \quad x = \frac{a}{b}.$$

Zero population is certainly an equilibrium population. If the population is initially zero, it will stay zero. The other equilibrium, $x = a/b$, is the largest population which the environment can sustain without loss, since the growth rate will be negative if $x > \frac{a}{b}$. This equilibrium $x = \frac{a}{b}$ is called the **carrying capacity** of the environment.

The stability of the equilibrium populations can be determined by a linearized stability analysis. The nonlinear differential equation $\frac{dx}{dt} = f(x) = ax - bx^2$ can be approximated by its linearization in the neighborhood of each equilibrium,

$$\frac{dy}{dt} = f'(x_e)y,$$

where $y = x - x_e$ is the displacement from an equilibrium and x_e denotes an equilibrium. In this case, $f'(x) = a - 2bx$. The zero population is unstable since $f'(0) = a > 0$. The small populations we consider here grow exponentially. However, the environment's carrying capacity $x = \frac{a}{b}$ is stable since $f'\left(\frac{a}{b}\right) = a - 2b \cdot \frac{a}{b} = -a < 0$.

To obtain the qualitative behavior of the populations away from the equilibria, we first sketch the one-dimensional phase line in figure 5.4.1. The derivative $\frac{dx}{dt}$ is a function of x, and the graph of this dependence is a parabola with intercepts at the

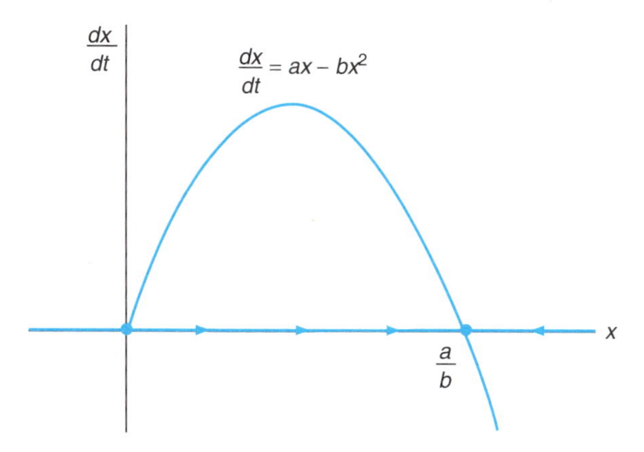

Figure 5.4.1 Phase line.

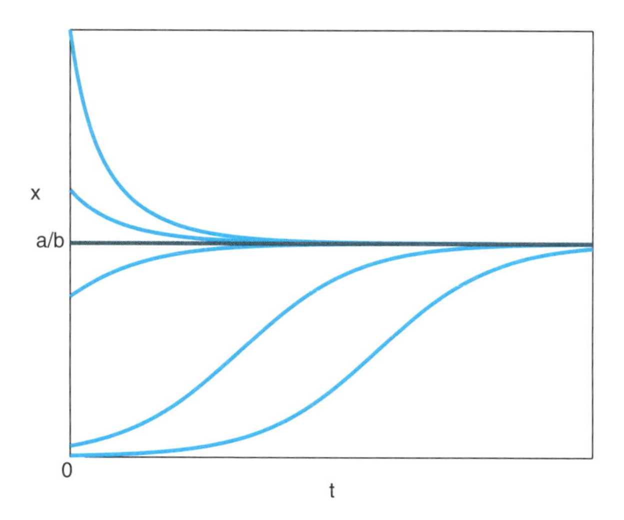

Figure 5.4.2 Solutions of the logistic equation.

equilibria $x = 0$ and $x = \frac{a}{b}$. Since x is a population, we are only concerned with $x \geq 0$. Arrows are introduced, indicating how the solution changes in time. In the upper half-plane, arrows point to the right since x is increasing there. In the lower half-plane arrows, point to the left. The instability of $x = 0$ and the stability of $x = a/b$ are seen. For populations less than the carrying capacity, the population increases as a function of time toward the carrying capacity (but the population never actually reaches the carrying capacity). For initial populations greater than the carrying capacity, the population decreases toward the carrying capacity without reaching it. The qualitative sketch of the population as a function of time is shown in figure 5.4.2 for various initial populations. Note that the equilibrium populations $x = 0$ and $x = \frac{a}{b}$ are horizontal lines (since they are constant). The analysis of possible inflection points is left for an exercise.

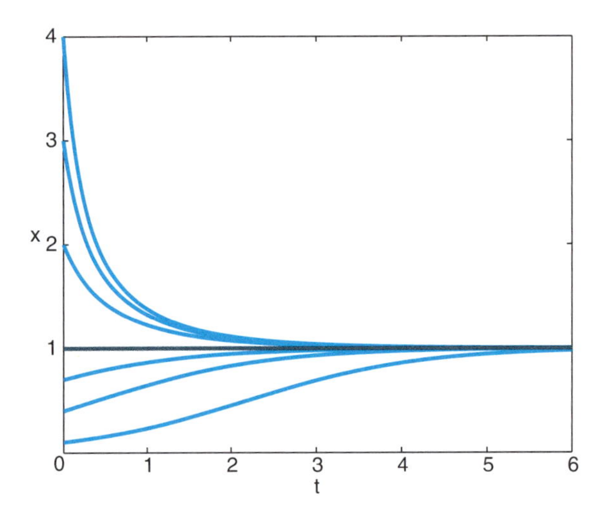

Figure 5.4.3 Graph of (2) with $a = b = 1$.

EXPLICIT SOLUTION OF THE LOGISTIC EQUATION BY SEPARATION. An explicit solution of the logistic equation can be obtained since the equation is separable. After using partial fractions and a considerable amount of algebra, we find that if $x(0) \neq 0$, then

$$x(t) = \frac{\frac{a}{b}}{1 + \left[\frac{a - bx(0)}{bx(0)}\right] e^{-at}}, \qquad (2)$$

in terms of the initial population $x(0)$. In the exercises, you are asked to show how this explicit solution verifies the qualitative results obtained from the one-dimensional phase line. Specific logistic curves depend on three parameters a, b, and $x(0)$. A few examples are sketched in figure 5.4.3 and are seen to be consistent with the qualitative behavior shown in figure 5.4.2.

Exercises

1. Consider $\frac{dx}{dt} = ax - bx^3$ (with $a > 0$, $b > 0$, and $x \geq 0$).
   a) How does the growth rate depend on the population?
   b) Sketch the one-dimensional phase line.
   c) Sketch the time-dependent solution qualitatively.
2. Suppose that we have a model of population growth which has a carrying capacity, in the sense that the growth rate is negative for populations greater than the carrying capacity and the growth rate is positive for populations less than the carrying capacity. Show that the qualitative behavior of this model will be the same as the qualitative behavior of the quadratic logistic model analyzed in this section.
3. Derive the general solution (2) from the differential equation (1) by separation.
4. Show from the exact solution (2) that by rescaling the population there are only two parameters.

5. Show from the differential equation (1) that by rescaling the population there are only two parameters.
6. Consider $\frac{dx}{dt} = ax^2 - bx$ (with $a > 0$, $b > 0$, and $x \geq 0$).
a) How does the growth rate depend on the population?
b) Sketch the one-dimensional phase line.
c) Sketch the time-dependent solution qualitatively.
d) Obtain the exact solution.
e) Show that the exact solution has the qualitative behavior obtained in (c).
7. Consider $\frac{dx}{dt} = ax + bx^2$ (with $a > 0$, $b > 0$, and $x \geq 0$).
a) How does the growth rate depend on the population?
b) Sketch the one-dimensional phase line.
c) Sketch the time-dependent solution qualitatively.
d) From the exact solution, show that the population reaches infinity in a finite time; what might be called a population explosion.
8. From the differential equation (1), investigate the concavity of $x(t)$. Where will an inflection point occur? Compare to the sketches in figure 5.4.2.
9. The logistic curve (2) is sometimes referred to as an S-curve for reasons to be described.
a) Show that

$$x(t) = \alpha + \beta \frac{e^{\frac{a}{2}(t-t_0)} - e^{-\frac{a}{2}(t-t_0)}}{e^{\frac{a}{2}(t-t_0)} + e^{-\frac{a}{2}(t-t_0)}}. \tag{3}$$

What are α, β, and t_0? (Hint: put (3) over a common denominator. Multiply numerator and denominator by $e^{-\frac{a}{2}(t-t_0)}$.)
b) Recall that the hyperbolic functions are defined as follows:

$$\sinh t = \frac{e^t - e^{-t}}{2}, \qquad \cosh t = \frac{e^t + e^{-t}}{2}, \qquad \tanh t = \frac{\sinh t}{\cosh t}.$$

Show that $x(t) = \alpha + \beta \tanh \frac{a}{2}(t - t_0)$.
c) Sketch $\tanh t$ as a function of t. Show that it might be called S-shaped. Consider the asymptotic behavior as $t \to \pm\infty$.
d) Now sketch the logistic curve.
e) Show that $\alpha + \beta = \frac{a}{b}$ and $\alpha - \beta = 0$.
10. Let $Q(t)$ be the population of a particular species at time t. Suppose that the rate of population growth depends only on the population size, so that

$$\frac{dQ}{dt} = f(Q). \tag{4}$$

Explain, in biological terms, why each of the following three statements could be a reasonable assumption.
a) $f(0) = 0$.
b) $f'(Q) > 0$ for small positive Q.
c) $f'(Q) < 0$ for large Q.
11. Suppose that $f(Q) = a + bQ + cQ^2$ is a second-degree polynomial in Q. Show that if f satisfies (a), (b), and (c) of Exercise 10, then $a = 0$, $b > 0$, and $c < 0$, so that (4) becomes the logistic equation.

CHAPTER 6

Nonlinear Systems of Differential Equations in the Plane

6.1 Introduction

In Chapter 5, we introduced the analysis of a single first-order differential equation $\frac{dx}{dt} = f(x)$. We discussed equilibrium and the stability of the equilibrium in that chapter. The current chapter will discuss similar questions for systems of **autonomous differential equations** of the form

$$\frac{dx}{dt} = f(x, y), \tag{1a}$$

$$\frac{dy}{dt} = g(x, y). \tag{1b}$$

This chapter is self-contained in the sense that the material from Chapter 5 is not necessary. However, Chapter 4 is essential for our presentation, in which we discussed the linear systems

$$\frac{dx}{dt} = ax + by \tag{2a}$$

$$\frac{dy}{dt} = cx + dy \tag{2b}$$

and the phase plane for linear systems. In Section 6.2, we will show that the phase plane near an equilibrium for the nonlinear system (1a)–(1b) usually looks like the phase plane for the corresponding linear system.

Before beginning this analysis in the next section, we shall give several examples of nonlinear systems, some of which will be examined more carefully later in this chapter.

Electric Circuits

(Note Sections 1.10 and 2.8.) In many devices, such as diodes, the voltage v and current i satisfy a nonlinear relationship. That is, the device has a nonlinear v-i characteristic. If the device is current-controlled, that is, if the voltage drop is a

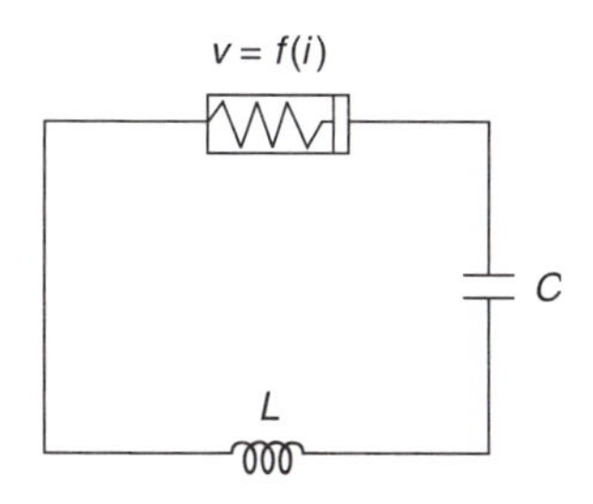

Figure 6.1.1

function of current, we get circuit models of the form shown in figure 6.1.1 which, by Kirchoff's voltage law, lead to the differential equation

$$L\frac{di}{dt} + f(i) + \frac{1}{C}q = 0, \tag{3}$$

or equivalently, the system

$$\frac{di}{dt} = -\frac{1}{L}f(i) - \frac{1}{LC}q, \tag{4a}$$

$$\frac{dq}{dt} = i, \tag{4b}$$

where i is the current in the loop and q is the charge on the capacitor.

Mechanical Systems

(Note Sections 1.11, 2.5, 2.7.) If we no longer assume small velocities and small displacements, we expect the resistance and spring force to vary in a nonlinear manner. This would lead to a spring-mass system with the differential equation

$$m\frac{d^2x}{dt^2} + f\left(\frac{dx}{dt}\right) + g(x) = 0, \tag{5}$$

or the equivalent system

$$\frac{dx}{dt} = y, \tag{6a}$$

$$\frac{dy}{dt} = -\frac{1}{m}f(y) - \frac{1}{m}g(x). \tag{6b}$$

Chemical Reactions or Populations

Chemical reactions involving two substances with concentrations x and y, which can not only combine but also disassociate, lead to differential equations of the form

$$\frac{dx}{dt} = ax + bxy + cy + d, \tag{7a}$$

$$\frac{dy}{dt} = ex + fxy + gy + h, \tag{7b}$$

where a, b, c, d, e, f, g, h are constants.

A closely related idea is that of populations (which could be nonbiological) that are either competing with or feeding on each other. Under the appropriate modeling assumptions, these lead to differential equations of the general form

$$\frac{dx}{dt} = ax + bxy + cx^2, \tag{8a}$$

$$\frac{dy}{dt} = cy + exy + fy^2, \tag{8b}$$

with a, b, c, d, e, f constants. Specific population examples will be given later.

AIDS Models of Perelson

Alan Perelson has developed some mathematical models using differential equations for how the HIV virus develops in humans. Perelson gained understanding over a long period of time by being exceptionally perceptive and by understanding numerous human experiments. He presented some of his work in 1999 with Patrick W. Nelson in "Mathematical Analysis of HIV-I Dynamics in Vivo", *SIAM Review* 41, 1(1999); 3–44. We give some of his basic ideas. Perelson continues to work on AIDS, so the interested reader should look at his more recent work.

Perelson has over many years developed many models for AIDS. A philosophical difficulty he faced is that the more accurate his mathematical model, the more complicated it was, and the less likely the best doctors in the world would follow it. So Perelson delicately and successfully walked this tightrope. We have changed only a few details for simplicity in our presentation. One of the simplest (yet effective) models for humans at first involves three populations:

$$\frac{dT}{dt} = s + aT - rT^2 - kVT, \tag{9a}$$

$$\frac{dT^*}{dt} = kVT - bT^*, \tag{9b}$$

$$\frac{dV}{dt} = NT^* - cV. \tag{9c}$$

In the presence of HIV, there are two types of target cells, uninfected and productively infected. Here T is the number of uninfected cells, T^* is the number of productively infected cells, V is the population of the HIV virus, and s represents the rate at which uninfected target cells are created from sources within the body, such as the thymus. The constant coefficients a and r are not well understood for humans, but represent proliferation of the logistic type described in Section 5.4. The most common way of modeling infection is by "mass-action," in which the uninfected cells diminish at a rate proportional to the number of infected cells (the more infected cells, the greater the decrease of uninfected cells), with proportionality constant k. The constant c is the natural death rate of the virus. The virus population grows proportional (with proportionality constant N) to the number of infected cells.

Perelson has shown the uninfected cells change very slowly compared to the virus and infected cells. So as a simplification, in some circumstances, Perelson replaces (9) with a simpler system with just two populations, where the number of uninfected cells in a person is approximated by a constant T_0:

$$\frac{dT^*}{dt} = kVT_0 - bT^*, \tag{10a}$$

$$\frac{dV}{dt} = NT^* - cV. \tag{10b}$$

This a linear system, whereas other models of his are nonlinear. These equations for humans have four parameters, kT_0, b, c, N. Unfortunately, none of the parameters are known for individual humans, and they vary between individual humans. However, these parameters have been determined by Perelson by finding the best fit between experimental measurements and numerical solutions of these two equations. The comparison for individual patients is so good that it has given Perelson confidence in the validity of these equations.

This is a prelude to the main contribution of Perelson. One goal is to understand the effectiveness of drugs (for HIV or any other disease). The main question is to determine whether a given drug is effective. The traditional way drug companies do this is by doing tests on a large number of people. This takes a great deal of time and is very risky to patients in the experimental stage, especially for drugs that are determined not effective in this way. Drug trials on humans are only attempted after trials with animals.

Perelson has realized that a different approach can be applied to a small number of individual infected patients. Experiments are run before a drug is introduced to determine the individual patient's parameters. Roughly speaking, after a patient takes a drug, the differential equations should be modified to be

$$\frac{dT^*}{dt} = kVT_0 - bT^*, \tag{11a}$$

$$\frac{dV}{dt} = NT^* - (c+d)V, \tag{11b}$$

where d is the rate at which the drug is effective at killing the virus. Unfortunately, we don't know the effectiveness d of the drug. Perelson determines effectiveness by doing experiments on a patient with the drug, and making a best fit with only one unknown parameter d between the experimental measurement and numerical solutions of (11). Perelson believes drug makers in the future will use this method to test effectiveness of new drugs. Drugs don't always work mathematically as we have described, so Perelson and others investigate more realistic models. We use Perelson's AIDS models as one motivation for our study of nonlinear systems of differential equations.

6.2 Equilibria of Nonlinear Systems, Linear Stability Analysis of Equilibrium, and the Phase Plane

In this section, we study the nonlinear autonomous system of differential equations

$$\frac{dx}{dt} = f(x, y), \tag{1a}$$

$$\frac{dy}{dt} = g(x, y). \tag{1b}$$

Examples have been given in the introduction to this chapter.

Equilibrium Solutions

If $x(t)$, $y(t)$ is an **equilibrium** or constant solution of (1a)–(1b), then $x(t)=r$, $y(t)=s$ for constants r, s. Substituting into (1a)–(1b), we find that the equilibrium solutions r, s must satisfy

$$0=f(r,s), \tag{2a}$$
$$0=g(r,s). \tag{2b}$$

These nonlinear algebraic equations may be used to determine the equilibrium for a given system of differential equations. For linear systems $x=0$, $y=0$ is an equilibrium, so that is not much of an issue.

Example 6.2.1 *Equilibrium*

Find all equilibria of

$$\frac{dx}{dt}=-x+xy, \tag{3a}$$
$$\frac{dy}{dt}=-4y+8xy. \tag{3b}$$

• SOLUTION. Let $x(t)=r$, $y(t)=s$, where r, s are constants. Thus (3a)–(3b) becomes

$$0=-r+rs=r(-1+s), \tag{4a}$$
$$0=-4s+8rs=4s(-1+2r). \tag{4b}$$

Equation (4a) implies $r=0$ or $s=1$. Then

$$\text{If } r=0,\text{ (4b) implies that } s=0.$$
$$\text{If } s=1,\text{ (4b) implies } -1+2r=0,\text{ so that } r=\tfrac{1}{2}$$

Thus there are two equilibria for (3a)–(3b):

$$x=0,\ y=0 \quad \text{and} \quad x=\tfrac{1}{2},\ y=1. \tag{5}$$

♦

6.2.1 Linear Stability Analysis and the Phase Plane

Once an equilibrium is found, we will develop here ideas that determine whether the equilibrium is stable or unstable. We now wish to determine the behavior of the solutions of (1a) and (1b) near an equilibrium. Suppose that $x(t)=r$, $y(t)=s$ is an equilibrium. In Chapter 5 we analyzed first-order nonlinear autonomous differential equations near an equilibrium and used a Taylor series of one variable to determine the stability of an equilibrium. For our system (1a)–(1b), we use the two-dimensional version of Taylor's approximations learned in calculus, and we get

$$f(x,y)\approx f(r,s)+f_x(r,s)(x-r)+f_y(r,s)(y-s), \tag{6a}$$
$$g(x,y)\approx g(r,s)+g_x(r,s)(x-r)+g_y(r,s)(y-s), \tag{6b}$$

which is **only valid as an approximation near the equilibrium** $x = r$, $y = s$. This provides the linearization of a function of two variables. It is very important that the partial derivatives are evaluated at the equilibrium, and that they are just constants. The subscript notation for partial derivatives has again been used. For example, $f_x = \frac{\partial f(x,y)}{\partial x}$. The approximation (6) uses only the linear or first-order terms. A more accurate approximation can be gotten by using terms with higher powers of $x - r$ and $y - s$. We neglect these higher-order nonlinear terms for now and discuss later the importance of the neglected nonlinear terms. Since (r, s) is an equilibrium, it satisfies (2a) and (2b), $f(r, s) = 0$, $g(r, s) = 0$. This suggests that, near the equilibrium, the solutions of (1a) and (1b) can be approximated and resemble those of the following **linearized system of differential equations**:

$$\frac{dx}{dt} = f_x(r, s)(x - r) + f_y(r, s)(y - s), \tag{7a}$$

$$\frac{dy}{dt} = g_x(r, s)(x - r) + g_y(r, s)(y - s). \tag{7b}$$

We introduce the displacement from the equilibrium (as we did for first-order equations):

$$z = x - r, \tag{8a}$$

$$w = y - s \tag{8b}$$

(which translates the equilibrium point r, s to the origin $z = 0$, $w = 0$). In this way, since r and s are constants, (7a) and (7b) become

$$\frac{dz}{dt} = az + bw, \tag{9a}$$

$$\frac{dw}{dt} = cz + dw, \tag{9b}$$

where a, b, c, d are constants given by

$$a = f_x(r, s), b = f_y(r, s), c = g_x(r, s), d = g_y(r, s). \tag{10}$$

But (9a)–(9b) is a linear homogeneous system like that discussed in Chapter 4.

JACOBIAN MATRIX. Using matrix multiplication, the system (9a)–(9b) can be written as

$$\frac{d}{dt}\begin{bmatrix} z \\ w \end{bmatrix} = \begin{bmatrix} f_x(r, s) & f_y(r, s) \\ g_x(r, s) & g_y(r, s) \end{bmatrix} \begin{bmatrix} z \\ w \end{bmatrix}. \tag{11}$$

Matrix notation is particularly effective here, and we introduce the matrix of first derivatives called the **Jacobian matrix**:

$$\text{Jacobian matrix} = \begin{bmatrix} \frac{\partial f}{\partial x} & \frac{\partial f}{\partial y} \\ \frac{\partial g}{\partial x} & \frac{\partial g}{\partial y} \end{bmatrix}. \tag{12}$$

The matrix must be evaluated at the equilibrium $x = r$, $y = s$. Introducing the vector displacement from equilibrium $\mathbf{z} = (z, w) = (x - r, y - s)$ whose components are z and w, equations (11) are more conveniently written using matrix notation:

$$\frac{d\mathbf{z}}{dt} = \mathbf{A}\mathbf{z}, \tag{13}$$

where $\mathbf{A}$ is a matrix of constants obtained by evaluating the Jacobian matrix at the equilibrium:

$$\mathbf{A} = \begin{bmatrix} f_x(r, s) & f_y(r, s) \\ g_x(r, s) & g_y(r, s) \end{bmatrix} = \begin{bmatrix} a & b \\ c & d \end{bmatrix}. \tag{14}$$

The Phase Plane

The phase plane that we discussed with many examples in Chapter 4 for linear systems can also be introduced for nonlinear systems:

$$\frac{dx}{dt} = f(x, y), \tag{15a}$$

$$\frac{dy}{dt} = g(x, y). \tag{15b}$$

Orbits or trajectories can be introduced and solutions graphed in the x, y-plane. Some procedures for the phase plane of nonlinear systems will be discussed in the next subsection. Near an equilibrium, the nonlinear differential equation can be approximated by its associated linearized system. Thus, we expect that the phase plane for the nonlinear system near an equilibrium can be approximated in some sense by the phase plane of the corresponding linear system near the equilibrium. We will try to be fairly precise and state a theorem below. The claim we wish to make is that in most cases the phase plane for the nonlinear system in the neighborhood of an equilibrium resembles the phase plane for the corresponding linear system (the linearized system of differential equations defined above).

STABILITY OF AN EQUILIBRIUM AND PHASE PLANE NEAR AN EQUILIBRIUM. We wish to describe the relationship between the nonlinear system (15a)–(15b) near an equilibrium and its linearized system of differential equations (9a)–(9b). We are interested in the stability of an equilibrium, and we are interested in the phase plane of the nonlinear system in the neighborhood of an equilibrium. The following theorem is motivated but not proved from the linearized system.

THEOREM 6.2.1 *Suppose that (r, s) is an equilibrium of (15a) and (15b). Define a, b, c, d by (10) or (14). Let λ_1, λ_2 be the eigenvalues of the matrix A (14) which satisfy the characteristic equation derived from the determinant condition:*

$$\det \begin{bmatrix} a - \lambda & b \\ c & d - \lambda \end{bmatrix} = \lambda^2 - (a + d)\lambda + ad - bc = 0. \tag{16}$$

Recall that the eigenvalues λ of the matrix correspond to the time dependency $e^{\lambda t}$. To be more precise, we recall that the solutions of the linear system are linear *combinations of solutions* $z = e^{\lambda t}\mathbf{v}$, *where* $\mathbf{v}$ *is an eigenvector of the matrix. Not all solutions are of this form. The stability of the equilibrium and the phase plane in the neighborhood of the equilibrium are usually determined by the linearization:*

1. *If* $\lambda_1 > 0, \lambda_2 > 0$, *the equilibrium of the nonlinear system is unstable, and the local phase plane will be an unstable node resembling figure 4.3.9 of Chapter 4.*
2. *If* $\lambda_1 < 0, \lambda_2 < 0$, *the equilibrium of the nonlinear system is asymptotically stable, and the local phase plane will be a stable node resembling figure 4.3.11 of Chapter 4.*
3. *If* λ_1, λ_2 *are nonzero and of opposite signs, the equilibrium is unstable, and the local phase plane will be a saddle point resembling figure 4.3.13 of Chapter 4.*
4. *If* $\lambda_1 = \lambda_2 > 0$ $(\lambda_1 = \lambda_2 < 0)$, *then the equilibrium is unstable (stable), but we do not discuss the local phase plane.*
5. *If* $\lambda_1 = 0$ *and* $\lambda_2 > 0$, *then the equilibrium is unstable. If* $\lambda_1 = 0$ *and* $\lambda_2 < 0$, *then the equilibrium for the nonlinear system may be stable or unstable depending on the neglected nonlinear terms. We do not discuss the local phase plane in either case.*
6. *If* $\lambda_1 = \alpha + i\delta, \lambda_2 = \alpha - i\delta$, *with* $\delta \neq 0$ *and* $\alpha > 0$, *the equilibrium is unstable, and the local phase plane with be an unstable spiral (see a and b of figure 4.3.16 of Chapter 4.)*
7. *If* $\lambda_1 = \alpha + i\delta, \lambda_2 = \alpha - i\delta$, *with* $\delta \neq 0$ *and* $\alpha < 0$, *the equilibrium is asymptotically stable, and the local phase plane with be a stable spiral (see c and d of figure 4.3.16 of Chapter 4.)*
8. *If* $\lambda_1 = i\delta, \lambda_2 = -i\delta$, *then the linearization has oscillatory solutions. The phase plane of the linearization is a center (see figure 4.3.17 of Chapter 4). The phase plane of the nonlinear system often looks like the same center, but* ***it is not guaranteed to look like a center.*** *The equilibrium for the nonlinear system will be stable or unstable depending on the nonlinear terms neglected in the linearization. These will be discussed later. Only cases 5 and 8 have this potential difficulty.*

Local means that the linearized phase plane is valid only near the equilibrium of the nonlinear system.

Resembles means that the axis may be bent and the phase plane somewhat distorted in the neighborhood of the equilibrium.

Example 6.2.2 *Phase Plane Near Equilibria*

Determine the phase plane of example (3a)–(3b),

$$\frac{dx}{dt} = f(x, y) = -x + xy, \tag{17a}$$

$$\frac{dy}{dt} = g(x, y) = -4y + 8xy, \tag{17b}$$

near the equilibria $(0, 0)$ and $\left(\frac{1}{2}, 1\right)$ found in Example 6.2.1.

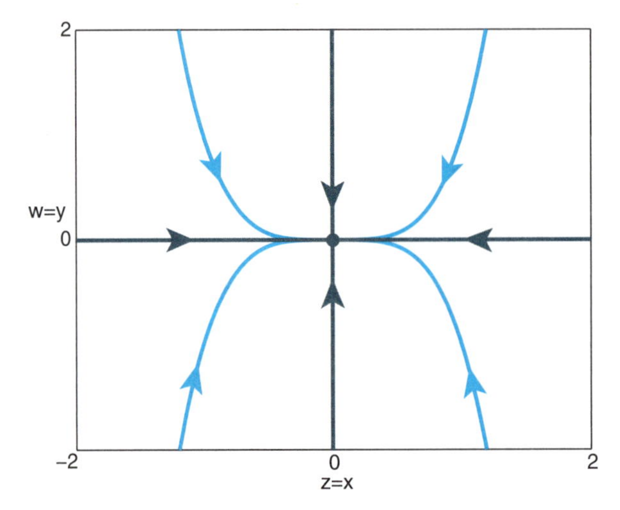

Figure 6.2.1 Stable node at (0, 0).

• SOLUTION. Since

$$\frac{\partial f}{\partial x} = -1 + y, \ \frac{\partial f}{\partial y} = x, \ \frac{\partial g}{\partial x} = 8y, \ \frac{\partial g}{\partial y} = -4 + 8x, \tag{18}$$

the matrix for this example is

$$\begin{bmatrix} -1+y & x \\ 8y & -4+8x \end{bmatrix}. \tag{19}$$

EQUILIBRIUM (0, 0). Consider first the equilibrium $x = 0$, $y = 0$. In this case the matrix is $\mathbf{A} = \left[\begin{smallmatrix} -1 & 0 \\ 0 & -4 \end{smallmatrix}\right]$. By Theorem 6.2.1, the behavior of (17a)–(17b) near (0, 0) is that of (9a)–(9b), which is

$$\frac{dz}{dt} = -z, \tag{20a}$$

$$\frac{dw}{dt} = -4w, \tag{20b}$$

where here $z = x - r = x$ and $w = y - s = y$. In this example, the linear system of differential equations decouples into two first-order differential equations, so that from our study in Chapter 1 of first-order equations (with constant coefficients) it is seen that eigenvalues (roots of the characteristic equation) are $-1, -4$. Thus, (0, 0) is a stable node as studied in Chapter 4. The eigenvalues λ and eigenvectors (u, v) satisfy the linear system

$$\begin{bmatrix} -1-\lambda & 0 \\ 0 & -4-\lambda \end{bmatrix} \begin{bmatrix} u \\ v \end{bmatrix} = \begin{bmatrix} 0 \\ 0 \end{bmatrix}. \tag{21}$$

The determinant condition is $\det \left[\begin{smallmatrix} -1-\lambda & 0 \\ 0 & -4-\lambda \end{smallmatrix}\right] = 0$. Thus, the characteristic equation is $(\lambda + 1)(\lambda + 4) = 0$ with roots (eigenvalues) $\lambda = -1, -4$ (as already determined),

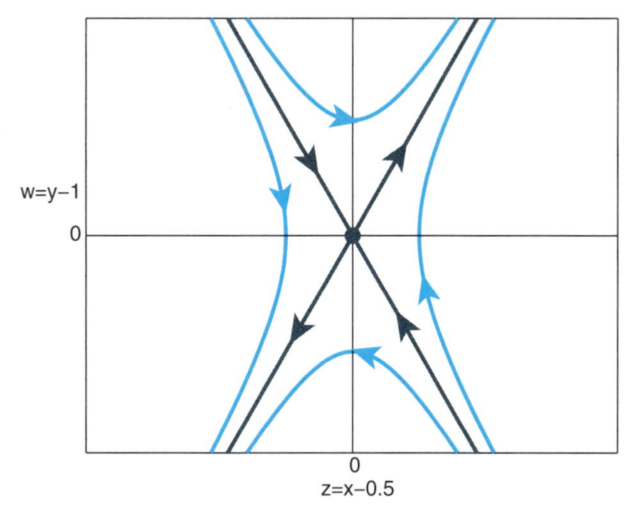

Figure 6.2.2 Saddle point at $\left(\frac{1}{2}, 1\right)$.

so that the equilibrium (0, 0) is a stable node. From (21), for $\lambda = -1$, since $-3v = 0$, the corresponding eigenvector is any multiple of $\left[\begin{smallmatrix}1\\0\end{smallmatrix}\right]$, the x-axis, while for $\lambda = -4$, the corresponding eigenvector is any multiple of $\left[\begin{smallmatrix}0\\1\end{smallmatrix}\right]$, since $3u = 0$ is in the direction of the y-axis. Near (0, 0) the phase plane of the nonlinear system (17a)–(17b) will resemble that of its linearization, (20a)–(20b), which is shown in figure 6.2.1. ♦

EQUILIBRIUM $\left(\frac{1}{2}, 1\right)$. For the equilibrium $x = \frac{1}{2}$, $y = 1$, the linearization involves the displacement from the equilibrium, so that here $z = x - r = x - \frac{1}{2}$ and $w = y - s = y - 1$. For this equilibrium, $(\frac{1}{2}, 1)$, the matrix is

$$\mathbf{A} = \begin{bmatrix} 0 & \frac{1}{2} \\ 8 & 0 \end{bmatrix}. \tag{22}$$

The behavior of the nonlinear system (17a)–(17b)near the equilibrium $(\frac{1}{2}, 1)$ can be approximated by its linearized system

$$\frac{dz}{dt} = \frac{1}{2}w, \tag{23a}$$

$$\frac{dw}{dt} = 8z. \tag{23b}$$

The eigenvalues λ and eigenvectors (u, v) satisfy the linear system

$$\begin{bmatrix} -\lambda & \frac{1}{2} \\ 8 & -\lambda \end{bmatrix} \begin{bmatrix} u \\ v \end{bmatrix} = \begin{bmatrix} 0 \\ 0 \end{bmatrix}. \tag{24}$$

The determinant condition is

$$\det \begin{bmatrix} -\lambda & \frac{1}{2} \\ 8 & -\lambda \end{bmatrix} = 0. \tag{25}$$

Thus, the characteristic equation is $\lambda^2 - 4 = 0$ with roots (eigenvalues) $\lambda = \pm 2$, so that the equilibrium $\left(\frac{1}{2}, 1\right)$ is a saddle point. From (24), for $\lambda = 2$, since $-2u + \frac{1}{2}v = 0$, the

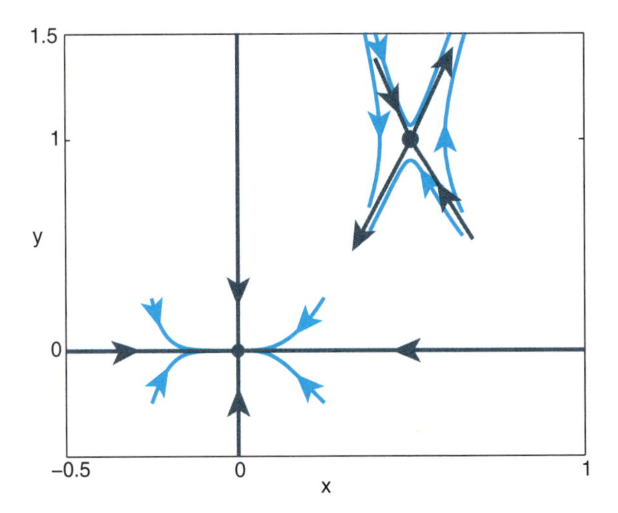

Figure 6.2.3 Phase plane using only linearizations.

corresponding eigenvector is any multiple of $\left[\begin{smallmatrix}1\\4\end{smallmatrix}\right]$, while for $\lambda=-2$, the corresponding eigenvector is any multiple of $\left[\begin{smallmatrix}1\\-4\end{smallmatrix}\right]$ since $2u+\frac{1}{2}v=0$. Thus, the local phase plane is shown in figure 6.2.2, with the corresponding eigenvectors being the stable and unstable directions for the saddle point $\left(\frac{1}{2}, 1\right)$.

PHASE PLANE OF A NONLINEAR SYSTEM. It is important to keep in mind that Theorem 6.2.1 describes only the behavior near each equilibrium, as shown in figure 6.2.3.

6.2.2 Nonlinear Systems: Summary, Philosophy, Phase Plane, Direction Field, Nullclines

Let us summarize the method for analyzing nonlinear systems:

1. Find equilibrium.
2. Linearize in the neighborhood of each equilibrium.
3. For each linearization, find eigenvalues. Find eigenvectors for the real eigenvalues.
4. For the usual cases described above, sketch the phase plane for the nonlinear system near each equilibrium by using the phase plane of the corresponding linear system.
5. Determine trajectories in the phase plane away from equilibrium by using
   a) direction field (or numerical solutions) from graphics software, described below;
   b) optional method of nullclines, described shortly, to understand 5(a).

In this subsection, we will do some examples to explain these ideas. More examples will be done in other sections of this chapter.

Direction Field

Here we will present some basic results for nonlinear systems of differential equations of the form

$$\frac{dx}{dt} = f(x, y), \tag{26a}$$

$$\frac{dy}{dt} = g(x, y). \tag{26b}$$

One of the beauties of mathematics is that we can determine the **direction field** for the phase plane directly from the differential equation. The direction field is a vector field in the tangential direction defined by the differential equation $\left(\frac{dx}{dt}, \frac{dy}{dt}\right) = (f(x, y), g(x, y))$. It is not necessary to determine the phase plane from brute force numerical solutions of the differential equation. Sometimes the following observation is useful.

For solutions that are not equilibria, solutions in the phase plane will be curves. Trajectories or orbits of the solution will move in time. From the differential equation (26a), we see that if $f(x, y) > 0$ then $\frac{dx}{dt} > 0$, which means x increases in time, and we introduce arrows moving toward the right ($\rightarrow$) in the x, y-plane. Similarly, if $f(x, y) < 0$ then x decreases in time, and we introduce arrows moving to the left ($\leftarrow$). From the other differential equation, (26b),we see that if $g(x, y) > 0$ then $\frac{dy}{dt} > 0$, which means y increases in time, and we introduce arrows moving upward ($\uparrow$) in the x, y-plane. And finally, if $g(x, y) < 0$, then $\frac{dy}{dt} < 0$, and we introduce downward arrows ($\downarrow$) in the x, y-plane. This is similar to drawing lines indicating which direction (northeast or southeast) the wind is going.

By using the chain rule, we obtain from the system (26a)–(26b),

$$\frac{dy}{dx} = \frac{dy/dt}{dx/dt} = \frac{g(x, y)}{f(x, y)}, \tag{27}$$

from which we may determine the phase plane. Equation (27) is a first-order differential equation, and we showed in Chapter 1 how easy it is for computers to sketch its slope field. It is somewhat subtle that the direction field is the slope field supplemented by the vector direction.

Example 6.2.3 *Direction Field and Phase Plane*

Find the direction field and phase plane for system (17a)–(17b),

$$\frac{dx}{dt} = -x + xy, \tag{28a}$$

$$\frac{dy}{dt} = -4y + 8xy. \tag{28b}$$

• SOLUTION. The direction field is shown in figure 6.2.4. Arrows have been included to make it a direction field. The phase plane for this example is shown in figure 6.2.5. It is obtained by combining information from the direction field with our knowledge of each equilibrium. The equilibrium (0, 0) is a stable node with eigenvalue -1 with corresponding eigenvector $\begin{bmatrix} 1 \\ 0 \end{bmatrix}$, and with eigenvalue -4 with corresponding

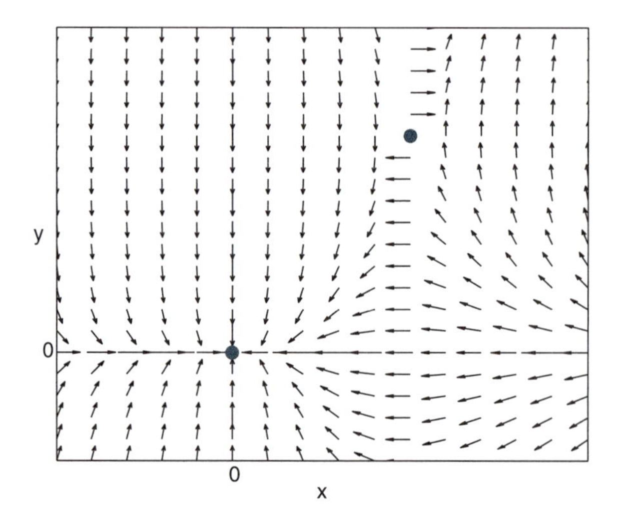

Figure 6.2.4 Direction field for (28a)–(28b).

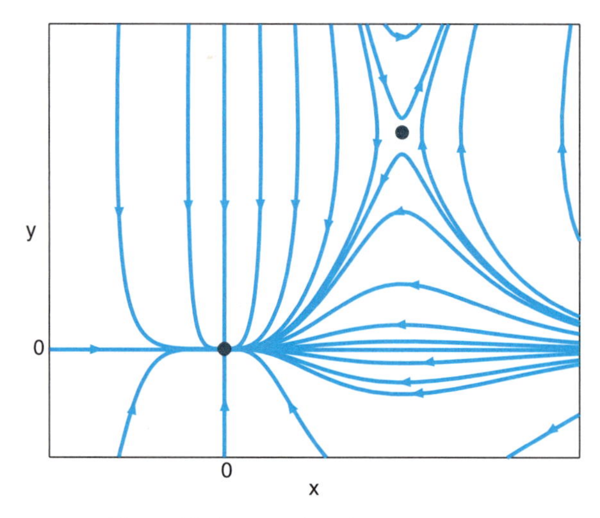

Figure 6.2.5 Phase plane for (28a)–(28b).

eigenvector $\begin{bmatrix} 0 \\ 1 \end{bmatrix}$. The other equilibrium $\left(\frac{1}{2}, 1\right)$ is an (unstable) saddle point with eigenvalue 2 with corresponding unstable eigenvector direction $\begin{bmatrix} 1 \\ 4 \end{bmatrix}$ and with eigenvalue -2 with corresponding stable eigenvector direction $\begin{bmatrix} 1 \\ -4 \end{bmatrix}$.

Method of Nullclines

It is easy for computers to sketch the phase plane (with or without the direction field) of a given nonlinear system of differential equations,

$$\frac{dx}{dt} = f(x, y), \tag{29a}$$

$$\frac{dy}{dt} = g(x, y). \tag{29b}$$

However, the method of nullclines can be used to understand why solutions behave the way they appear. The phase plane satisfies the slope field of the nonlinear first-order differential equation obtained by dividing the two equations to give

$$\frac{dy}{dx} = \frac{g(x, y)}{f(x, y)}. \tag{30}$$

First graph $f(x, y) = 0$. Along this curve from (30), $\frac{dy}{dx} = \infty$, so we draw short vertical dashes (see figure 6.2.6) since the trajectories have vertical tangents there. Similarly, along $g(x, y) = 0$, $\frac{dy}{dx} = 0$, so we draw short horizontal dashes. Both curves are called **nullclines**. Interestingly enough, the intersection of nullclines will provide a graphical determination of the equilibria. More important, we note that in the region in which $f(x, y) > 0$, $\frac{dx}{dt} > 0$, so that x increases as a function of time which we can mark in the phase plane with a right arrow $\rightarrow$ (and vice versa, $f(x, y) < 0$ corresponds to x decreasing in time, $\leftarrow$). Similarly, regions with $g(x, y) > 0$ correspond to y increasing in time, which we indicate in the phase plane with an upper arrow $\uparrow$ (and vice versa meaning $g(x, y) < 0$ corresponds to y decreasing in time, $\downarrow$). We combine the arrows, as shown in the following example, so that we can predict and mark the direction of the flow in the phase plane directly from the differential equation.

Example 6.2.4 *Method of Nullclines (for Example 6.2.2)*

Sketch the phase plane for the nonlinear example (28a)–(28b) using the method of nullclines and the phasc plane of the linearizations. The equations are

$$\frac{dx}{dt} = x(y - 1), \tag{31a}$$

$$\frac{dy}{dt} = 4y(2x - 1). \tag{31b}$$

• SOLUTION. For the method of nullclines, the trajectories in the phase plane satisfy the first-order nonlinear differential equation

$$\frac{dy}{dx} = \frac{4y(2x - 1)}{x(y - 1)}. \tag{32}$$

The nullclines $y = 1$ and $x = 0$ are sketched in figure 6.2.7, along which $\frac{dy}{dx} = \infty$, marked with vertical dashes. The nullcline $x = 0$ is parallel to the vertical dashes, so that $x = 0$ is a solution, as can be seen from the differential equation. The nullclines $y = 0$ and $x = \frac{1}{2}$ are also sketched, along which $\frac{dy}{dx} = 0$, marked with horizontal dashes, so that $y = 0$ is a solution. Only the intersections of different families are equilibria, so that the two equilibria are $x = 0$, $y = 0$ and $x = \frac{1}{2}$, $y = 1$. The nullclines break the phase plane into nine regions, and we must determine what happens in each. If $x > 0$ and $y > 1$, then $\frac{dx}{dt} > 0$, and hence x increases as a function of time, which we indicate by $\rightarrow$. Similarly, $x = 0$ and $y = 1$ divide the x, y-plane into four regions, and the arrows in the x-direction can be determined in the other three regions as shown in figure 6.2.7. If $y > 0$, and $x > \frac{1}{2}$, we have from the differential equation $\frac{dy}{dt} > 0$, and thus y increases, which we mark by $\uparrow$. There are four regions of different y behavior.

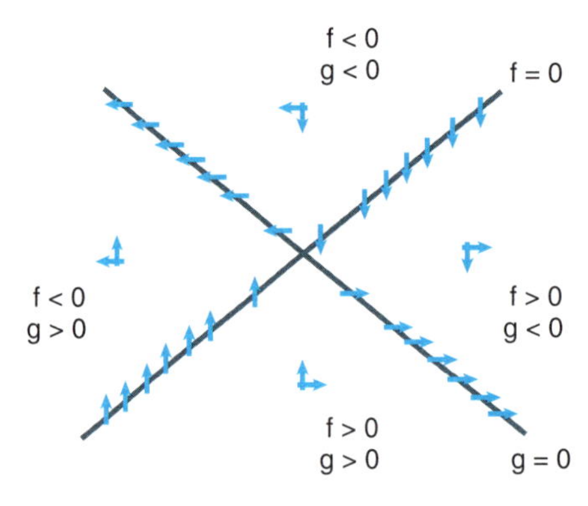

Figure 6.2.6 Method of Nullclines.

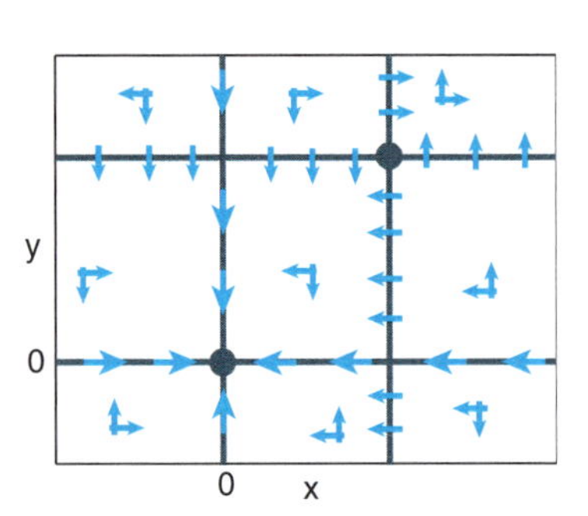

Figure 6.2.7 Method of nullclines for (31a)–(31b).

In the figure, we have indicated left-right and up-down arrows differently in each of the nine regions. On each nullcline we can determine more information. For example, on the horizontal dashes on the nullcline $x = \frac{1}{2}$, we introduce left arrows for $y < 1$ and right arrows for $y > 1$. We note that for the solution $x = 0$, down arrows occur for $y > 0$ and up arrows occur for $y < 0$. Similarly, for the solution $y = 0$, left arrows occur for $x > 0$ and right arrows occur for $x < 0$. We can now sketch trajectories by hand in the phase portrait, but it is best also to use our knowledge of the local phase plane near each equilibrium. Trajectories do not cross the solutions $x = 0$ and $y = 0$. Trajectories must have zero slope when they cross the nullcline $x = \frac{1}{2}$ and infinite slope when they cross the nullcline $y = 1$.

The method of nullclines helps us understand why the phase plane looks they way it is shown in figure 6.2.8. We also have noted in figure 6.2.8 that $x = 0$, $y = 0$ is a stable node with eigenvectors that we have determined and used. Similarly, $x = \frac{1}{2}$, $y = 1$ is a saddle point, and we know and use the eigenvectors carefully. Note that time-dependent solutions $x(t)$ may have a local maximum or minimum when $y = 1$ and $y(t)$ may have a local maximum or minimum when $x = \frac{1}{2}$. The phase plane using computer software is shown in figure 6.2.9 and can be compared to the other results. ♦

Example 6.2.5 *Method of Nullclines for a Linear Example*

Sketch the phase plane for the following interesting linear example using the method of nullclines:

$$\frac{dx}{dt} = -x + y, \tag{33a}$$

$$\frac{dy}{dt} = -y. \tag{33b}$$

• SOLUTION. The trajectories in the phase plane satisfy the first- order nonlinear differential equation

$$\frac{dy}{dx} = \frac{-y}{y - x}. \tag{34}$$

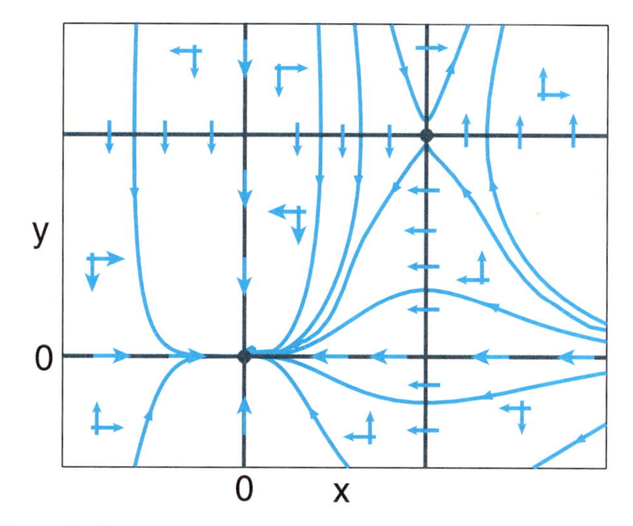

Figure 6.2.8 Method of nullclines and linear systems for (31a)–(31b).

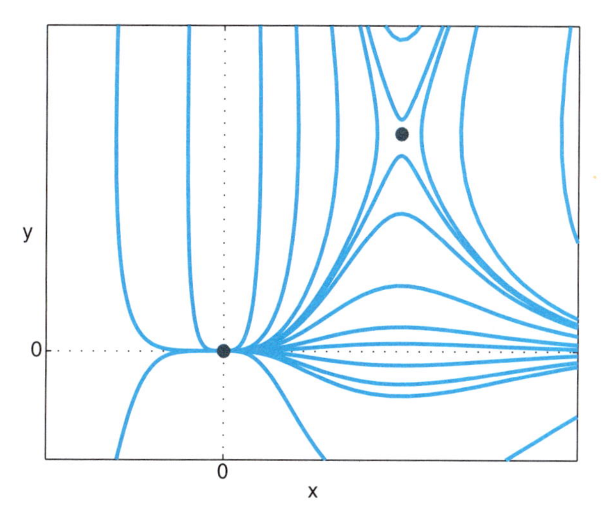

Figure 6.2.9 Solutions in the phase plane using computer software for (31a)–(31b).

The nullcline $y = x$ is sketched in figure 6.2.10, along which $\frac{dy}{dx} = \infty$, marked with vertical dashes. The nullcline $y = 0$ is also sketched, along which $\frac{dy}{dx} = 0$ marked with horizontal dashes. However, the nullcline $y = 0$ is itself horizontal, so $y = 0$ is itself a solution, as can be verified from (33b). The nullclines intersect at the origin (0, 0), which is the only equilibrium for (33a)–(33b). The nullclines break the phase plane into four regions. We find the general direction of the trajectories in these four regions. If $y > x$, then $\frac{dx}{dt} > 0$ and hence x increases as a function of time, which we indicate by $\rightarrow$; if $y < x$, $\frac{dx}{dt} < 0$ and hence x decreases as a function of time, $\leftarrow$. If $y > 0$, we have $\frac{dy}{dt} < 0$ and y decreases, which we mark by $\downarrow$, and if $y < 0$, we

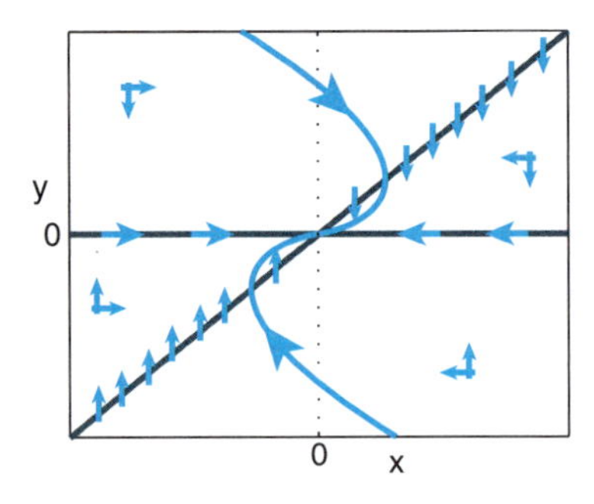

Figure 6.2.10

have $\frac{dy}{dt} > 0$ and y increases in time, ↑. In the figure, we have indicated left-right and up-down arrows in each of the four regions. We now sketch trajectories by hand by being consistent with the flow direction and remembering that trajectories must have infinite slope when they cross the nullcline $y = x$. In this problem, since $y = 0$ is a nullcline and part of a constant solution, the trajectories do not cross it. ♦

Exercises

In Exercises 1–12,

(a) Determine all equilibria and classify (node, saddle, spiral, center, stable or unstable).
(b) If an eigenvalue is real, find the eigenvectors.
(c) Graph the phase plane using the phase plane of linearized system.
(d) Graph the phase plane using the direction field from software.
(e) Graph the phase plane using the method of nullclines and part (c).

1. $\frac{dx}{dt} = x + xy,\ \frac{dy}{dt} = 2y - 4xy.$

2. $\frac{dx}{dt} = -x + xy,\ \frac{dy}{dt} = -2y + 8xy.$

3. $\frac{dx}{dt} = 2x - 2xy,\ \frac{dy}{dt} = y - xy.$

4. $\frac{dx}{dt} = 1 - x^2,\ \frac{dy}{dt} = y + 1.$

5. $\frac{dx}{dt} = x - y,\ \frac{dy}{dt} = -2x + 2xy.$

6. $\frac{dx}{dt} = y^3 + 1,\ \frac{dy}{dt} = x^2 + y.$

7. $\frac{dx}{dt} = 1 - y^2,\ \frac{dy}{dt} = 1 - x^2.$

8. $\frac{dx}{dt} = x(1 - y^2),\ \frac{dy}{dt} = x + y.$

9. $\dfrac{dx}{dt} = x - y + x^2, \ \dfrac{dy}{dt} = x + y.$

10. $\dfrac{dx}{dt} = 2x - y - xy, \ \dfrac{dy}{dt} = x + 2y.$

11. $\dfrac{dx}{dt} = -x - 2y, \ \dfrac{dy}{dt} = 2x - y + xy^2.$

12. $\dfrac{dx}{dt} = 2x + y, \ \dfrac{dy}{dt} = -x - 2y + y^3.$

Exercises 13–17 refer to $\frac{dx}{dt} = x - xy + \gamma x^2$, $\frac{dy}{dt} = -y + xy$. In each case find all equilibria and classify (node, saddle, spiral, center, stable or unstable).

13. $\gamma = -8.$

14. $\gamma = -\dfrac{1}{3}.$

15. $\gamma = \dfrac{1}{3}.$

16. $\gamma = 1.$

17. $\gamma = 8.$

6.3 Population Models

One of the many interesting applications of systems has been in developing population models for interacting species. The species can be chemical, economic, or quantities in Perelson's HIV models. However, we shall consider two species and speak of them as **biological species**.

Assume that there are two species whose populations at time t are given by $x(t)$, $y(t)$. We assume the rate of change of each population depends just on that of the other. Often, it is the growth rate of each population that depends on the other, so that we have

$$\frac{dx}{dt} = f(x, y) = xu(x, y), \tag{1a}$$

$$\frac{dy}{dt} = g(x, y) = yv(x, y), \tag{1b}$$

where $u(x, y)$ is the growth rate of species x and $v(x, y)$ is the growth rate of species y. See Chapter 1 for a discussion of constant growth rate models. It should be stressed that merely writing (1a) and (1b) down has ruled out effects due to time (seasonal variations), delay effects, and external factors such as a fluctuating food supply. However, models like (1a)–(1b) do provide some insight into the dynamics of populations. Two special cases will illustrate how a model might be developed. In all our examples we will assume $x \geq 0$, $y \geq 0$.

6.3.1 Two Competing Species

In this example, we consider two populations that are competing for the same limited food and shelter. In this situation, the simplest assumptions concerning the growth rates are

$$u(x, y) = a - by - cx, \tag{2a}$$

$$v(x, y) = q - rx - sy, \tag{2b}$$

where a, b, c, q, r, s are positive constants, and a and q represent the growth rates if there were no competition. Here we assume that the growth rates of both species diminishes as either species gets larger due to finite resources for both.

These assumptions lead to the nonlinear system

$$\frac{dx}{dt} = x(a - by - cx) = ax - bxy - cx^2, \tag{3a}$$

$$\frac{dy}{dt} = y(q - rx - sy) = qy - rxy - sy^2. \tag{3b}$$

We note that if the population of one species is zero, $y(t) = 0$, then species x satisfies the equation we called in Chapter 5, the **logistic equation**,

$$\frac{dx}{dt} = x(a - cx) = ax - cx^2. \tag{4}$$

Similarly, if $x(t) = 0$, then species y satisfies the different logistic equation

$$\frac{dy}{dt} = y(q - sy) = qy - sy^2. \tag{5}$$

The x- and y-axes contain trajectories. Equations (4) and (5) are first-order nonlinear equations analyzed in Chapter 5 by one-dimensional phase lines. The logistic equation in x, for example, has the property that there is a carrying capacity to the environment $x = \frac{a}{c}$. Populations less than $\frac{a}{c}$ increase and approach $\frac{a}{c}$. Populations greater than $\frac{a}{c}$ decrease and and approach $\frac{a}{c}$. When we do a phase plane for our two-dimensional systems, we will obtain the same result.

EQUILIBRIA. The equilibria of (3a)–(3b) are given by

$$x = 0 \text{ or } a - by - cx = 0 \tag{6}$$

and

$$y = 0 \text{ or } q - rx - sy = 0, \tag{7}$$

so that with some effort the four equilibria are

$$(0, 0), \left(0, \frac{q}{s}\right), \left(\frac{a}{c}, 0\right), (\alpha, \beta), \tag{8}$$

where (α, β) is the solution of

$$cx + by = a, \tag{9a}$$

$$rx + sy = q. \tag{9b}$$

(We assume $cs - br \neq 0$, so that the point (α, β) exists and is unique.) Thus

$$\alpha = \frac{as - bq}{cs - br}, \quad \beta = \frac{cq - ar}{cs - br}. \tag{10}$$

We note that $(0, 0)$ is zero population of both species and $\left(0, \frac{q}{s}\right)$ and $\left(\frac{a}{c}, 0\right)$ are the environment's carrying capacity for each species. The fourth equilibrium only corresponds to nonnegative populations if both $\alpha \geq 0$ and $\beta \geq 0$.

Example 6.3.1 *Competing Species*

Suppose $b = q = r = s = 1$, $c = 3$, and $a = 2$ and carry out the analysis.

• SOLUTION. Then (3a)–(3b) is the nonlinear system for the competing species:

$$\frac{dx}{dt} = f(x, y) = x(2 - y - 3x) = 2x - xy - 3x^2, \tag{11a}$$

$$\frac{dy}{dt} = g(x, y) = y(1 - x - y) = y - xy - y^2. \tag{11b}$$

First we determine the equilibria, which satisfy

$$x(2 - y - 3x) = 0, \tag{12a}$$

$$y(1 - x - y) = 0. \tag{12b}$$

From (12a), one possibility is $x = 0$, in which case (12b) gives $y = 0$ and $y = 1$. This corresponds so far to two equilibria, $(0, 0)$ and $(0, 1)$. If $2 - y - 3x = 0$, then (12b) gives the equilibrium $\left(\frac{2}{3}, 0\right)$ as well as a fourth equilibrium satisfying

$$2 - y - 3x = 0, \tag{13a}$$

$$1 - x - y = 0, \tag{13b}$$

which can be solved to show that $(\frac{1}{2}, \frac{1}{2})$ is an equilibrium. In summary the four equilibria are

$$(0, 0), (0, 1), \left(\frac{2}{3}, 0\right), \left(\frac{1}{2}, \frac{1}{2}\right). \tag{14}$$

LINEARIZATIONS. We recall that near each equilibrium the solution (and the trajectories in the phase plane) can be approximated by the corresponding linearized system with the following matrix

$$\frac{d\mathbf{z}}{dt} = \mathbf{A}\,\mathbf{z}. \tag{15}$$

We compute the matrix in general for this example:

$$\mathbf{A} = \begin{bmatrix} \frac{\partial f}{\partial x} & \frac{\partial f}{\partial y} \\ \frac{\partial g}{\partial x} & \frac{\partial g}{\partial y} \end{bmatrix} = \begin{bmatrix} 2 - y - 6x & -x \\ -y & 1 - x - 2y \end{bmatrix}. \tag{16}$$

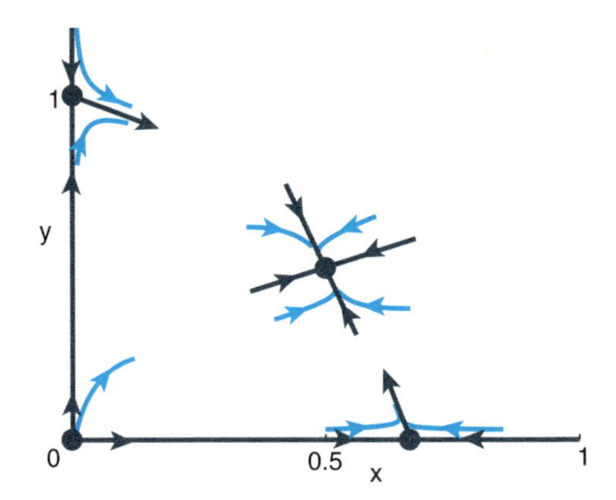

Figure 6.3.1 Phase plane for competing species model using linearized systems.

We recall that $\mathbf{z} = \left[\begin{smallmatrix} x-r \\ y-s \end{smallmatrix}\right]$ is the displacement from the equilibrium. The matrix given by (16) must be evaluated at each equilibrium. Each of the four equilibria must be analyzed separately.

EQUILIBRIUM (0, 0). The matrix evaluated at (0, 0) is

$$\mathbf{A} = \begin{bmatrix} 2 & 0 \\ 0 & 1 \end{bmatrix}. \tag{17}$$

In this case the equilibrium is the origin (0, 0), so that the Taylor series or linearization corresponds just to ignoring the nonlinear terms, $\frac{dx}{dt} = 2x$, $\frac{dy}{dt} = y$. Since the matrix is diagonal, the eigenvalues are just $\lambda = 2, 1$. Thus, the origin is an unstable node. For $\lambda = 1$, the eigenvector is $\left[\begin{smallmatrix} 0 \\ 1 \end{smallmatrix}\right]$, while for $\lambda = 2$, the eigenvector is $\left[\begin{smallmatrix} 1 \\ 0 \end{smallmatrix}\right]$. Observe that the trajectories are tangent to the y-axis as $t \to -\infty$, and the trajectories move away from the origin as time increases. In the competing species figure 6.3.1, we have sketched the phase plane diagram in the first quadrant near the origin.

EQUILIBRIUM (0, 1). The matrix (16) evaluated at (0, 1) is

$$\mathbf{A} = \begin{bmatrix} 1 & 0 \\ -1 & -1 \end{bmatrix}. \tag{18}$$

The eigenvalues λ and eigenvectors (u, v) satisfy the linear system

$$\begin{bmatrix} 1-\lambda & 0 \\ -1 & -1-\lambda \end{bmatrix} \begin{bmatrix} u \\ v \end{bmatrix} = \begin{bmatrix} 0 \\ 0 \end{bmatrix}. \tag{19}$$

The eigenvalues are clearly $\lambda = 1, -1$ determined from the determinant condition $(\lambda + 1)(\lambda - 1) = 0$. Hence the equilibrium (0, 1) is a (unstable) saddle point. From (19), for $\lambda = -1$, the eigenvector is $\left[\begin{smallmatrix} 0 \\ 1 \end{smallmatrix}\right]$, while for $\lambda = +1$, the eigenvector is any multiple of $\left[\begin{smallmatrix} 2 \\ -1 \end{smallmatrix}\right]$ since $-u - 2v = 0$, which has been graphed in competing species figure 6.3.1.

EQUILIBRIUM $(\frac{2}{3}, 0)$. The matrix evaluated at $\left(\frac{2}{3}, 0\right)$ is

$$\mathbf{A} = \begin{bmatrix} -2 & -\frac{2}{3} \\ 0 & \frac{1}{3} \end{bmatrix}. \tag{20}$$

The eigenvalues λ and eigenvectors (u, v) satisfy the linear system

$$\begin{bmatrix} -2-\lambda & -\frac{2}{3} \\ 0 & \frac{1}{3}-\lambda \end{bmatrix}\begin{bmatrix} u \\ v \end{bmatrix} = \begin{bmatrix} 0 \\ 0 \end{bmatrix}. \tag{21}$$

The eigenvalues are clearly $\lambda = -2, \frac{1}{3}$ determined from the determinant condition $(\lambda - \frac{1}{3})(\lambda + 2) = 0$. Hence the point $(\frac{2}{3}, 0)$ is also an (unstable) saddle point. From (21), For $\lambda = -2$, the eigenvector is $\left[\begin{smallmatrix} 1 \\ 0 \end{smallmatrix}\right]$ while for $\lambda = \frac{1}{3}$, the eigenvector is any multiple of $\left[\begin{smallmatrix} -2 \\ 7 \end{smallmatrix}\right]$ since $-\frac{7}{3}u - \frac{2}{3}v = 0$, as has also been sketched in figure 6.3.1.

EQUILIBRIUM $(\frac{1}{2}, \frac{1}{2})$. The matrix evaluated at $(\frac{1}{2}, \frac{1}{2})$ is

$$\mathbf{A} = \begin{bmatrix} -\frac{3}{2} & -\frac{1}{2} \\ -\frac{1}{2} & -\frac{1}{2} \end{bmatrix}. \tag{22}$$

The eigenvalues λ and eigenvectors (u, v) satisfy the linear system

$$\begin{bmatrix} -\frac{3}{2}-\lambda & -\frac{1}{2} \\ -\frac{1}{2} & -\frac{1}{2}-\lambda \end{bmatrix}\begin{bmatrix} u \\ v \end{bmatrix} = \begin{bmatrix} 0 \\ 0 \end{bmatrix}. \tag{23}$$

The eigenvalues can be computed from the determinant condition $\lambda^2 + 2\lambda + \frac{3}{4} - \frac{1}{4} = \lambda^2 + 2\lambda + \frac{1}{2} = 0$, and we obtain the eigenvalues $\lambda = -1 \pm \frac{\sqrt{2}}{2}$. Hence the equilibrium $(\frac{1}{2}, \frac{1}{2})$ is a stable node. From (23), for $\lambda = -1 + \frac{\sqrt{2}}{2}$, since $(-1 - \sqrt{2})u - v = 0$ the eigenvector is any multiple of $\left[\begin{smallmatrix} -1 \\ 1+\sqrt{2} \end{smallmatrix}\right]$, while for $\lambda = -1 - \frac{\sqrt{2}}{2}$, the eigenvector is any multiple of $\left[\begin{smallmatrix} -1 \\ 1-\sqrt{2} \end{smallmatrix}\right]$ since $(-1 + \sqrt{2})u - v = 0$. As $t \to +\infty$, it is most important that the trajectories go to the equilibrium $(\frac{1}{2}, \frac{1}{2})$. We note, in addition, that as $t \to +\infty$, the trajectories approach the equilibrium $(\frac{1}{2}, \frac{1}{2})$ tangent to the eigenvector $\left[\begin{smallmatrix} -1 \\ 1+\sqrt{2} \end{smallmatrix}\right]$ associated with the eigenvalue $-1 + \frac{\sqrt{2}}{2}$, as shown in figure 6.3.1.

PHASE PLANE. In figure 6.3.1, we have combined the phase planes in the neighborhood of each of the four equilibria.

METHOD OF NULLCLINES AND LINEARIZATIONS. To understand the phase plane, especially away from any equilibrium, we first apply the method of nullclines to this example. Then we will also consider the linearizations. The method of nullclines should give a graph very quickly. There is a huge amount of information in the method of nullclines. A picture is worth a thousand words. We are considering the specific system of differential equations corresponding to a competing species model

$$\frac{dx}{dt} = f(x, y) = x(2 - y - 3x) = 2x - xy - 3x^2, \tag{24a}$$

$$\frac{dy}{dt} = g(x, y) = y(1 - x - y) = y - xy - y^2. \tag{24b}$$

Since it is a population model, we only consider the first quadrant and include the usual x- and y-axes. Most of the results come from (24a) and (24b), but it is sometimes helpful to note that

$$\frac{dy}{dx} = \frac{y(1 - x - y)}{x(2 - y - 3x)}. \tag{25}$$

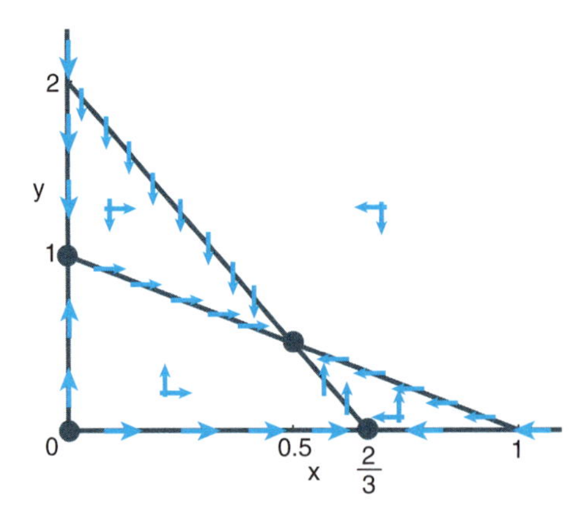

Figure 6.3.2 Method of nullclines for competing species model.

1. Nullclines are curves along which $\frac{dy}{dx}=0$ or $\frac{dy}{dx}=\infty$. We first draw the nullclines, which in this case include the axes. The x-axis, $y=0$, is a nullcline and a horizontal line solution in the phase plane along which the slope is zero; and the y-axis, $x=0$, is a nullcline and a vertical line solution along which the slope is infinite. Now we determine and plot the other nullclines. Along $x+y=1$, the slope is zero, and we place short horizontal lines along that nullcline. Along $3x+y=2$, the slope is infinite, and we place short vertical lines along that nullcline; see figure 6.3.2.
2. We also determine the equilibria by intersection of the different families of nullclines. In this example, carefully note that there are four equilibria (marked with large dot) $(0, 0)$, $(0, 1)$, $\left(\frac{2}{3}, 0\right)$, $\left(\frac{1}{2}, \frac{1}{2}\right)$.
3. In this example, these nullclines divide the first quadrant into four regions. We must be prepared to calculate whether x and y are increasing or decreasing as a function of time t in each region. Let us do x first. From (24a), we see that if $y>2-3x$, then $\frac{dx}{dt}<0$, and thus x decreases as a function of t and we put left arrows $(\leftarrow)$ in the two regions. In addition, we put left arrows $(\leftarrow)$ along the appropriate part of $y=0\left(y>\frac{2}{3}\right)$ and $x+y=1$. We put right arrows $(\rightarrow)$ in the other regions, and also along other portions of the nullcline with $\frac{dy}{dx}=0$.
4. For the time dependence of y, we use (24b). In this case, if $y>1-x$, then $\frac{dy}{dt}<0$, y decreases as a function of time t, and we introduce downward arrows $(\downarrow)$ in the appropriate two regions. In addition, downward arrows are placed along $x=0$ for $y<1$ and along the appropriate part of the nullcline $2=y+3x$. Upward arrows $(\uparrow)$ are placed everywhere else.
5. The next step is to improve the solutions obtained by the method of nullclines by using the phase plane behavior near each equilibrium, using knowledge of the theory of linear systems (nodes, saddles, spirals) and using the eigenvectors, which should have been calculated. Briefly, for example, the equilibrium $\left(\frac{2}{3}, 0\right)$ is a a saddle point, the stable direction is quite clear (x-axis), but the unstable direction must be consistent with the northeasterly flow in the region near the equilibrium determined by the method of nullclines, which is confirmed by the specific eigenvector previously calculated $\begin{bmatrix} -2 \\ 7 \end{bmatrix}$. For the other saddle point $(0, 1)$, the unstable eigenvector direction was $\begin{bmatrix} 2 \\ -1 \end{bmatrix}$,

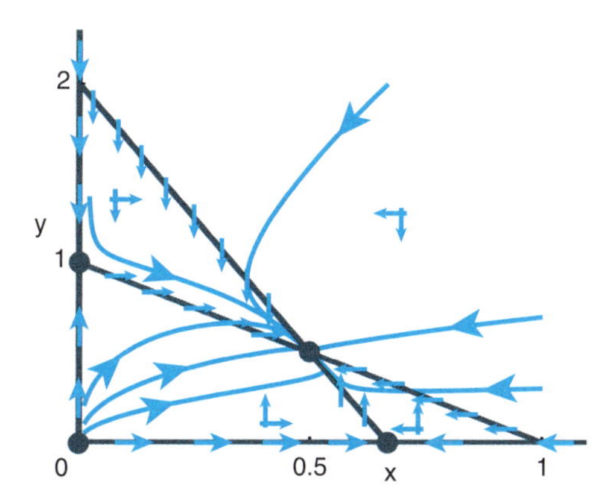

Figure 6.3.3 Trajectories using method of nullclines for competing species model.

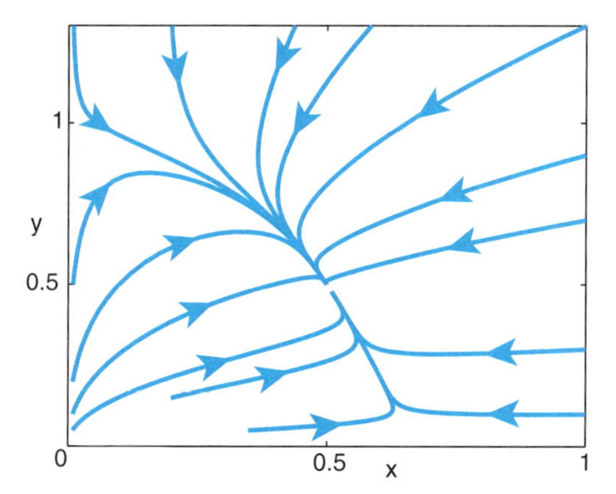

Figure 6.3.4 Trajectories (using software) for competing species model.

and we include that information as well. The equilibrium $\left(\frac{1}{2}, \frac{1}{2}\right)$ is a stable node, and the directions of the two eigenvectors must be consistent with the direction of solutions within each quadrant near that equilibrium determined by the method of nullclines. One should note that even some hand calculators can find eigenvalues and eigenvectors of 2×2 matrices.

6. The eigenvector solutions which appear to be straight near the equilibrium will curve (as shown) to be consistent with the information learned about the direction field from the method of nullclines. We include trajectories that are the continuation of the eigenvectors at each equilibrium.
7. Finally, we sketch a few representative solutions in the phase plane, making sure we are consistent with the linear phase plane and with the direction field determined by the method of nullclines. The result is shown in figure 6.3.3.

THE PHASE PLANE. We can obtain trajectories (using software) for the competing species model, which we present in figure 6.3.4. In figure 6.3.3, we have combined the phase planes in the neighborhood of each of the four equilibria and sketched the phase plane diagram in the first quadrant for the system (11a)–(11b). This specific competing species model has the property that zero population as well as both single species carrying capacities (0, 1) and $\left(\frac{2}{3}, 0\right)$ are unstable. Only the competing equilibrium population $\left(\frac{1}{2}, \frac{1}{2}\right)$ is stable. Furthermore, from the phase plane we see that nearly all initial conditions approach this stable competing equilibrium population $\left(\frac{1}{2}, \frac{1}{2}\right)$. This model runs counter to **niche theory**, which says that, given two essentially similar species, one will eventually replace the other. Other competing species models are considered in the exercises.

6.3.2 Predator-Prey Population Models

Now suppose we have a **prey** species x (rabbits) that is eaten by a **predator** species y (foxes). If there are no predators, we assume the prey has sufficient sources of food. On the other hand, if there are no prey, we assume the predators would die off. In this situation, the simplest assumptions concerning the growth rates are

$$u(x, y) = a - by - cx, \tag{26a}$$

$$v(x, y) = -q + rx - sy, \tag{26b}$$

where a, b, c, q, r, s are positive constants, a is the growth rate of prey without predators, and q is the death rate of predators without prey. The interaction terms are such that increasing the predator y diminishes the growth rate of the prey x, while increasing the prey x increases the growth rate of the predator y. The growth rates of both species diminish as either species gets large due to finite resources for both:

$$\frac{dx}{dt} = x(a - by - cx) = ax - bxy - cx^2, \tag{27a}$$

$$\frac{dy}{dt} = y(-q + rx - sy) = -qy + rxy - sy^2. \tag{27b}$$

If there are no predators, $y(t) = 0$, we find that the prey $x(t)$ satisfies the **logistic equation** discussed in Chapter 5:

$$\frac{dx}{dt} = x(a - cx) = ax - cx^2. \tag{28}$$

Without predators, the prey grows or decays to its environmental carrying capacity $x = \frac{a}{c}$. If there are no prey, the predators die off:

$$\frac{dy}{dt} = y(-q - sy) = -qy - sy^2. \tag{29}$$

These one-dimensional phase lines (of Chapter 5) will correspond to trajectories in the phase plane along the x- and y-axes.

EQUILIBRIA. The equilibria of (27a)–(27b) are given by

$$x = 0 \quad \text{or } a - by - cx = 0 \tag{30}$$

and

$$y = 0 \quad \text{or } -q + rx - sy = 0, \tag{31}$$

so that there are four equilibria:

$$(0,0),\ \left(0,-\frac{q}{s}\right),\ \left(\frac{a}{c},0\right),\ (\alpha,\beta). \tag{32}$$

Here (α, β) is the solution of

$$cx+by=a, \tag{33a}$$

$$rx-sy=q, \tag{33b}$$

so that

$$\alpha=\frac{as+bq}{cs+br},\quad \beta=\frac{-cq+ar}{cs+br}. \tag{34}$$

We note that $-\frac{q}{s}$ is nonphysical because of the negative y-value. Certainly meaningful are $(0, 0)$, the zero population of both species, and $\left(\frac{a}{c}, 0\right)$ the carrying capacity for the prey. The other equilibrium corresponds to nonnegative populations if $\beta \geq 0$, since it is guaranteed that $\alpha \geq 0$.

Example 6.3.2 *Predator-Prey*

Suppose that $b=q=r=s=1$, $c=2$, and $a=3$, so that (27a)–(27b) is the nonlinear system for the predatory-prey system:

$$\frac{dx}{dt}=f(x,y)=x(3-y-2x)=3x-xy-2x^2, \tag{35a}$$

$$\frac{dy}{dt}=g(x,y)=y(-1+x-y)=-y+xy-y^2. \tag{35b}$$

- SOLUTION. First we determine the equilibria, which satisfy

$$x(3-y-2x)=0, \tag{36a}$$

$$y(-1+x-y)=0. \tag{36b}$$

From the first equation, one possibility is $x=0$, in which case the second equation gives $y=0$ and $y=-1$. If $3-y-2x=0$, then the second equation gives the equilibrium $\left(\frac{3}{2}, 0\right)$ as well as a fourth equilibrium satisfying

$$3-y-2x=0, \tag{37a}$$

$$-1+x-y=0, \tag{37b}$$

which can be solved to show that $\left(\frac{4}{3}, \frac{1}{3}\right)$ is an equilibrium. In summary, the equilibria are

$$(0,0),\ (0,-1),\ \left(\frac{3}{2},0\right),\ \left(\frac{4}{3},\frac{1}{3}\right). \tag{38}$$

The equilibrium $(0, -1)$ is not physical, and we do not use it.

LINEARIZATIONS. The solution and phase plane near each equilibrium can be approximated by the linearized system of differential equations:

$$\frac{d\mathbf{z}}{dt}=\mathbf{A}\,\mathbf{z}. \tag{39}$$

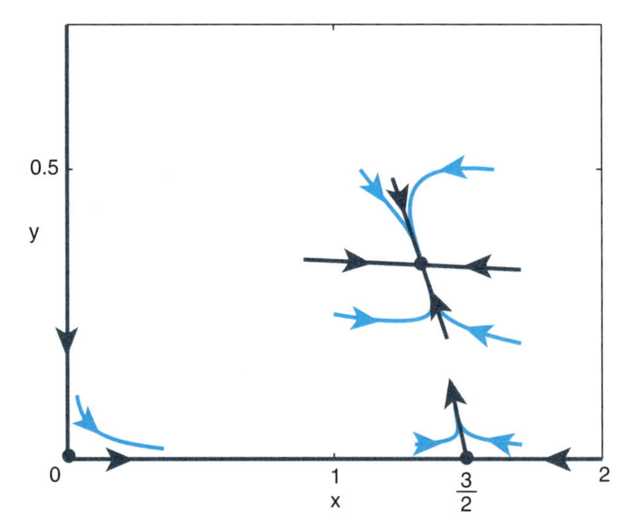

Figure 6.3.5 Phase plane just based on linearizations,

For this example, the matrix is

$$\mathbf{A} = \begin{bmatrix} \frac{\partial f}{\partial x} & \frac{\partial f}{\partial y} \\ \frac{\partial g}{\partial x} & \frac{\partial g}{\partial y} \end{bmatrix} = \begin{bmatrix} 3 - y - 4x & -x \\ y & -1 + x - 2y \end{bmatrix}. \tag{40}$$

We recall that $\mathbf{z} = \left[\begin{smallmatrix} x-r \\ y-s \end{smallmatrix}\right]$ is the displacement from the equilibrium. The matrix must be evaluated at the equilibrium. Each of the equilibrium must be analyzed separately.

EQUILIBRIUM (0, 0). The matrix evaluated at (0, 0) is

$$\mathbf{A} = \begin{bmatrix} 3 & 0 \\ 0 & -1 \end{bmatrix}. \tag{41}$$

Since the matrix is diagonal, the eigenvalues are just $\lambda = 3, -1$. Thus, the origin is an (unstable) saddle point. A saddle point has stable and unstable directions, which are the eigenvectors. From the one-dimensional problem, it is known that y decreases in time ($\downarrow$) and x increases ($\rightarrow$) near the origin; see figure 6.3.5.

EQUILIBRIUM $(\frac{3}{2}, 0)$. The matrix evaluated at $\left(\frac{3}{2}, 0\right)$ is

$$\mathbf{A} = \begin{bmatrix} -3 & -\frac{3}{2} \\ 0 & \frac{1}{2} \end{bmatrix}. \tag{42}$$

The eigenvalues λ and eigenvectors (u, v) satisfy the linear system

$$\begin{bmatrix} -3 - \lambda & -\frac{3}{2} \\ 0 & \frac{1}{2} - \lambda \end{bmatrix} \begin{bmatrix} u \\ v \end{bmatrix} = \begin{bmatrix} 0 \\ 0 \end{bmatrix}. \tag{43}$$

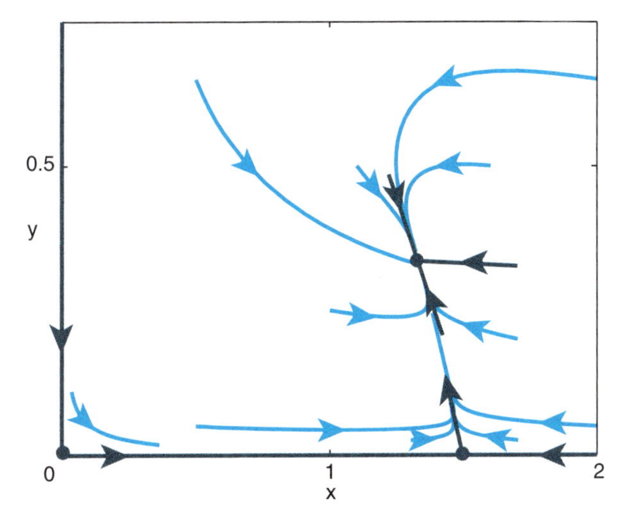

Figure 6.3.6 Phase plane for based on linearization and direction field for predator-prey model.

The eigenvalues are $\lambda = -3, \frac{1}{2}$ from the determinant condition $\left(\lambda - \frac{1}{2}\right)(\lambda + 3) = 0$. Since the eigenvalues have different signs, the equilibrium $\left(\frac{2}{3}, 0\right)$ is also an (unstable) saddle point. From the one-dimensional phase line, the stable direction for $\lambda = -3$ is x. The unstable eigenvector direction for $\lambda = \frac{1}{2}$ is $\left[\begin{smallmatrix} -3 \\ 7 \end{smallmatrix}\right]$ from $-7u - 3v = 0$. This is graphed in figure 6.3.5.

EQUILIBRIUM $\left(\frac{4}{3}, \frac{1}{3}\right)$. The matrix evaluated at $\left(\frac{4}{3}, \frac{1}{3}\right)$ is

$$\mathbf{A} = \begin{bmatrix} -\frac{8}{3} & -\frac{4}{3} \\ \frac{1}{3} & -\frac{1}{3} \end{bmatrix}. \tag{44}$$

The eigenvalues λ and eigenvectors (u, v) satisfy the linear system

$$\begin{bmatrix} -\frac{8}{3} - \lambda & -\frac{4}{3} \\ \frac{1}{3} & -\frac{1}{3} - \lambda \end{bmatrix} \begin{bmatrix} u \\ v \end{bmatrix} = \begin{bmatrix} 0 \\ 0 \end{bmatrix}. \tag{45}$$

From the determinant condition, the eigenvalues satisfy $\lambda^2 + 3\lambda + \frac{8}{9} + \frac{4}{9} = \lambda^2 + 3\lambda + \frac{4}{3} = 0$. The eigenvalues are $\lambda = -\frac{3}{2} \pm \frac{1}{2}\sqrt{\frac{11}{3}}$, so that $\lambda_1 \approx -0.54$, $\lambda_2 \approx -2.45$. The equilibrium $\left(\frac{4}{3}, \frac{1}{3}\right)$ is a stable node. For $\lambda_1 \approx -0.54$, the corresponding eigenvector is $\left[\begin{smallmatrix} -0.62 \\ 1 \end{smallmatrix}\right]$, while for $\lambda_2 \approx -2.45$ the eigenvector is $\left[\begin{smallmatrix} -6.35 \\ 1 \end{smallmatrix}\right]$. As $t \to +\infty$, solutions tend to the stable node in the direction $\left[\begin{smallmatrix} -0.62 \\ 1 \end{smallmatrix}\right]$. This is also graphed in figure 6.3.5.

THE PHASE PLANE. In figure 6.3.6, we have graphed the phase plane based on the linearizations and the direction field. We have noted that the equilibrium $\left(\frac{2}{3}, 0\right)$ is a saddle point, and we have included the trajectory that begins in the direction of the unstable direction (eigenvector) of that saddle point. We see from the phase plane that most initial conditions approach the equilibrium $\left(\frac{4}{3}, \frac{1}{3}\right)$, which is a stable node. Our

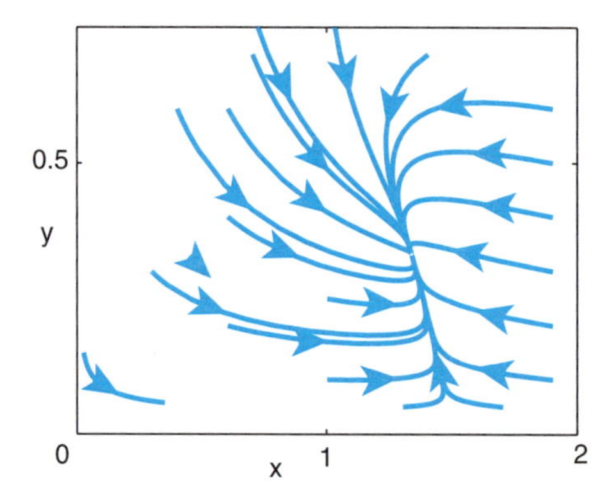

Figure 6.3.7 Computer-drawn phase plane for predator-prey model.

understanding of the phase plane and the direction field can be improved using the method of nullclines (as we did with the competing species example). In figure 6.3.7 we show a phase plane for this model drawn using a computer. ♦

Exercises

In Exercises 1–17, the x- and y-axes contain trajectories, and you are only to consider the trajectories for $x \geq 0$, $y \geq 0$.
In Exercises 1–9, determine and classify all equilibria (node, saddle, spiral, center, stable or unstable).

Two Competing Species Models (3a)–(3b)

1. Consider the competing species model with $a = 2$, $b = c = 1$, $q = r = 6$, $s = 0$.
a) Determine and classify all equilibria (node, saddle, spiral center, stable or unstable).
b) Sketch the phase portrait using the linear systems (do not calculate the eigenvectors).
c) Sketch the phase portrait using the method of nullclines.
d) Improve the phase plane using eigenvectors if the eigenvalues are real.
e) Sketch the phase plane using software.
2. Consider the competing species model with $a = 3$, $b = 2$, $c = 1$, $q = 2$, $r = s = 1$.
a) Determine and classify all equilibria (node, saddle, spiral center, stable or unstable).
b) Sketch the phase portrait using the linear systems (do not calculate the eigenvectors).
c) Sketch the phase portrait using the method of nullclines.
d) Improve the phase plane using eigenvectors if the eigenvalues are real. What happens to the competing populations as time increases?
e) Sketch the phase plane using software.
3. In Example 6.3.1, reduce the inherent growth rate of x by changing a to $\frac{3}{2}$, increase the effect of competition on x by changing b to 2, and increase the resource limitations on x by changing c to 4, so that $b = 2$, $c = 4$, $q = r = s = 1$, $a = \frac{3}{2}$.

a) Determine and classify all equilibria (node, saddle, spiral center, stable or unstable).
b) Sketch the phase portrait using the linear systems (do not calculate the eigenvectors).
c) Sketch the phase portrait using the method of nullclines. Explain in biological terms what the method of nullclines represents.
d) Improve the phase plane using eigenvectors if the eigenvalues are real.
e) Sketch the phase plane using software.

4. In the system (3a)–(3b), if there are ample food and shelter, then one might set $c = 0$, $s = 0$. For example, consider, $b = a = 2$ and $q = r = 1$.

a) Determine and classify all equilibria (node, saddle, spiral center, stable or unstable).
b) Sketch the phase portrait using the linear systems (including eigenvectors if the eigenvalues are real).
c) Sketch the phase portrait using the method of nullclines.
d) Sketch the phase plane using software.

Predator-Prey Problems (27a)–(27b)

5. Consider the predator-prey model with $a = 3$, $b = c = 1$, $q = r = 1$, $s = 0$.

a) Determine and classify all equilibria (node, saddle, spiral, center, stable or unstable).
b) Sketch the phase portrait using the linear systems (do not calculate the eigenvectors).
c) Sketch the phase portrait using the method of nullclines.
d) Improve the phase plane using eigenvectors if the eigenvalues are real.
e) Sketch the phase plane using software.

6. Consider the predator-prey model with $a = 3$, $b = 1$, $c = 6$, $q = r = s = 1$.

a) Determine and classify all equilibria (node, saddle, spiral, center, stable or unstable).
b) Sketch the phase portrait using the linear systems (do not calculate the eigenvectors). What happens to the competing populations as time increases?
c) Sketch the phase portrait using the method of nullclines.
d) Improve the phase plane using eigenvectors if the eigenvalues are real.
e) Sketch the phase plane using software.

7. In Example 6.3.2, remove the limitation on prey population by setting c to 0, so that $a = 3$, $c = 0$, $q = r = s = b = 1$

a) Determine and classify all equilibria (node, saddle, spiral center, stable or unstable).
b) Sketch the phase portrait using the linear systems (do not calculate the eigenvectors).
c) Sketch the phase portrait using the method of nullclines.
d) Improve the phase plane using eigenvectors if the eigenvalues are real.
e) Sketch the phase plane using software.

8. In Example 6.3.2, remove the limitation on predator and prey populations by setting $c = s = 0$, so that $a = 3$ and $q = r = b = 1$.

a) Determine and classify all equilibria (node, saddle, spiral center, stable or unstable).

b) Sketch the phase plane using the method of nullclines.
c) To show that periodic solutions exist, integrate the direction field equation.
9. In Example 6.3.2, increase the effect of the predator on the prey by increasing b to 6, so that $a=3, b=6, c=2, q=r=s=1$.
a) Determine and classify all equilibria (node, saddle, spiral center, stable or unstable).
b) Sketch the phase portrait using the linear systems (including eigenvectors if the eigenvalues are real).
c) Sketch the phase portrait using the method of nullclines.
d) Sketch the phase plane using software.

In exercises 10–12, consider a predator-prey model in which we assume there is unlimited food and shelter available, so that, left to itself, the prey will grow exponentially. Assume that, without the prey, the predator will die out slowly. (Perhaps there is an alternative food supply, but it lacks the proper nutrients.) Instead of a constant multiple of xy, use the following more realistic uptake function:

$$\frac{dx}{dt} = x(a - b\frac{y}{1+x}) = ax - b\frac{x}{1+x}y, \tag{46a}$$

$$\frac{dy}{dt} = y(-q + r\frac{x}{1+x}) = -qy + r\frac{x}{1+x}y. \tag{46b}$$

Note that again the x- and y-axes contain trajectories, and we need only consider $x \geq 0$, $y \geq 0$.

10. Let $a=b=1, q=2, r=1$ in (46a)–(46b).
a) Determine and classify all equilibria (node, saddle, spiral center, stable or unstable).
b) Sketch the phase portrait using the linear systems (do not calculate the eigenvectors).
c) Sketch the phase portrait using the method of nullclines.
d) Improve the phase plane using eigenvectors if the eigenvalues are real.
e) Sketch the phase plane using software.
11. Let $a=b=1, q=0.5, r=1$ in (46a)–(46b).
a) Determine and classify all equilibria (node, saddle, spiral center, stable or unstable).
b) Sketch the phase portrait using the linear systems (do not calculate the eigenvectors).
c) Sketch the phase portrait using the method of nullclines.
d) Improve the phase plane using eigenvectors if the eigenvalues are real.
e) Sketch the phase plane using software.
12. Let $a=b=1, q=1, r=2$ in (46a)–(46b).
a) Determine and classify all equilibria (node, saddle, spiral center, stable or unstable).
b) Sketch the phase portrait using the linear systems (do not calculate the eigenvectors).
c) Sketch the phase portrait using the method of nullclines.
d) Improve the phase plane using eigenvectors if the eigenvalues are real.
e) Sketch the phase plane using software.

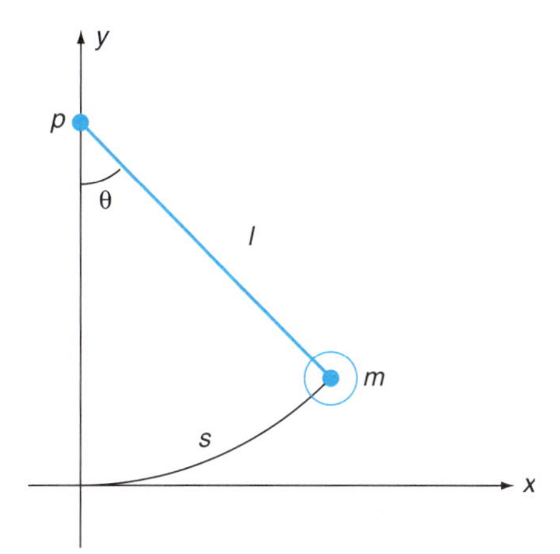

Figure 6.4.1 Nonlinear pendulum.

In Exercises 13–16, consider Perelson's linear model(10a)–(10b). Do the following: (a) determine all equilibria and classify (node, saddle, spiral, center), and stable or unstable. (b) Sketch the linear system. If an eigenvalue is real, find the eigenvectors. (c) Graph the phase plane using method of nullclines. (d) Graph the phase plane using direction field from software.

13. $kT_0 = \frac{3}{4}, b = 1, N = 1, c = 2.$

14. $kT_0 = \frac{5}{4}, b = 1, N = 1, c = 3.$

15. $kT_0 = \frac{9}{4}, b = 1, N = 1, c = 1.$

16. $kT_0 = \frac{15}{4}, b = 1, N = 1, c = 2.$

6.4 Mechanical Systems

Most mechanics problems are linear only if one assumes that there are "small displacements" or "limited variations in velocity." In this section, we shall analyze nonlinear problems, including one specific physical problem to motivate the subject.

6.4.1 Nonlinear Pendulum

The problem to be considered is a rigid pendulum of length l (see Figure 6.4.1). It is free to rotate around the point P, which is $(0, l)$. We suppose that the mass of the arm of the pendulum is negligible with respect to the mass m at the end of the pendulum. The mass moves along a circle.

Let (x, y) be the location of the mass at time t, s the distance along the circle from the origin to the point (x, y), and θ the angle the pendulum makes with the $y-$axis. At a given s, the velocity along the circle is $\frac{ds}{dt}$, in a direction tangential to the circle the mass is following. The component of the force of gravity in this direction is $-mg\sin\theta$. As in Section 2.5, we assume the resistance is linear and in the opposite direction to the velocity, so that

$$\text{mass times acceleration} = \text{total force} = \text{gravity} + \text{damping}$$

or

$$m\frac{d^2s}{dt^2} = -mg\sin\theta - \delta\frac{ds}{dt}. \tag{1}$$

If θ is measured in radians, then $s = l\theta$ and equation (1) becomes $ml\frac{d^2\theta}{dt^2} = -mg\sin\theta - \delta l\frac{d\theta}{dt}$ or

$$ml\frac{d^2\theta}{dt^2} + \delta l\frac{d\theta}{dt} + mg\sin\theta = 0. \tag{2}$$

This is the equation of a nonlinear pendulum, which we use as motivation for the consideration of a nonlinear system. We analyze important special cases later in this section. Now, we will use x instead of θ, so that the angle is x and satisfies

$$ml\frac{d^2x}{dt^2} + \delta l\frac{dx}{dt} + mg\sin x = 0. \tag{3}$$

6.4.2 Linearized Pendulum

If the angle x is small, then $\sin x \approx x$, and (3) is often approximated by the linear constant coefficient differential equation

$$ml\frac{d^2x}{dt^2} + \delta l\frac{dx}{dt} + mgx = 0, \tag{4}$$

which we studied in Chapter 2. If there is no damping, $\delta = 0$, and we find that (4) predicts oscillations with frequency $\frac{1}{2\pi}\sqrt{\frac{g}{l}}$ as shown in Section 2.5.

6.4.3 Conservative Systems and the Energy Integral

If there is no damping, so that $\delta = 0$, then the differential equation for the nonlinear pendulum is

$$\frac{d^2x}{dt^2} + \frac{g}{l}\sin x = 0. \tag{5}$$

This is an example of a conservative system. In physics, by Newton's law of motion, $F = ma$, where $F =$ forces, and the acceleration is given by $a = \frac{d^2x}{dt^2}$. A system is said to be **conservative** if it conserves energy. We show the system conserves energy if the forces depend only on position,

$$m\frac{d^2x}{dt^2} = f(x) \text{ or } m\frac{d^2x}{dt^2} - f(x) = 0. \tag{6}$$

DERIVATION OF CONSERVATION OF ENERGY. Conservation of energy is a very useful principle in physics for many purposes. We will find that it is very useful for finding the phase planes of conservative systems. There are many ways to derive conservation of energy. Conservation of energy follows from the differential equation

$$m\frac{d^2x}{dt^2} = f(x). \tag{7}$$

We write the second-order differential equation as a system in the usual way. We introduce $y = \frac{dx}{dt}$, so that our two variables are x, position, and y, velocity. The system is

$$\frac{dx}{dt} = y, \tag{8a}$$

$$\frac{dy}{dt} = \frac{d^2x}{dt^2} = \frac{1}{m}f(x), \tag{8b}$$

using (7). Conservation of energy can be derived from the slope field equations

$$\frac{dy}{dx} = \frac{\frac{dy}{dt}}{\frac{dx}{dt}} = \frac{1}{m}\frac{f(x)}{y}. \tag{9}$$

The first-order slope field equation separates:

$$my\,dy - f(x)dx = 0. \tag{10}$$

Integrating this equation yields a constant we call E:

$$\frac{1}{2}my^2 - \int f(x)dx = E. \tag{11}$$

Equation (11) is called conservation of energy, and we can write it in different ways. We note that $y = \frac{dx}{dt} = v =$ velocity, so that (11) can be written as

$$\frac{1}{2}m\left(\frac{dx}{dt}\right)^2 - \int f(x)dx = E. \tag{12}$$

The first term in (12) is called the kinetic energy and the second term the potential energy. Thus, the sum of the kinetic energy plus the potential energy is the constant total energy E:

$$\text{kinetic energy} = \frac{1}{2}m\left(\frac{dx}{dt}\right)^2 = \frac{1}{2}mv^2, \tag{13}$$

$$\text{potential energy} = -\int f(x)dx = V(x). \tag{14}$$

We note that the **potential energy is defined as minus the integral of the force**:

$$\frac{1}{2}m\left(\frac{dx}{dt}\right)^2 + V(x) = E. \tag{15}$$

With this definition of the potential energy, we note that **the derivative of the potential energy is minus the force**:

$$V'(x) = -f(x). \tag{16}$$

ALTERNATE DERIVATION I OF CONSERVATION OF ENERGY. If both sides of the differential equation (7) are multiplied by $\frac{dx}{dt}$, we obtain

$$m\frac{dx}{dt}\frac{d^2x}{dt^2} - f(x)\frac{dx}{dt} = 0. \tag{17}$$

We note this is equivalent to

$$\frac{d}{dt}\left[\frac{1}{2}m\left(\frac{dx}{dt}\right)^2 - \int f(x)dx\right] = 0, \tag{18}$$

since by the chain rule,

$$\frac{d}{dt}\int f(x)dx = \frac{d}{dx}\int f(x)dx\frac{dx}{dt} = f(x)\frac{dx}{dt}.$$

Integrating (18) yields conservation of energy:

$$\frac{1}{2}m\left(\frac{dx}{dt}\right)^2 - \int f(x)dx = E = \text{constant}. \tag{19}$$

ALTERNATE DERIVATION II OF CONSERVATION OF ENERGY. If the energy is defined as the sum of the kinetic and the potential energy, can we prove that energy is constant in time? The energy could depend on time t, since $x(t)$ from the differential equation (5) or (7):

$$E(t) \equiv \frac{1}{2}m\left(\frac{dx}{dt}\right)^2 + V(x). \tag{20}$$

Taking the derivative of the total energy with respect to t yields, using the chain rule,

$$\frac{dE}{dt} = \left[m\frac{d^2x}{dt^2} + V'(x)\right]\frac{dx}{dt}. \tag{21}$$

Since $V'(x) = -f(x)$, we obtain

$$\frac{dE}{dt} = \left[m\frac{d^2x}{dt^2} - f(x)\right]\frac{dx}{dt} = 0, \tag{22}$$

using the differential equation (7). Since $\frac{dE}{dt} = 0$, we have proved using the differential equation that the energy $E(t)$ is a constant.

Example *Determine the Phase Plane of the Linearized Pendulum without Damping Using the Energy Equation.*

The linearized pendulum without damping ($\delta = 0$) satisfies (4)

$$l\frac{d^2x}{dt^2} + gx = 0. \tag{23}$$

Letting $y = \frac{dx}{dt}$, the slope field equations are

$$\frac{dy}{dx} = \frac{\frac{dy}{dt}}{\frac{dx}{dt}} = -\frac{gx}{ly}, \tag{24}$$

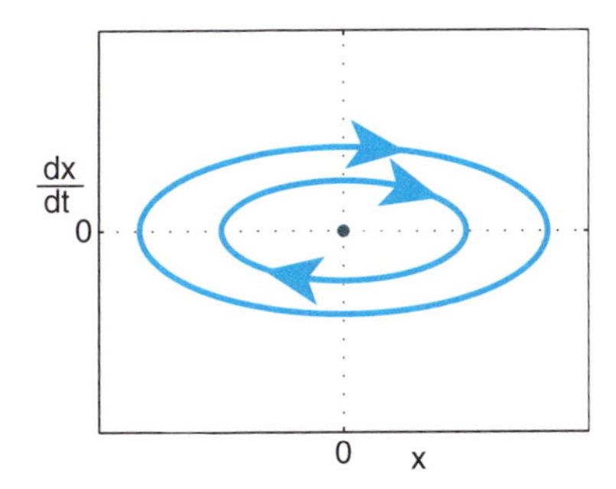

Figure 6.4.2 Phase plane for linearized pendulum (without damping).

since $\frac{dy}{dt} = \frac{d^2x}{dt^2} = -\frac{g}{l}x$. Equation (24) is separable, and we find

$$lydy = -gxdx. \tag{25}$$

Integrating this shows that the solution in the phase plane are ellipses, as shown in figure 6.4.2:

$$\frac{1}{2}ly^2 + \frac{1}{2}gx^2 = C. \tag{26}$$

Equation (26) is a form of conservation of energy for the linearized pendulum:

$$\frac{1}{2}l\left(\frac{dx}{dt}\right)^2 + \frac{1}{2}gx^2 = C, \tag{27}$$

where $\frac{1}{2}l\left(\frac{dx}{dt}\right)^2 = \frac{1}{2}lv^2$is the kinetic energy and $\frac{1}{2}gx^2$ is the potential energy. The solutions are confined to these ellipses, and hence flow neither toward nor away from the equilibrium. The equilibrium $x = 0$, $y = 0$ is said to be **stable**, and its type a **center**. Counterclockwise arrows are added, since in the upper half-plane $y = \frac{dx}{dt} > 0$ and hence x increases ($\rightarrow$) in time. Similarly, in the lower half-plane $y = \frac{dx}{dt} < 0$ and hence x decreases ($\leftarrow$) in time. We have previously (Example 4.3.1) determined the phase plane for a center by using the explicit time-dependent sinusoidal solution to derive ellipses.

6.4.4 The Phase Plane and the Potential

We reconsider second-order conservative systems (6)

$$m\frac{d^2x}{dt^2} = f(x). \tag{28}$$

Sometimes it is convenient to write (28) as a system:

$$\frac{dx}{dt} = f_1(x, y) = y, \tag{29a}$$

$$\frac{dy}{dt} = f_2(x, y) = \frac{f(x)}{m}. \tag{29b}$$

We will use conservation of energy,

$$E = \frac{1}{2}my^2 + V(x), \tag{30}$$

where the potential is related to the force through

$$V(x) = -\int f(x)dx \text{ and } V'(x) = -f(x). \tag{31}$$

EQUILIBRIUM AND POTENTIAL. Equilibria for the system are those values of x where the force vanishes, $f(x) = 0$ and $y = \frac{dx}{dt} = 0$. Often the potential is known and easy to graph. We wish to determine properties of our system in terms of the potential. **Equilibrium points are equivalent to critical points of the potential, since** V′(x)=0 **there.** In calculus, critical points (candidates for maxima and minima of the potential) are places where the derivative is zero (the slope of the potential is zero).

CENTERS AND SADDLES FOR THE PHASE PLANE AND THE POTENTIAL. We will show in three different ways that a **minimum of the potential is a center** and a **maximum of the potential is a saddle point**.

DERIVATION I. Suppose there is an equilibrium $x = a$ determined from $f(a) = 0$. From the differential equation (28), $m\frac{d^2x}{dt^2} - f(x) = 0$, we may approximate the differential equation in the neighborhood of an equilibrium. We obtain, using Taylor series for the function $f(x)$ or tangent line approximation or linearization,

$$m\frac{d^2x}{dt^2} - f(a) - f'(a)(x - a) = 0. \tag{32}$$

Introducing the displacement from equilibrium, $z = x - a$, and noting $f(a) = 0$, since a is an equilibrium, we obtain

$$m\frac{d^2z}{dt^2} + \alpha z = 0, \tag{33}$$

where α is a constant, $\alpha = -f'(a) = V''(a)$. This second-order constant coefficients linear differential equation is easy to solve by the methods of Chapter 2. Letting $z = e^{\lambda t}$, we find the characteristic equation, $m\lambda^2 + \alpha = m\lambda^2 + V''(a) = 0$. If $V''(a) > 0$, so that the **potential has a minimum**, the roots of the characteristic equation are imaginary, the solution to the linear differential equation is sinusoidal, and **the phase plane is a center**. If $V''(a) < 0$, so that the **potential has a maximum**, one of the roots of the characteristic equation is positive and the other negative, the solution to the linear differential equation is a linear combination of plus and minus exponentials, and **the phase plane is a saddle point**. We restrict our attention to $V''(a) \neq 0$.

DERIVATION II. The same result can be obtained directly from the energy integral,

$$E = \frac{1}{2}mv^2 + V(x). \tag{34}$$

We can approximate the potential near the equilibrium $V(x) = V(a) + V'(a)(x - a) + \frac{V''(a)}{2}(x - a)^2$ using a Taylor series. Since for an equilibrium $V'(a) = 0$ we obtain

$$E - V(a) = \frac{1}{2}mv^2 + \frac{V''(a)}{2}(x - a)^2. \tag{35}$$

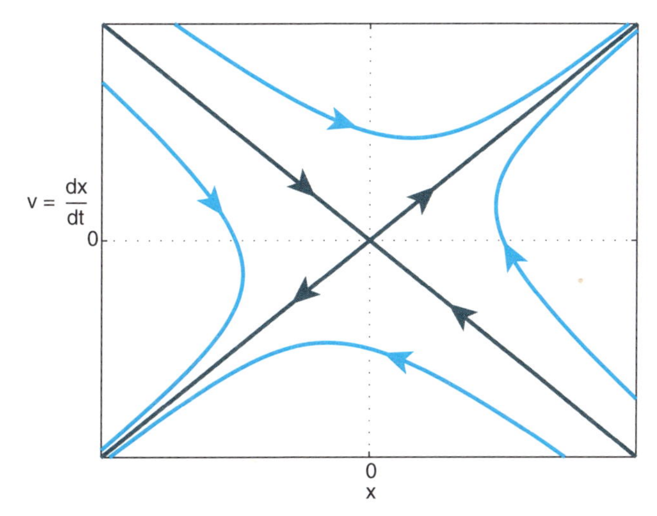

Figure 6.4.3 Phase plane of a saddle point.

Since E and $V(a)$ are constants, we see that the **phase plane will be ellipses (center) if $V''(a) > 0$. If $V''(a) < 0$, the phase plane will be hyperbolas (saddle point)**, graphed in figure 6.4.3, and in particular at energy $E = V(a)$ the phase plane near the equilibrium will have two intersecting straight lines corresponding to the eigenvectors of the saddle point. We add arrows to the right ($\rightarrow$) in the upper half-plane $y = \frac{dx}{dt} > 0$ since x increases in time there. Similarly, in the lower half-plane $y = \frac{dx}{dt} < 0$ and hence x decreases ($\leftarrow$) in time.

DERIVATION III. Let us obtain the same result using matrices and their eigenvalues of the matrix. For the system (29a)–(29b), the matrix is

$$\begin{bmatrix} \frac{\partial f_1}{\partial x} & \frac{\partial f_1}{\partial v} \\ \frac{\partial f_2}{\partial x} & \frac{\partial f_2}{\partial y} \end{bmatrix} = \begin{bmatrix} 0 & 1 \\ \frac{f'(a)}{m} & 0 \end{bmatrix}. \tag{36}$$

The eigenvalues of the matrix, obtained from $\det \begin{bmatrix} -\lambda & 1 \\ \frac{f'(a)}{m} & -\lambda \end{bmatrix} = 0$, satisfy $\lambda^2 = \frac{f'(a)}{m} = -\frac{V''(a)}{m}$. Thus, if $V''(a) > 0$, corresponding to a minimum of the potential, the eigenvalues are purely imaginary and the phase plane is a center, while if $V''(a) < 0$ corresponding to a maximum of the potential, one eigenvalue is positive and one eigenvalue is negative, and the phase plane is a unstable saddle point.

CENTERS. When the linear analysis has purely imaginary eigenvalues and the phase plane for the linear system are centers, this is the case in which the linear theory does not guarantee that the equilibrium is stable for the nonlinear system. However, we see from the examples to follow that the phase plane of the nonlinear system is a center and that the equilibrium is indeed a stable center for conservative systems if the potential has a local minimum at the equilibrium.

THE PHASE PLANE FROM A POTENTIAL USING CONSERVATION OF ENERGY. We claim that the phase plane x, $y = \frac{dx}{dt}$ can be determined by graphing the potential and using various constant values of the energy,

$$E = \frac{1}{2}my^2 + V(x). \tag{37}$$

The energy equation defines y as a function of x which is the phase plane. First we obtain

$$\frac{1}{2}my^2 = E - V(x). \tag{38}$$

From this, we learn that the **total energy must be greater than the potential energy** since kinetic energy is nonnegative. This will help our phase plane graph. Solving for y gives two symmetric curves since $y = \pm$. The method is best explained with examples. We immediately do one example, and later we will graph the phase plane for the nonlinear pendulum in this way. There is a different curve in the phase plane for each value of the constant energy E.

Example 6.4.1 *Conservative System*

Consider the conservative system

$$\frac{d^2x}{dt^2} = f(x) = x(6 - 3x) = 6x - 3x^2. \tag{39}$$

Find equilibria (classify) and graph the phase plane using the graph of the potential.

• SOLUTION. There are two equilibria, $x = 0, 2$. The potential satisfies

$$V(x) = -\int f(x)dx = -3x^2 + x^3. \tag{40}$$

The cubic potential is graphed in figure 6.4.4. The cubic has two critical points: $x = 0$ is a local maximum and $x = 2$ is a local minimum. Thus from this section, $x = 0$ is an (unstable) saddle point and $x = 2$ is a center. We will graph the phase plane x, $y = \frac{dx}{dt}$ directly underneath the potential with the same x variable. It is not necessary to write the energy equation, but it is being used throughout and is

$$E = \frac{1}{2}y^2 - 3x^2 + x^3. \tag{41}$$

The procedure is more straightforward, with some practice, than our written description. We introduce a few allowable constant values of E on the graph of the potential; see Figure 6.4.4. One important value of E corresponds to the local maximum of the potential at $x = 0$, $E = V(0) = 0$. At this value of E, there is a maximum allowable value of x, which we can compute (but is not necessary) to graph the qualitative features of the phase plane. We draw vertical lines down to the phase plane at $x = 0$ and the maximum value of x. Now we go down to the phase plane and draw the phase plane diagram for that value of E. At $x = 0$ and at the maximum value of x, E equals the the potential energy $V(x)$, so that the velocity $y = \frac{dx}{dt} = 0$ at these two points. Here $x = 0$, $y = 0$ is the equilibrium which we know to be a saddle, while the largest value of x (which also has $y = 0$) is not an equilibrium. We draw the $E = 0$ portion of the phase

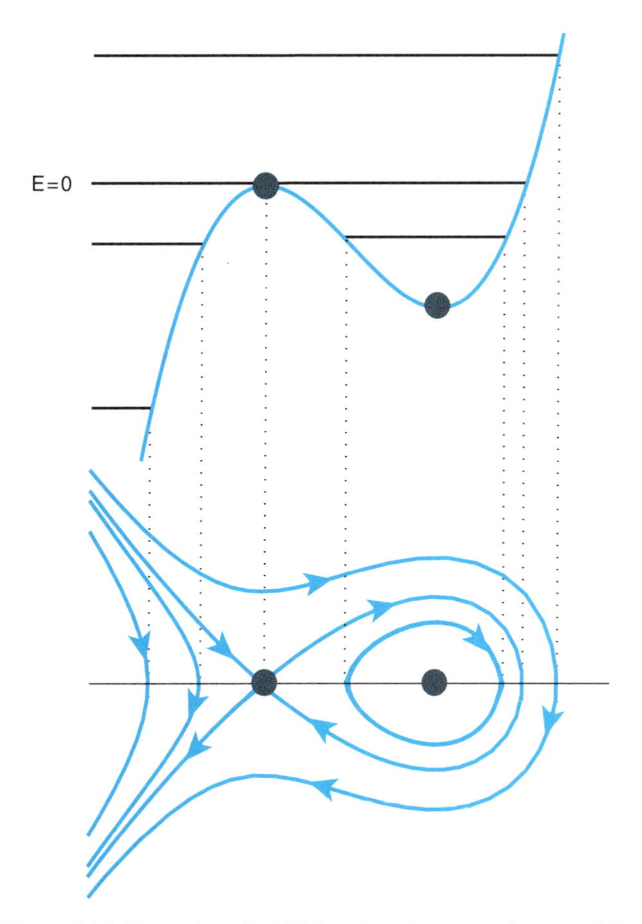

Figure 6.4.4 Phase plane for (39) based on the potential energy (40).

plane near the saddle point which is two straight lines near the saddle. We continue drawing a small portion of the phase plane with this one fixed constant E. First we do larger and larger values of x starting from the saddle point $x = 0$. Since E is getting larger than the potential $V(x)$, the velocity in the phase plane $y = \frac{dx}{dt}$ will continue to increase until its maximum (which occurs at the center $x = 2$) then decrease to the maximum value of x. Since the curve is symmetric in y, we have drawn one closed curve in this way which looks like an oval pointed near the saddle point. We must also graph the phase plane for the same constant value of $E = 0$, but include x less than the saddle point $x = 0$. As x diminishes away from $x = 0$, $V(x)$ continues to get smaller than E so $y = \frac{dx}{dt} \to \pm\infty$ as $x \to -\infty$, as shown in the figure pendulum, with the continuation from the straight line associated with the saddle point when $E = 0$. The center $x = 2$ corresponds to $E = V(2) = -12 + 8 = -4$. Other values of $-4 < E < 0$ give smooth closed curves inside the pointed loop. For $E < 0$, there are other orbits (see figure 6.4.4), which turn around at their maximum and $y = \frac{dx}{dt} \to \pm\infty$. Orbits for positive values of E are also sketched. They have a maximum value of x, follow "near" the pointed loop for a while, and then going off to $x = -\infty$. The phase plane must be symmetric in $y = \frac{dx}{dt}$, so two curves must be drawn in all cases. Note the saddle point at $x = 0$, $y = 0$ and the center at $x = 2$, $y = 0$. **For all conservative**

systems, the center (in this example $x=2,\ y=0$) is really stable, even though our theorem about the linear center did not guarantee it would be stable.

After the phase plane has been drawn, arrows must be added. **For all conservative systems arrows move to the right ($\rightarrow$)in the upper half-plane (since there $y=\frac{dx}{dt}>0$) and to the left ($\leftarrow$) in the lower half-plane (since $y=\frac{dx}{dt}<0$).** To understand the solutions, we see that each closed orbit inside the pointed loop is a periodic solution. We will show how to calculate its period. The pointed loop is called a **separatix** because it separates solutions with different kinds of behavior. ♦

NONLINEAR PENDULUM WITHOUT DAMPING. We return to the nonlinear pendulum without damping, (3), $ml\frac{d^2x}{dt^2}+mg\sin x=0$. Dividing by ml yields

$$\frac{d^2x}{dt^2}=f(x)=-\beta\sin x, \tag{42}$$

where $\beta=\frac{g}{l}>0$. Some may prefer to make a scaling of time, so that the equation would be equivalent to $\beta=1$, but we leave β. Converting to the usual system gives

$$\frac{dx}{dt}=y=f_1(x,y), \tag{43a}$$

$$\frac{dy}{dt}=-\beta\sin x=f_2(x,y). \tag{43b}$$

The equilibria are $\sin x=0,\ y=0$, or

$$y=0,\ x=n\pi \qquad \text{for } n=0,\pm1,\pm2,\ldots \tag{44}$$

Note that the differential equation (42) is the same when shifted by $x=2\pi$, so that the phase plane will repeat itself every 2π units in x. Note that x is the angle of the pendulum, and the physical problem is the same when x is increased by 2π. The equilibrium $x=0$ is called the **natural position** of the pendulum, and it is the same as $x=2n\pi$. The equilibrium $x=\pi$ is called the **inverted position** of the pendulum, and it is the same as $x=\pi+2n\pi$. To determine the behavior near the equilibrium, we need the eigenvalues of the linearized coefficient matrix

$$\begin{bmatrix} \frac{\partial f_1}{\partial x} & \frac{\partial f_1}{\partial v} \\ \frac{\partial f_2}{\partial x} & \frac{\partial f_2}{\partial y} \end{bmatrix}=\begin{bmatrix} 0 & 1 \\ -\beta\cos x & 0 \end{bmatrix}. \tag{45}$$

The eigenvalues of the matrix, obtained from $\det\begin{bmatrix} -\lambda & 1 \\ -\beta\cos x & -\lambda \end{bmatrix}=0$, satisfy the characteristic equation

$$\lambda^2+\beta\cos x=0. \tag{46}$$

There are two equilibria.

NATURAL POSITION ($x=0$). The equilibrium is $x=\pi,\ y=\frac{dx}{dt}=0$. Then the characteristic equation is $\lambda^2+\beta=0$. The eigenvalues (roots) are $\lambda=\pm i\sqrt{\beta}$. This suggests the possibility of, but does not prove that, there are periodic solutions. To determine what solutions look like near the equilibriums look like, we will graph the phase plane shortly.

INVERTED POSITION ($x=\pi$). The equilibrium is $x=\pi$, $y=\frac{dx}{dt}=0$. Then the characteristic equation is $\lambda^2-\beta=0$ with real roots $\lambda=\pm\sqrt{\beta}$ of opposite signs. These equilibria, which correspond to the pendulum delicately balanced sticking straight up, are unstable (saddles).

ENERGY EQUATION. We determine the phase plane and determine the behavior near and far from the natural equilibrium position $x=0$ using the energy equation. The kinetic energy is $\frac{1}{2}y^2$, and the potential energy is $V(x)=-\int f(x)=\beta\int\sin x=-\beta\cos x$. Thus,

$$E=\frac{1}{2}y^2-\beta\cos x. \tag{47}$$

The energy equation can be derived from the slope field equation.

PHASE PLANE FROM POTENTIAL ENERGY. To find the phase plane, we first graph the potential energy and various values of E, where

$$V(x)=-\beta\cos x. \tag{48}$$

There are a number of cases (see figure 6.4.5). The lowest energy, $E=-\beta$, corresponds only to the equilibrium at the natural position $x=0$, $x=2n\pi$. If $-\beta<E<+\beta$, then there are a minimum and maximum value of x which the time-dependent solution periodically oscillates between, and the phase plane consists of oval-shaped closed curves (marked C in the phase plane) surrounding the natural position $x=0$. We put in arrows counterclockwise since arrows move to the right ($\rightarrow$)in the upper half-plane (since there $y=\frac{dx}{dt}>0$) and to the left ($\leftarrow$) in the lower half-plane (since $y=\frac{dx}{dt}<0$). This proves that the natural position is stable, and nearby trajectories look like a center. Here x is the angle of the pendulum, so that the angle of the pendulum oscillates periodically around the natural position. The energy level $E=\beta$ corresponds to the sequence of identical saddle points at the inverted equilibrium position, for example, $x=\pi$, $y=\frac{dx}{dt}=0$, but there are orbits that go from one saddle point to the next: see orbits marked B in figure 6.4.5. The trajectories B correspond to the situation when the initial energy is just right, so that the pendulum approaches the unstable inverted equilibrium. If $E>\beta$, there is no minimum or maximum value of x. We put in arrows which move right ($\rightarrow$)in the upper half-plane (since there $y=\frac{dx}{dt}>0$) and left in the lower half-plane. Trajectories A in figure 6.4.5 correspond to the situation when the pendulum is given enough initial velocity to keep swinging around the pivot point of the pendulum, and the angle of the pendulum keeps increasing (upper half-plane) or decreasing (lower half-plane). The trajectories B separate the trajectories C, where the pendulum oscillates around the stable natural equilibrium position from those like A, where it keeps whirling around the pivot point. For this reason, trajectories B are called **separatrices**.

PERIOD OF PERIODIC ORBITS. A closed orbit (that does not have an equilibrium on it) in the phase plane corresponds to a periodic solution. To show that it is periodic, consider a specific initial condition. Since the orbit is closed, after some time (the period) the solution returns to that point. Now the differential equation can be considered to have the same initial condition, and hence, by the uniqueness theory of the initial value problem, the time-dependent solution will be the same again and again.

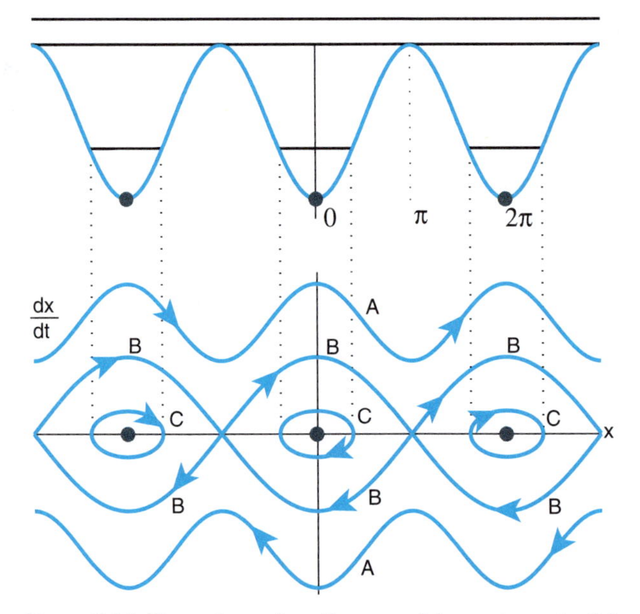

Figure 6.4.5 Phase plane of nonlinear pendulum using potential.

PERIOD. The system of equations (1a) and (1b) from section 6.1 can be solved for dt:

$$dt = \frac{dx}{f(x, y)} = \frac{dy}{g(x, y)}. \tag{49}$$

The **period** T may be computed from either of the following equations (obtained by integrating (49) around the orbit):

$$T = \oint \frac{dx}{f(x, y)} = \oint \frac{dy}{g(x, y)}. \tag{50}$$

Sometimes the x-integral is easier. If the orbit is symmetric, as it is for conservative systems, then the period is twice the integral from the minimum x to the maximum x:

$$\text{period} = T = 2\int_{x_{\min}}^{x_{\max}} \frac{dx}{f(x, y)}, \tag{51}$$

assuming $\frac{dx}{dt} \geq 0$.

PERIOD OF A LINEAR SPRING-MASS SYSTEM (LINEAR OSCILLATOR). Consider the linear spring-mass system without damping satisfying the differential equation

$$m\frac{d^2x}{dt^2} + kx = 0. \tag{52}$$

First we obtain the period using the general solution $x = c_1 \cos \omega t + c_2 \sin \omega t$, where $\omega = \sqrt{k/m}$. The period is $T = 2\pi/\omega = 2\pi\sqrt{m/k}$. One property of linear oscillators is that the period is a constant and does not depend on the amplitude or the energy of the oscillation. The phase plane consists of a family of concentric ellipses (see Chapter 4). One way to obtain the ellipses is directly from conservation of energy:

$$E = \frac{1}{2}m\left(\frac{dx}{dt}\right)^2 + \frac{1}{2}kx^2. \tag{53}$$

To obtain the period from (53), we note $dt = \frac{dx}{(dx/dt)}$ and solve for $\frac{dx}{dt} = \pm\sqrt{\frac{k}{m}}\sqrt{\frac{2E}{k} - x^2}$ from the energy equation. In the upper half-plane $\left(\frac{dx}{dt} > 0\right)$ we choose the plus sign. It is also necessary to know that $x_{\text{min}} = -\sqrt{\frac{2E}{k}}$ and $x_{\text{max}} = \sqrt{\frac{2E}{k}}$ determined from (53), since $\frac{dx}{dt} = 0$ at maxima and minimas. Using (51) and evaluating this integral yield

$$T = 2\sqrt{\frac{m}{k}}\int_{-\sqrt{\frac{2E}{k}}}^{\sqrt{\frac{2E}{k}}} \frac{dx}{\sqrt{\frac{2E}{k} - x^2}} = 2\sqrt{\frac{m}{k}}\sin^{-1}\left(\frac{x}{\sqrt{\frac{2E}{k}}}\right)\Bigg|_{-\sqrt{\frac{2E}{k}}}^{\sqrt{\frac{2E}{k}}} = 2\pi\sqrt{\frac{m}{k}}, \tag{54}$$

The answer appears to depend on the energy. However, by evaluating the integrals above we see that the period does not depend on energy, as we obtained more easily from the solution of the differential equation.

PERIOD OF A NONLINEAR OSCILLATOR (CONSERVATIVE SYSTEM). Let us determine the period for conservative systems satisfying

$$m\frac{d^2x}{dt^2} = f(x). \tag{55}$$

Conservation of energy is quite useful:

$$E = \frac{m}{2}\left(\frac{dx}{dt}\right)^2 + V(x), \tag{56}$$

where the potential energy satisfies $V(x) = -\int f(x)dx$. We assume the potential is shaped like a well somewhere along the x-axis. The phase planes will consist of regions with closed curves corresponding to periodic solutions. The period can be computed from $dt = \frac{dx}{(dx/dt)} = \frac{dx}{v}$, where $\frac{dx}{dt} = v = \pm\sqrt{\frac{2}{m}}\sqrt{E - V(x)}$ is the velocity determined from the energy equation. We can use (51). In this way, by integrating, we obtain

$$\text{period} = T(E) = 2\sqrt{\frac{m}{2}}\int_{x_{\text{min}}(E)}^{x_{\text{max}}(E)} \frac{dx}{\sqrt{E - V(x)}}. \tag{57}$$

We have noted that the minimum and maximum of x depend on the energy E, as can be seen in figure 6.4.4. In general, the period $T(E)$ for the nonlinear oscillator will depend on the energy E.

NONLINEAR PENDULUM WITH DAMPING. This is a nonconservative system and is briefly discussed in the exercises.

Exercises

In Exercises 1–6, the potentials are given. Find and classify the equilibria (node, saddle, spiral, center, stable, or unstable), and graph the phase plane:

1.

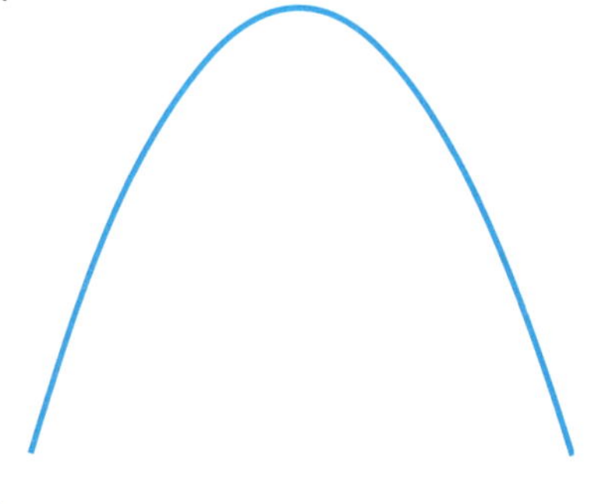

2.

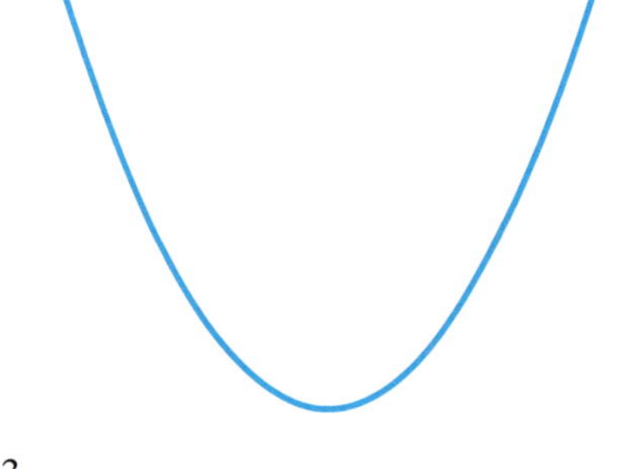

3.

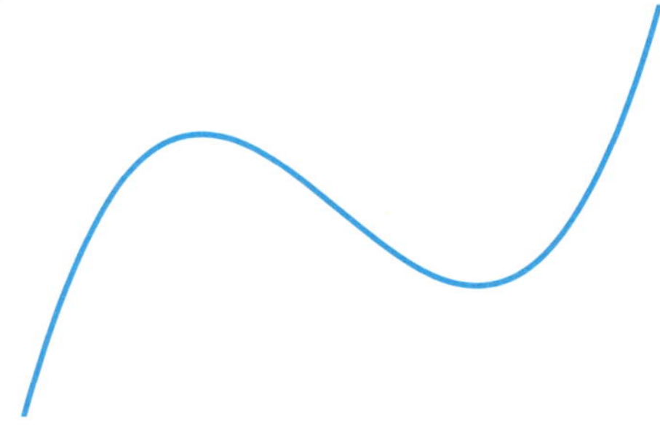

4.

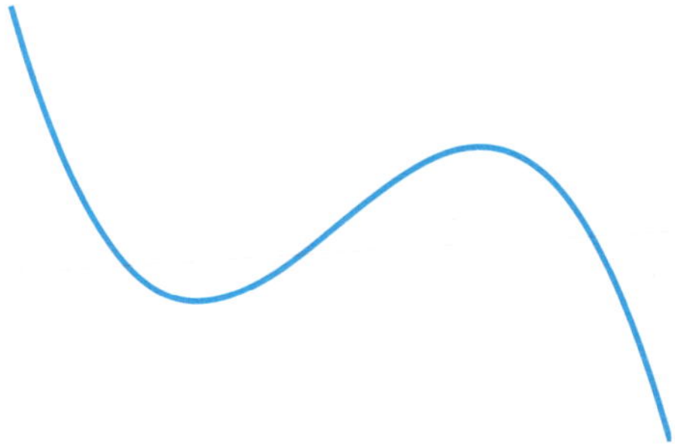

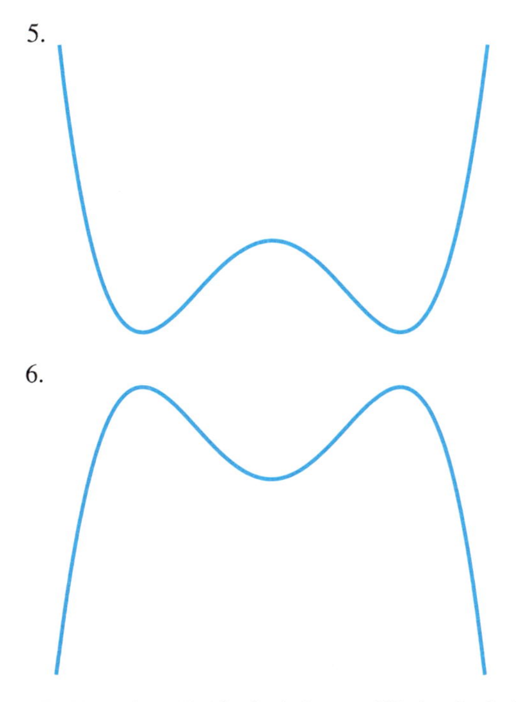

In Exercises 7–12, find the equilibria, find the potential, classify the equilibria (node, saddle, spiral, center, stable or unstable), and graph the phase plane using the potential

7.
$$\frac{d^2x}{dt^2} = x(1+x^2). \tag{58}$$

8.
$$\frac{d^2x}{dt^2} = -x(1+x^2). \tag{59}$$

9.
$$\frac{d^2x}{dt^2} = x(1-x^2). \tag{60}$$

10.
$$\frac{d^2x}{dt^2} = -x(1-x^2). \tag{61}$$

11.
$$\frac{d^2x}{dt^2} = -x(6-3x) = -6x+3x^2. \tag{62}$$

12.
$$\frac{d^2x}{dt^2} = (1-x^2)(4-x^2). \tag{63}$$

13. Derive conservation of energy for the pendulum without damping using the slope field equation.
14. Derive conservation of energy for a general conservative system using the slope field equation.
15. Derive conservation of energy for a general conservative system by multiplying the differential equation by $\frac{dx}{dt}$.
16. Derive conservation of energy for a general conservative system by showing that the derivative of energy is zero.

In exercises 17–19, suppose that in the pendulum there is a resistive force of $-\delta\frac{dx}{dt}$. Determine the nature of the equilibrium under the given parameters.

17. $l = 32$ ft, $m = 1$slug, $\delta = 1$ slug/s.

18. $l = 32$ ft, $m = \frac{1}{2}$slug, $\delta = 1$ slug/s.

19. $l = 32$ ft, $m = \frac{1}{3}$slug, $\delta = 1$ slug/s.

20. Determine the values of δ for which $(0, 0)$ is a stable spiral and stable node. Interpret the answer in terms of under-damping and over-damping.

In Exercises 21–25, consider the damped pendulum $\frac{d^2x}{dt^2} + \sin x = -\delta\frac{dx}{dt}$. Determine the phase plane using software or numerically determined solutions.

21. $\delta = 0$.
22. $\delta = 0.5$.
23. $\delta = 0.25$.
24. $\delta = 2.5$.
25. $\delta = 2.0$.

ANSWERS TO ODD-NUMBERED EXERCISES

Chapter 1: First-Order Differential Equations and their Applications

1.2 The Definite Integral and the Initial Value Problem

1–7. Substitute the expression for x into the differential equation.

9. $x = 3e^t + c.$

11. $x = -\frac{5}{6}\sin 6t + c.$

13. $x = 8\int_0^t \cos\bar{t}^{-1/2} d\bar{t} + c.$

15. $x = \int_0^t \ln(4 + \cos^2\bar{t})d\bar{t} + c.$

17. $x = \frac{1}{5}t^5 - \frac{17}{5}.$

19. $x = \int_2^t \frac{\ln\bar{t}}{4+\cos^2\bar{t}} d\bar{t} + 5.$

21. $x = \int_1^t \frac{e^{\bar{t}}}{1+\bar{t}} d\bar{t} + 3.$

23. $v_0 = \sqrt{2250}$ m/ sec $= 15\sqrt{10}$ m/ sec.

25. $x = 0.72$ km.

27. $x = \frac{v_0^2}{5000}$ km.

29. $x = 2\left(\frac{62}{3}\right)^{\frac{3}{2}} - 116.$

31. (b) $x = \frac{1}{k}\ln\frac{4}{3}$ m.

33. $v_0 = \sqrt{1960}$ m/ sec.

35. $t = \frac{20}{\sqrt{9.8}}$ sec.

37. $c = x_0 - \int_0^{t_0} f(\bar{t})d\bar{t}.$

1.3 First-Order Separable Differential Equations

1. $\ln|x+1| = \ln|t| + C$ or $x = ct - 1.$

3. $x = e^t + c.$

5. $x = ce^{(t^2/4)+4t} - 3.$

7. $x = 3t + c.$

9. $x = (9 - 4t)^{-1/4}.$

11. $$x = \frac{1}{1 - \int_0^t \cos\bar{t}^2 d\bar{t}}.$$

13. $$\int_2^x \frac{d\bar{t}}{\cos\bar{t}^{-1/2}} = \frac{1}{2}(t^2 - 1).$$

15. $u^3 + 12u = t^3 + 3t + 13$.

17. $x = \tan\left(\frac{t^3}{3} + t + \tan^{-1} 2\right)$.

19. $-\ln|x| + \ln|x-1| = x + c$; $x = (1 - ke^t)^{-1}$ and $x = 0$.

21. $\ln|x-1| - \ln|x-2| - (x-2)^{-1} = t + c$; $x = 1, x = 2$.

23. $te^t xe^x = \bar{c}$.

25. $\left(\frac{z-3}{z+3}\right) = \bar{c}\left(\frac{t+2}{t-2}\right)^{3/2}$.

27. $-e^{-t} - \frac{1}{4}e^{-4x} = c$; $x = \frac{1}{4}\ln(-4e^{-t} + k)$.

29. $z' = bf(z) + a$.

31. $\frac{1}{2}\tan^{-1}(2t + 8x - 2) = t + c$; $x = \frac{1}{4}\left(1 - t + \frac{1}{2}\tan(2t + k)\right)$.

33. $(t + x + 1)e^{-t-x} = -t + c$.

1.4 Direction Fields

1.

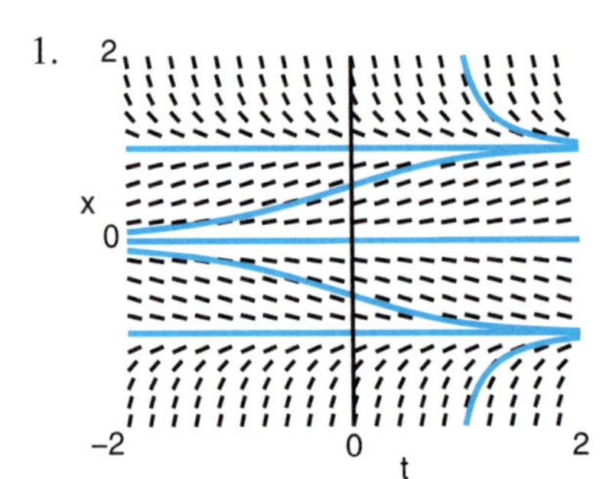

3.

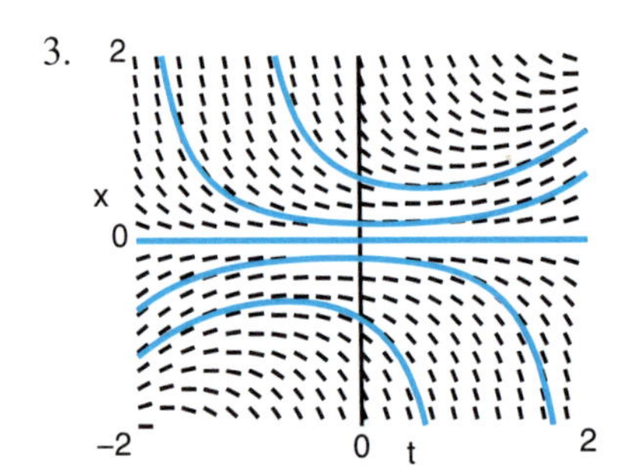

5.

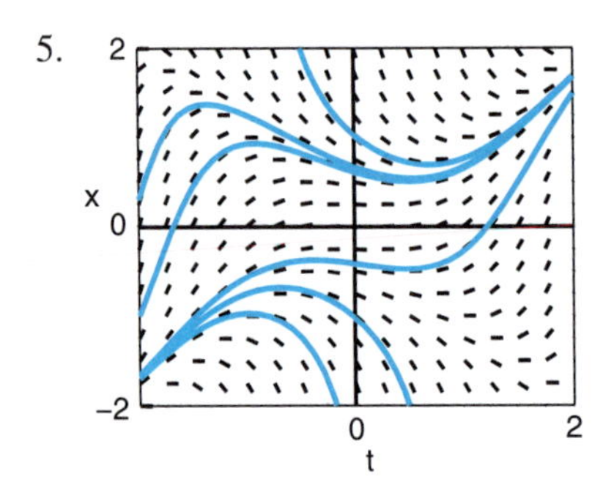

1.4.1. Existence and Uniqueness

1. All (t_0, x_0).
3. All (t_0, x_0).
5. $t_0 + x_0 \neq 0$.
7. $x_0 \neq 1$.
9. $t_0^2 + 2x_0^2 < 1$.
11. $f(t,x) = t^{-1/3}$, $f_x = 0$ (if $t \neq 0$); both continuous if $t \neq 0$.
13. $x_0 = 0$.
15. $x_0 = 0$.
17. $t_0 = n\pi, n = 0, \pm 1, \pm 2, \ldots$.
19. None.
21. $x = 1$ and $x = \left(\frac{4}{5}(t - t_0)\right)^{5/4} + 1$, f_x not continuous at $x_0 = 1$.

1.5 Euler's Numerical Method

1. 3
3. 0.
5. 5.
7. 0.
9. a) 59, 049.
11. a) 4.15×10^{-20}.
13. 4.109, 6.129, 30.390, 193.137.
15. $x(2) = 551, 627$. Appears to be going from -5 to infinity.
17. $x_N = x_{20} = 0.601$ with $h = 0.1$.

1.6 First-Order Linear Differential Equations

1.6.1 Solutions of Homogeneous First-Order Linear Differential Equations

1. $x = ce^{3t}$.
3. $x = c\exp(t^2)$.
5. $x = ct^{-1/2}$.
7. $x = ce^{-\sin t}$.
9. $x = c\exp\left(-\int_0^t \cos s^{-1/2} ds\right)$.
11. $x = 9e^{-5t}$.
13. $x = 7e^{9(t-3)}$.
15. $x = 10\exp\left(-\int_5^t \frac{\sin s}{4+e^s} ds\right)$.
17. $x = 3\exp\left(\frac{1}{t} - 1\right)$.

1.6.2 Integrating Factors for First-Order Linear Differential Equations

19. $x = t^{-1}e^t + (1-e)t^{-1}$.

21. $x = 3e^t + c$ (note: This can be done by just antidifferentiating both sides).

23. $x = t(t^2+1)^{-1} + c(t^2+1)^{-1}$.

25. $x = \frac{1}{4}t - \frac{1}{16} + \frac{1}{16}e^{-4t}$.

27. $x = 4t^2$.

29. $x = t^{-1}\ln t + ct^{-1}$.

31. $x = \frac{1}{4}t + ct^{-3}$. One solution continuous at $(0, 0)$; for the rest, $|x| \to \infty$ as $t \to 0$.

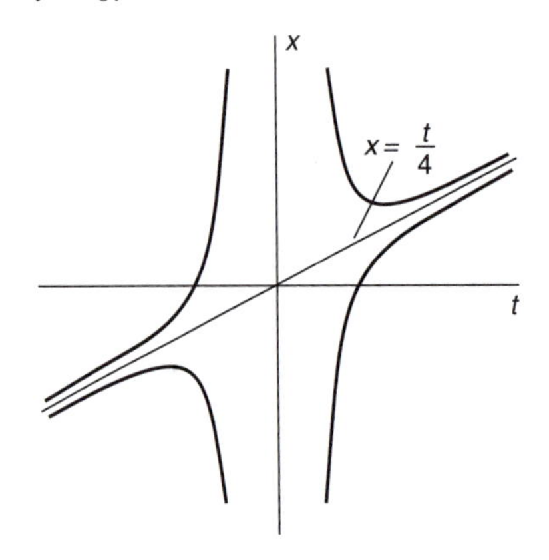

33. $x = t^2 + 3t$. All solutions continuous and pass through $(0, 0)$.

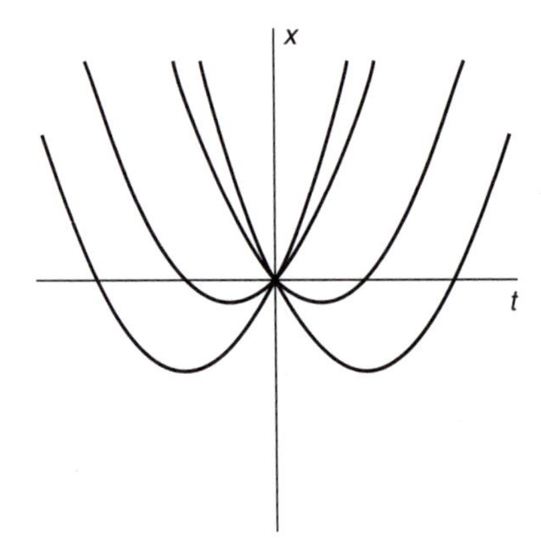

35. $x = e^{-t^2}\int_0^t e^{s^2}ds + ce^{-t^2}$.

37. $x = e^{-t^3/3}\int_0^t e^{s^3/3}s\,ds + ce^{-t^3/3}$.

39. $x = e^{-e^t}\int_0^t 3e^{e^s}ds + ce^{-e^t}$.

41. $x = e^{-t}\int_2^t \frac{e^s}{s+1}ds + e^{-(t-2)}$.

43. $x = \frac{1}{3}t^{1/3}\int_5^t s^{-\frac{1}{3}}\sin s\,ds$.

45. $x = \frac{1}{7}t^{-\frac{1}{7}}\int_0^t s^{-\frac{6}{7}}e^s ds + ct^{-\frac{1}{7}}$.

47. $t = -x - 1 + ce^x$.
49. Hint. $e^{\int p(t)dt + c_1} = e^{\int p(t)dt} e^{c_1} = ue^{c_1}$, so the new integrating factor is just a constant times the old one.
51. $(ux)' = u(x' + px) \longleftrightarrow (kux)' = ku(x' + px)$.
53. $G(t, \bar{t}) = \exp\left(\int_t^{\bar{t}} \frac{\sin s}{s} ds\right) = \exp\left(-\int_{\bar{t}}^{t} \frac{\sin s}{s} ds\right)$.
55. $\frac{du}{dt} = qe^{\int pdt}$, $u = \int qe^{\int pdt} dt$, $x = e^{-\int pdt} \int qe^{\int pdt} dt$.

1.7 Linear First-Order Differential Equations with Constant Coefficients with a Constant Input

1. $x = ce^{8t}$.
3. $x = ce^{-2t}$.
5. $x = ce^{-7t}$.
7. $x = ce^{-t}$.
9. $x = ce^{5t}$.
11. $x = \frac{8}{3} + ce^{-3t}$.
13. $x = \frac{9}{4} + ce^{4t}$.
15. $x = -\frac{15}{4} + ce^{4t/5}$.
17. $x = -2 + ce^{-2t/3}$.
19. $x = -9 + ce^{2t}$.
21. $x = -\frac{17}{3} + ce^{-t}$.
23. $x = \frac{8}{3}e^{-4t} + ce^{-7t}$.
25. $x = \frac{3}{7}e^{-5t} + ce^{2t}$.
27. $x = \frac{3}{8}e^{4t} + ce^{-4t}$.
29. $x = e^{-2t} + ce^{-3t}$.
31. $x = \frac{5}{7}t + \frac{16}{49}$.
33. $x = 7t - \frac{7}{2}$.
35. $x = 2t^2 + t - 9$.
37. $x = \frac{1}{3}t^3 - \frac{1}{3}t^2 + \frac{2}{9}t - \frac{2}{27}$.
39. $x = \frac{1}{5}t - \frac{1}{25}$.
41. $x = -\frac{9}{20}\cos 6t + \frac{3}{20}\sin 6t$.
43. $x = -\frac{5}{13}\cos t + \frac{1}{13}\sin t$.
45. $x = \frac{1}{10}\cos 2t + \frac{13}{10}\sin 2t$.
47. $x = \frac{6}{37}\cos t + \frac{1}{37}\sin t - \frac{5}{61}\cos 5t + \frac{6}{61}\sin 5t$.
49. $x = -\frac{5}{9} - \frac{1}{15}\cos 3t - \frac{1}{5}\sin 3t$.
51. $x = \frac{1}{2}e^{3t} - \frac{1}{2}\cos t + \frac{1}{2}\sin t$.
53. $x = 8te^{-3t} + ce^{-3t}$.

55. $\frac{dv}{dt} = 7$, $v = 7t + c$, $x = 7te^{2t} + ce^{2t}$.

57. $x = 4te^{t} + ce^{t}$.

59. $x = 5te^{-t} + ce^{-t}$.

61. $x = 8te^{7t} + ce^{7t}$.

63. $x = \begin{cases} 2e^{-2t}, & 0 \le t \le 1, \\ 2e^{-1}e^{-t} = 2e^{-(t+1)}, & 1 \le t \le 2. \end{cases}$

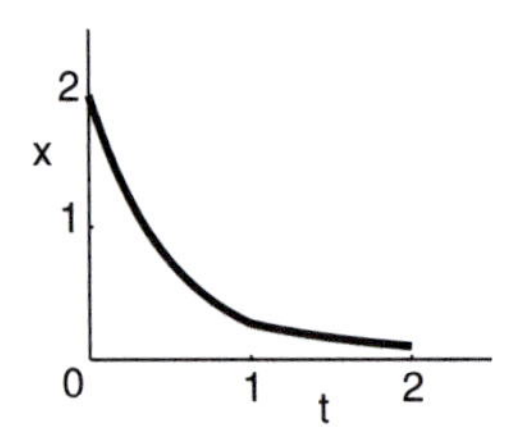

65. $x = \begin{cases} t, & 0 \le t \le 1, \\ 2 - t, & 1 \le t \le 2. \end{cases}$

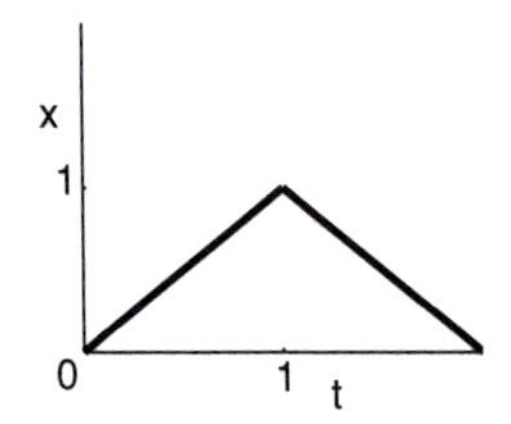

67. $x = \begin{cases} 2e^{-t}, & 0 \le t \le 1, \\ t + 2e^{-1} - 1, & 1 \le t \le 2. \end{cases}$

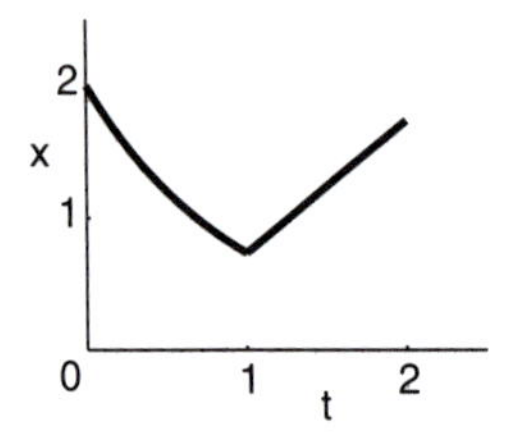

1.8 Growth and Decay Problems

1. $t_{\text{doubling}} = \frac{\ln 2}{.015} \approx 46.2$ years.

3. $k = 3\ln 2 \approx 2.08 = 208\%$ per day.

5. $x = 1500\left(\frac{4}{3}\right)^4$.

7. 9.97 years.

9. $Q' = k_1 Q - k_2 Q + k$.

11. 3% per year implies $t_{\text{doubling}} = \frac{\ln 2}{0.03} \approx 23.10$ years,
3% yield implies $t_{\text{doubling}} = \frac{\ln 2}{\ln(1.03)} \approx 23.45$ years.

13. $t_{\text{doubling}} = \frac{\ln 2}{\ln(1.064)} \approx 11.17$ years.

15. $k = \ln(1.1) \approx 0.0953 = 9.53\%$ per year.

17. 110.04 g.

19. a) 31.063 yr, b) 75.018 yr.

21. a) $\frac{10 \ln 17}{\ln 17 - \ln 15} \approx 226.4$ min.

b)

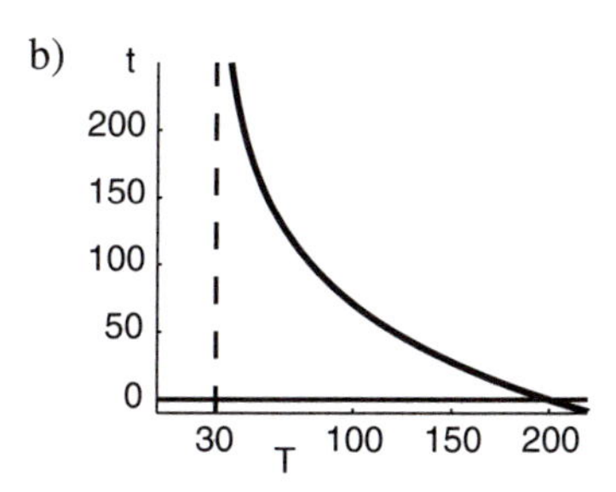

$t(T) = \frac{10}{\ln 15 - \ln 17} \ln\left(\frac{T-30}{170}\right)$.

23. $k = -\frac{1}{10} \ln 3 \approx -0.1099$.

25. a) $\frac{dT}{dt} = -k(T - Q_0) = -k[T - (20 + 10t)]$.

b) $T = 10 + 10t + 30e^{-t}$.

c)

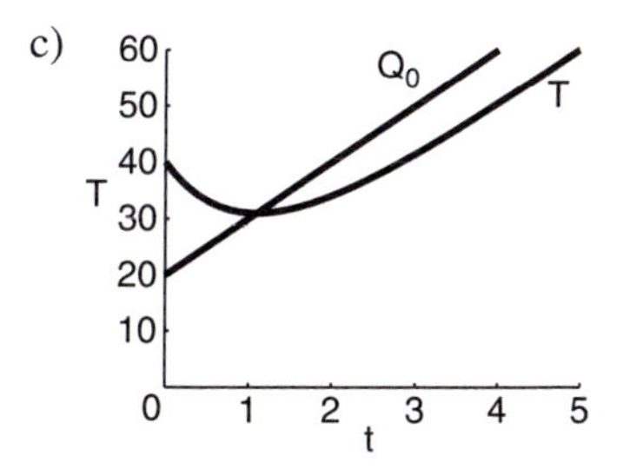

27. a) $\frac{dA}{dt} = 0.06(A - 500)$, $A(0) = 2000$.

b) $A = 500 + 1500e^{0.06t}$, \$3233.18.

c) \$411.06.

29. $Q' = 0.08Q + 365B$, $Q(0) = 1000$.

a) $B = \$3.79$.

b) $B(t) = (800 - 80e^{0.08t})(365e^{0.08t} - 365)^{-1}$.

31. $Q' = 0.2Q + 400 \cos 2\pi t$, $Q(0) = 100$.

a) $Q = \dfrac{1}{\pi^2 + 0.01}(-20 \cos 2\pi t + 200\pi \sin 2\pi t) + \left(100 + \dfrac{20}{\pi^2 + 0.01}\right) e^{0.2t}$.

b)

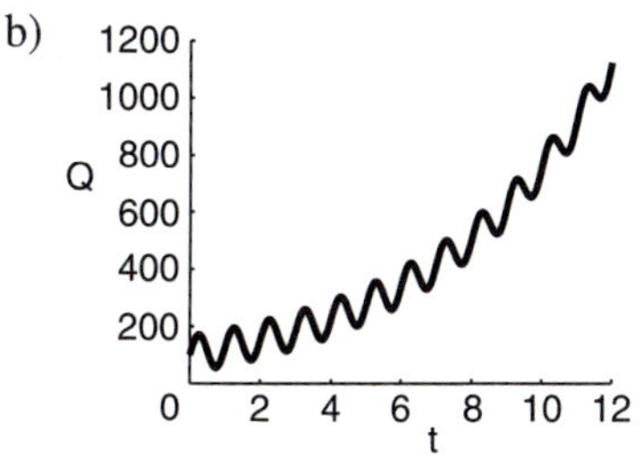

1.9 Mixture Problems

1. a) $Q(t) = 120 - 60e^{-t/150}$ lb,
   $c(t) = \frac{Q}{300} = \frac{2}{5} - \frac{1}{5}e^{-t/150}$ lb/gal.
   b) $t = -150\ln(0.5) \approx 104$ min.
3. a) $Q(t) = 25 - e^{-t/25}15$, $c(t) = 0.25 - e^{-t/25}0.15$,
   b) $\lim_{t\to\infty} c(t) = 0.25 > 0.2$.
5. a) and b)

$$Q(t) = \begin{cases} 10\left(\dfrac{100}{100+t}\right)^{1/5}, & 0 \le t \le 100, \\ 10e^{3(100-t)/500}(2^{-1/5}), & 100 \le t, \end{cases}$$

$$c(t) = \begin{cases} \dfrac{Q(t)}{500+5t}, & 0 \le t \le 100, \\ \dfrac{Q(t)}{1000}, & 100 \le t. \end{cases}$$

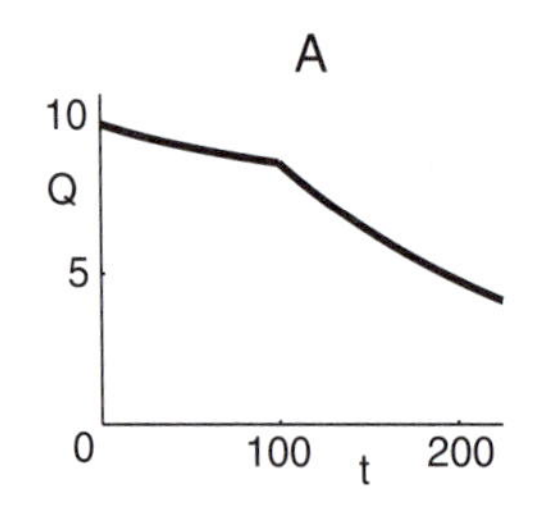

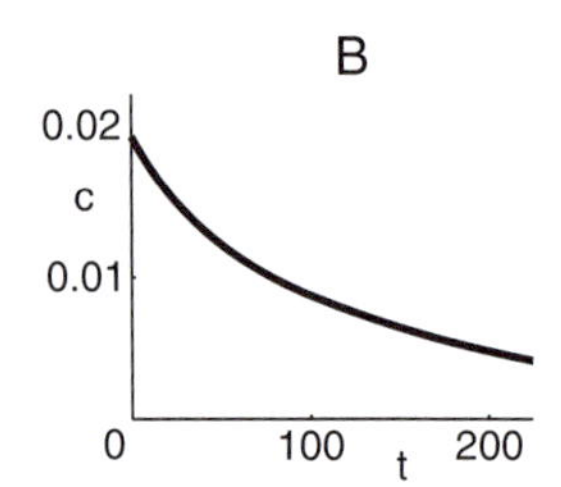

7. $Q(t) = 7(1000+3t) - 500000(1000+3t)^{-2/3}$,
   $c(t) = 7 - 500000(1000+3t)^{-5/3}$.
9. $\dfrac{dS_1}{dt} = 13 \cdot 3 - 7\dfrac{S_1}{150+6t}, \dfrac{dS_2}{dt} = 7\dfrac{S_1}{150+6t} - 28\dfrac{S_2}{250-21t}$.
11. $\dfrac{dS_1}{dt} = 21 \cdot 5 - 18\dfrac{S_1}{230+3t}, \dfrac{dS_2}{dt} = 9\dfrac{S_1}{230+3t} - 4\dfrac{S_2}{275+5t}$.
13. $\dfrac{dS_1}{dt} = 11 \cdot 5 - 18\dfrac{S_1}{100-7t}, \dfrac{dS_2}{dt} = 18\dfrac{S_1}{100-7t} - 18\dfrac{S_2}{200}, \dfrac{dS_3}{dt} = 18\dfrac{S_2}{200}$.

1.10 Electronic Circuits

1. $\dfrac{di}{dt} + 2i - 1 = 0$. Equilibrium is $i = \frac{1}{2}$, $i = \frac{1}{2} + ce^{-2t} = \frac{1}{2} + \left[i(0) - \frac{1}{2}\right]e^{-2t}$.
3. $\dfrac{di}{dt} + i - \sin t = 0$, $i = \frac{1}{2}\sin t - \frac{1}{2}\cos t + \left[i(0) + \frac{1}{2}\right]e^{-t}$.
5. $i(t) = \begin{cases} 9(1-e^{-t}) & 0 \le t \le 10, \\ 9(e^{10}-1)e^{-t} & t \ge 10. \end{cases}$
7. $\dfrac{dq}{dt} + q = 3\sin t$, $q(0) = 1$, $q = \frac{3}{2}(\sin t - \cos t) + \frac{5}{2}e^{-t}$.

1.11 Mechanics II: Including Air Resistance

1. a) $v = -1960(1 - e^{-t/2})$.
   b) $v = -1947$.
   c) -1960.
3. $-14,416$ cm/s.
5. a) $v = -\sqrt{32}\dfrac{(1 + ke^{-2\sqrt{32}t})}{(1 - ke^{-2\sqrt{32}t})}$, $k = (1000 + \sqrt{32})(1000 - \sqrt{32})^{-1}$,
   b) $v = -\sqrt{32}$.

1.12 Orthogonal Trajectories

1. $x = -t + c_2$.

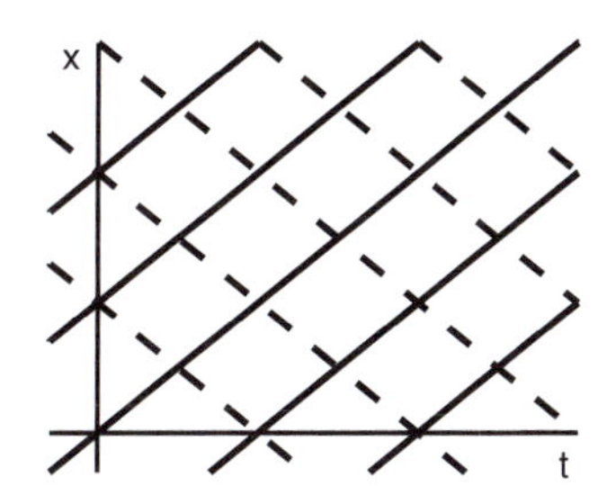

3. $x^2 - t^2 = c_2$.

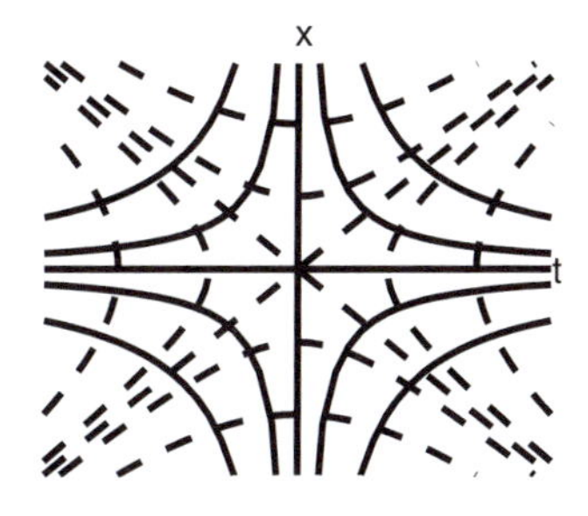

5. $2x^2 \ln|x| - x^2 = t^2 - 2t^2 \ln|t| + c_2$.
7. $x = -\dfrac{\ln|t|}{2} + c_2$.
9. $x = \pm\frac{4}{3}t^{3/2} + c_2$.
11. $x + \dfrac{x^3}{3} = -t + c_2$.
13. $x^2 = 2\ln|\sin t| + c_2$.
15. $2x^2 = -t^2 + c_2$.
17. $x = \left(c_2 - \frac{3}{2}\ln|t|\right)^{-1}$.

Chapter 2: Linear Second- and Higher-Order Differential Equations

2.1 General Solution of Second-Order Linear Differential Equations

1. $x = c_1 \sin t + c_2 \cos t + 1$, $x = 1 - \cos t$.
3. $x = c_1 e^t + c_2 e^{2t} + t + \frac{3}{2}$, $x = -\frac{1}{2}e^{2t} + t + \frac{3}{2}$.
5. $x = c_1 e^{-t} \cos t + c_2 e^{-t} \sin t + 3$, $x = -2e^{-t} \cos t - e^{-t} \sin t + 3$.
7. $x = c_1 \cos t + c_2 \sin t + t \sin t$, $x = \cos t - \sin t + t \sin t$.
9. b) $\tilde{x} = 2e^t + e^{2t} + e^{-t} = e^t + e^{2t} + 2\cosh t = x$.

2.2 Initial Value Problem for Homogeneous Equations

1. Yes, $W[\sin t, \cos t] = -1 \neq 0$.
3. t, no, only one solution.
5. a) $W[x_1, x_2](1) = -1 \neq 0$ b) $x_3 = 2x_1 + 2x_2$.
7. $W[x_1, x_2](t_0) = 1$.
9. $W = x_1 x_2' - x_2 x_1'$, $\frac{dW}{dt} = 0$.
11. $W(t) = 1$; $p(t) = 0$, $W(t) = W(t_0) = 1$.
13. $W(t) = e^{3t}$; $p(t) = -3$, $W(t) = e^{3t_0} e^{3(t-t_0)} = e^{3t}$.
15. $W(t) = -t^{-4}$, $p(t) = \frac{4}{t}$, $W(t) = -t_0^{-4}\,(t_0/t)^4 = -t^{-4}$.

2.3 Reduction of Order

1. $x = c_1 t^{-1} \ln t + c_2 t^{-1}$.
3. $x = c_1 e^{-5t} + c_2 t e^{-5t}$.
5. $x = c_1(t-1) + c_2 e^{-t}$.
7. $x = c_2 t \int_1^t \frac{1}{s^2} e^{-\frac{1}{2}s^2}\, ds + c_1 t$.
9. $r = 2$, $x = c_1 t^2 + c_2 t^2 \ln t$.
11. $r = -3$, $x = c_1 t^{-3} + c_2 t^{-3} \ln t$.
13. $r = 2$, $x = c_1 \sin 2t + c_2 \cos 2t$.
15. $r = 2$, $x = c_1 e^{2t} + c_2 t e^{2t}$.
17. $r = -1$, $x = c_1 e^{-t} + c_2 t e^{-t}$.
19. $r = \pm 2$, $x = c_1 e^{2t} + c_2 e^{-2t}$.
21. Use Wronskian.

2.4 Homogeneous Linear Constant Coefficient Differential Equations (Second-Order)

1. $x = c_1 e^{2t} + c_2 e^{-3t}$.
3. $x = c_1 \cos t + c_2 \sin t$.

5. $x = c_1 e^{-2t} \cos t + c_2 e^{-2t} \sin t$.
7. $x = c_1 e^{t} + c_2 e^{2t}$.
9. $x = c_1 + c_2 e^{t}$.
11. $x = c_1 + c_2 t$.
13. $x = c_1 e^{-t} + c_2 e^{t/3}$.
15. $x = \frac{1}{3} e^{t} - \frac{1}{3} e^{-2t}$.
17. $x = c_1 \cos \sqrt{2}t + c_2 \sin \sqrt{2}t$.
19. $x = -e^{-3t} + 3e^{-t}$.
21. $x = c_1 e^{-5t} + c_2 t e^{-5t}$.
23. $x = c_1 e^{7t} + c_2 t e^{7t}$.
25. $x = c_1 e^{3t} \cos 4t + c_2 e^{3t} \sin 4t$.
27. $x = c_1 e^{\sqrt{12}t} + c_2 e^{-\sqrt{12}t}$.
29. $x = c_1 e^{-2t} \cos 2t + c_2 e^{-2t} \sin 2t$.
31. $x = c_1 \cos \sqrt{8}t + c_2 \sin \sqrt{8}t$.
33. $x = e^{-3t}(c_1 e^{\sqrt{2}t} + c_2 e^{-\sqrt{2}t})$.
35. $W[e^{\alpha t} \cos \beta t, e^{\alpha t} \sin \beta t] = \beta e^{2\alpha t} \neq 0$ if $\beta \neq 0$.
37. $b = -2ar_1, c = ar_1^2$.
39. No answer required.
41. $x'' + 3x' + 2x = 0$.
43. $x'' - 6x' + 9x = 0$.
45. $x'' + 16x = 0$.
47. $x'' - 2x' + 2x = 0$.
49. $x'' = 0$.

2.4.1 Homogeneous Linear Constant Coefficient Differential Equations (nth-Order)

51. $x = c_1 e^{2t} + c_2 t e^{2t} + c_3 t^2 e^{2t}$.
53. $x = c_1 e^{t} + c_2 e^{-t} + c_3 e^{2t} + c_4 e^{-2t}$.
55. $x = c_1 + c_2 e^{t} + c_3 e^{-2t}$.
57. $x = c_1 e^{-t} + c_2 t e^{-t} + c_3 t^2 e^{-t} + c_4 t^3 e^{-t}$.
59. $x = c_1 e^{3t}$.
61. $x = c_1 e^{-t} \cos t + c_2 e^{-t} \sin t + c_3 e^{t}$.
63. $x = c_1 e^{-t} \cos t + c_2 e^{-t} \sin t + c_3 t e^{-t} \cos t + c_4 t e^{-t} \sin t$.
65. $x = c_1 e^{t} + c_2 t e^{t} + c_3 t^2 e^{t} + c_4 t^3 e^{t}$.
67. $x = c_1 e^{t} + c_2 t e^{t} + c_3 t^2 e^{t} + c_4 e^{-t} + c_5 t e^{-t} + c_6 t^2 e^{-t}$.
69. $x = c_1 e^{t} + c_2 e^{-t} + c_3 \cos t + c_4 \sin t$.
71. $x = c_1 \cos 5t + c_2 \sin 5t + c_3 t \cos 5t + c_4 t \sin 5t$.
73. $x = c_1 e^{t} + c_2 e^{-2t} \cos t + c_3 e^{-2t} \sin t$.

75. $x''' - 4x'' + 4x' = 0;\ r(r-2)^2 = 0.$

77. $x'''' + 50x'' + 625x = 0;\ (r^2+25)^2 = r^2 + 50r^2 + 625 = 0.$

79. $x''' + 6x'' + 12x' + 8x = 0;\ (r+2)^3 = 0.$

81. $x'''' - 4x''' + 8x'' - 8x' + 4x = 0;\ (r^2 - 2r + 2)^2 = 0.$

83. $x'''' + 2x'' + x = 0;\ (r^2+1)^2 = 0.$

85. $x = e^{8t}(c_1 + c_2t + c_3t^2 + c_4t^3) + e^{8t}(c_5 \cos 7t + c_6 \sin 7t).$

87. $x = c_1 + c_2t + c_3t^2 + e^{4t}(c_4 \cos 5t + c_5 \sin 5t) + (c_6 + c_7t)\cos 5t$
$+ (c_8 + c_9t)\sin 5t.$

89. $x = e^{2t}[(c_1 + c_2t + c_3t^2)\cos t + (c_4 + c_5t + c_6t^2)\sin t].$

91. $x = e^{2t}(c_1 \cos 6t + c_2 \sin 6t) + e^{-2t}(c_3 \cos 6t + c_4 \sin 6t) +$
$(c_5 + c_6t)\cos 6t + (c_7 + c_8t)\sin 6t.$

2.5 Mechanical Vibrations I: Formulation and Free Response

No Friction

1. $x(t) = 3\cos 5t - 7\sin 5t = R\cos(5t - \phi)$; the answer is $R = \sqrt{58}$, $\tan\phi = -\frac{7}{3}$, 4th quadrant, $\phi = \arctan -\frac{7}{3} = -1.1659$ radians.

3. $x(t) = \sqrt{3}\cos 14t + \sin 14t = R\cos(14t - \phi)$; the answer is $R = 2$, $\tan\phi = \frac{1}{\sqrt{3}}$, 1st quadrant, $\phi = \frac{\pi}{6}$ radians.

5. $x(t) = -6\cos 5t + 6\sin 5t = R\cos(5t - \phi)$; the answer is $R = 6\sqrt{2}$, $\tan\phi = -1$, 2nd quadrant, $\phi = \frac{3\pi}{4}$ radians.

7. $x(t) = \sqrt{3}\cos 6t - \sin 6t = R\cos(6t - \phi)$; the answer is $R = 2$, $\tan\phi = -\frac{1}{\sqrt{3}}$, 4th quadrant, $\phi = -\frac{\pi}{6}$ radians.

9. $x(t) = -4\cos 2t + 4\sqrt{3}\sin 2t = R\cos(2t - \phi)$; the answer is $R = 8$, $\tan\phi = -\sqrt{3}$, 2nd quadrant, $\phi = \frac{2\pi}{3}$ radians.

11. $30x'' + 1470x = 0,\ x(0) = 10,\ x'(0) = 0,\ x = 10\cos 7t.$

13. $2x'' + 128x = 0,\ x(0) = 2,\ x'(0) = 1,\ x = 2\cos 8t + \frac{1}{8}\sin 8t.$

15. $k = 1000\pi^2 \text{lb/ft}.$

17. $x = c_1\cos 2t + c_2 \sin 2t,\ c_1 = R\cos\phi = 1,\ c_2 = R\sin\phi = \sqrt{3},\ x(0) = c_1 = 1,$
$x'(0) = 2c_2 = 2\sqrt{3}.$

19. a) $x = 10\sqrt{\frac{m}{k}}\sin\sqrt{\frac{k}{m}}t.$

    b) Amplitude $= 10\sqrt{\frac{m}{k}}$,

    c) decreases,

    d) increases

21. a) $x = \cos\sqrt{\frac{k}{m}}t + \sqrt{\frac{m}{k}}\sin\sqrt{\frac{k}{m}}t.$

    b) Amplitude $= \sqrt{1 + \frac{m}{k}}.$

    c) Increasing m increases amplitude, increasing k decreases amplitude toward 1.

23. $\int 0dt = \int (mx''x' + kxx')dt$. Integrating gives $c = \frac{1}{2}m(x')^2 + \frac{1}{2}x^2$.
25. $x(t) = c_1 \cos\sqrt{7}t + c_2 \sin\sqrt{7}t$.
27. $x(t) = c_1 e^{\sqrt{7}t} + c_2 e^{-\sqrt{7}t}$.
29. $x(t) = 2\cos\sqrt{5}t + \frac{3}{\sqrt{5}}\sin\sqrt{5}t = R\cos(\sqrt{5}t - \phi)$, where $R = \sqrt{\frac{29}{5}}$, $\tan\phi = \frac{3}{2\sqrt{5}}$, $\phi = .59087$ (1st quadrant).
31. If the constant $= c$, then $\frac{c}{2\pi} =$ cycles per second. Thus $c =$ cycles per 2π seconds or radians per second (the circular frequency). $x(t) = r\cos\theta$, where $\theta = ct + \theta_0$ and r is a constant.

With Friction

33. $10x'' + 40x' + 30x = 0$, $x(0) = 3$, $x'(0) = -5$, $x = e^{-3t} + 2e^{-t}$, overdamped.
35. $x'' + 49x = 0$, $x(0) = 1$, $x'(0) = 7$, $x = \sin 7t + \cos 7t$, $x = \sqrt{2}\cos\left(7t - \frac{\pi}{4}\right)$, harmonic.
37. $x'' + 7x' + 12x = 0$, $x(0) = -1$, $x'(0) = 1$, $x = 2e^{-4t} - 3e^{-3t}$, overdamped.
39. $x'' + 4x' + 5x = 0$, $x(0) = 2$, $x'(0) = 2$, $x = 4e^{-2t}\sin t + 2e^{-2t}\cos t$, damped oscillation, $x = e^{-2t}\sqrt{20}\cos(t - \phi)$, $\phi = \tan^{-1} 2 = 1.107$ radian.
41. Differentiate x, set equal to zero, and show that there is at most one solution.
43. $0 < m < \delta^2/4k$, overdamped; $m = \delta^2/4k$, critically dampled; $\delta^2/4k < m$ is underdampled oscillation. As $m \to +\infty$, solution decays slower, period $\to \infty$, frequency $\to 0$.
45. $\delta = 12\pi$.
47. $x(t) = e^{-t}\left(c_1 \cos\frac{\sqrt{2}}{2}t + c_2 \sin\frac{\sqrt{2}}{2}t\right)$, underdamped.
49. $x(t) = e^{-5t/6}\left(c_1 \cos\frac{\sqrt{23}}{6}t + c_2 \sin\frac{\sqrt{23}}{6}t\right)$, underdamped.
51. $x(t) = c_1 e^{-2t} + c_2 e^{-2t/3}$, overdamped.
53. $x(t) = c_1 e^{-2t/3} + c_2 t e^{-2t/3}$, critically damped.
55. $x(t) = c_1 e^{-t} + c_2 e^{-t/3}$, overdamped.
57. $t = \dfrac{1}{r_1 - r_2}\ln\left(\dfrac{1 - \frac{v_0}{r_1}}{1 - \frac{v_0}{r_2}}\right)$.

2.6 Method of Undetermined Coefficients

1. Yes.
3. No, $\ln|t|$.
5. No, $\dfrac{\sin t}{\cos t}$ not allowed.
7. No, not constant coefficients.
9. No, negative power of t.
11. Yes, $\sinh 3t = \frac{1}{2}e^{3t} - \frac{1}{2}e^{-3t}$.
13. $x = \frac{2}{3} - \frac{2}{27}t + \frac{1}{9}t^3 + c_1\cos 3t + c_2 \sin 3t$.

15. $x = \frac{81}{64}t + \frac{7}{16}t^2 + c_1 + c_2e^{-8t}$.

17. $x = \frac{5}{2}e^{2t} + c_1e^{3t} + c_2e^{4t}$, $c_1 = -1$, $c_2 = -\frac{1}{2}$.

19. $x = -2te^t + c_1e^{2t} + c_2e^t$.

21. $x = e^t + c_1e^t \cos 2t + c_2e^t \sin 2t$.

23. $x = -\frac{5}{6}te^{-3t} + c_1e^{3t} + c_2e^{-3t}$.

25. $x = \frac{3}{5}\sin t - \frac{3}{10}\cos t + c_1e^{-t}\cos 2t + c_2e^{-t}\sin 2t$, $c_1 = \frac{13}{10}$, $c_2 = \frac{17}{20}$.

27. $x = -\frac{2}{3}t\cos 3t + c_1\cos 3t + c_2\sin 3t$.

29. $x = -\frac{1}{3}\cos 2t + c_1\cos t + c_2\sin t$, $c_1 = \frac{1}{3}$, $c_2 = 2$.

31. $x = \frac{1}{5}te^{3t} - \frac{6}{25}e^{3t} + c_1e^{2t} + c_2e^{-2t}$.

33. $x_p = (A_0t + A_1t^2)e^t$, $x = -\frac{1}{4}te^t - \frac{1}{4}t^2e^t + c_1e^{3t} + c_2e^t$.

35. $x = \frac{3}{8}t\sin 4t + c_1\cos 4t + c_2\sin 4t$.

37. $x_p = A_1e^t\cos 3t + A_2e^t\sin 3t$, $x = -\frac{1}{5}e^t\cos 3t + c_1e^t\cos 2t + c_2e^t\sin 2t$.

39. $x_p = t(A_0 + A_1t + A_2t^2) + A_3e^t$, $x = \frac{3}{8}t - \frac{3}{4}t^2 + \frac{1}{5}e^t + \frac{29}{32} - \frac{17}{160}e^{-4t}$.

41. $x_p = At^2e^{-t}$, $x = \frac{3}{2}t^2e^{-t} + c_1e^{-t} + c_2te^{-t}$.

43. $x_p = A_1e^t + A_2te^t + A_3e^{3t}$, $x = e^t + te^t + 2e^{3t} + c_1e^{2t} + c_2te^{2t}$.

45. $x_p = A_0 + A_1t + A_2t^2 + A_3\cos 2t + A_4\sin 2t$,
$x = 6 - 5t + 2t^2 + \frac{1}{5}\sin 2t + c_1e^{-t} + c_2e^{-4t}$.

47. $x_p = A_0 + A_1t + A_2t^2$, $x = \frac{11}{27} - \frac{2}{9}t + \frac{1}{3}t^2 + ce^{-3t}$.

49. $x_p = At\sin 2t + Bt\cos 2t$, $x = -\frac{1}{4}t\cos 2t + c_1\cos 2t + c_2\sin 2t$.

51. $x_p = (A_1 + A_2t)e^t$, $x = -3e^t + te^t + ce^{2t/3}$.

53. $x_p = A_0 + A_1t + A_2t^2 + A_3t^3 + A_4t^4 + A_5t^5$.

55. $x_p = t(A_0 + A_1t + A_2t^2 + A_3t^3)$.

57. $x_p = Ae^{4t}$.

59. $x_p = Ate^{2t}$.

61. $x_p = At^2e^{-3t}$.

63. $x_p = t^2(A_0 + A_1t + A_2t^2 + A_3t^3)e^{-t}$.

65. $x_p = t(A_0 + A_1t + A_2t^2 + A_3t^3 + A_4t^4 + A_5t^5)$.

67. $x_p = t^2(A_0 + A_1t + A_2t^2 + A_3t^3 + A_4t^4)e^{3t}$.

69. $x_p = t(A_0 + A_1t + A_2t^2)e^{2t} + A_3e^{5t}$.

71. $x_p = A\cos 5t + B\sin 5t$.

73. $x_p = (A_0 + A_1t + A_2t^2)\cos 4t + (B_0 + B_1t + B_2t^2)\sin 4t$.

75. $x_p = t(A\cos 5t + B\sin 5t)$.

77. $x_p = Ate^t\sin 2t + Bte^t\cos 2t$.

79. $x_p = (A_0 + A_1t)e^{-t}\sin 2t + (B_0 + B_1t)e^{-t}\cos 2t$.

81. $x_p = (A_0 + A_1t + A_2t^2)e^t\cos t + (B_0 + B_1t + B_2t^2)e^t\sin t$.

83. $x_p = A_1te^{-t}\cos t + B_1te^{-t}\sin t + A_2e^t\cos t + B_2e^t\sin t$.

85. $x_p = Ate^{-t} + Bte^t + e^t(D\cos t + E\sin t)$.

87. $x_p = t(A_0 + A_1 t)\cos 4t + t(B_0 + B_1 t)\sin 4t + A_2 e^{-t}\sin 4t + B_2 e^{-t}\cos 4t + A_3 e^{-4t}$.

89. $x_p = (A_0 + A_1 t + A_2 t^2 + A_3 t^3)e^{-t}\sin t + (A_4 + A_5 t + A_6 t^2 + A_7 t^3)e^{-t}\cos t + A_8 te^t \sin t + A_9 te^t \cos t$.

91. $x_p = (A_0 + A_1 t)e^t\cos 3t + (B_0 + B_1 t)e^t \sin 3t$.

93. $x_p = t(A_0 + A_1 t + A_2 t^2)e^{-2t}\cos 2t + t(B_0 + B_1 t + B_2 t^2)e^{-2t}\sin 2t$.

95. $x_p = t(A_0 + A_1 t + A_2 t^2 + A_3 t^3)e^{-2t}\cos 3t + t(B_0 + B_1 t + B_2 t^2 + B_3 t^3)e^{-2t}\sin 3t$.

97. (a) $x_p = t^2(A_0 + A_1 t + A_2 t^2 + A_3 t^3)$ (b) $x = \frac{1}{20}t^5 + \frac{7}{6}t^3 - t^2$.

99. $x = e^{2t} + c_1 e^t + c_2 te^t + c_3 t^2 e^t$.

101. $x = \frac{1}{2}\cos t + c_1 + c_2 e^t + c_3 e^{-t}$.

103. $x_p = At\cos t + Bt\sin t + Ct$, $x = -t\cos t + 3t - 2\sin t$.

105. $x = -\frac{4}{45}e^t - \frac{1}{3}te^t + c_1 e^{2t} + c_2 e^{-2t} + c_3\cos 2t + c_4 \sin 2t$.

107. $x = -\frac{1}{40}e^{3t} + \frac{1}{12}te^{2t} + c_1 e^t + c_2 e^{-t} + c_3 e^{2t} + c_4 e^{-2t}$.

109. $x_p = t^3(A_0 + A_1 t + A_2 t^2)e^t$.

111. $x_p = t^4(A_0 + A_1 t + A_2 t^2 + A_3 t^3)e^t + (A_4 + A_5 t + A_6 t^2)e^{-t}$.

113. $x_p = t(Ae^{-t}\cos t + Be^{-t}\sin t)$.

115. $r^4 + 4r^3 + 8r^2 + 8r + 4 = (r^2 + 2r + 2)^2$, $x_p = t^2(Ae^{-t}\cos t + Be^{-t}\sin t)$.

2.7 Mechanical Vibrations II: Forced Response

Friction is Absent ($\delta = 0$)

1. $m = \frac{1}{400\pi^2}$.

3. $k = 144(22)^2\pi^2$.

5. $\frac{10}{(140)^2\pi^2} < m < \frac{10}{(20)^2\pi^2}$.

7. $\sqrt{\frac{8}{15}}\frac{1}{2\pi}$.

9. $m = (240\pi^2)^{-1}g$.

13. $mg = kd$, $mx'' = F_T = -k(x + d) + mg = -kx$, $mx'' + kx = 0$.

15. $x = \frac{F}{k - m\omega^2}(\cos\omega t - \cos\omega_0 t) + x_0\cos\omega_0 t$.

17. $x = \frac{F}{k - m\omega^2}(\sin\omega t - \frac{\omega}{\omega_0}\sin\omega_0 t) + x_0 \cos\omega_0 t$.

19. $x = \frac{F}{2m\omega}t\sin\omega t$.

21. $x = \frac{8}{21}\cos 5t + c_1\cos 2t + c_2 \sin 2t$.

23. $x = -\frac{8}{29}\cos 5t + c_1 e^{2t} + c_2 e^{-2t}$.

25. $x = \frac{3}{4}t\sin 2t + c_1\cos 2t + c_2\sin 2t$.

27. $x = \frac{F}{2m\omega}t\sin\omega t + c_1\cos\omega t + c_2\sin\omega t$.

Friction is Present ($\delta > 0$)

29. $x(t) = \frac{3}{73}\cos 4t + \frac{8}{73}\sin 4t = \frac{\sqrt{73}}{73}\cos(4t - \phi)$, where $\tan\phi = \frac{8}{3}$, $\phi = 1.212$.
31. $x(t) = \frac{9}{29}\cos 2t + \frac{8}{29}\sin 2t = \frac{\sqrt{145}}{29}\cos(2t - \phi)$, where $\tan\phi = \frac{8}{9}$, $\phi = 0.72661$.
33. $x(t) = -\frac{3}{208}\cos t + \frac{15}{208}\sin t = \frac{3\sqrt{26}}{208}\cos(t - \phi)$, where $\tan\phi = -5$, $\phi = 1.7682$.
35. $x(t) = -\frac{3}{10}\cos t + \frac{1}{10}\sin t = \frac{\sqrt{10}}{10}\cos(t - \phi)$, $\tan\phi = -\frac{1}{3}$, $\phi = 2.8198$.
37. $x(t) = -\frac{2}{5}\cos t + \frac{1}{5}\sin t = \frac{\sqrt{5}}{5}\cos(t - \phi)$, $\tan\phi = -\frac{1}{2}$, $\phi = 2.6779$.
39. $x(t) = -\frac{1}{2}\cos t = \frac{1}{2}\cos(t - \phi)$, $\tan\phi = 0$, $\phi = \pi$.
41. If $\gamma < \frac{\sqrt{2}}{2}$, maximum occurs at $z = 1 - 2\gamma^2$ or $\omega = \omega_0\sqrt{1 - 2\gamma^2}$. As $\gamma \to 0$, $\omega \to \omega_0$. If $\gamma \geq \frac{\sqrt{2}}{2}$, the maximum occurs at $\omega = 0$ since $\frac{dy}{d\omega} \neq 0$ for all $\omega > 0$ and $y \to 0$ as $\omega \to \infty$.

2.8 Linear Electric Cicuits

1. $q = \frac{6}{13}\cos t + \frac{9}{13}\sin t - \frac{1}{2}e^{-t} + \frac{1}{26}e^{-5t}$,
   $i = -\frac{6}{13}\sin t + \frac{9}{13}\cos t + \frac{1}{2}e^{-t} - \frac{5}{26}e^{-5t}$.
3. $q = 3 + 6e^{-0.5t} - 7e^{-t}$, $i = -3e^{-0.5t} + 7e^{-t}$.
5. $L < 10(60\pi)^{-2}$ or $L > 10(40\pi)^{-2}$.
7. $q(t) = \begin{cases} \frac{1}{2} - \frac{1}{2}e^{-t}\cos t - \frac{1}{2}e^{-t}\sin t, & 0 \leq t \leq \pi, \\ \frac{1}{2}(e^{-\pi} + 1)e^{-t}(\cos t + \sin t), & \pi \leq t \leq 2\pi. \end{cases}$

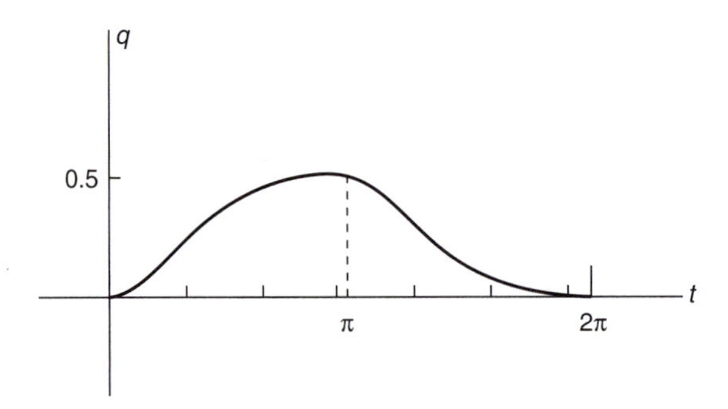

9. $\frac{dE}{dt} = Lii' + \frac{1}{c}qq' = (Li' + \frac{1}{c}q)i = 0i = 0$.
11. $q(0) = -\frac{4}{35}$, $i(0) = 0$.
17. $I = \frac{3}{\sqrt{340}}$.

2.9 Euler Equation

1. $x = c_1 t + c_2 t^{-1}$.
3. $x = c_1 t^{-1} + c_2 t^{-1}\ln t$.

5. $x = c_1 t^{-1/2} + c_2 t^{-1/2} \ln t$.
7. $x = 2t^{-1} - t^{-2}$.
9. $x = c_1 \cos(2 \ln t) + c_2 \sin(2 \ln t)$.
11. $x = t^{-1}[c_1 \cos(\sqrt{7} \ln t) + c_2 \sin(\sqrt{7} \ln t)]$.
13. $x = c_1 t^r + c_2 t^r \ln t$.
15. a) $t\frac{dx}{dt} = t\frac{dx}{dy}\frac{dy}{dt} = t\frac{dx}{dy}\frac{1}{t} = \frac{dx}{dy}$ (etc.).
17. $x = c_1 t + c_2 t^{-1} + c_3 \cos(\ln t) + c_4 \sin(\ln t)$.
19. $x = c_1 t + c_2 t \ln t + c_3 t (\ln t)^2$.

2.10 Variation of Parameters (Second-Order)

1. $x = \frac{1}{3}e^{2t} + c_1 e^t + c_2 e^{-t}$, yes.
3. $x = (\sin t) \ln |\sin t| - t \cos t + c_1 \sin t + c_2 \cos t$, no.
5. $x = [-\ln |\sec t + \tan t| + \sin t] \cos t + (-\cos t) \sin t + c_1 \cos t + c_2 \sin t = (-\cos t) \ln |\sec t + \tan t| + c_1 \cos t + c_2 \sin t$, no.
7. $x = -\frac{1}{2}e^{-2t}\ln(1 + e^{2t}) + e^{-t} \tan^{-1} e^t + c_1 e^{-2t} + c_2 e^{-t}$, no.
9. $x = -\frac{3}{7}t^{7/2}e^{3t} + \frac{2}{5}t^{5/2}te^{3t} + c_1 e^{3t} + c_2 te^{3t} = \frac{4}{35}t^{7/2}e^{3t} + c_1 e^{3t} + c_2 te^{3t}$, no.
11. $x = -\frac{\ln t}{4}e^{-t/2} - \frac{t^{-1}}{4}te^{-t/2} + c_1 e^{-t/2} + c_2 te^{-t/2} = -\frac{\ln t}{4}e^{-t/2} + \widetilde{c}_1 e^{-t/2} + c_2 te^{-t/2}$, no.
13. $x = (-\frac{1}{25}e^{-5t})e^{3t} - \frac{1}{5}te^{-2t} + c_1 e^{3t} + c_2 e^{-2t} = -\frac{1}{5}te^{-2t} + c_1 e^{3t} + \widetilde{c}_2 e^{-2t}$, yes.
15. $x = \frac{t^3}{2} + c_1 t + c_2 t^2$.
17. $x = -t^{-1} - t^{-1} \ln t + c_1 + c_2 t^{-1} = -t^{-1} \ln t + c_1 + \widetilde{c}_2 t^{-1}$.
19. Use formulas for v_1' and v_2' with $v_i = \int_0^t v_i'(s)ds$ and $x_p = v_1 x_1 + v_2 x_2$.
21. $\int_0^t \sinh(t - s)e^{-s^2} ds + c_1 e^t + c_2 e^{-t}$, $\sinh z = \frac{1}{2}e^z - \frac{1}{2}e^{-z}$.
23. $-\int_0^t [e^{3(s-t)} - e^{2(s-t)}]\frac{1}{s+1} ds + c_1 e^{-2t} + c_2 e^{-3t}$.
25. $y = v_1 t + v_2 t^{-1}$, where $v_1 = \frac{1}{2}\int_1^t \frac{e^s}{s^2} ds + c_1$, $v_2 = -\frac{1}{2}e^t + c_2$.

2.11 Variation of Parameters (nth Order)

1. $x = \left(-\frac{e^{2t}}{2}\right) 1 + \left(\frac{e^t}{2}\right) e^t + \left(\frac{e^3 t}{6}\right) e^{-t} + c_1 + c_2 e^t + c_3 e^{-t} = \frac{e^{2t}}{6} + c_1 + c_2 e^t + c_3 e^{-t}$.
3. $x = \left(-\frac{t^4}{8} - \frac{t^2}{2} + \frac{t^3}{3}\right) 1 + \left(\frac{t^3}{3} - \frac{t^2}{2}\right) t - \frac{t^2}{4}t^2 - (t + 1)e^{-t}e^t + c_1 + c_2 t + c_3 t^2 + c_4 e^t = -\frac{t^4}{24} - \frac{t^3}{6} + \tilde{c}_1 + \tilde{c}_2 t + \tilde{c}_3 t^2 + c_4 e^t$.
7. $x = -\frac{t}{2}e^t - \frac{1}{3}e^{-t}e^{2t} + \frac{e^{2t}}{12}e^{-t} + c_1 e^t + c_2 e^{2t} + c_3 e^{-t} = -\frac{t}{2}e^t + \tilde{c}_1 e^t + c_2 e^{2t} + c_3 e^{-t}$.
9. $x = -\frac{1}{6}e^{-t}e^t - \frac{1}{2}e^t e^{-t} + \frac{1}{6}e^{2t}e^{-2t} + c_1 e^t + c_2 e^{-t} + c_3 e^{-2t} = -\frac{1}{2} + c_1 e^t + c_2 e^{-t} + c_3 e^{-2t}$.

11. $x = \frac{t}{8}e^t - \frac{1}{8}e^{2t}e^{-t} + \frac{1}{32}e^{4t}e^{-3t} + c_1e^t + c_2e^{-t} + c_3e^{-3t} = \frac{1}{8}te^t + \tilde{c}_1e^t + c_2e^{-t} + c_3e^{-3t}$.

15. $x = \frac{1}{2}\ln te^{-t} + t^{-1}(te^{-t}) - \frac{1}{4}t^{-t}t^2e^{-t} + c_1e^{-t} + c_2te^{-t} + c_3t^2e^{-t} = \frac{1}{2}\ln te^{-t} + \tilde{c}_1e^{-t} + c_2te^{-t} + c_3t^2e^{-t}$.

17. $x = \frac{1}{13}t^{13/2}e^{2t} - \frac{2}{11}t^{11/2}te^{2t} + \frac{1}{9}t^{9/2}t^2e^{2t} + c_1e^{2t} + c_2te^{2t} + c_3t^2e^{2t} = \frac{8}{1287}t^{13/2}e^{2t} + c_1e^{2t} + c_2te^{2t} + c_3t^2e^{2t}$.

19. $x = \frac{1}{12}t^3e^t - \frac{1}{4}t^2te^t + \frac{1}{4}tt^2e^t + c_1e^t + c_2te^t + c_3t^2e^t = \frac{1}{12}t^3e^t + c_1e^t + c_2te^t + c_3t^2e^t$.

21. $x = \frac{1}{2}t^41 + \frac{2}{3}t^3t - \frac{8}{7}t^{7/2}t^{1/2} + c_1 + c_2t + c_3t^{1/2} = \frac{1}{42}t^4 + c_1 + c_2t + c_3t^{1/2}$.

Chapter 3: The Laplace Transform

3.1 Definition and Basic Properties

1. $\lim_{t\to 2^+} f(t)$ does not exist.
3. Continuous except at $t = 1$; $\lim_{t\to 1^+} f(t)$, $\lim_{t\to 1^-} f(t)$ exist.
5. $(1 - e^{-s})/s$.
7. $(1 - 2e^{-s} + e^{-2s})/s^2$.
9. $\int_0^\infty e^{-st}\sin at\,dt = \lim_{B\to\infty}(s^2+a^2)^{-1}(-ae^{-st}\cos at - se^{-st}\sin at)\big|_{t=0}^{B}$.

15. $\frac{3s}{s^2-4}$.

17. $\frac{30}{s^2+36}$.

19. $-\frac{1}{s^2} + \frac{3}{s}$.

21. $\frac{2}{s} + \frac{s}{s^2+25}$.

23. $\frac{3}{s^2-9}$.

25. $\frac{2}{s+1} + \frac{6}{s-3}$.

27. $\frac{3}{s^2} - \frac{1}{s} + \frac{s}{s^2-4}$.

29. $F(s) = \frac{42}{s^4} + \frac{11}{s^2} + \frac{8}{s}$.

31. $F(s) = \frac{3\cdot 4!}{s^5} + \frac{24}{s^4}$.

33. $F(s) = \frac{5}{(s-3)^2+25}$.

35. $F(s) = \frac{s-4}{(s-4)^2+49}$.

37. $F(s)=\frac{5}{(s+3)^2+25}$.

39. $F(s)=\frac{s+4}{(s+4)^2+49}$.

41. $F(s)=\frac{6!}{(s-2)^7}$.

43. $F(s)=\frac{6!}{(s+2)^7}$.

45. $F(s)=\frac{5!}{(s-3)^6}+\frac{6}{(s-3)^4}+\frac{1}{s-3}$.

47. $t-1$.

49. $\frac{1}{3}\sin 3t$.

51. $1+t$.

53. $3-7t+19\sin t$.

55. $3\cos 4t+\frac{7}{4}\sin 4t$.

57. $f(t)=5e^{-3t}+7e^{5t}$.

59. $f(t)=\frac{1}{5!}t^5$.

61. $f(t)=\frac{2}{3}\sin 3t$.

63. $f(t)=\frac{3}{9!}t^9$.

65. $f(t)=\frac{3}{2}\sqrt{\frac{2}{7}}\sin\sqrt{\frac{7}{2}}t$.

67. $f(t)=\frac{1}{2}t^2e^{4t}$.

69. $f(t)=e^{5t}\cos 3t$.

71. $f(t)=\frac{7}{4}e^{7t}\sin 4t$.

73. $f(t)=e^{-3t}\cos\sqrt{5}t$.

77. Yes.

79. $|f(t)e^{-st}|\leq Me^{\alpha t}e^{-st}=Me^{(\alpha-s)t}$ and $\int_0^\infty e^{(\alpha-s)t}dt$ converges if $s>\alpha$.

3.1.2 Derivative Theorem (Optional)

81. $F(s)=\frac{1}{(s-5)^2}$.

83. $F(s)=\frac{s^2-25}{(s^2+25)^2}$.

85. $F(s)=\frac{2s}{(s^2+9)^2}-\frac{72s}{(s^2+9)^3}$.

87. $F(s)=\frac{6(s-5)}{[(s-5)^2+9]^2}$.

89. $F(s)=\frac{(s+4)^2-25}{[(s+4)^2+25]^2}$.

91. $f(t)=\frac{1}{5!}t^5e^{-3t}$.

93. $f(t)=\frac{5}{6}t\sin 3t$.

95. $f(t)=t\cos 3t$.

97. $f(t)=\frac{5}{18}\left(\frac{1}{3}\sin 3t-t\cos 3t\right)$.

99. $f(t) = \frac{1}{2}(\sin t - t\cos t) + t\sin t$.

101. Use change of variables $\tau = ct$.

103. If $F(s) = \frac{1}{(s-a)}$, then $F^{(n)}(s) = (-1)^n \frac{n!}{(s-a)^{n+1}}$.

3.2 Inverse Laplace Transforms (Roots, Quadratics, and Partial Fractions)

1. $\frac{1}{6}(e^{3t} - e^{-3t})$.

3. $5\cos\sqrt{7}t - \frac{1}{\sqrt{7}}\sin\sqrt{7}t$.

5. $\cosh t + 2\sinh t$ or $\frac{3}{2}e^t - \frac{1}{2}e^{-t}$.

7. $1 + e^t$.

9. $-\frac{1}{3}e^{-2t} - \frac{2}{3}e^t$.

11. $-\frac{4}{5}e^{-3t} + \frac{4}{5}e^{2t}$.

13. $e^t\cos 5t + \frac{1}{5}e^t\sin 5t$.

15. $2e^{3t}\cos 3t + \frac{11}{3}e^{3t}\sin 3t$.

17. $3e^{5t}\cos t - 17e^{-5t}\sin t$.

19. $3e^{3t} - 2e^{2t}$.

21. $\frac{1}{2} + \frac{1}{2}e^{2t} - e^t$.

23. $-\frac{2}{7} + \frac{1}{6}e^{-t} + \frac{47}{42}e^{-7t}$.

25. $-3 + 2e^t + e^{-t}$.

27. $\frac{2s+1}{s^3+4s^2+13s} = \frac{1}{13s} - \frac{1}{13}\frac{-22+s}{s^2+4s+13}$, $f(t) = \frac{1}{13} - \frac{1}{13}e^{-2t}(-8\sin 3t + \cos 3t)$.

29. $\frac{s^2-3}{s^3+2s^2+26s} = -\frac{3}{26s} + \frac{1}{26}\frac{6+29s}{s^2+2s+26}$, $f(t) = -\frac{3}{26} + \frac{1}{26}e^{-t}\left(-\frac{23}{5}\sin 5t + 29\cos 5t\right)$.

31. $\frac{s-8}{(s-5)(s^2+4)} = -\frac{3}{29(s-5)} + \frac{1}{29}\frac{44+3s}{s^2+4}$, $f(t) = -\frac{3}{29}e^{5t} + \frac{1}{29}(22\sin 2t + 3\cos 2t)$.

33. $\frac{s^2}{(s-3)(s^2-2s+26)} = \frac{9}{29(s-3)} + \frac{2}{29}\frac{39+10s}{s^2-2s+26}$, $f(t) = \frac{9}{29}e^{3t} + \frac{2}{29}e^t\left(\frac{49}{5}\sin 5t + 10\cos 5t\right)$.

35. $\frac{1}{24}\sin 3t + \frac{9}{8}\cos 3t - \frac{1}{8}\sin t - \frac{1}{8}\cos t$.

37. $\frac{4}{3}\cosh 2t - \frac{1}{3}\cosh t = \frac{2}{3}e^{2t} + \frac{2}{3}e^{-t} - \frac{1}{6}e^t - \frac{1}{6}e^{-t}$.

39. $\frac{1}{8}\cosh 2t - \frac{1}{8}\cos 2t = \frac{1}{16}e^{2t} + \frac{1}{16}e^{-2t} - \frac{1}{8}\cos 2t$.

41. $\frac{4}{5!}t^5e^{-3t} = \frac{1}{30}t^5e^{-3t}$.

43. $\frac{1}{3!3^5}t^4e^{-t/3}$.

45. $\frac{s^2}{(s+3)^2(s-3)^2} = \frac{1}{4(s+3)^2} - \frac{1}{12(s+3)} + \frac{1}{4(s-3)^2} + \frac{1}{12(s-3)}$, $f(t) = \left(\frac{1}{4}t - \frac{1}{12}\right)e^{-3t} + \left(\frac{1}{4}t + \frac{1}{12}\right)e^{3t}$.

47. $\frac{s}{(s+1)^2(s^2+1)} = -\frac{1}{2(s+1)^2} + \frac{1}{2(s^2+1)}$, $f(t) = -\frac{1}{2}te^{-t} + \frac{1}{2}\sin t$.

49. $\frac{s+5}{(s+1)(s-1)^3} = -\frac{1}{2(s+1)} + \frac{3}{(s-1)^3} - \frac{1}{(s-1)^2} + \frac{1}{2(s-1)}$, $f(t) = -\frac{1}{2}e^{-t} + e^t\left(\frac{1}{2} - t + \frac{3}{2}t^2\right)$.

51. $\frac{s^2-s}{(s^2+4)^2}=\frac{1}{s^2+4}-\frac{s+4}{(s^2+4)^2}=\frac{1}{2}\frac{2}{s^2+4}-\frac{1}{4}\frac{2\cdot 2s}{(s^2+4)^2}-\frac{1}{2}\frac{2\cdot 4}{(s^2+4)^2}$, $f(t)=\frac{1}{2}\sin 2t-\frac{1}{4}t\sin 2t-\frac{1}{2}\left(\frac{1}{2}\sin 2t-t\cos 2t\right)=\frac{1}{4}\sin 2t-\frac{1}{4}t\sin 2t+\frac{1}{2}t\cos 2t$.

53. $\frac{s^3-1}{(s^2+9)^2}=\frac{s^3+9s-9s-1}{(s^2+9)^2}=\frac{s}{s^2+9}-\frac{9s+1}{(s^2+9)^2}$, $f(t)=\cos 3t-\frac{3}{2}t\sin 3t-\frac{1}{18}\left(\frac{1}{3}\sin 3t-t\cos 3t\right)$.

55. $\frac{s^3-s^2}{(s^2+36)^2}=\frac{-36s+36}{(s^2+36)^2}+\frac{s-1}{s^2+36}$, $f(t)=\frac{1}{2}\left(\frac{1}{6}\sin 6t-t\cos 6t\right)-3t\sin 6t-\frac{1}{6}\sin 6t+\cos 6t$.

57. $\frac{3s-1}{(s^2+2s+10)^2}$, $f(t)=-\frac{2}{25}e^{-t}\left(\frac{1}{5}\sin 5t-t\cos 5t\right)+\frac{3}{10}te^{-t}\sin 5t$.

59. $\frac{s^2}{(s^2+6s+25)^2}=\frac{-6s-25}{(s^2+6s+25)^2}+\frac{1}{s^2+6s+25}$, $f(t)=-\frac{7}{32}e^{-3t}\left(\frac{1}{4}\sin 4t-t\cos 4t\right)-\frac{3}{4}te^{-3t}\sin 4t+\frac{1}{4}e^{-3t}\sin 4t$.

61. $F(s)=\frac{s+5}{(s+1)(s-1)^3}=-\frac{1}{2(s+1)}+\frac{3}{(s-1)^3}-\frac{1}{(s-1)^2}+\frac{1}{2(s-1)}$, $f(t)=-\frac{1}{2}e^{-t}+e^t\left(\frac{1}{2}-t+\frac{3}{2}t^2\right)$.

63. $F(s)=\frac{12}{s(s-8)^6}=\frac{3}{65536s}+\frac{3}{2(s-8)^6}-\frac{3}{16(s-8)^5}+\frac{3}{128(s-8)^4}-\frac{3}{1024(s-8)^3}+\frac{3}{8192(s-8)^2}-\frac{3}{65536(s-8)}$, $f(t)=\frac{12}{8^6}+e^{8t}\left(-\frac{12}{8^6}+\frac{12}{8^5}t-\frac{3}{2048}t^2+\frac{1}{256}t^3-\frac{1}{128}t^4+\frac{1}{80}t^5\right)$.

65. $F(s)=\frac{s+5}{(s+1)^2(s-1)^2}=\frac{1}{(s+1)^2}+\frac{5}{4(s+1)}+\frac{3}{2(s-1)^2}-\frac{5}{4(s-1)}$, $f(t)=e^{-t}\left(\frac{5}{4}+t\right)+e^t\left(-\frac{5}{4}+\frac{3}{2}t\right)$.

67. $F(s)=\frac{s}{(s^2+1)(s-7)^4}=\frac{7}{50(s-7)^4}-\frac{12}{625(s-7)^3}+\frac{161}{2^2 5^6(s-7)^2}-\frac{527}{2^2 5^8(s-7)}+\frac{1}{2^2 5^8}\frac{-336+527s}{s^2+1}$, $f(t)=\frac{1}{2^2 5^8}\left[527\cos t-336\sin t+e^{7t}\left(-527+5^2\cdot 161t-15000t^2+\frac{5^6 7}{3}t^3\right)\right]$.

3.3 Initial Value Problems for Differential Equations

1. $-\frac{1}{4}+\frac{3}{8}e^{2t}-\frac{1}{8}e^{-2t}$.
3. e^t.
5. $\frac{1}{6}+\frac{5}{2}e^{-2t}-\frac{5}{3}e^{-3t}$.
7. $\frac{1}{2}+\frac{3}{2}e^{2t}-2e^t$.
9. $\frac{1}{10}e^{-t}+\frac{7}{10}\sin 3t+\frac{9}{10}\cos 3t$.
11. $\frac{2}{13}+\frac{22}{39}e^{-2t}\sin 3t+\frac{11}{13}e^{-2t}\cos 3t$.
13. $-1+\frac{9}{4}e^t-\frac{1}{4}e^{-t}+2\cos t+\frac{5}{2}\sin t$.
15. $t-e^t$.
17. $\frac{1}{2}t^2e^{-t}+te^{-t}$.
19. $\frac{1}{2}e^{-t}\cos t+\frac{1}{2}e^{-t}\sin t-\frac{1}{2}e^{-t}$.

21. $y(t) = -\frac{1}{2}t\cos t + \frac{1}{2}\sin t$.
23. $y(t) = \frac{1}{6}t\sin 3t + \frac{1}{3}\sin 3t$.
25. $y(t) = \frac{1}{4}t\sin 2t$.
27. $y(t) = -\frac{1}{2}t\cos t + \frac{1}{2}t\sin t + \frac{1}{2}\sin t$.

3.4 Discontinuous Forcing Functions

1. $f(t) = 3[H(t-2) - H(t-5)] + t[H(t-5)] = 3H(t-2) + (t-3)H(t-5)$.

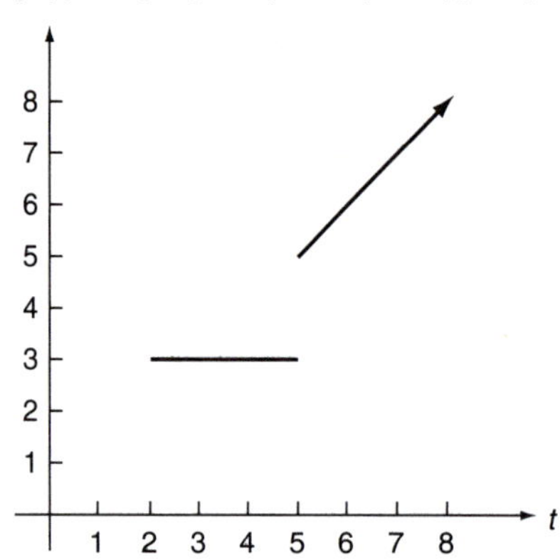

3. $f(t) = \sin t[1 - H(t-\pi)] + \sin t[H(t-2\pi) - H(t-3\pi)]$.

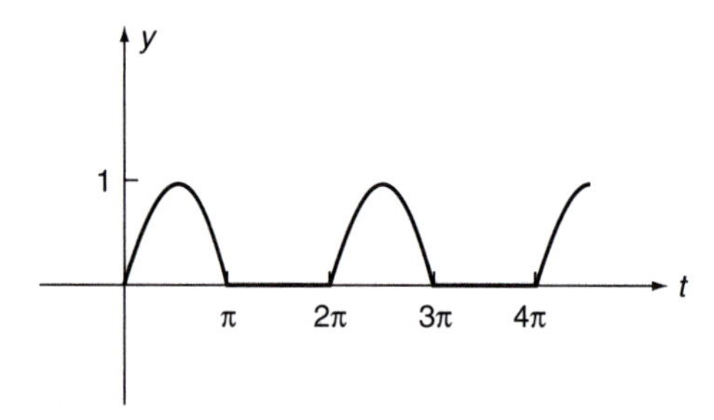

5. $f(t) = 1[1 - H(t-1)] + 1[H(t-2) - H(t-3)] = 1 - H(t-1) + H(t-2) - H(t-3)$.
7. $f(t) = (t-1)[H(t-1) - H(t-2)] + [H(t-2) - H(t-3)] + (4-t)[H(t-3) - H(t-4)] = (t-1)H(t-1) + (2-t)H(t-2) + (3-t)H(t-3) - (4-t)H(t-4)$.
9. $f(t) = (2-t)[H(t-1) - H(t-3)] + (t-4)[H(t-3) - H(t-5)]$.
11. $Y(s) = e^{-2s}\left[\frac{1}{s^2} + \frac{2}{s}\right]$.
13. $Y(s) = e^{-s}\left[\frac{6}{s^4} + \frac{6}{s^3} + \frac{3}{s^2} + \frac{2}{s}\right]$.
15. $Y(s) = e^{-\pi s}\left[-\frac{1}{s^2+1}\right]$.
17. $Y(s) = e^{-2\pi s}\left[\frac{s}{s^2+1}\right]$.

19. $Y(s) = e^{-3s}\left[\frac{e^6}{s-2}\right]$.

21. $Y(s) = e^{-2s}e^{10}\left[\frac{1}{(s-5)^2} + \frac{2}{s-5}\right]$.

23. $y(t) = [-e^{-(t-2)} + 1]H(t-2) + [-e^{-(t-3)} + 1]H(t-3) =$ $[1 - e^{2-t}]H(t-2) + [1 - e^{3-t}]H(t-3)$.

25. $y(t) = e^{-(t-3)}\sin(t-3)H(t-3)$.

27. $y(t) = \sin(t-1)H(t-1) - \frac{1}{2}\sin(2(t-2))H(t-2)$.

29. $y(t) = [4 + 6\cos(3(t-1))]H(t-1)$.

31. $y(t) = [1 - \cos(t-2)]H(t-2) - [1 - \cos(t-5)]H(t-5)$.

33. $y(t) = e^{3t} + \left[-\frac{1}{9} - \frac{t}{3} + \frac{1}{9}e^{3t}\right] - 2\left[-\frac{1}{9} - \frac{(t-1)}{3} + \frac{1}{9}e^{3(t-1)}\right]H(t-1) +$ $\left[-\frac{1}{9} - \frac{(t-2)}{3} + \frac{1}{9}e^{3(t-2)}\right]H(t-2)$.

35. $y(t) = \left[\frac{1}{10} + \frac{9}{10}e^{-t}\cos 3t + \frac{3}{10}e^{-t}\sin 3t\right] - \left[\frac{1}{10} - \frac{1}{10}e^{-(t-3)}\cos 3(t-3) -\right.$ $\left.\frac{1}{10}e^{-(t-3)}\sin 3(t-3)\right]H(t-3)$.

37. $y(t) = \frac{1}{4}[1 - \cos 2t] - \frac{1}{4}[1 - \cos(2(t-1))]H(t-1) +$ $\frac{1}{4}[1 - \cos(2(t-2))]H(t-2) - \frac{1}{4}[1 - \cos(2(t-3))]H(t-3)$.

39. $y(t) = \frac{1}{9} - \frac{1}{9}\cos 3t - \left[\frac{1}{9} - \frac{1}{9}\cos 3(t-2)\right]H(t-2) + \frac{1}{10}e^{-2}\left[e^{-(t-2)} +\right.$ $\left.\frac{1}{3}\sin 3(t-2) - \cos 3(t-2)\right]H(t-2)$

41. $y(t) = -\frac{1}{2}e^{3t} + \frac{1}{2}e^{5t} + e^{12}\left(\frac{1}{2}e^{3(t-4)} - \frac{1}{2}e^{5(t-4)}\right)H(t-4)$.

43. $y(t) = -H\left(t - \frac{\pi}{2}\right)\left(\frac{1}{10}e^{-3(t-\frac{\pi}{2})} - \frac{1}{10}\cos\left(t - \frac{\pi}{2}\right) + \frac{3}{10}\sin\left(t - \frac{\pi}{2}\right)\right) =$ $-H\left(t - \frac{\pi}{2}\right)\left(\frac{1}{10}e^{-3(t-\frac{\pi}{2})} - \frac{1}{10}\sin t - \frac{3}{10}\cos t\right)$.

45. $y(t) = \frac{1}{5}H(t-3)[(2\cos 3 + \sin 3)\sin(t-3) + (-\cos 3 + 2\sin 3)\cos(t-3)$ $-\frac{1}{5}e^{-2(t-3)}(-\cos 3 + 2\sin 3)]$.

47. $y(t) = \frac{1}{8}\sin t - \frac{1}{24}\sin 3t + \frac{1}{8}H(t-4)\left[\frac{1}{3}\cos 4\sin 3(t-4) + \sin 4\cos 3(t-4) -\right.$ $\left.\cos 4\sin(t-4) - \sin 4\cos(t-4)\right]$.

49. $y(t) = \frac{1}{4}[1 - \cos 2t] + \frac{1}{2}\sum_{n=1}^{\infty}(-1)^n[1 - \cos(2(t-n))]H(t-n)$.

3.5 Periodic Functions

1. $G(s) = \frac{1+e^{-\pi s}}{(s^2+1)(1-e^{-\pi s})}$.

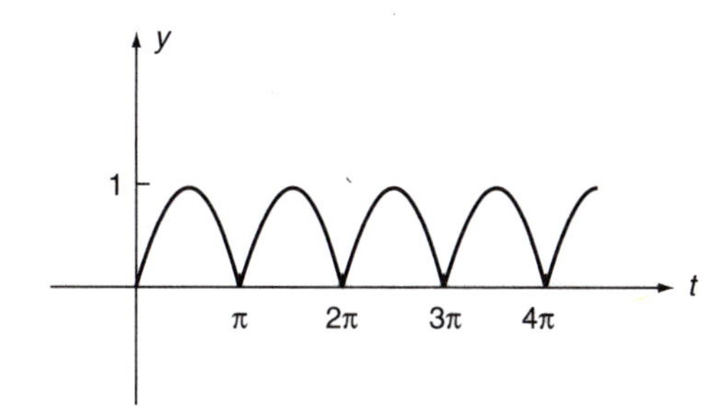

3. $G(s) = \left[\frac{2}{s^3} - \left(\frac{4}{s^3} + \frac{4}{s^2}\right)e^{-s} + \left(\frac{2}{s^3} + \frac{4}{s^2} + \frac{2}{s}\right)e^{-2s}\right]\frac{1}{1-e^{-2s}}$.

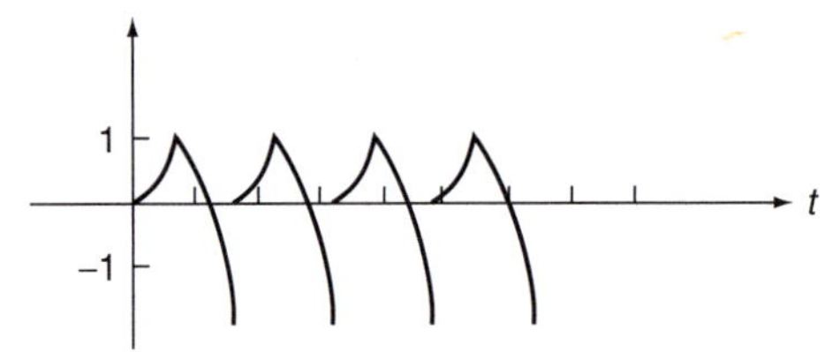

5. $G(s) = \frac{e^{1-s}-1}{(1-s)(1-e^{-s})}$.

7. $G(s) = \left[\frac{1}{s} - \frac{2}{s}e^{-s} + \frac{1}{s}e^{-2s}\right]\frac{1}{1-e^{-2s}}$.

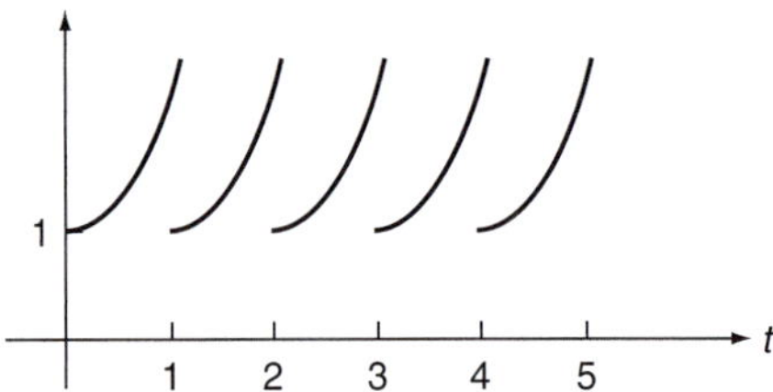

9. $\displaystyle\sum_{n=0}^{\infty}\left[(t-n) + \frac{(t-n)^2}{2}\right]H(t-n)$.

11. $\displaystyle\sum_{n=0}^{\infty}\cos(2(t-n))H(t-n)$.

13. $$\sum_{n=0}^{\infty}\left[(t-2n) + \frac{(t-1-2n)^2}{2}H(t-1-2n)\right]H(t-2n)$$
$$= \sum_{n=0}^{\infty}(t-2n)H(t-2n) + \frac{(t-1-2n)^2}{2}H(t-1-2n).$$

15. $$\sum_{n=0}^{\infty}\left[(1 + \sin\left(t - \tfrac{\pi}{2} - n\pi\right)H\left(t - \tfrac{\pi}{2} - n\pi\right)\right]H(t-n\pi)$$
$$= \sum_{n=0}^{\infty}H(t-n\pi) + \sin\left(t - \tfrac{\pi}{2} - n\pi\right)H\left(t - \tfrac{\pi}{2} - n\pi\right).$$

17. $\displaystyle\sum_{n=0}^{\infty}(-1)^n\left[\tfrac{1}{2}(t-5n)^2H(t-5n) + \tfrac{1}{6}(t-5n-2)^2H(t-5n-2)\right]$.

19. $G(s) = \frac{1}{s^2}\operatorname{csch} s = \frac{2}{s^2(e^s-e^{-s})} = \frac{2e^{-s}}{s^2(1-e^{-2s})}$.
$$g(t) = \sum_{n=0}^{\infty}2(t-2n-1)H(t-2n-1).$$

23. $$\frac{1}{3}\sum_{n=0}^{\infty}\left\{\left[2\sin\left(t-\frac{n\pi}{2}\right) - \sin(2t-n\pi)\right]H\left(t-\frac{n\pi}{2}\right)\right.$$
$$\left. + \left[2\sin\left(t-\frac{\pi}{2}-\frac{n\pi}{2}\right) - \sin(2t-\pi-n\pi)\right]H\left(t-\frac{\pi}{2}-\frac{n\pi}{2}\right)\right\}.$$

25. $e^{-t}+\sum_{n=0}^{\infty}\sinh(t-n)H(t-n)-e\sinh(t-n-1)H(t-n-1).$

27. $\frac{1}{4}\sum_{n=0}^{\infty}[\cosh(2(t-2n))-1]H(t-2n)-2[\cosh(2(t-2n-1))-1]\times$
$H(t-2n-1)+[\cosh(2(t-2n-2))-1]H(t-2n-2).$

29. $\sum_{n=0}^{\infty}\frac{1}{s^{n+1}}=\frac{1}{s}\sum_{n=0}^{\infty}\frac{1}{s^n}=\frac{1}{s}\frac{1}{1-\frac{1}{s}}=\frac{1}{s-1}.$

3.6 Integrals and the Convolution Theorem

1. $t*e^t=-t-1+e^t.$
3. $t.$
5. $t-\sin t.$
7. $y(t)=\int_0^t f(\tau)\left(\frac{1}{3}e^{4(t-\tau)}-\frac{1}{3}e^{t-\tau}\right)d\tau.$
9. $y(t)=\int_0^t f(\tau)\left(\frac{1}{6}e^{3(t-\tau)}-\frac{1}{6}e^{-3(t-\tau)}\right)d\tau.$
11. $y(t)=\int_0^t f(\tau)\frac{1}{2}\sin 2(t-\tau)d\tau+\frac{7}{2}\sin 2t.$
13. $y(t)=\int_0^t f(\tau)e^{-7(t-\tau)}d\tau+2e^{-7t}.$
15. $y(t)=\int_0^t f(\tau)\frac{1}{3}e^{t-\tau}\sin 3(t-\tau)d\tau.$
17. $y(t)=\int_0^t f(\tau)\frac{1}{3}e^{-2(t-\tau)}\sin 3(t-\tau)d\tau.$
19. $X(s)=\frac{1}{s^2-s-1},\ x(t)=\frac{2}{\sqrt{3}}e^{t/2}\sin\left(\frac{\sqrt{3}}{2}t\right).$
21. $X(s)=\frac{2(s+1)}{s^2},\ x(t)=2+2t.$
23. $X(s)=\frac{s^2+4}{s^2(s^2+1)},\ x(t)=4t-3\sin t.$
25. $x(t)=3-\frac{3}{2}t^2.$
27. $|F(s)|\le\int_0^\infty e^{-st}|f(t)|dt\le\int_0^\infty e^{-st}Me^{\alpha t}dt=\frac{M}{s-\alpha}$ for $s>\alpha.$

3.7 Impulses and Distributions

1. $y(t)=e^{-8(t-1)}H(t-1)+e^{-8(t-2)}H(t-2)=e^{8-8t}H(t-1)+$
$e^{16-8t}H(t-2).$
3. $y(t)=\frac{1}{10}e^{-3(t-1)}\sin(10(t-1))H(t-1)-$
$\frac{1}{10}e^{-3(t-7)}\sin(10(t-7))H(t-7).$
5. $y(t)=\frac{1}{3}(1-e^{-3t})+\frac{1}{2}[e^{-(t-3)}-e^{-3(t-3)}]H(t-3).$
7. $y(t)=1+\sin(t-2\pi)H(t-2\pi)=1+\sin tH(t-2\pi).$
9. (a) The following are graphs of the solutions of Exercises 1–4.

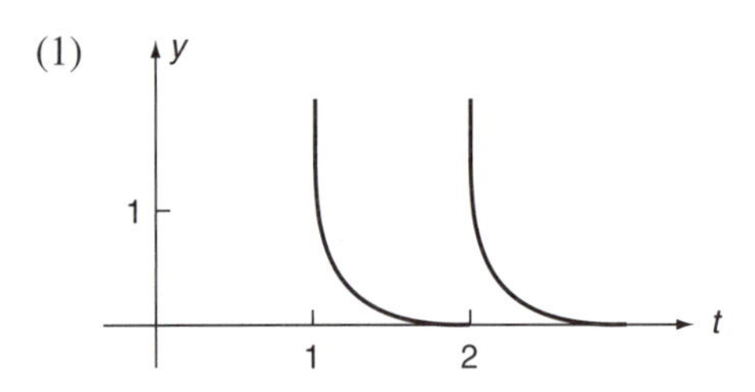

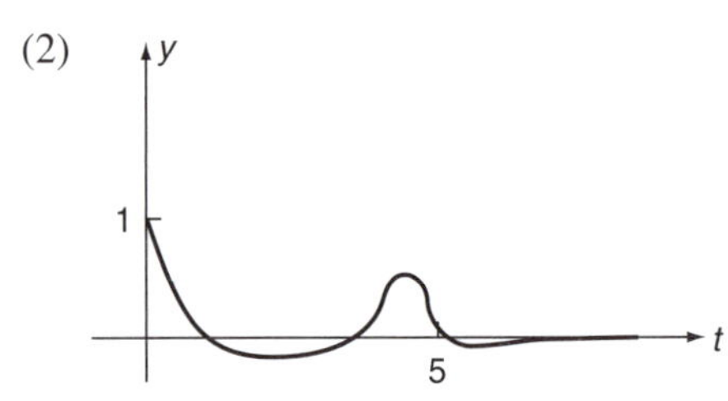

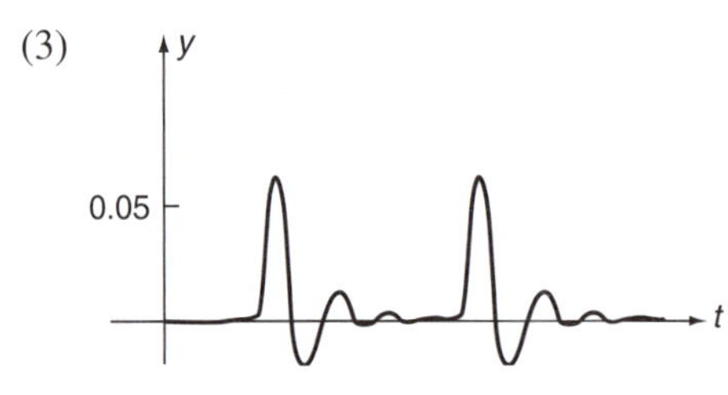

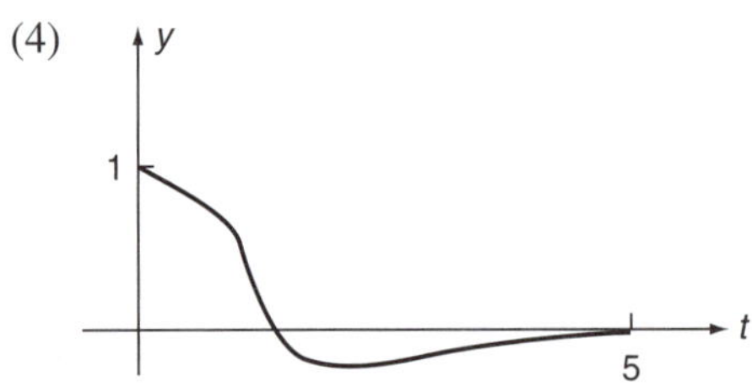

11. $\displaystyle\int_0^\infty \delta^{(n)}(t-a)e^{-st}dt = (-1)^n \left[\frac{d^n(e^{-st})}{dt^n}\right]\Bigg|_{t=a} = (-1)^n(-s)^n e^{-as}.$

Chapter 4: Linear Systems of Differential Equations and their Phase Plane

4.2 Introduction to Linear Systems of Differential Equations

1. $\lambda = \pm 2i$.
3. $1, \begin{bmatrix} -1 \\ 1 \end{bmatrix}$ and $3, \begin{bmatrix} 1 \\ 1 \end{bmatrix}$.
5. $\lambda = \dfrac{5}{2} \pm \dfrac{\sqrt{3}}{2} i$.
7. $-3, \begin{bmatrix} 0 \\ 1 \end{bmatrix}$ and $1, \begin{bmatrix} 4 \\ 1 \end{bmatrix}$.

9. $1, \begin{bmatrix} 1 \\ 1 \end{bmatrix}$ and $3, \begin{bmatrix} 3 \\ 1 \end{bmatrix}$.

11. $3, \begin{bmatrix} 1 \\ 0 \end{bmatrix}$ and $4, \begin{bmatrix} 2 \\ 1 \end{bmatrix}$.

13. $\lambda^2 + 4$, $\lambda = \pm 2i$, trace $= 0 + 0 = 2i - 2i$, det $= 4 = 2i(-2i)$.

15. $\lambda^2 - 4\lambda + 3$, $\lambda = 1, 3$. trace $= 2 + 2 = 1 + 3$, det $= 3 = 1 \cdot 3$.

17. $\lambda = \pm 2i$.

19. $x = c_1 e^{3t} \begin{bmatrix} 1 \\ 1 \end{bmatrix} + c_2 e^{t} \begin{bmatrix} 1 \\ -1 \end{bmatrix}$.

21. $\lambda = \frac{5}{2} \pm \frac{\sqrt{3}}{2} i$.

23. $x = c_1 e^{t} \begin{bmatrix} 4 \\ 1 \end{bmatrix} + c_2 e^{-3t} \begin{bmatrix} 0 \\ 1 \end{bmatrix}$.

25. $x = c_1 e^{t} \begin{bmatrix} 1 \\ 1 \end{bmatrix} + c_2 e^{3t} \begin{bmatrix} 1 \\ 3 \end{bmatrix}$.

27. $x = c_1 e^{3t} \begin{bmatrix} 1 \\ 0 \end{bmatrix} + c_2 e^{4t} \begin{bmatrix} 2 \\ 1 \end{bmatrix}$.

29. $x = c_1 e^{t} \begin{bmatrix} -1 \\ 1 \end{bmatrix} + c_2 e^{2t} \begin{bmatrix} 0 \\ 1 \end{bmatrix}$.

31. $\lambda = \pm 2i$.

33. $\lambda = 1 \pm i$.

35. $x = c_1 e^{-t} \begin{bmatrix} -1 \\ 1 \end{bmatrix} + c_2 e^{-3t} \begin{bmatrix} -2 \\ 1 \end{bmatrix}$.

37. $x = c_1 e^{2t} \begin{bmatrix} 1 \\ 2 \end{bmatrix} + c_2 e^{-t} \begin{bmatrix} 1 \\ -1 \end{bmatrix}$.

39. $x = c_1 e^{-2t} \begin{bmatrix} 1 \\ -3 \end{bmatrix} + c_2 e^{-4t} \begin{bmatrix} 1 \\ -1 \end{bmatrix}$.

41. $\lambda = -1 \pm 4i$.

43. $x = c_1 e^{7t} \begin{bmatrix} 1 \\ 1 \end{bmatrix} + c_2 e^{t} \begin{bmatrix} 1 \\ -1 \end{bmatrix}$.

45. $\alpha = 0, \beta = 3, \mathbf{a} = \begin{bmatrix} 2 \\ 0 \end{bmatrix}, \mathbf{b} = \begin{bmatrix} 0 \\ 1 \end{bmatrix}$,

$$x = c_1 e^{0t} \left(\cos 3t \begin{bmatrix} 2 \\ 0 \end{bmatrix} - \sin 3t \begin{bmatrix} 0 \\ 1 \end{bmatrix} \right) + c_2 e^{0t} \left(\cos 3t \begin{bmatrix} 0 \\ 1 \end{bmatrix} + \sin 3t \begin{bmatrix} 2 \\ 0 \end{bmatrix} \right)$$
$$= c_1 \begin{bmatrix} 2\cos 3t \\ -\sin 3t \end{bmatrix} + c_2 \begin{bmatrix} 2\sin 3t \\ \cos 3t \end{bmatrix}.$$

47. $\alpha = 1, \beta = 2, \mathbf{a} = \begin{bmatrix} 1 \\ 0 \end{bmatrix}, \mathbf{b} = \begin{bmatrix} 1 \\ 3 \end{bmatrix}$,

$$x = c_1 e^{t} \left(\cos 2t \begin{bmatrix} 1 \\ 0 \end{bmatrix} - \sin 2t \begin{bmatrix} 1 \\ 3 \end{bmatrix} \right) + c_2 e^{t} \left(\cos 2t \begin{bmatrix} 1 \\ 3 \end{bmatrix} + \sin 2t \begin{bmatrix} 1 \\ 0 \end{bmatrix} \right)$$
$$= c_1 e^{t} \begin{bmatrix} \cos 2t - \sin 2t \\ -3\sin 2t \end{bmatrix} + c_2 e^{t} \begin{bmatrix} \cos 2t + \sin 2t \\ 3\cos 2t \end{bmatrix}.$$

49. $\alpha = 0, \beta = 5, \mathbf{a} = \begin{bmatrix} 1 \\ 0 \end{bmatrix}, \mathbf{b} = \begin{bmatrix} 1 \\ 1 \end{bmatrix},$

$$x = c_1 e^{0t}\left(\cos 5t \begin{bmatrix} 1 \\ 0 \end{bmatrix} - \sin 5t \begin{bmatrix} 1 \\ 1 \end{bmatrix}\right) + c_2 e^{0t}\left(\cos 5t \begin{bmatrix} 1 \\ 1 \end{bmatrix} + \sin 5t \begin{bmatrix} 1 \\ 0 \end{bmatrix}\right)$$

$$= c_1 \begin{bmatrix} \cos 5t - \sin 5t \\ -\sin 5t \end{bmatrix} + c_2 \begin{bmatrix} \cos 5t + \sin 5t \\ \cos 5t \end{bmatrix}.$$

4.3 Phase Plane for Linear Systems of Differential Equations

1. 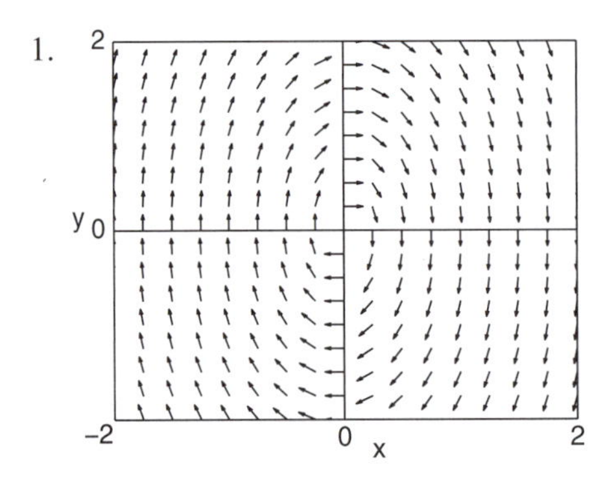

3. 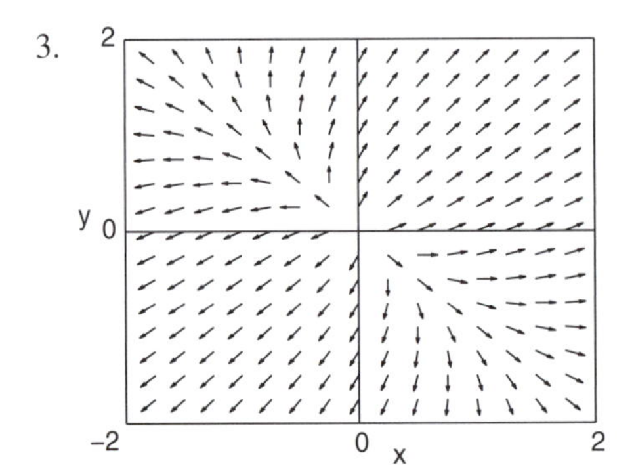

5. 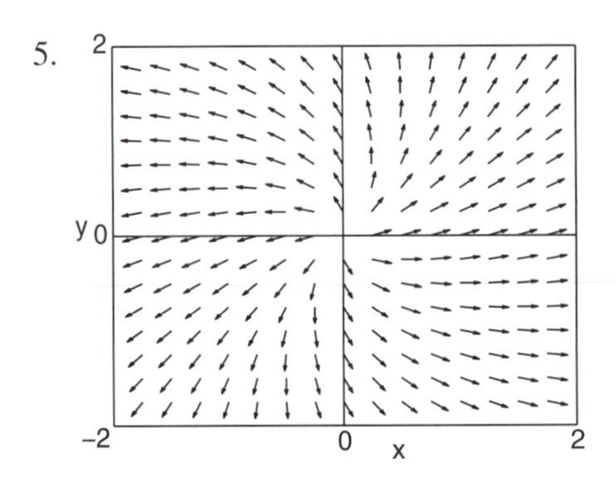

7. $\lambda_1 = 1, \begin{bmatrix} -1 \\ 1 \end{bmatrix}, \ \lambda_2 = 2, \begin{bmatrix} 0 \\ 1 \end{bmatrix}$, unstable node.

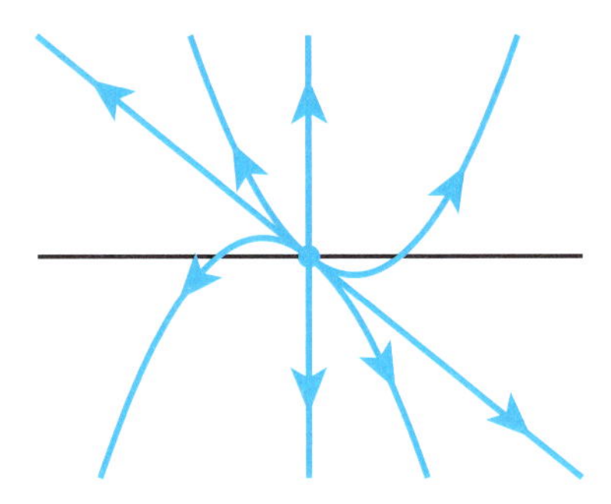

9. $\lambda = \pm 2i$, center.

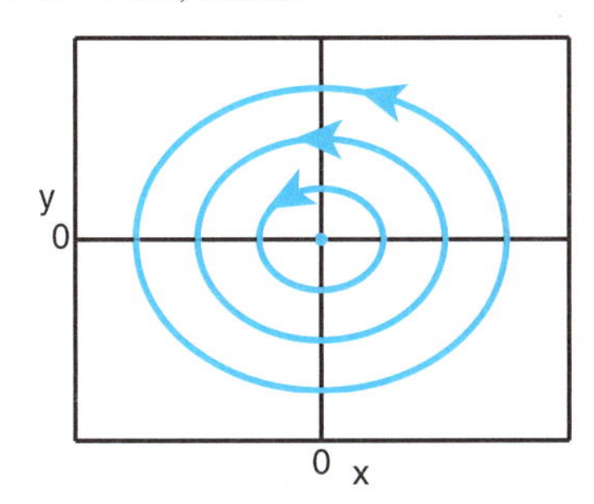

11. $\lambda = 1 \pm i$, unstable spiral.

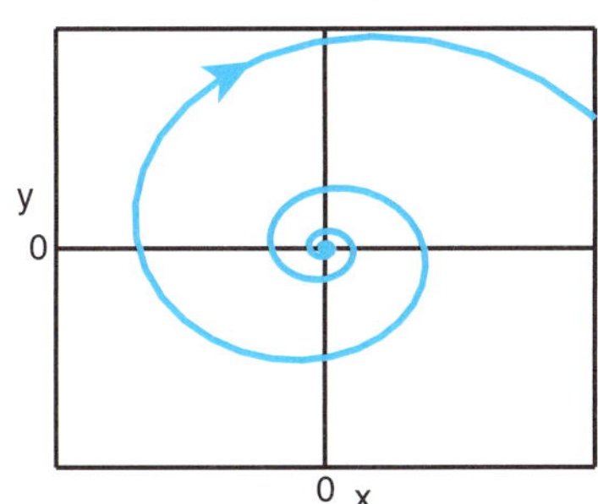

13. $\lambda_1 = -1, \begin{bmatrix} -1 \\ 1 \end{bmatrix}, \ \lambda_2 = -3, \begin{bmatrix} -2 \\ 1 \end{bmatrix}$, stable node.

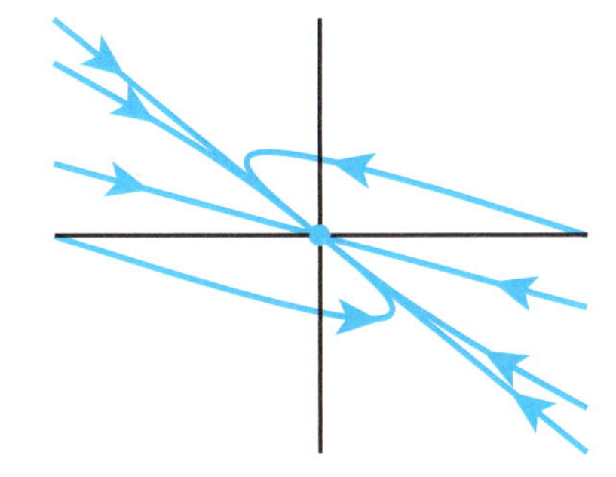

15. $\lambda_1 = 2, \begin{bmatrix} 1 \\ 2 \end{bmatrix}, \ \lambda_2 = -1, \begin{bmatrix} 1 \\ -1 \end{bmatrix}$, saddle point.

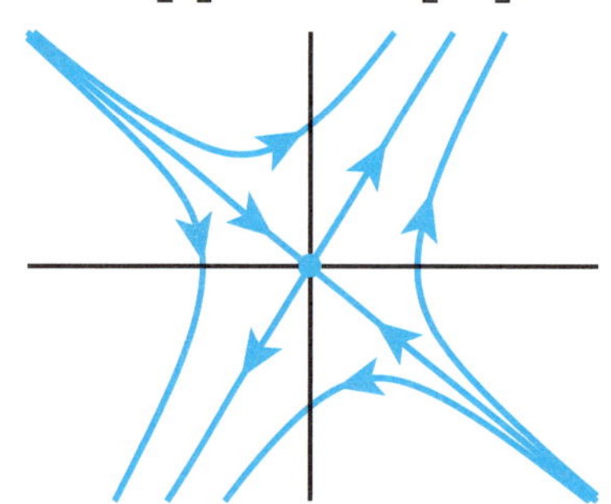

17. $\lambda_1 = -2, \begin{bmatrix} 1 \\ -3 \end{bmatrix}, \ \lambda_2 = -4, \begin{bmatrix} 1 \\ -1 \end{bmatrix}$, stable node.

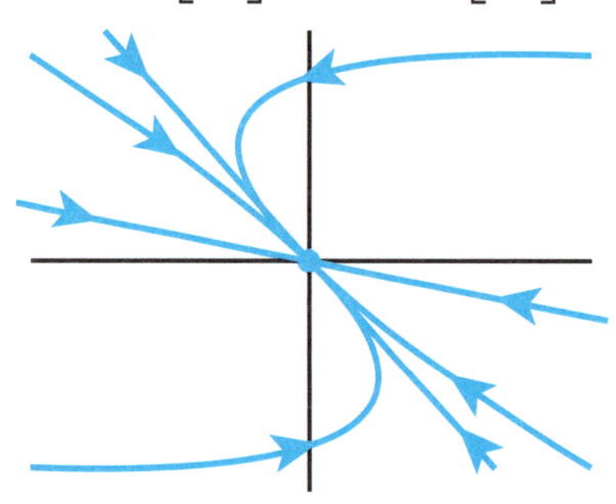

19. $\lambda = -1 \pm 4i$, stable sprial.

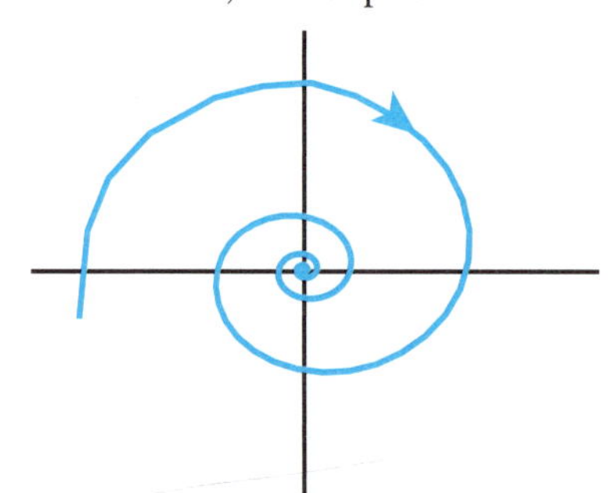

21. $\lambda_1 = 1, \begin{bmatrix} -1 \\ 1 \end{bmatrix}, \ \lambda_2 = 7, \begin{bmatrix} 1 \\ 1 \end{bmatrix}$, unstable node.

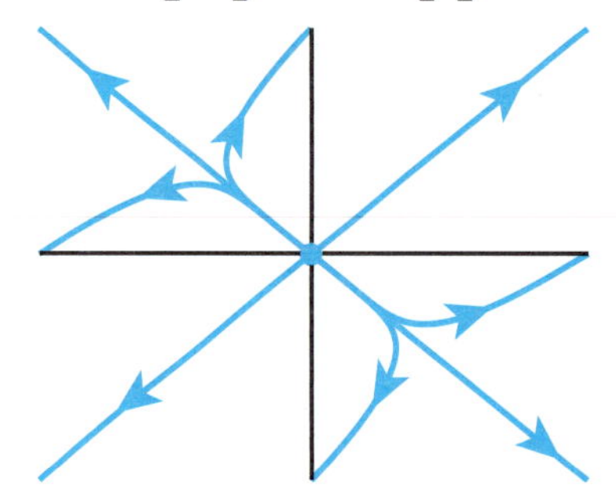

23. $\lambda = \pm 2i$, center

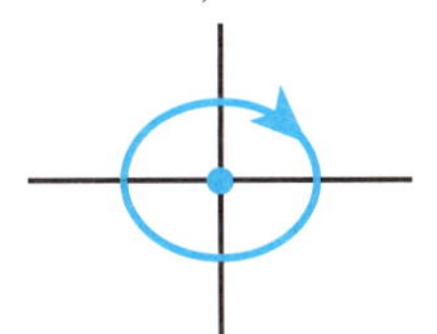

25. $\lambda_1 = 3, \begin{bmatrix} 1 \\ 1 \end{bmatrix}, \lambda_2 = 1, \begin{bmatrix} 1 \\ -1 \end{bmatrix}$, unstable node.

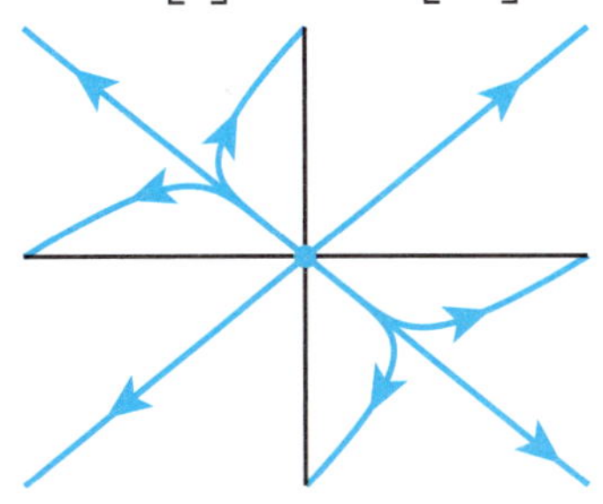

27. $\lambda = \frac{5}{2} \pm i\frac{\sqrt{3}}{2}$, unstable spiral.

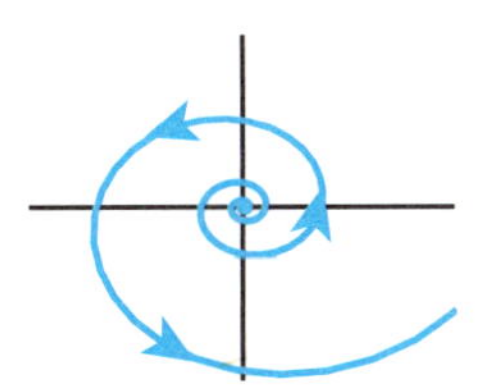

29. $\lambda_1 = 1, \begin{bmatrix} 4 \\ 1 \end{bmatrix}, \lambda_2 = -3, \begin{bmatrix} 0 \\ 1 \end{bmatrix}$, saddle point.

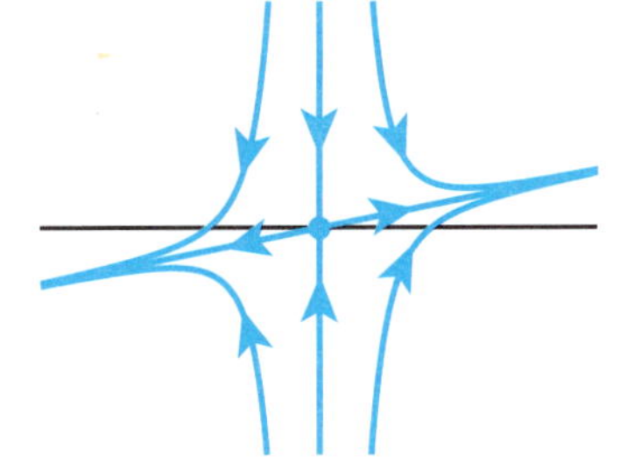

31. $\lambda_1 = 1, \begin{bmatrix} 1 \\ 1 \end{bmatrix}, \lambda_2 = 3, \begin{bmatrix} 1 \\ 3 \end{bmatrix}$, unstable node.

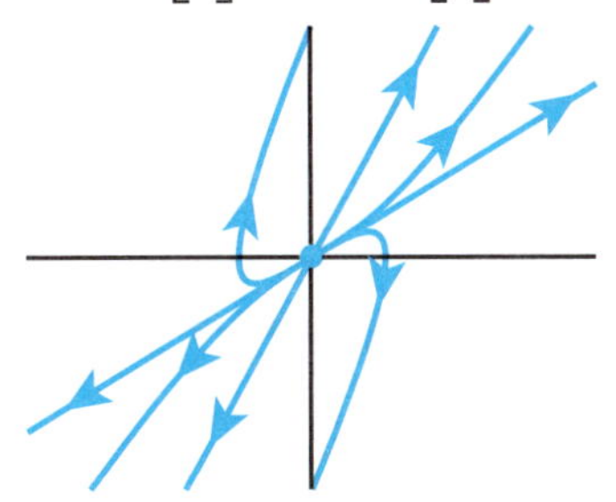

33. $\lambda_1 = \lambda_2 = 1$. Not two distinct eigenvalues.

35. $\lambda_1 = 1, \begin{bmatrix} 1 \\ 0 \end{bmatrix}, \lambda_2 = -3, \begin{bmatrix} 0 \\ 1 \end{bmatrix}$, saddle point.

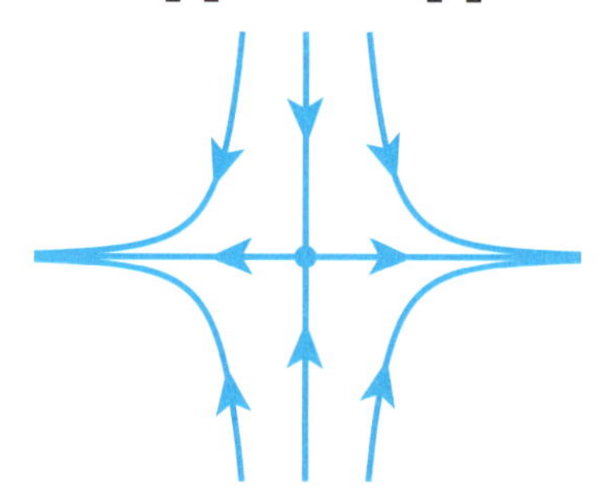

39.

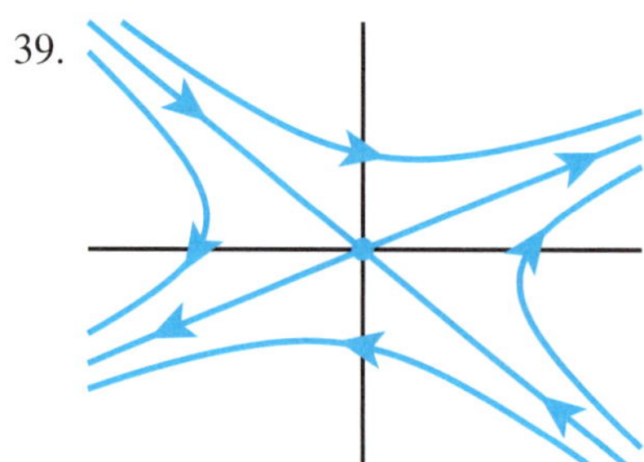

41.

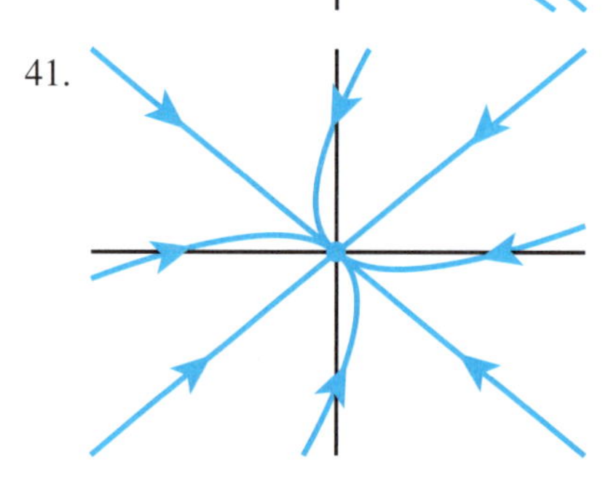

Chapter 5: Mostly Nonlinear First-Order Equations

5.2 Equilibria and Stability

1. $x = 0, 1$.
3. $x = 1, 2$.
5. $x = n\pi$ for $n = 0, \pm 1, \pm 2, \ldots$ (n any integer).
7. $x = 0$ and $x = a/b$ if $b \neq 0$.
9. $x = 0$ is stable, for $x = 1$ the linear analysis is inconclusive.
11. $x = -1, \sqrt{3}$ are unstable, $x = 1, -\sqrt{3}$ are stable.
13. $x = n\pi$, any integer n, unstable.
15. $x = -\frac{1}{3} \ln 5$, stable.

5.3 One-Dimensional Phase Lines

1. $x = 0$ is stable, $x = 1$ is unstable.

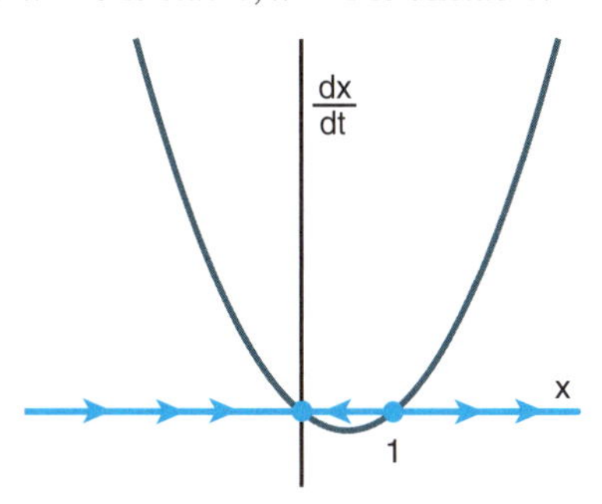

3. $x = 1$ is unstable, for $x = 2$ the linear analysis is inconclusive (but $x = 2$ is unstable).

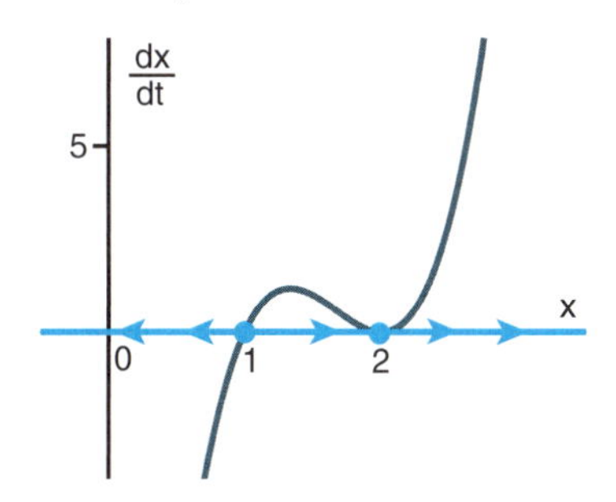

5. $x = n\pi$ is unstable if n is even and stable if n is odd.

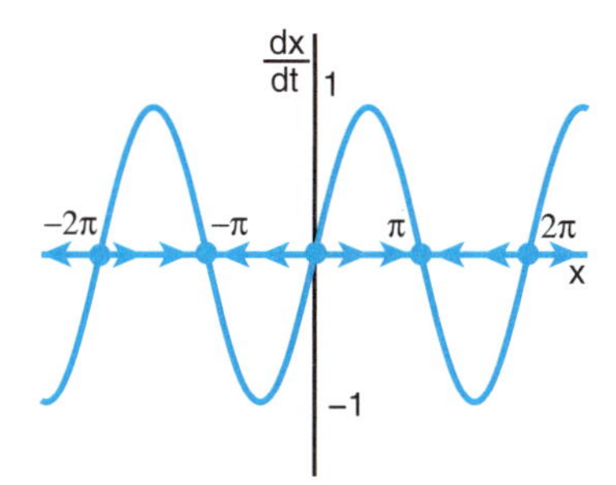

7. No equilibria.

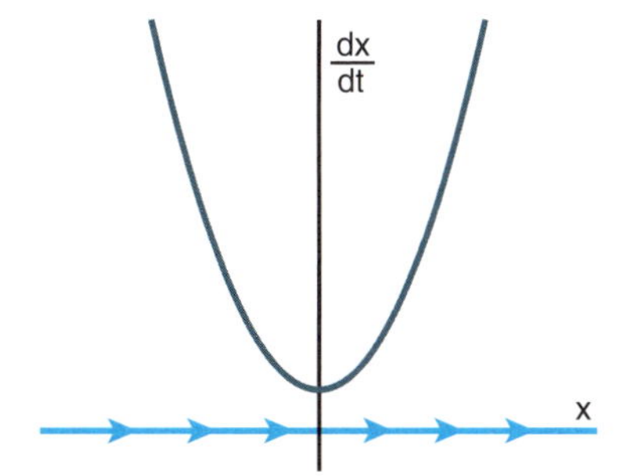

9. For $x = 0$ the linear analysis is inconclusive (but $x = 0$ is unstable).

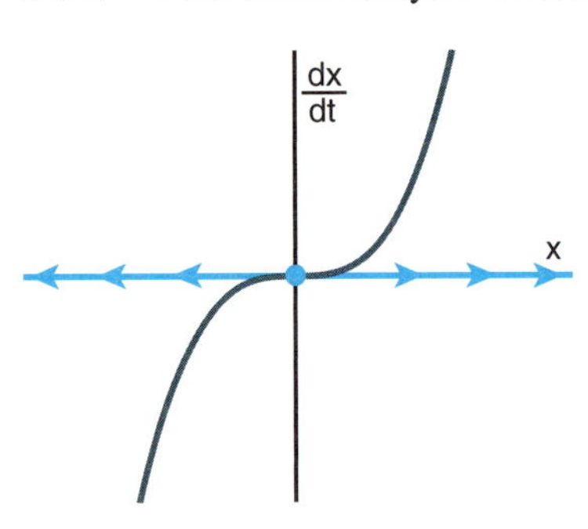

11. For $x = 3$ the linear analysis is inconclusive (but $x = 3$ is unstable).

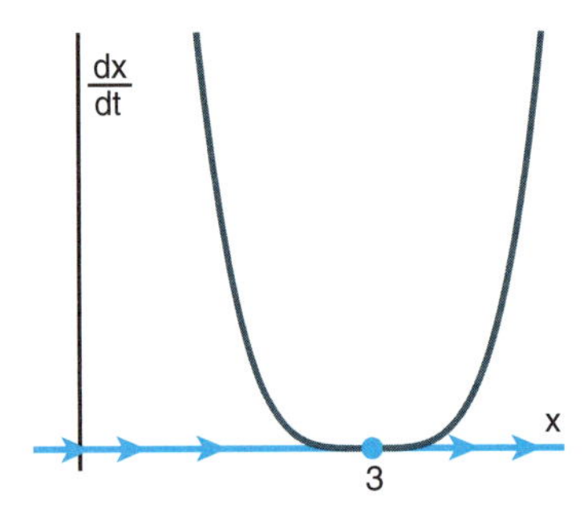

13. $x = 1, 3$ are stable and $x = 2, 4$ are unstable.

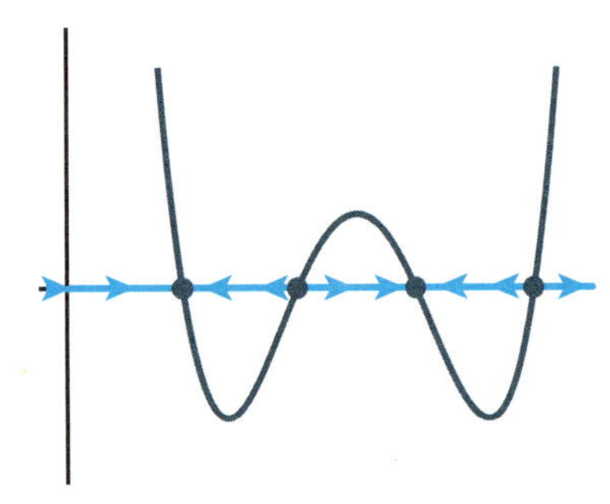

15. x

1

0

t

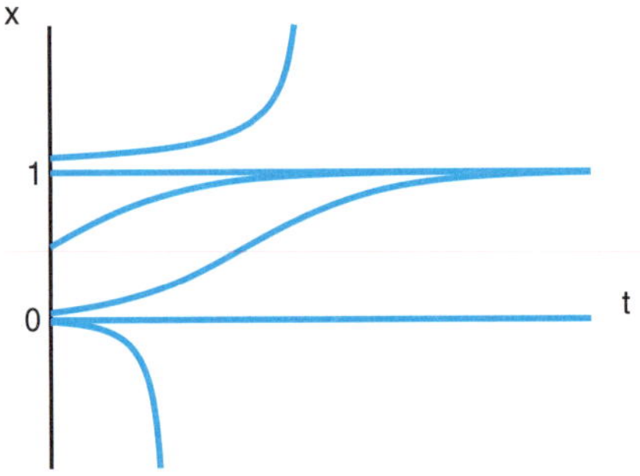

17.

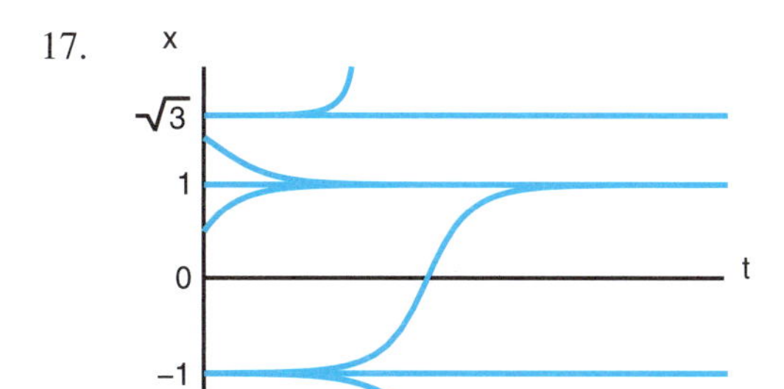

19. $\int \frac{dx}{x^2} = \int dt \Longrightarrow -\frac{1}{x} + c$, $x = -\frac{1}{t+c}$. Note that $c > 0$ if $x(0) < 0$ and $c < 0$ if $x(0) > 0$. If $x(0) > 0$ and $x(0)$ is close to zero, then $x(t)$ increases, so not stable.

21. From Exercise 19, $x(t) = -\frac{1}{t+c}$ and $c > 0$ if $x(0) < 0$. Thus $\lim_{t\to\infty} x(t) = 0$. $x(t)$ is defined for all positive t and never reaches zero in finite time.

5.4 Applications to Population Dynamics: The Logistic Equation

1. a) growth rate $= a - bx^2$. The growth rate is positive for $x < \sqrt{\frac{a}{b}}$ and the growth rate is negative for $x > \sqrt{\frac{a}{b}}$.

b)

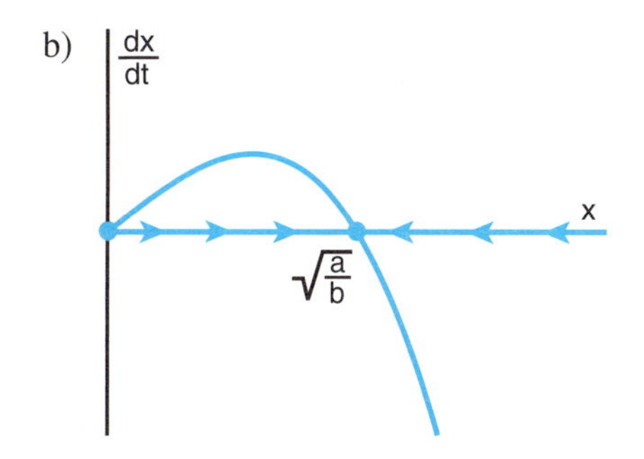

c)

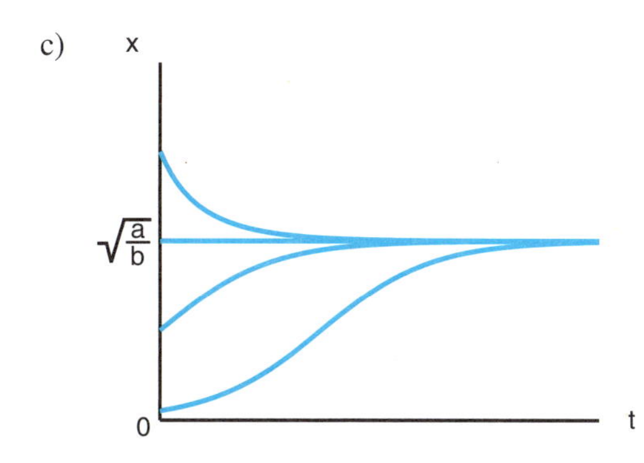

7. a) growth rate $= a + bx$. The growth rate increases as x increases.

b)

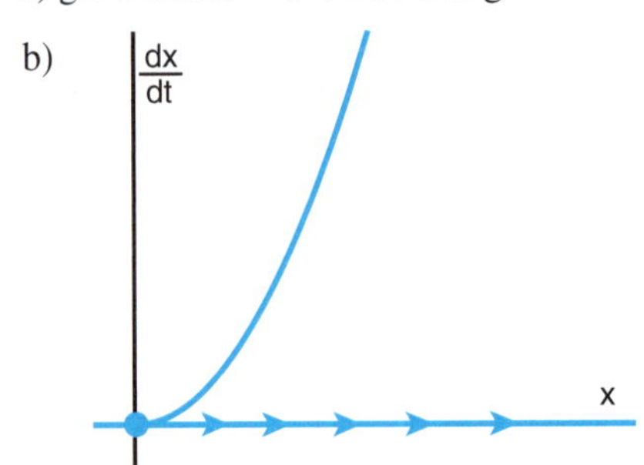

c)

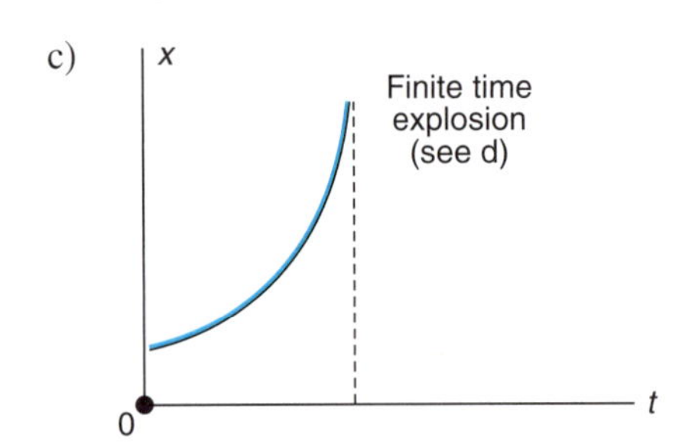

d) $x(t) = \dfrac{-\frac{a}{b}x_0}{x_0 - e^{-at}\left(x_0 + \frac{a}{b}\right)}$.

9. a) $\alpha = \beta = \dfrac{a}{2b}$, $t_0 = \dfrac{1}{a} \ln \dfrac{a - bx_0}{bx_0}$.

c) $\lim_{t\to\infty} \tanh t = 1$, $\lim_{t\to-\infty} \tanh t = -1$.

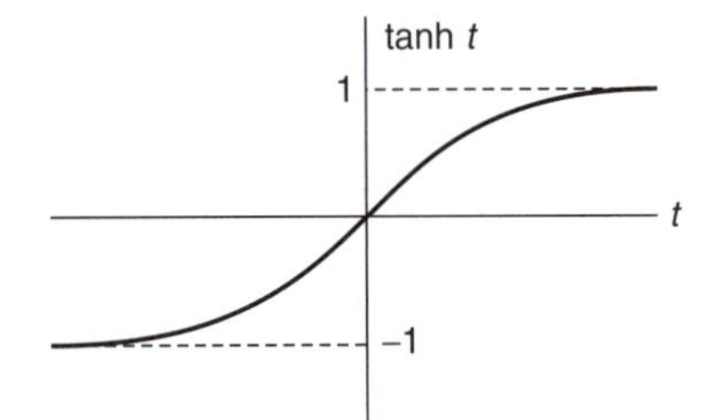

d) and e)

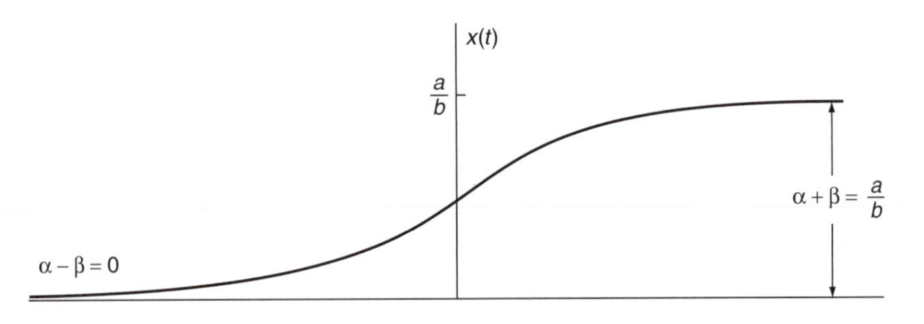

11. $f(0) = a \Rightarrow a = 0$. $f'(Q) = b + 2cQ$. Condition (c) $\Rightarrow c < 0$. Condition (b) $\Rightarrow b > 0$. Thus $\frac{dQ}{dt} = rQ - kQ^2$ where $r = b$, $k = -c$.

Chapter 6: Nonlinear Systems of Differential Equations in the Plane

6.2 Equilibria of Nonlinear Systems, Linear Stability Analysis of Equilibrium, and Phase Plane

1. (a) and (b) Equilibrium (0, 0): $A = \begin{bmatrix} 1 & 0 \\ 0 & 2 \end{bmatrix}$, $\lambda = 1, 2$ so unstable node.

   $\lambda = 1$, $\begin{bmatrix} 1 \\ 0 \end{bmatrix}$, $\lambda = 2$, $\begin{bmatrix} 0 \\ 1 \end{bmatrix}$,

   Equilibrium $(\frac{1}{2}, -1)$: $A = \begin{bmatrix} 0 & \frac{1}{2} \\ 4 & 0 \end{bmatrix}$, $\lambda = \pm\sqrt{2}$ so saddle point.

   $\lambda = \sqrt{2}$, $\begin{bmatrix} 1 \\ 2\sqrt{2} \end{bmatrix}$, $\lambda = -\sqrt{2}$, $\begin{bmatrix} 1 \\ -2\sqrt{2} \end{bmatrix}$,

   c)

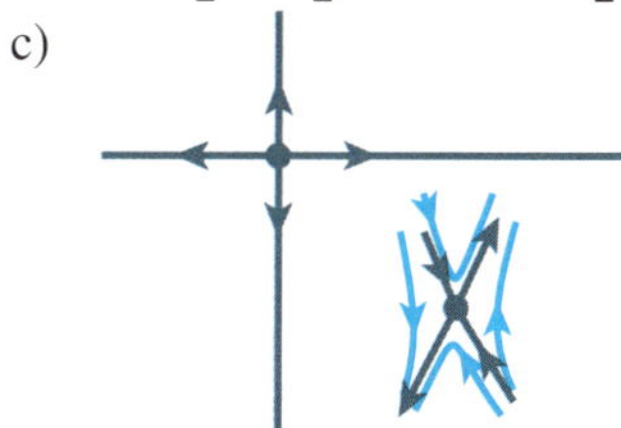

   e)

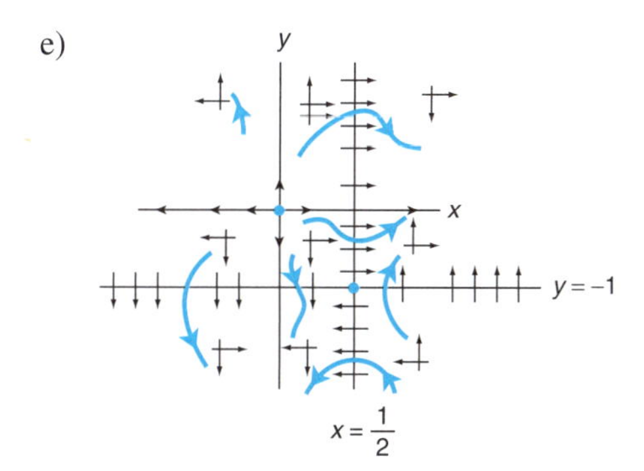

3. (a) and (b) Equilibrium (0, 0): $A = \begin{bmatrix} 2 & 0 \\ 0 & 1 \end{bmatrix}$, $\lambda = 2, 1$ so unstable node.

   $\lambda = 2$, $\begin{bmatrix} 1 \\ 0 \end{bmatrix}$, $\lambda = 1$, $\begin{bmatrix} 0 \\ 1 \end{bmatrix}$,

   Equilibrium (1, 1): $A = \begin{bmatrix} 0 & -2 \\ -1 & 0 \end{bmatrix}$, $\lambda = \pm\sqrt{2}$ so saddle point.

   $\lambda = \sqrt{2}$, $\begin{bmatrix} \sqrt{2} \\ -1 \end{bmatrix}$, $\lambda = -\sqrt{2}$, $\begin{bmatrix} \sqrt{2} \\ 1 \end{bmatrix}$,

   c)

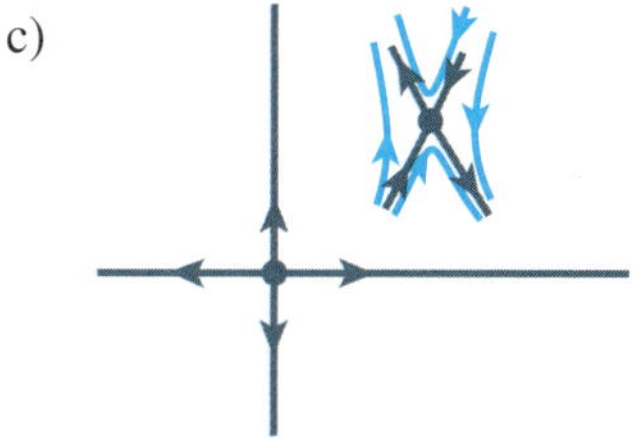

e)

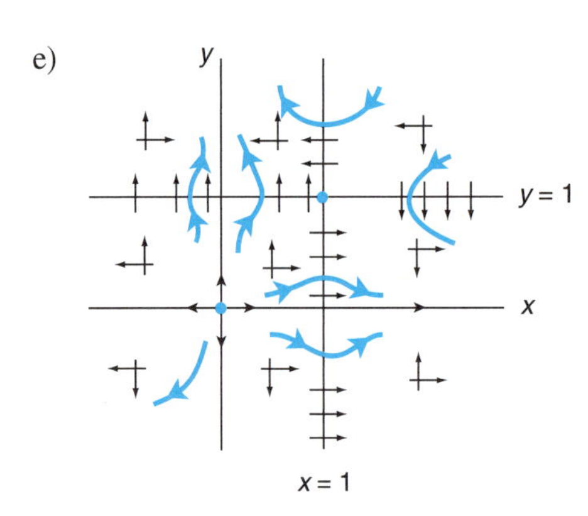

5. (a) and (b) Equilibrium (0, 0): $A=\begin{bmatrix}1 & -1\\ -2 & 2\end{bmatrix}$, $\lambda=0, 3$ so borderline case. Will not be stable. $\lambda=0$, $\begin{bmatrix}1\\1\end{bmatrix}$, $\lambda=3$, $\begin{bmatrix}1\\-2\end{bmatrix}$,

Equilibrium (1, 1): $A=\begin{bmatrix}1 & -1\\ 0 & 2\end{bmatrix}$, $\lambda=1, 2$ so unstable node. $\lambda=1$, $\begin{bmatrix}1\\0\end{bmatrix}$, $\lambda=2$, $\begin{bmatrix}1\\-1\end{bmatrix}$,

c)

e)

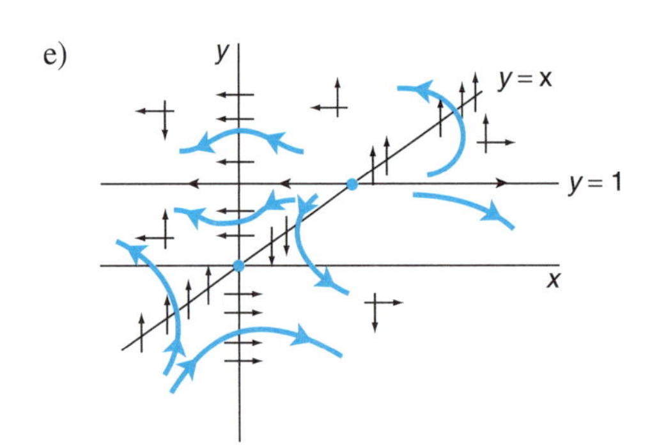

7. (a) and (b) Equilibrium (1, 1): $A=\begin{bmatrix}0 & -2\\ -2 & 0\end{bmatrix}$, $\lambda=2, -2$ saddlepoint. $\lambda=2$, $\begin{bmatrix}1\\-1\end{bmatrix}$, $\lambda=-2$, $\begin{bmatrix}1\\1\end{bmatrix}$,

Equilibrium $(-1, -1)$: $A = \begin{bmatrix} 0 & 2 \\ 2 & 0 \end{bmatrix}$, $\lambda = 2, -2$ saddlepoint. $\lambda = 2$, $\begin{bmatrix} 1 \\ 1 \end{bmatrix}$, $\lambda = -2$, $\begin{bmatrix} 1 \\ -1 \end{bmatrix}$,

Equilibrium $(1, -1)$: $A = \begin{bmatrix} 0 & 2 \\ -2 & 0 \end{bmatrix}$, $\lambda = \pm 2i$, center (nonlinear may be different).

Equilibrium $(-1, 1)$: $A = \begin{bmatrix} 0 & -2 \\ 2 & 0 \end{bmatrix}$, $\lambda = \pm 2i$, center (nonlinear may be different).

c) 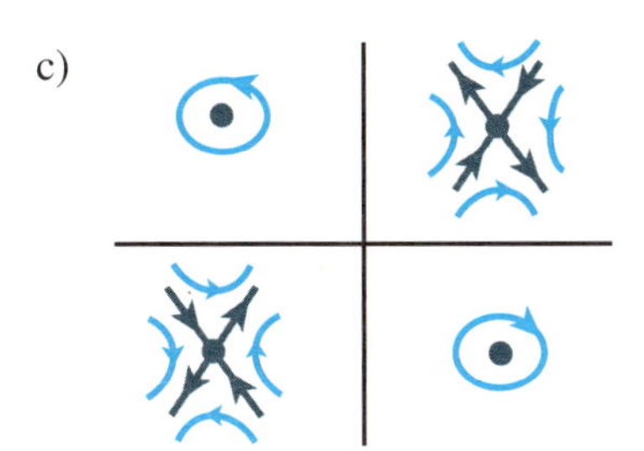

e) 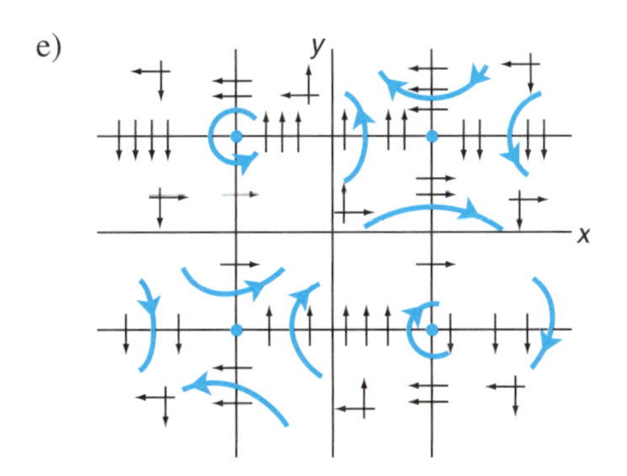

9. (a) and (b) Equilibrium $(0, 0)$: $A = \begin{bmatrix} 1 & -1 \\ 1 & 1 \end{bmatrix}$, $\lambda = 1 \pm i$, unstable spiral.

Equilibrium $(-2, 2)$: $A = \begin{bmatrix} -3 & -1 \\ 1 & 1 \end{bmatrix}$, $\lambda = -1 \pm \sqrt{3}$, saddlepoint.

$\lambda = -1 - \sqrt{3}$, $\begin{bmatrix} -2 + \sqrt{3} \\ 1 \end{bmatrix}$, $\lambda = -1 + \sqrt{3}$, $\begin{bmatrix} -2 - \sqrt{3} \\ 1 \end{bmatrix}$,

c) 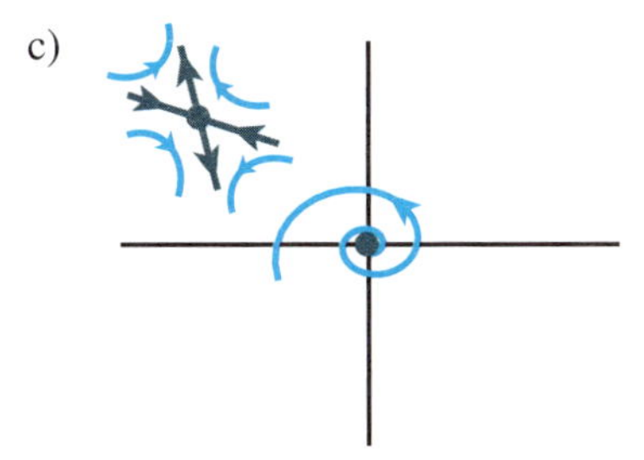

e)

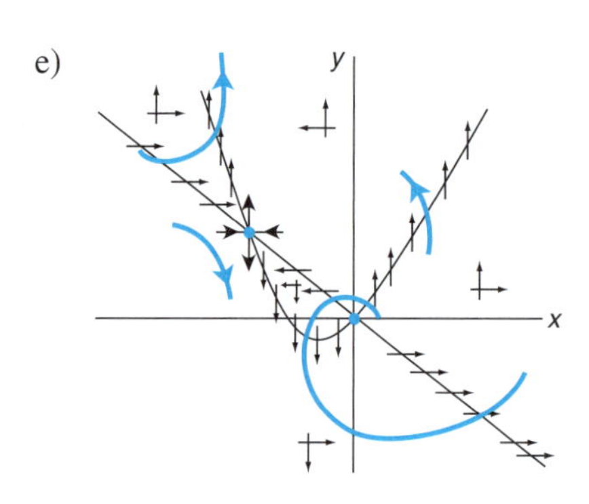

11. (a) and (b) Equilibrium (0, 0): $A = \begin{bmatrix} -1 & -2 \\ 2 & -1 \end{bmatrix}$, $\lambda = -1 \pm 2i$, stable spiral.

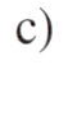
c)

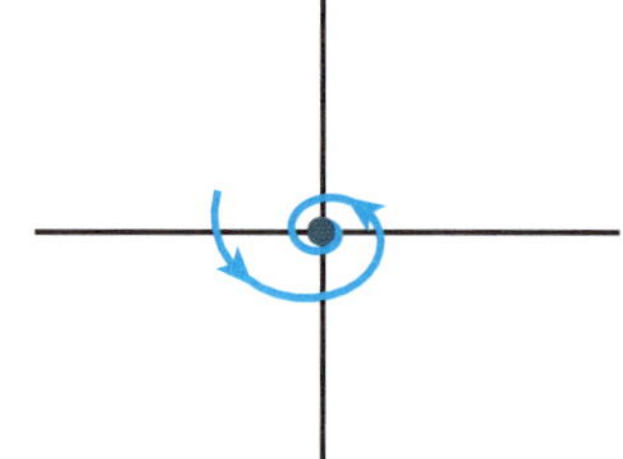

e)

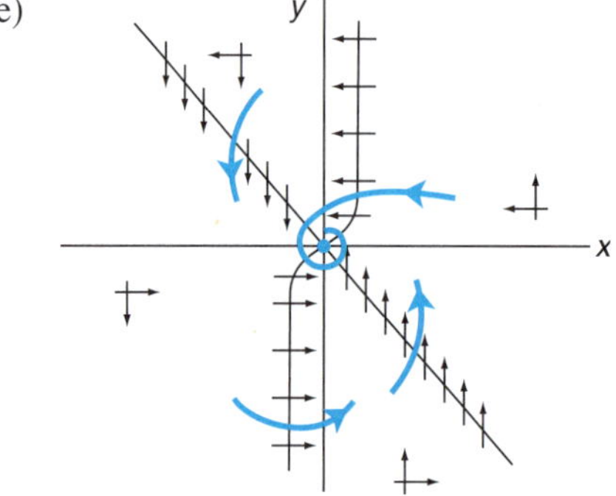

13. (0, 0), saddle; $\left(\frac{1}{8}, 0\right)$, stable node; (1, −7), saddle.
15. (0, 0), saddle; (−3, 0), stable node; $\left(1, \frac{4}{3}\right)$, unstable spiral.
17. (0, 0), saddle; $\left(-\frac{1}{8}, 0\right)$, stable node; (1, 9), unstable node.

6.3 Population Models

1. Equilibrium (0, 0): $A = \begin{bmatrix} 2 & 0 \\ 0 & 6 \end{bmatrix}$, $\lambda = 2, 6$, unstable node, $\lambda = 2$: $\begin{bmatrix} 1 \\ 0 \end{bmatrix}$, $\lambda = 1$: $\begin{bmatrix} 0 \\ 1 \end{bmatrix}$,

Equilibrium (2, 0): $A = \begin{bmatrix} -2 & -2 \\ 0 & -6 \end{bmatrix}$, $\lambda = -2, -6$, stable node, $\lambda = -2$: $\begin{bmatrix} 1 \\ 0 \end{bmatrix}$, $\lambda = -6$: $\begin{bmatrix} 1 \\ 2 \end{bmatrix}$,

Equilibrium (1, 1): $A = \begin{bmatrix} -1 & -1 \\ -6 & 0 \end{bmatrix}$, $\lambda = 2, -3$, saddlepoint, $\lambda = 2$: $\begin{bmatrix} 1 \\ -3 \end{bmatrix}$, $\lambda = -3$: $\begin{bmatrix} 1 \\ 2 \end{bmatrix}$,

b)

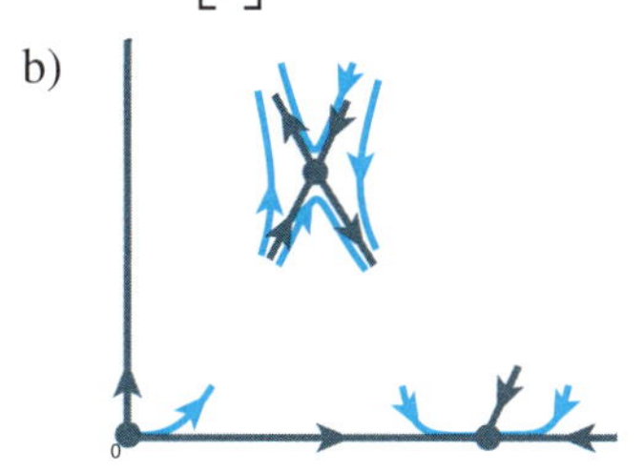

c)

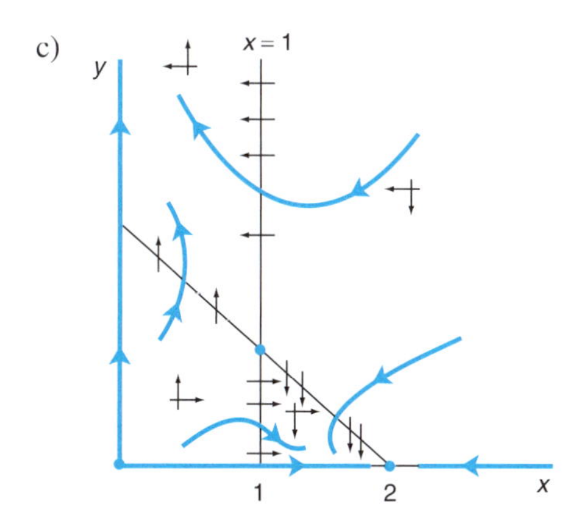

d)

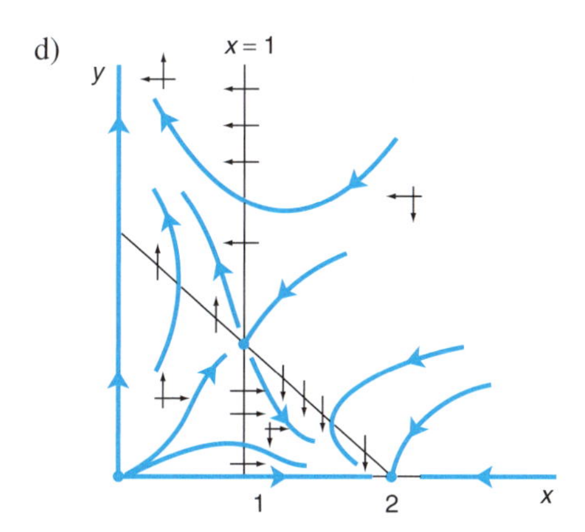

3. Equilibrium (0, 0): $A = \begin{bmatrix} \frac{3}{2} & 0 \\ 0 & 1 \end{bmatrix}$, $\lambda = \frac{3}{2}, 1$, unstable node, $\lambda = \frac{3}{2}$: $\begin{bmatrix} 1 \\ 0 \end{bmatrix}$, $\lambda = 1$: $\begin{bmatrix} 0 \\ 1 \end{bmatrix}$,

Equilibrium (0, 1): $A = \begin{bmatrix} -\frac{1}{2} & 0 \\ -1 & -1 \end{bmatrix}$, $\lambda = -\frac{1}{2}, -1$, stable node, $\lambda = -1$: $\begin{bmatrix} 0 \\ 1 \end{bmatrix}$, $\lambda = -\frac{1}{2}$: $\begin{bmatrix} 1 \\ -2 \end{bmatrix}$,

Equilibrium $\left(\frac{3}{8}, 0\right)$: $A = \begin{bmatrix} -\frac{3}{2} & -\frac{3}{4} \\ 0 & \frac{5}{8} \end{bmatrix}$, $\lambda = -\frac{3}{2}, \frac{5}{8}$, saddlepoint, $\lambda = -\frac{3}{2}$: $\begin{bmatrix} 1 \\ 0 \end{bmatrix}$, $\lambda = \frac{5}{8}$: $\begin{bmatrix} 6 \\ -17 \end{bmatrix}$,

b)

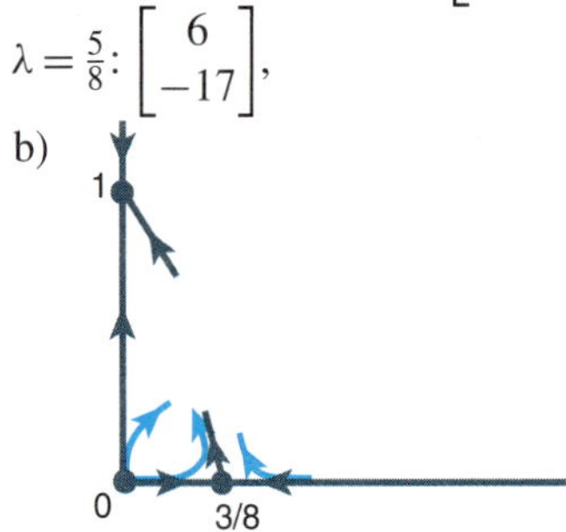

c) See combined figure for part d.

d)

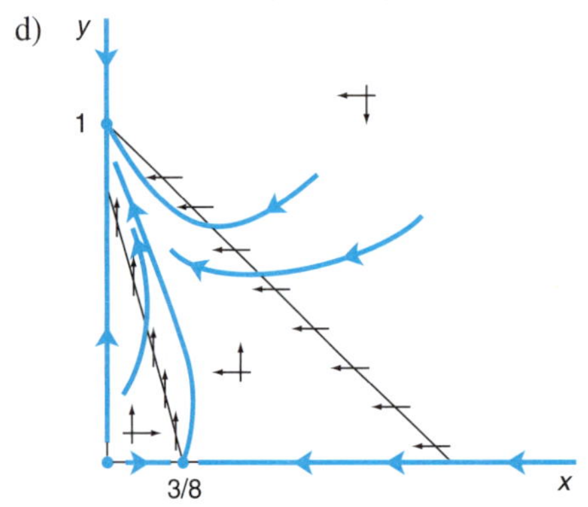

5. Equilibrium (0, 0): $A = \begin{bmatrix} 3 & 0 \\ 0 & -1 \end{bmatrix}$, $\lambda = 3, -1$, saddlepoint, $\lambda = 3$: $\begin{bmatrix} 1 \\ 0 \end{bmatrix}$, $\lambda = -1$: $\begin{bmatrix} 0 \\ 1 \end{bmatrix}$,

   Equilibrium (3, 0): $A = \begin{bmatrix} -3 & -3 \\ 0 & 2 \end{bmatrix}$, $\lambda = -3, 2$, saddlepoint, $\lambda = -3$: $\begin{bmatrix} 1 \\ 0 \end{bmatrix}$, $\lambda = 2$: $\begin{bmatrix} 3 \\ -5 \end{bmatrix}$,

   Equilibrium (1, 2): $A = \begin{bmatrix} -1 & -1 \\ 2 & 0 \end{bmatrix}$, $\lambda = (-1 \pm i\sqrt{7})/2$, stable spiral,

   b)

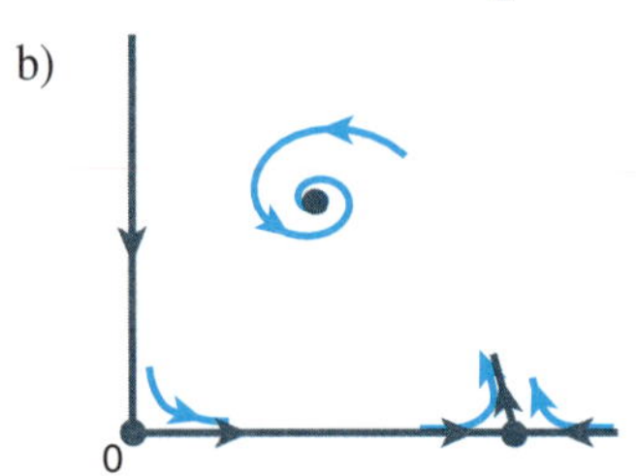

c) and d) combined

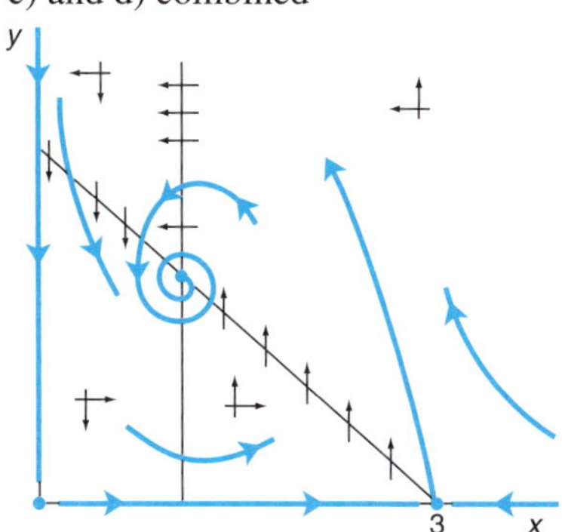

7. Equilibrium $(0, 0)$: $A = \begin{bmatrix} 3 & 0 \\ 0 & -1 \end{bmatrix}$, $\lambda = 3, -1$, saddlepoint, $\lambda = 3$: $\begin{bmatrix} 1 \\ 0 \end{bmatrix}$, $\lambda = -1$: $\begin{bmatrix} 0 \\ 1 \end{bmatrix}$,
Equilibrium $(4, 3)$: $A = \begin{bmatrix} 0 & -4 \\ 3 & -3 \end{bmatrix}$, $\lambda = (-3 \pm i\sqrt{39})/2$, stable spiral,

b)

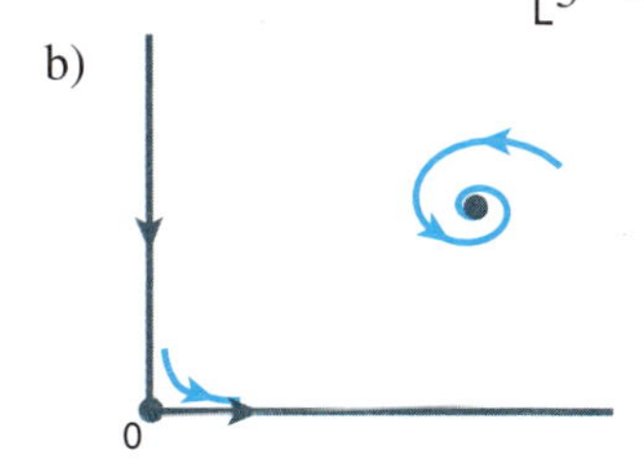

c)

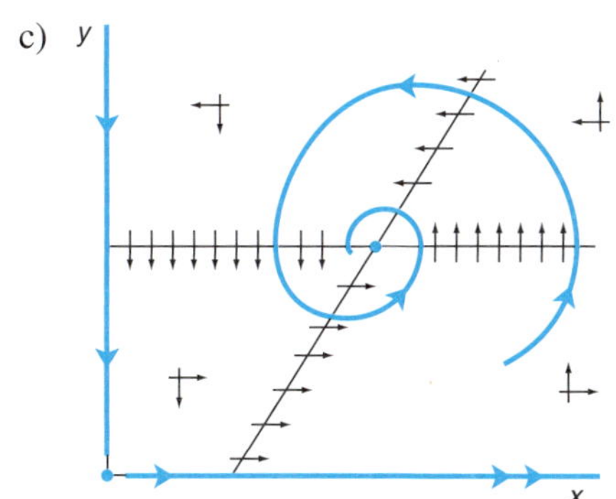

d)

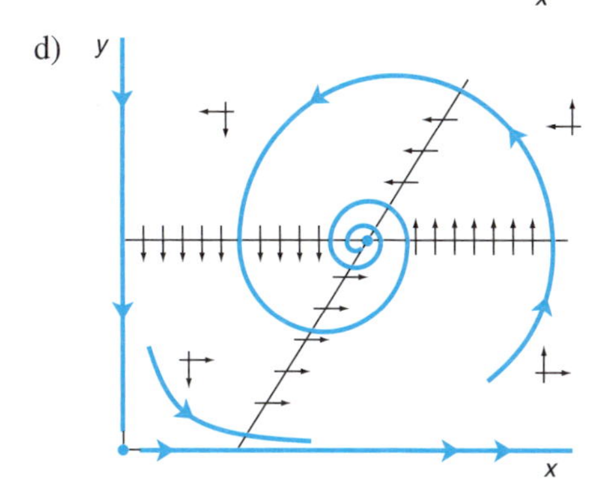

9. Equilibrium $(0,0)$: $A=\begin{bmatrix}3 & 0\\ 0 & -1\end{bmatrix}$, $\lambda=3,-1$, saddlepoint, $\lambda=3$: $\begin{bmatrix}1\\0\end{bmatrix}$, $\lambda=-1$: $\begin{bmatrix}0\\1\end{bmatrix}$,

Equilibrium $\left(\frac{3}{2},0\right)$: $A=\begin{bmatrix}-3 & -9\\ 0 & \frac{1}{2}\end{bmatrix}$, $\lambda=-3,\frac{1}{2}$, saddlepoint, $\lambda=-3$: $\begin{bmatrix}1\\0\end{bmatrix}$, $\lambda=\frac{1}{2}$: $\begin{bmatrix}18\\-7\end{bmatrix}$,

Equilibrium $\left(\frac{9}{8},\frac{1}{8}\right)$: $A=\begin{bmatrix}-\frac{9}{4} & -\frac{27}{4}\\ \frac{1}{8} & -\frac{1}{8}\end{bmatrix}$, $\lambda=\frac{-19\pm\sqrt{73}}{16}$, stable node,

b)

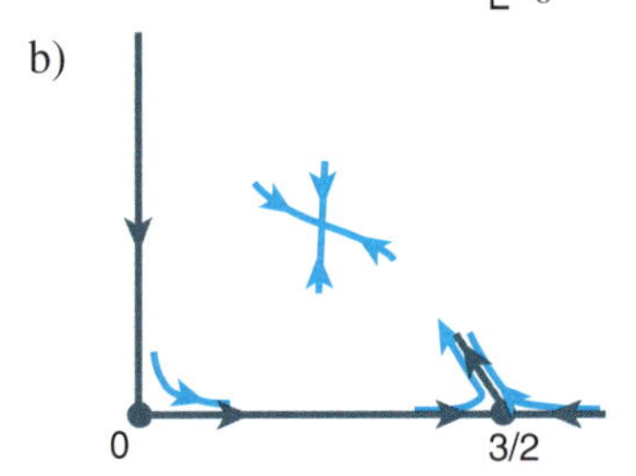

c)

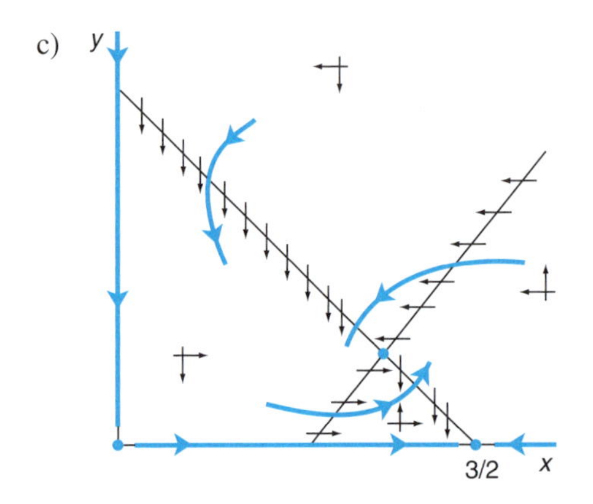

11. Equilibrium $(0,0)$: $A=\begin{bmatrix}1 & 0\\ 0 & -\frac{1}{2}\end{bmatrix}$, $\lambda=1,-\frac{1}{2}$, saddlepoint, $\lambda=1$: $\begin{bmatrix}1\\0\end{bmatrix}$, $\lambda=-\frac{1}{2}$: $\begin{bmatrix}0\\1\end{bmatrix}$,

Equilibrium $(1,2)$: $A=\begin{bmatrix}\frac{1}{2} & -\frac{1}{2}\\ \frac{1}{2} & 0\end{bmatrix}$, $\lambda=\frac{1}{4}\pm\frac{1}{2}i\sqrt{\frac{3}{4}}$, unstable spiral,

b)

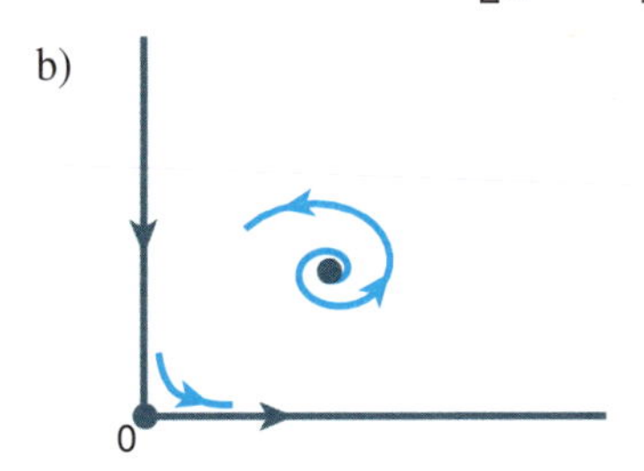

c) and d) combined

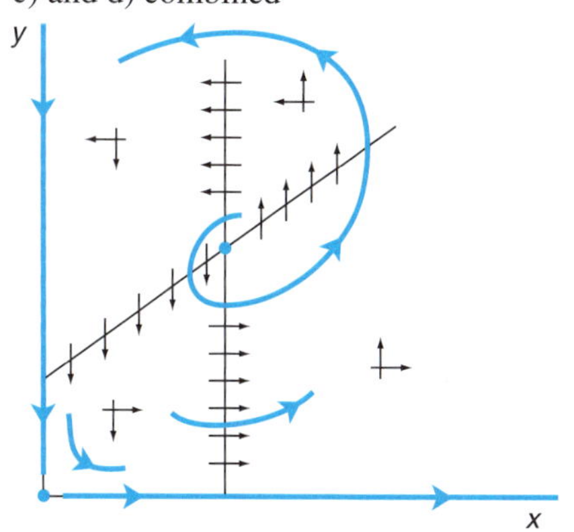

13. Equilibrium $T^* = 0, v = 0$; $A = \begin{bmatrix} -1 & \frac{3}{4} \\ 1 & -2 \end{bmatrix}$, $\lambda = -\frac{1}{2}, -\frac{5}{2}$, stable node, $\lambda = -\frac{1}{2}$: $\begin{bmatrix} 3 \\ 2 \end{bmatrix}$, $\lambda = -\frac{5}{2}$: $\begin{bmatrix} 1 \\ -2 \end{bmatrix}$,

b)

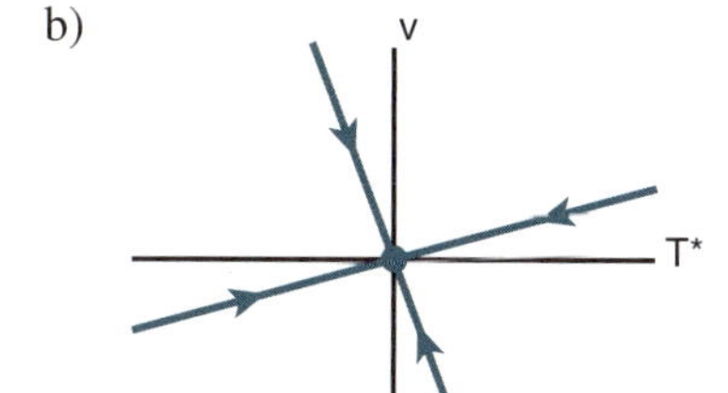

c)

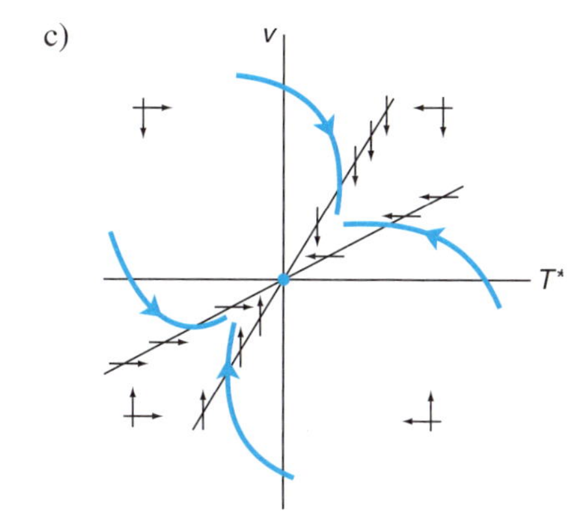

15. Equilibrium $T^* = 0, v = 0$; $A = \begin{bmatrix} -1 & \frac{9}{4} \\ 1 & -1 \end{bmatrix}$, $\lambda = -\frac{5}{2}, \frac{1}{2}$, saddle point, $\lambda = \frac{1}{2}$: $\begin{bmatrix} 3 \\ 2 \end{bmatrix}$, $\lambda = -\frac{5}{2}$: $\begin{bmatrix} 3 \\ -2 \end{bmatrix}$,

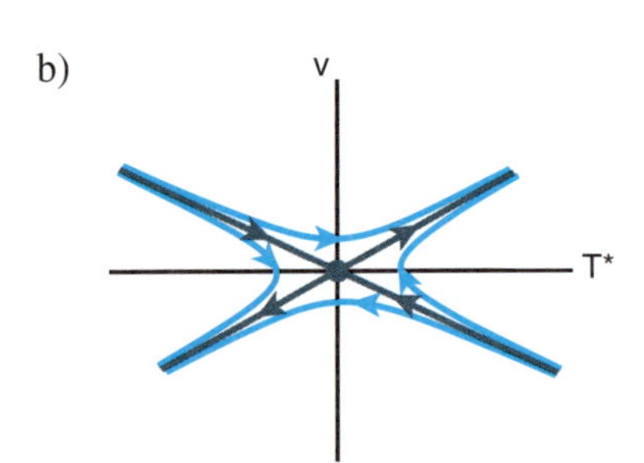

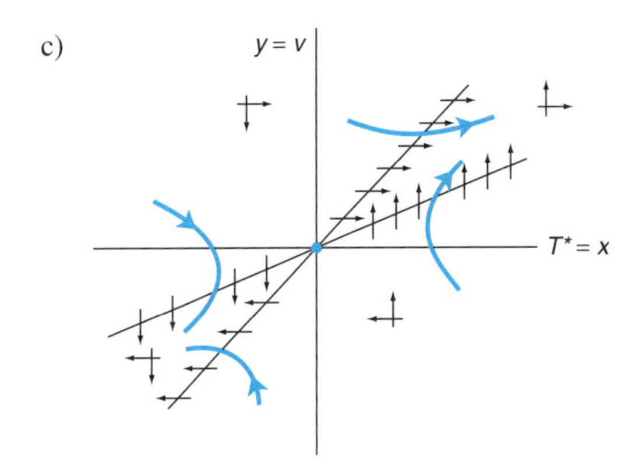

6.4 Mechanical Systems

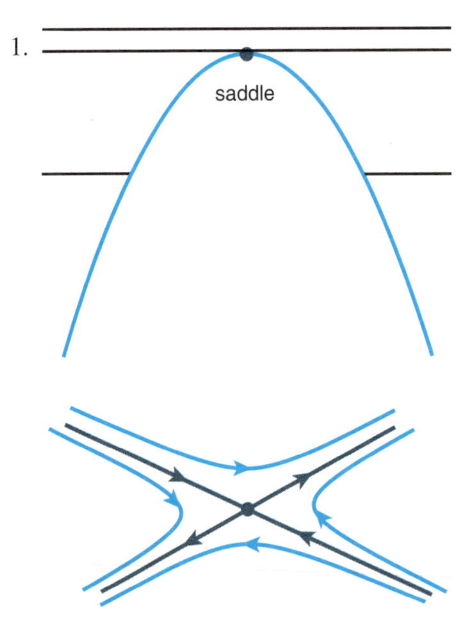

3.

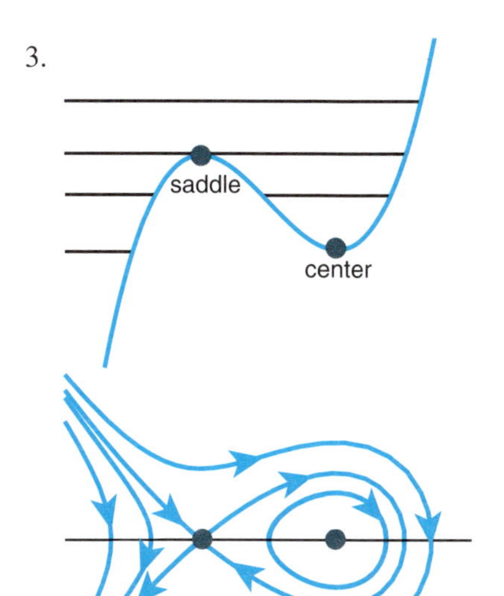

5.

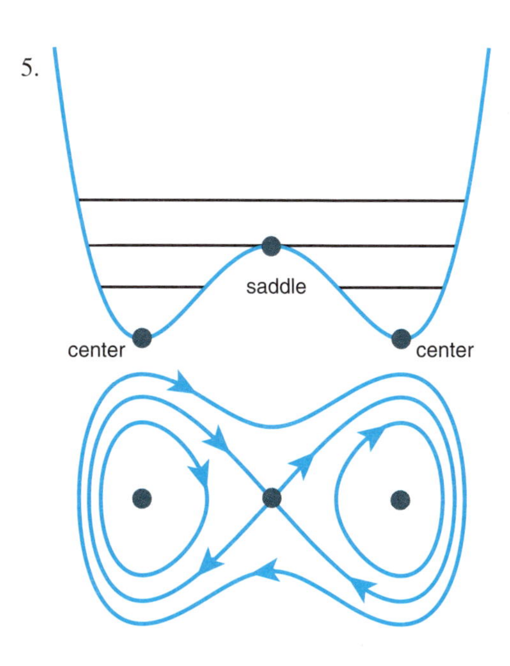

7. $V(x) = -\frac{x^2}{2} - \frac{x^4}{4}$. Equilibrium $x = 0$ is a saddlepoint.

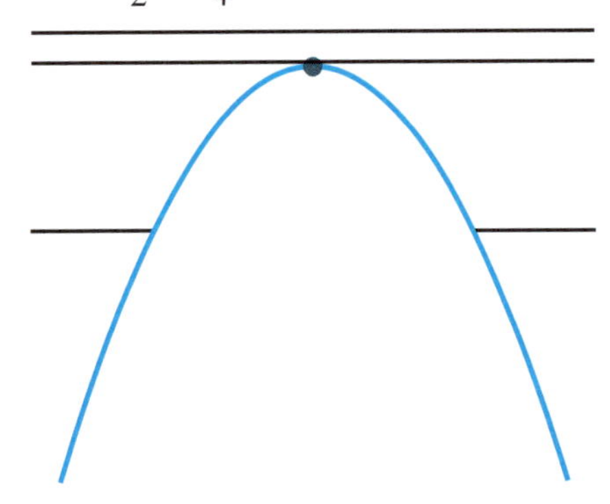

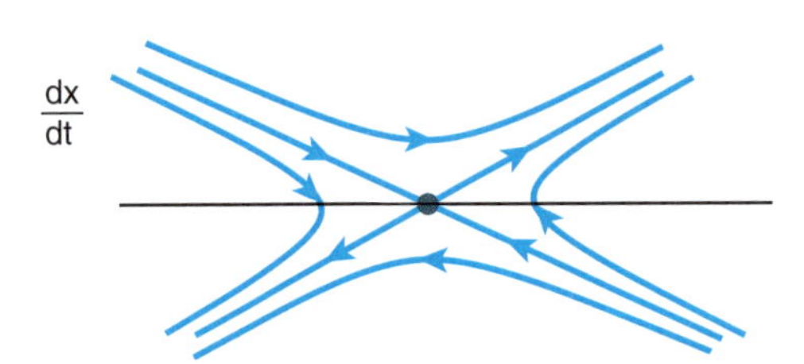

9. $V(x) = -\frac{x^2}{2} + \frac{x^4}{4}$. Equilibrium $x = 0$ is a saddlepoint while $x = \pm 1$ are centers.

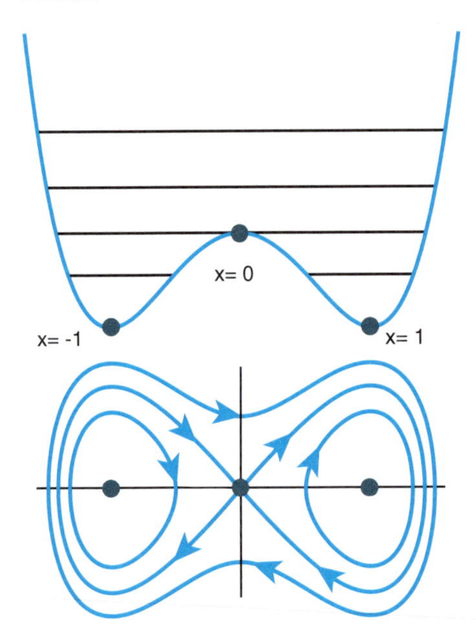

11. $V(x) = 3x^2 - x^3$. Equilibrium $x = 0$ is a center and $x = 2$ is a saddlepoint.

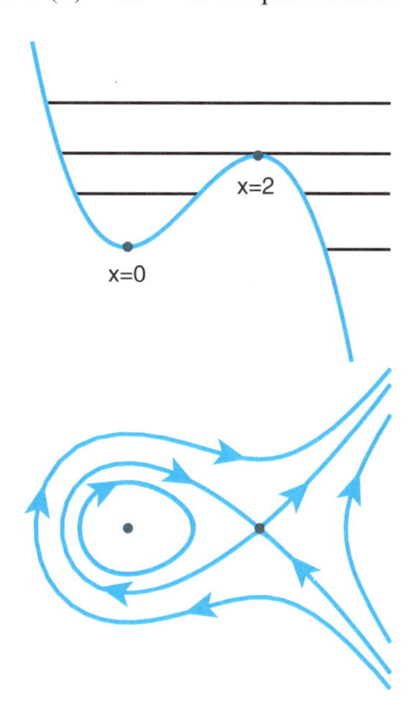

17. $x = 2n\pi$ are stable spirals, $x = \pi + 2n\pi$ are saddlepoints
19. $x = 2n\pi$ are stable nodes, $x = \pi + 2n\pi$ are saddlepoints
21. Same phase plane as in Figure 6.4.5.

23.

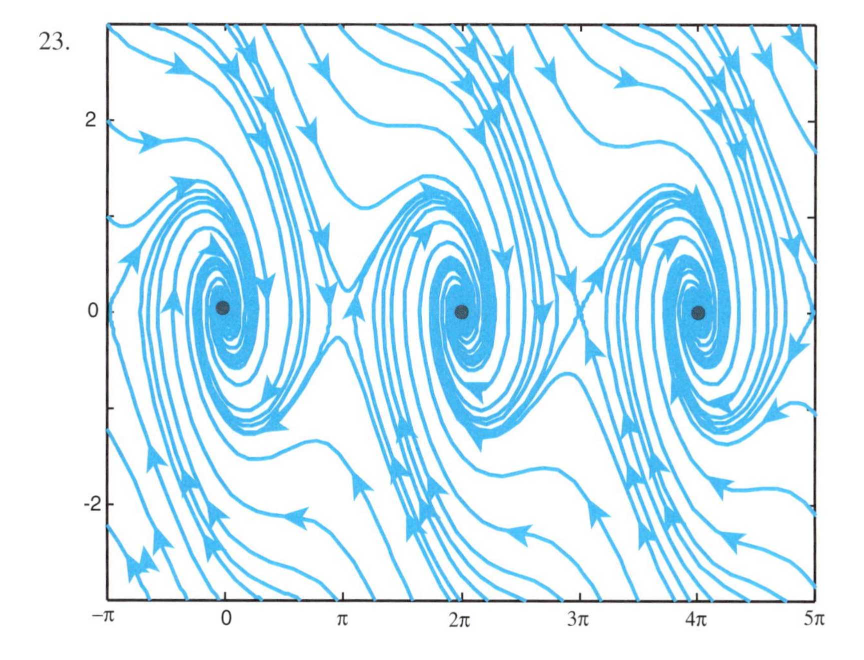

25.

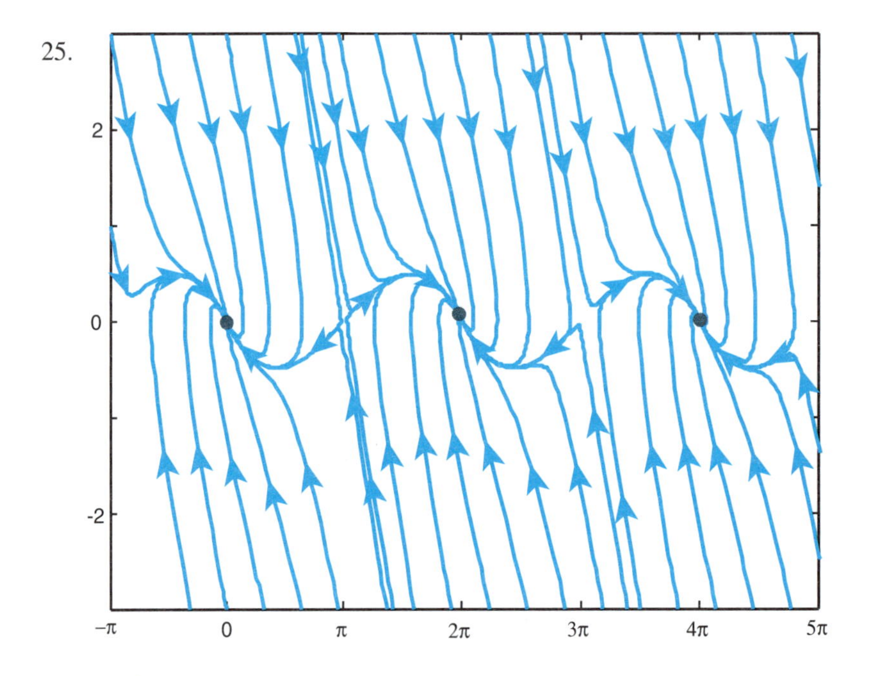

INDEX